W9-BWR-781

Encyclopedia
of Animal Behavior

Encyclopedia of Animal Behavior

Volume 2: D–P

Edited by
Marc Bekoff

Foreword by
Jane Goodall

GREENWOOD PRESS
Westport, Connecticut • London

Library of Congress Cataloging-in-Publication Data

Encyclopedia of animal behavior / edited by Marc Bekoff; foreword by Jane Goodall.
    p. cm
    Includes bibliographical references (p.)
    ISBN 0-313-32745-9 (set : alk. paper)—ISBN 0-313-32746-7 (vol. 1 : alk. paper)—
    ISBN 0-313-32747-5 (vol. 2 : alk. paper)—ISBN 0-313-33294-0 (vol. 3 : alk. paper)
    1. Animal behavior—Encyclopedias. I. Bekoff, Marc.
QL750.3.E53   2004
591.5'03—dc22   2004056073

British Library Cataloguing in Publication Data is available.

Library of Congress Catalog Card Number: 2004056073
ISBN: 0-313-32745-9 (set code)
        0-313-32746-7 (Vol. 1)
        0-313-32747-5 (Vol. 2)
        0-313-33294-0 (Vol. 3)

First published in 2004

Greenwood Press, 88 Post Road West, Westport, CT 06881
An imprint of Greenwood Publishing Group, Inc.
www.greenwood.com

Printed in the United States of America

The paper used in this book complies with the
Permanent Paper Standard issued by the National
Information Standards Organization (Z39.48-1984).

10 9 8 7 6 5 4 3 2

*These volumes are dedicated to the memory
of one of the most amazing people
I have ever known—my mother Beatrice*

Contents

■ Alphabetical List of Entries

(S) indicates a side bar. It is listed under the entry within which it appears.

■ Guide to Related Topics

ANIMALS

Following is a list of entries that focus primarily on one type of animal. Most of the other entries in the *Encyclopedia of Animal Behavior* also discuss various species. Please see the index for additional references to specific animals.

Ants

Animal Architecture—*Subterranean Ant Nests*

Bats

Bats—*The Behavior of a Mysterious Mammal*
Communication—Auditory—*Bat Sonar*

Bears

Feeding Behavior—*Grizzly Foraging*
Human (Anthropogenic) Effects—*Bears: Understanding, Respecting, and Being Safe around Them*

Birds

Antipredatory Behavior—*Sentinel Behavior*
Behavioral Physiology—*Plumage Color and Vision in Birds*
Caregiving—*Brood Parasitism among Birds*
Cognition—*Grey Parrot Cognition and Communication*
Corvids—*The Crow Family*
Communication—Vocal—*Social System and Acoustic Communication in Spectacled Parrotlets*
Communication—Vocal—*Singing Birds: From Communication to Speciation*
Mimicry—*Magpies*
Play—*Birds at Play*
Reproductive Behavior—*Bowerbirds and Sexual Displays*
Tools—*Tool Use and Manufacture by Birds*
Welfare, Well-Being, and Pain—*Enrichment for Chickens*
Welfare, Well-Being, and Pain—*Feather Pecking in Birds*
Welfare, Well-Being, and Pain—*Rehabilitation of Raptors*

Cats

Cats—*Domestic Cats*
  Wild versus Domestic Behaviors: When Normal
  Behaviors Lead to Problems (S)

Fish

Communication—Auditory
    Noisy Herring (S)
Sharks
    Sharks: Mistaken Identity (S)
Caregiving—*Brood Parasitism in Freshwater Fish*
Caregiving—*Parental Care in Fish*
Communication—Visual—*Fish Display Behavior*
Human (Anthropogenic) Effects—*Genetically Modified Fish*
Methods
    Sonic Tracking of Endangered Atlantic Salmon (S)
Reproductive Behavior—*Alternative Male Reproductive Tactics in Fish*
Reproductive Behavior—*Sex Change in Fish*
Social Organization—*Shoaling Behavior in Fish*

Frogs

Communication—Vocal—*Choruses in Frogs*

Horses

Horses—*Behavior*
    Sleeping Standing Up (S)
Horses—*Horse Training*

Hyenas

Development—*Spotted Hyena Development*

Insects

Animal Architecture—*Subterranean Ant Nests*
Behavioral Physiology
    Colors—How Do Flowers and Bees Match? (S)
Behavioral Physiology—*Insect Vision and Behavior*
Caregiving—*Parental Care by Insects*
Communication—*Honeybee Dance Language*
Education—Classroom Activities—*Insects in the Classroom*
Predatory Behavior—*Praying Mantids*
Reproductive Behavior—*Sexual Behavior in Fruit Flies*—Drosophila
Reproductive Behavior—*Sexual Cannibalism*

Lemurs. *See also* Primates

Lemurs—*Behavioral Ecology of Lemurs*
Lemurs—*Learning from Lemurs*

Macaques

Cognition—*Tactical Deception in Wild Bonnet Macaques*
Social Organization—*Social Knowledge in Wild Bonnet Macaques*
Tools—*Tool Manufacture by a Wild Bonnet Macaque*

Cognition—*Talking Chimpanzees*
Culture—*Orangutan Culture*
    CULTURE IN OUR RELATIVES, THE ORANGUTANS (S)
Education—Classroom Activities—*All About Chimpanzees!*
Gorillas—*Gorillas / Koko*
Human (Anthropogenic) Effects—*Logging, Behavior, and the Conservation
    of Primates*
Learning—*Social Learning and Intelligence in Primates*
Lemurs—*Behavioral Ecology of Lemurs*
Lemurs—*Learning from Lemurs*
Methods—*Zen in the Art of Monkey Watching*
Personality and Temperament—*Personality in Chimpanzees*
Social Evolution—*Flowering Plants and the Origins of Primates*
Social Evolution—*Social Evolution in Bonnet Macaques*
Social Organization—*Cooperation and Aggression among Primates*
Social Organization—*Monkey Families and Human Families*
Social Organization—*Social Knowledge in Wild Bonnet Macaques*
Tools—*Tool Manufacture by a Wild Bonnet Macaque*
Welfare, Well-Being, and Pain—*Enrichment for Monkeys*
    FORAGING PRIMATES (S)
Welfare, Well-Being, and Pain—*Enrichment for New World Monkeys*
Welfare, Well-Being, and Pain—*Primate Rescue Groups*

Rodents

Communication—Auditory—*Ultrasound in Small Rodents*
Communication—Vocal—*Vocalizations of Northern Grasshopper Mice*
Conservation and Behavior
    PREBLE'S MEADOW JUMPING MOUSE AND CULVERTS (S)
Reproductive Behavior—*Shredding Behavior of Rodents*

Sharks

Sharks
    SHARKS: MISTAKEN IDENTITY (S)

Snakes

Welfare, Well-Being, and Pain—*Experiments in Enriching Captive Snakes*

Spiders

Predatory Behavior—*Orb-Web Spiders*

Turtles

Behavioral Physiology—*Turtle Behavior and Physiology*
Navigation—*Natal Homing and Mass Nesting in Marine Turtles*
Reproductive Behavior—*Marine Turtle Mating and Evolution*
Welfare, Well-Being, and Pain—*Back Scratching and Enrichment
    in Sea Turtles*

Behavioral Physiology

Color Vision in Animals
  COLORS—HOW DO FLOWERS AND BEES MATCH? (S)
  DO SQUID MAKE A LANGUAGE ON THEIR SKIN? (S)
Insect Vision and Behavior
Plumage Color and Vision in Birds
Thermoregulation
Thermoregulatory Behavior
Turtle Behavior and Physiology
Visual Perception Mechanisms
  VISION, SKULL SHAPE, AND BEHAVIOR IN DOGS (S)

Behaviorism

Burrowing Behavior

Caregiving

Attachment Behaviors
Brood Parasitism among Birds
Brood Parasitism in Freshwater Fish
Fostering Behavior
How Animals Care for their Young
  HELPERS IN COMMON MARMOSETS (S)
Incubation
Mother–Infant Relations in Chimpanzees
Non-Offspring Nursing
Parental Care
  PARENTAL BEHAVIOR IN MARSUPIALS (S)
Parental Care and Helping Behavior
Parental Care by Insects
Parental Care in Fish
Parental Investment
  PARENTS DESERT NEWBORNS AFTER BREAK-IN! MOTHER GOES
  FOR FOOD AS SIBLINGS BATTLE TO THE DEATH! FATHER EATS BABIES! (S)

Cats

Domestic Cats
  WILD VERSUS DOMESTIC BEHAVIORS: WHEN NORMAL BEHAVIORS
  LEAD TO PROBLEMS (S)

Cephalopods

Octopuses, Squid, and other Mollusks
  ROVING OCTUPUSES (S)

BEHAVIOR OF ANIMALS *(continued)*

Cognition

Animal Consciousness
Animal Languages, Animal Minds
Audience Effect in Wolves
Behavior, Archaeology, and Cognitive Evolution
Cache Robbing
Caching Behavior
Categorization Processes in Animals
Cognitive Ethology: The Comparative Study of Animal Minds
Concept Formation
Deception
Dogs Burying Bones: Unraveling a Rich Action Sequence
Domestic Dogs Use Humans as Tools
Equivalence Relations
    FAIRNESS IN MONKEYS (S)
Food Storing
Grey Parrot Cognition and Communication
Imitation
Limited Attention and Animal Behavior
Mirror Self-Recognition
Mirror Self-Recognition and Kinesthetic–Visual Matching
Social Cognition in Primates and other Animals
Tactical Deception in Wild Bonnet Macaques
Talking Chimpanzees
Theory of Mind

Communication—Auditory

Acoustic Communication in Extreme Environments
Audition
    NOISY HERRING (S)
Bat Sonar
Long Distance Calling, the Elephant's Way
Ultrasound in Small Rodents

Communication

Electrocommunication
    ELECTRIC FISH (S)
Honeybee Dance Language
Modal Action Patterns

Communication—Olfaction

Chemical Communication
    DOG SCENTS AND "YELLOW SNOW" (S)
Mammalian Olfactory Communication

BEHAVIOR OF ANIMALS (continued)

Unisexual Vertebrates

Welfare, Well-Being, and Pain

Animal Welfare and Reward
Back Scratching and Enrichment in Sea Turtles
Behavior and Animal Suffering
Behavioral Assessment of Animal Pain
Behavioral Correlates of Animal Pain and Distress
Carnivores in Captivity
Enrichment for Chickens
Enrichment for Monkeys
Enrichment for New World Monkeys
Experiments in Enriching Captive Snakes
Feather Pecking in Birds
Obsessive–Compulsive Behaviors
Primate Rescue Groups
Psychological Well-Being
Rehabilitation of Marine Mammals
Rehabilitation of Raptors
Sanctuaries
Stress in Dolphins
Veterinary Ethics and Behavior
Wildlife Trapping, Behavior, and Welfare

Wildlife Management and Behavior

Wolf Behavior

Learning to Live in Life or Death Situations

CAREERS IN ANIMAL BEHAVIOR

Careers

Animal Behavior and the Law
Animal Tracking and Animal Behavior
Animal-Assisted Psychotherapy
Applied Animal Behavior
Careers in Animal Behavior Science
Mapping their Minds: Animals on the Other Side of the Lens
Recording Animal Behavior Sounds: The Voice of the Natural World
Significance of Animal Behavior Research
    UNCONVENTIONAL USES OF ANIMAL BEHAVIOR (S)
Veterinary Practice Opportunities for Ethologists
Wildlife Filmmaking
Wildlife Photography
Writing about Animal Behavior: The Animals Are Also Watching Us

HUMANS AND ANIMALS (continued)

Careers

Animal Behavior and the Law
Animal Tracking and Animal Behavior
Animal-Assisted Psychotherapy
Applied Animal Behavior
Careers in Animal Behavior Science
Mapping their Minds: Animals on the Other Side of the Lens
Recording Animal Behavior Sounds:The Voice of the Natural World
Significance of Animal Behavior Research
Veterinary Practice Opportunities for Ethologists
Wildlife Filmmaking
Wildlife Photography
Writing about Animal Behavior: The Animals Are Also Watching Us

Cognition

Domestic Dogs Use Humans as Tools
Grey Parrot Cognition and Communication

Conservation and Behavior

Species Reintroduction
    PREBLE'S MEADOW JUMPING MOUSE AND CULVERTS (S)
Wildlife Behavior as a Management Tool in United States National Parks

Ecopsychology

Human—Nature Interconnections

Education

Classroom Activities in Behavior
    CLASSROOM RESEARCH (S)

Education—Classroom Activities

All About Chimpanzees!
Insects in the Classroom
"Petscope"
Planarians

History

History of Animal Behavior Studies
    SOME LEADERS IN ANIMAL BEHAVIOR STUDIES (S)
Niko Tinbergen and the "Four Questions" of Ethology

Human (Anthropogenic) Effects

Bears: Understanding, Respecting, and Being Safe around Them
Edge Effects and Behavior
The Effect of Roads and Trails on Animal Movement

Environmentally Induced Behavioral Polymorphisms
Genetically Modified Fish
Human (Anthropogenic) Effects on Animal Behavior
Logging, Behavior, and the Conservation of Primates
Pollution and Behavior
Urban Wildlife Behavior

Religion and Animal Behavior

Robotics
Animal Robots
Artificial Pets

Telepathy and Animal Behavior

Welfare, Well-being, and Pain
Animal Welfare and Reward
Back Scratching and Enrichment in Sea Turtles
Enrichment for Chickens
Enrichment for Monkeys
Enrichment for New World Monkeys
Experiments in Enriching Captive Snakes
Primate Rescue Groups
Psychological Well-Being
Rehabilitation of Marine Mammals
Rehabilitation of Raptors
Sanctuaries
Veterinary Ethics and Behavior
Wildlife Trapping, Behavior, and Welfare

Wildlife Management and Behavior

Zoos and Aquariums
Animal Behavior Research in Zoos
Giant Pandas in Captivity
Studying Animal Behavior in Zoos and Aquariums

LITERATURE, ARTS, AND RELIGION AND ANIMAL BEHAVIOR

Animals in Myth and Lore
Animals in Native American Lore
Fairy Tales and Myths of Animal Behavior
Mythology and Animal Behavior
The Wolf in Fairy Tales (S)

LITERATURE, ARTS, AND RELIGION AND ANIMAL BEHAVIOR *(continued)*

Art and Animal Behavior

Careers

Mapping their Minds: Animals on the Other Side of the Lens
Recording Animal Behavior Sounds: The Voice of the Natural World
Wildlife Filmmaking
Wildlife Photography
Writing about Animal Behavior: The Animals Are Also Watching Us

METHODS OF STUDY. *(See* RESEARCH, EXPLANATION, AND METHODS OF STUDY)

PEOPLE IN ANIMAL BEHAVIOR RESEARCH

Craig, Wallace (1876–1954)

Darwin, Charles (1809–1882)

Frisch, Karl von (1886–1982)

Griffin, Donald Redfield (1915–2003)

Hamilton, William D. III (1936–2000)

History

History of Animal Behavior Studies
    SOME LEADERS IN ANIMAL BEHAVIOR STUDIES (S)
Niko Tinbergen and the "Four Questions" of Ethology

Lorenz, Konrad Z. (1903–1989)

Maynard Smith, John (1920–2004)

Nobel Prize

1973 Nobel Prize for Medicine or Physiology

Tinbergen, Nikolaas (1907–1988)
A Visit with Niko and Lies Tinbergen (S)

Turner, Charles Henry (1867–1923)

Washburn, Margaret Floy (1871–1939)

Whitman, Charles Otis (1842–1910)

RESEARCH, EXPLANATION, AND METHODS OF STUDY

Behaviorism

Careers
Significance of Animal Behavior Research

Comparative Psychology

Development
Behavioral Stages

Ecopsychology

Frisch, Karl von
Decoding the Language of the Bees

History
A History of Animal Behavior Studies
Human Views of the Great Apes from Ancient Greece to the Present

Levels of Analysis in Animal Behavior

Lorenz, Konrad Z.
Analogy as a Source of Language

Methods
Computer Tools for Measurement and Analysis of Behavior
Deprivation Experiments

Darwin, Charles
(1809–1882)

Charles Darwin's discoveries and theories were of such importance that it is hard to imagine how people perceived things before they were made. Modern evolutionary biology, molecular sciences, and many other new subfields of science would be unthinkable without Darwin's ideas. The idea that humans have evolved is now also very well supported by fossil evidence of hominid predecessors. Conversely, many of Darwin's ideas that were speculative at his own time have now been shown to be true. One such basic tenet was that all organisms are individually unique; that is, they vary from one another by inherited traits. This is *diversity* and diversity has now been shown to exist even at molecular and genetic (chromosomal) levels to an extent totally unknown and unsuspected at Darwin's time. More off-

spring are produced (by any species) than can survive to adulthood and reproduce, so that only certain traits and characteristics survive—known as *natural selection*. And most importantly, all organisms are connected through a process of evolution: Selection processes favor some variants over others, and their accumulation may lead to new species.

Charles Darwin was born in 1809 in Shrewsbury, England, and died in Downe, near Croydon, England, in 1882. Darwin, as was typical of his class and background, was a man of privilege who did not have to work for a living. He could devote his life to his studies and, because of his social position, was able to converse with and move in circles of scholars without any specific university affiliations. His formal studies of medicine (in Edinburgh) and of theology (in Cambridge) came to nothing. He did not complete his course in Edinburgh and, in Cambridge, Darwin spent probably more time dabbling in scientific hobbies than attending to his lecturers. He never became a man of the cloth as he was meant to do and, instead, escaped on a sea voyage. Just as he had finally completed his Cambridge studies, he was offered a place as a companion to Captain Fitzroy of the H.M.S. *Beagle*. Fitzroy was just 23 years of age when he was given command over the *Beagle*, and Darwin was a mere 22 years old

Charles Darwin.
Courtesy of the Library of Congress.

when they sailed from England in 1831. The journey, with the clear task to establish better naval charts, eventually extended over 5 years in a complete circumnavigation of the world. In those years, Darwin turned himself from social companion into a de facto naturalist of the ship, noting down his observations along the way. His meticulousness and his passion for natural history led him to make many discoveries. Among these were the seminal studies of Galapagos finches. He discovered 14 species and drew them all separately, noting their feeding preferences and beak shapes in particular. Influenced by a new spirit of science that began to ask for reasons (causes) of events, forms, and structures, particularly by authors such as Charles Lyell and his book *Principles of Geology* (1830–1833), Darwin was slowly beginning to doubt that species were immutable and unchanging (see below). The finches seemed to show that variability and adaptability were crucial factors in survival. From these observations, he began to shape his theories and, on return to England in 1836, took another 20 years to perfect them.

The Theory of Evolution and Natural Selection

Before the publication of Darwin's theory of evolution, the study of animal behavior had been dominated by the idea that the species were immutable, or unchangeable. The perception was that God had created living organisms in different but static forms. The scientific study of animals, hence, focused on examining the anatomies of different species. Baron Cuvier, head of the Académie des Sciences in Paris, was one of the most influential advocates of this approach and he collected enormous amounts of laboratory-based data. His ideas and influences suffocated any other more unifying approach to the study of animals and any attempts to develop unifying theories. This led to the famous debates of the 1830s between Cuvier and Geoffroy St. Hilaire, who embraced the trends seen in animals in nature. Within the next two decades, the debates, often bitter, led to the establishment of two disciplines: animal psychology, focused on laboratory testing and analysis as emphasized by Cuvier, and ethology, termed such by the son of Geoffroy St. Hilaire and focused on studying the characters of animals in their natural habitats. On this background, and in 1859, Darwin published the *Origin of the Species*, formulating a unifying theory based on the mutability of species.

Darwin's theory of evolution led to a revolutionary change in the study of animals. For the first time, it made research on animals a natural science relevant to understanding humans and, within the unifying framework of evolution, the behavior of one animal species was seen to be relevant to another. This laid the foundations for comparative studies of animal behavior. By postulating that one species could evolve into another by the gradual accumulation of chance variations in *phenotype* (physical form) and the process of natural selection, Darwin changed our view of ourselves in the context of all animal life. The earlier notion that humans were entirely different from animals, separated from them by an insurmountable barrier, was replaced by the view that the difference between humans and animals was a matter of degree and not of kind. Although natural selection was seen to act on phenotype, the structure of the sensory organs and the brain itself was seen as part of the phenotype, and these aspects are expressed in behavior. For example, in support of Darwin, Thomas Huxley studied the structural differences between the brains of different higher primates and interpreted his findings from an evolutionary perspective.

By the end of the nineteenth century, interest in Darwin's theory of evolution had declined. The theory was seen as largely disproven because geological and paleontological evidence, showing that life had existed on earth for only a short time, indicated that the gradual accumulation of chance variations took too long to account for the appearance of each species. Even Darwin's formulation of the concept of *sexual selection* (referring to preferences in mate choice for particular phenotypes, hence *genotypes*) failed to convince his critics at the time. Study of animal psychology developed at this time when Darwin's theory of natural selection was seen to be of little merit, and it emphasized the value of laboratory experimentation. Not until the 1930s did natural selection move back into a central place in the theory of evolution. Then, as it had done before and also in more recent times, it was also used by emerging fascist groups and individuals as a political force to oppress Jews, blacks and others. Hence, Darwin's theories have had very seminal and revolutionary impact on the study of animal behavior but, at the same time, some disastrously negative outcomes in the study and theoretical development of research of human behavior (sexism, racism and many other discriminatory theories of present day, as seen, for instance, in the relatively new field of evolutionary psychology) have laid claim to Darwin's theories.

The 1970s, for instance, saw the emergence of sociobiology as a mainstay of Darwin's ideas, and, while its attempts to explain human behavior in divisive genetic terms were hotly contested, its application to animal behavior gained a large following, as is still the case today. At the same time *ethology*, the scientific study of animal behavior, fragmented into neuroethology, behavioral ecology and other subdisciplines. Sociobiologists were entirely focused on genetic explanations of animal behavior, as a direct application of Darwin's theories (natural selection and sexual selection), and they declared large parts of ethology to be irrelevant, especially research on development and causation. In time, sociobiology spawned evolutionary psychology which, by and large, has remained a crude attempt to explain much of human behavior in terms of genetic causes.

Both sociobiology and evolutionary psychology are, it is argued, characterized by reductionist thinking, reducing complex behavior to unitary genetic causes, and they are both used frequently in the service of conservative social and political forces. Although there is a lineage of these ideas directly back to Darwin, the study of behavior has almost disappeared in these accounts, other than as a by-product of the genes. While innovative (but highly problematic in their consequences on the social and political arena of human society), some of these theoretical developments have done little to foster our understanding of animal behavior in the context of its natural environment. In the supposed application of Darwin's ideas, they have often degenerated into pseudoscientific discussions, betraying their political convictions rather than scientific rigor and in that they are very "un-Darwinian."

Darwin and Ethology

The field of ethology has been referred to by Niko Tinbergen, one of its chief modern exponents, as "the biological study of behavior." Many ethologists study the behavior of animals in the wild. Others observe them carefully in captive environments or in the laboratory, always taking into account the known behavior of the species in its natural environment. Ethology is just one approach to the study of animal behavior, and Darwin is not always regarded as the first ethologist. Indeed, evaluations of Darwin in the history of science have at times placed him in the prehistory of ethology. Credit for ethology has been placed either much more recently in the direction of Karl von Frisch, Konrad Lorenz, and Tinbergen, the winners of the Nobel Prize in 1973, or in the hands of their relatively unacknowledged predecessors such as O. Heinroth or C. O. Whitman, who undertook behavioral observations much earlier than the Nobel Prize winners. However, in the English tradition, Darwin has maintained a founding role of the modern discipline of animal behavior, although he is by no means considered the only founder: G. M. Romanes (1848–1894) and C. L. Morgan (1852–1936) have been accorded similar founding status, while the role of Alfred Wallace, despite Darwin's and Wallace's first joint paper on natural selection (1844), was largely eclipsed by Darwin's *Origins* 15 years later. To further complicate matters, some early researchers of animal behavior were decidedly against Darwin (such as J. von Uexküll).

Darwin's own and detailed descriptions of animal behavior are closely linked to animal psychology and usually formulated with reference to human emotions and feelings. Both Romanes and Darwin have been criticized by some writers for a hopelessly entangled anthropomorphizing of animals. On the other hand, recent investigators of higher cognition in animals have seen such approach as less problematic and even valuable.

In the English tradition Darwin's role in ethology is considered central. Although the term ethology had been used over more than a hundred years prior to Darwin's time, it had been used to refer to the study of the ethics or science of building character in humans. It is generally accepted in this tradition that the use of "ethology" as referring to the study of animal behavior began with the publication of Darwin's *The Expression of Emotions in Man and Animals* in 1872, although it was not until well after Darwin that the label ethology was used consistently to refer to a particular way of studying animal behavior. Also, the origin of ethology could be attributed equally to the publication of Darwin's books *Variation of Animals and Plants under Domestication*, published in 1868, and *The Descent of Man*, published in 1871. In all of these he made comparisons between the mental processes of humans and other animals. Further impetus to the study of animal behavior came from the publication, in 1882, of the book *Animal Intelligence* by Romanes, a strong follower of Darwin. Romanes is seen by many as being the key figure in placing the study of animal behavior within an evolutionary framework and making it a comparison between species. He made a considerable move away from the Cartesian view and saw anthropomorphism as a means of understanding the minds and emotions of animals, thus foregrounding the debates that led ethologists of the mid-twentieth century to adopt the opposite position and avoid any hint of anthropomorphism.

Following the influence of Darwin, early ethologists studied animals in their natural habitats, and this focus persisted in the ethologists of the 1940s in Europe and the United States of America, as a clear distinction from laboratory-based comparative psychology. Also, as a direct line of influence from Darwin, the ethologists were largely concerned with instinctual behavior (inherited behavior). Darwin had argued against the division made between animals as being largely controlled by instincts, whereas humans are controlled by reason. His view was that all behavior is partly instinctive and partly dependent on past experience (learning), and it is not a simple matter of balancing off one by the other. In other words, the same animal may display many instinctive patterns of behavior as well as rapid and obvious learning.

Darwin's pivotal role in modern studies of animal behavior was his original combination of evolutionary theory with theories of instinct, even though some of these relationships have never been solved to Darwin's own satisfaction. Social insects, such as ants and bees, posed a major difficulty in his theory of natural selection because a colony contains large numbers of sterile workers that still show a rich diversity of adaptations and hence a complex division of labor. Yet variation in individual reproductive success is the lynch pin in the theory of natural selection, and this is a problem that appeared not to fit the theory (particularly in his detailed observations of the behaviors of ants). While Darwin speaks at length of instincts in the *Origin of the Species*, he points out repeatedly that "instinct" has no clear definition, and that learning by imitation and/or experience may modify inherited patterns of behavior.

Hence, even in the first writings of evolutionary theory (and its postulation of heritability of traits, physical and behavioral), there was room for learning and development—a point that Tinbergen's formulations acknowledged and maintained. In Tinbergen's model, the task of the ethologist is four-fold: namely to address the *ontogeny* (the development of an animal), the *phylogeny* (the evolution of the animal), the cause, and the function of the animal's behavior. By erecting these four pillars of investigation, ethology had turned away from subjective states, cognitive abilities, or emotions of animals, and hence away from presumed unscientific ways of engaging with animals. Cartesian philosophy, from Descartes onward with its mechanistic reflex theories of animal behavior, thus had strangely combined with the calls of Tinbergen and colleagues for a rigorous science, to maintain a solid taboo over the question of animal intelligence and emotions.

Ethologists studied a wide range of species in their natural environment and emphasized the comparisons between species, the most famous exponent of this approach being Konrad Lorenz. Ethologists saw value in studying each species in its own right. Their approach was a direct application of Darwin's theories, and it contrasted with the approach of comparative psychologists, who focused on experimenting with just a few selected species and conducted laboratory studies on learning, usually with the aim of understanding human behavior rather than animals per se. Some syntheses were attempted between the two directions. Particularly in the second half of the twentieth century, it was discovered that there are methodological ways of studying areas of animal behavior that had been thought to escape scientific scrutiny. Today, there is evidence that many mammals may have higher cognitive abilities that, for a long time, were thought to be uniquely human or, just by proximity to humans, partially present in apes.

Emotions and Cognitive Abilities in Animals

And it is at this point that one ought to return to Darwin and examine what he attributed to animals. In *The Descent of Man*, particularly in chapters 2 and 3, called "Mental Powers of Man and the Lower Animals," Darwin makes a number of observations that, despite certain shortcomings in interpretation (but only when judged against current scientific knowledge), illuminate a field of studies in animal behavior that has been in ascent ever since the 1980s. We shall address his catalog of behavioral attributes in animals in the order in which Darwin described and named them in these chapters (note that page numbers in brackets refer to *The Descent of Man*).

Feelings

First, Darwin addresses the question of feeling in animals. He argues that animals feel pleasure, pain, misery and fear. These claims are now largely supported by research and form the basis of animal ethics codes around the western world. He further claimed that certain species show strong individual differences in temperament. Today, we may not use Darwin's terminology, such as "timidity" or speak of "ill-tempered" or "sultry" animals, as Darwin did, but the study of individual temperament is now being investigated again.

Higher Mental Powers

In the same chapter, he makes an even more daring assumption that complex emotions are common to "higher animals" and humans. Complex emotions, in Darwin's catalog, include jealousy and modesty. This is something that we still have not proven, although owners of companion animals have long since believed that their pets can show such emotions and some researchers are investigating these questions.

Attention and Imitation

In Darwin's view, two particular abilities foster learning and therefore higher cognition: One is the "the power of attention" and the other the "power of imitation" (p.44). If an individual organism can attend to something then it is possible for that individual either to imitate what it has seen or to be taught to do something. Research into a large variety of species (including birds and primates) has shown that both can be present in one species and be part of learning, hence behavior is not based solely on instinct.

Reason

Today we might refer to the powers of reason Darwin described as problem-solving ability. Problems can be solved by trial and error, by learning from others (observation) or by insight. Many authors rather too hastily attributed the problem-solving abilities of apes merely to imitation learning, but there are many more recent reports indicating insight learning. It may well be that only a few individuals within a group or species may possess the power of insight and that others then learn from that individual, by imitation, but this does not negate this ability.

As an aspect of higher cognitive abilities, Darwin also mentions memory and imagination. The ability to recognize an individual after a considerable passage of time or the ability to dream indicated to Darwin that "higher animals," as he called them, possess the rudiments of mind.

Tool Use

Even in Darwin's time there was debate about the existence of tool using in animals; some denied its existence altogether, but Darwin was able to point to a few recorded and published examples of tool using in chimpanzees and baboons. He describes the case of a chimpanzee "in a state of nature" (i.e., not in captivity) using a stone to crack a nut and an orangutan using a stick as a lever or, baboons, in conflict with neighboring troops of another species, using stones in defense, hurling them down a slope against their assailants (p.51). Similarly, Darwin quotes reports by Wallace of orangutans throwing sticks and spiky durian fruit like missiles at the human observer. We now have further evidence that some species use tools, and that such tool using may be more versatile than had once been thought. Elephants are credited with tool using, and the tool using of the great apes is now very well documented. Moreover, tool using has been discovered more recently in birds.

Darwin, to some extent, disagreed with the Duke of Argyll that the fashioning of tools is altogether unique to humans by downplaying the human "invention" of using flint stones and how the first fashioning of such tools may have come about (p.52). Today, we have detailed research accounts of the fashioning of tools by some great apes and by one avian species, the New Caledonian crow, reported in *Nature* and other journals in the 1990s. Such fashioning of tools may be relatively simple—varying the length of a stick to make it fit a tree hole—but also more complex—varying the shape of a tool (creating multistep tools in the case of New Caledonian crows) in order to fulfill a particularly difficult extracting task—or one may think of the detailed descriptions of chimpanzees using an anvil and a rock to crack nuts. The anvil has to be carefully selected or even fashioned so that the nut stays in position while force is applied by a rock hammered from above.

As examples of tool using Darwin also referred to nest building and building of shelters, as had been observed of orangutans and baboons. Some of the examples of tool using today are still hotly debated in the sense that they may have come about as a result of proximity to humans and can, therefore, not be classified as something peculiar to the species. Darwin's point about tool using acquired because of the presence of humans is still noteworthy because the individuals in question continued to use the tools thereafter in appropriate ways, and sometimes in new and different situations, showing that they had not just mimicked a behavior but that they had understood the function and uses of the tool they were shown.

Property, Art, Culture, Perception of Beauty

Darwin's description of rudimentary presence of a sense of property, art and perception of beauty does not necessarily strike a cord today, certainly not in scientific study of animals. However, we cannot quite dismiss them outright either. He described, for instance, the episode of a chimpanzee hiding his favorite rock with which he liked to crack nuts. Darwin thought that this was evidence of a sense of property. The stone had no direct food value and other stones were available. Research refers to food storing in animals as *caching* and in many of these cases (known in species of various corvids and described in detail in woodpeckers and especially in the nutcracker), such caching becomes a vital survival task; in such cases, food has to be stored for the lean winter months ahead or the individual will starve. However, some crows and ravens simply do not pass up a food offer even when there is plenty about, and they cache it and retrieve it, sometimes on the same day. The important point here, relating to Darwin's claims of a sense of property ownership, is that the caching process in these crows and ravens is accompanied by repeated checking of the environment. If there is another crow that might have watched, the caching individual will at once retrieve the find and try to hide it somewhere else. In other words, there is some understanding that another might desire the same morsel and might take it. Guarding against this could certainly be regarded as taking possession of the food item. It may also indicate that the caching individual is aware of a state of mind of another and, evidently, capable of foreseeing that it could be stolen. Hence, a sense of property ownership might indeed be regarded as a higher cognitive ability.

Sexual Selection

Darwin refers here largely to sexual selection and female choice—the colorful displays of male birds or their complex tunes, in his view, are a sign that the female must have a sense of beauty and will choose the most beautiful one. Modern research regards certain features of display as signaling health, strength and perhaps even intelligence, but not "beauty." However, the case of the bowerbird is somewhat difficult to accommodate in the model of "health," and this is Darwin's clinching argument for a perception of beauty. After all, the bowers that various species of male bowerbirds build are not nest sites. Bowers are purely for display and they can be extremely elaborate structures that serve no known specific purpose other than to attract females. The bowers are often decorated with shapes and objects of bright colors. Some decorate only in one color, others in multiple arrangements. Females come and inspect them in detail and they may consent to a mating or fly off after such inspections. However, there is another way to argue this—a bower can certainly signal the age and maturity of a male bird. Immature and young adult male birds also build bowers, but these are not skillful, even to the human eye, and it takes bowerbirds a long time to perfect the art of building such a bower, usually by watching successfully breeding adult males build their bowers—the age and maturity of the courting male may be the trigger of which the bower is its visible signal.

Culture

Today, scientists take less, or no, offense when the argument is made that certain species of animal, foremost primates, possess something akin to "culture," as Darwin suggested. Culture in animal behavior studies refers to the presence of behaviors or habits that may be specific to a species in one geographical location but not in another, and persist over generations within the same area.

Sociability in Animals

Darwin argues that social animals in particular might well develop higher cognitive abilities and even a conscience, namely, altruism. He writes that animals living in groups tend to perform a number of services for each other, the most common of which is to respond as a group or on behalf of the group in the case of danger. A joint action might ward off a much larger predator. Another is to warn each other of danger. He cites groups of monkeys issuing not only cries of danger but also of safety so that group members can resume normal activities (p.74). He includes mutual touching, licking and grooming among the services, and even such cases as protecting group members when ill or even blind (p.77, final pages of Ch.3, *The Descent of Man*). The concept of altruism gave rise to many debates, especially in the 1970s and 1980s, since sociobiologists could not see how genes encoding altruistic behavior would be passed on from generation to generation.

Darwin's Contribution to Science

In Charles Darwin's theoretical mold, animals are not extraneous to us, in the sense that we are not separate from the animal world, but one of the evolutionary extant outcomes of this process. Darwin's influence on the study of animal behavior is as diffuse as it is distinct. His theories have been misused and taken at face value in alluring pseudoscientific justifications for a host of unethical human acts, beliefs and values. We cannot blame Darwin for the latter (even though his views on human cultures were circumscribed by nineteenth century understanding) because he presented us with a way of viewing all living things in such a new light that his contribution is equal to Galileo's. As in the case of the position of the earth in the solar system and the loss of the earth as the center of the universe, Darwin's view took away something of the centrality and importance of human beings. What has been gained instead is a new appreciation of the importance of animal life around us as something that matters for understanding life, including our life. That is part of the reason why we continue to have an abiding interest in animals and in animal behavior. More unsettling is the thought that evolution is an ongoing process and that living organisms will change in the future. The study of animal behavior today is also a study of how these processes have functioned and to discover laws about the existence, preservation or disappearance of certain traits. Yet it is wise to remember that Darwin's many theoretical contributions and astute observations are not easily fitted into a simple formula. It would be nice if we could crystallize Darwin's views on animal behavior into a few simple lines. Unfortunately, as Stephen J. Gould reminded us a decade ago, conceptual complexity is not reducible to a formula "as we taxonomists of life's diversity should know better than most" (Gould 1994, 6774).

See also Anthropomorphism

Cognition—*Cognitive Ethology: The Comparative Study of Animal Minds*

Cognition—*Theory of Mind*
Culture
Emotions—*Emotions and Affective Experiences*
Frisch, Karl von *(1886–1982)*
History—*History of Animal Behavior Studies*
Lorenz, Konrad Z. *(1903–1989)*
Sociobiology
Tinbergen, Nikolaas *(1907–1988)*
Tools—*Tool Use*

Further Resources

Barnett, M. 1981. *Modern Ethology*. London: Oxford University Press.

Boakes, R. 1984. *From Darwin to Behaviourism*. Cambridge, UK: Cambridge University Press.

Darwin, C. 1981. *The Descent of Man, and Selection in Relation to Sex*. Facsimile from 1871 edition held in the Firestone Library, Princeton University, Princeton University Press, Princeton, NJ.

Darwin, C. 1890. *The Origin of Species by Means of Natural Selection*. 6th edn., London: John Murray.

Darwin, C. 1904. *The Expression of the Emotions in Man and Animals*. London: John Murray.

Darwin, C. & Wallace, A. R. 1844. *Evolution by Natural Selection*, Cambridge, UK: Cambridge University Press.

Dewsbury, D. A. 1992. *Comparative psychology and ethology. A Reassessment*. American Psychologist, 47(2), 208–215.

Gould, S. J. 1994. *Tempo and mode in the macroevolutionary reconstruction of Darwinism*. Proceedings of the National Academy of Science USA, 91, 6764–6771.

Heinroth, O. 1985. *Contributions to the biology, especially the ethology and psychology of the Anatidae*. In: *Foundations of Comparative Ethology* (Ed. by G. M. Burghardt), pp. 246–301, New York: Van Nostrand Reinhold.
　　(original work: 1911, Ornithologen-Kongress, Berlin, pp. 589–702).

Hinde, R. A. 1966. *Animal Behaviour: A Synthesis of Ethology and Comparative Psychology*. London: MacGraw-Hill.

Huxley, J. 1942. *The Living Thoughts of Darwin*. London: Cassell.

Jaynes, J. 1969. *The historical origins of 'ethology' and 'comparative psychology'*. Animal Behaviour, 17, 601–606.

Kaplan, G. & Rogers, L.J. 2003. *Gene Worship*. New York: Other Press.

Lehrman, D. S. 1953. *A critique of Konrad Lorenz's theory of instinctive behavior*. Quarterly Review of Biology, 28, 337–363.

Lorenz, K. 1980. *A personal introductory history of ethology*. The Journal of Mind and Behaviour, 1(2), 171–122.

Page, R. E. Jr. 1997. *The evolution of insect societies*, Endeavour, 21 (3), 114–120.

Plotkin, H. C. (Ed.) 1988. *The Role of Behaviour in Evolution*. Cambridge, MA: MIT.

Rogers, L. J. & Kaplan, G. Eds. 2004. *Comparative Vertebrate Cognition: Are Primates Superior to Non-primates?* New York: Kluwer Academic /Plenum Publishers.

Thorpe, W. H. 1979. *The Origins and Rise of Ethology: The Science of the Natural Behaviour of Animals*. New York: Praeger.

Tinbergen, N. 1951. *The Study of Instinct*. Oxford, UK: Oxford University Press.

Whitman, C. O. 1898. *Animal Behaviour*. Biology Lectures of the Marine Biology Laboratory, Wood's Hole, MA, 285–338.

Gisela Kaplan & Lesley J. Rogers

■ Development
Adaptive Behavior and Physical Development

Body parts clearly serve adaptive functions, promoting individuals' survival and reproduction. But, appropriate patterns of behavior must also evolve for anatomical tools emerging during development to be implemented effectively. Gonads and genitalia allow reproduction, for example, only for animals able to identify and court appropriate partners as potential mates upon reaching adulthood. Overwhelmingly, in fact, animals restrict their sexual solicitations to prospective partners of their own kind and also avoid mating with close kin. Similarly, bodies that resemble thorns or leaves can suppress risk of predation principally for insects that also rest routinely where these illusions are credible. Some leaf-like species even simulate swaying in breezes. Wings coevolved with flight behavior. And, monarch butterflies and many northern birds would perish if individuals did not also feel compelled to fly hundreds of miles south as days grow shorter each autumn, to particular supportive winter habitats. Some squirrels and bears, instead, change their dietary preferences, increase their appetites, and lower their metabolisms to accumulate fat, before seeking shelters whose specific features promise good insulation. Thereafter, successful individuals reduce behavioral activity and maintain low levels throughout winter months.

Since, prereproductive conditions of existence necessarily differ from those engaged by the fully mature in all species, we should expect that patterns of behavior change in adaptive ways over the course of individual lives. Offspring must begin smaller than their parents, for example. So, faster changes of body temperature and higher numbers of predators will inevitably threaten immatures in comparison to adults. Any hard-to-process foods on which adults rely will generally be inaccessible to weaker youngsters. Juveniles lack experience with dangers and resources, and many must work to establish themselves on territories or other preferred habitat. To whatever extent immatures' constraints, challenges, and opportunities characteristically differ from those of adults, in any given species, distinct prereproductive life ways will have evolved, with success, as ever, depending on close, functional dovetailing of behavior with anatomy.

Consider any of the thousands of marine invertebrate species—starfish, corals, crabs, clams, sea urchins, whelks, barnacles, and more. All begin life as planktonic larval forms, comprehensively different from adult members of their species. After traveling passively with ocean currents for a week or more, surviving on yolk, individuals undertake one or more metamorphoses to reorganize anatomy and behavior for successive life ways. First, larvae seek to launch juvenile life phases on the ocean floor, responding to substrate type, local odors, chemical evidence of predators, and other cues to select habitats known (by scientists) to yield high rates of postsettlement survival. After more months or years, a second metamorphosis may occur, again pairing anatomical changes with vital behavioral adjustments. American lobsters (*Homarus americanus*), for instance, depart juvenile shelters to undertake adult lives entailing movement over 100 miles or more each year.

How about humans? Imagine what it would be like to be an adult just under 3 feet tall with a head 50% larger than normal, shockingly pudgy feet, and severely limited coordination. With these attributes, how might it feel to try and dash through a crowd up a long, steep stairwell to connect from your subway to the bus? To command attention at a public demonstration? To compete in 90 minutes of full-field soccer? Such anatomical proportions and capacities would turn these and many other activities into real struggles. Yet, these very constraints form the basis of daily experience for human 2-year-olds.

Having learned to walk just months earlier, these inexperienced and uncertain creatures are, in fact, dangerously top-heavy, with heads that are huge relative to overall body size.

Many animals develop in such a top-down or front-to-back manner. But, distinctions among taxa relate clearly to specific behavioral factors affecting chances for survival. Any infant mammals' abilities to breathe and swallow are obviously vital. But also, in nature, newborns that disproportionately develop their front ends first, like humans, need not move around much on their own early in life. By contrast, open savannas and pervasive predator pressure caused East Africa's antelopes to evolve their amazing abilities to run adroitly at full speed within minutes of birth. Understandably, percentages of full neonatal stature comprising posterior body parts (e.g., legs) are much greater in these species than in humans, most rodents, and other den and nest-building animals.

Other behavioral factors pertaining to human infants' uniquely large skulls are suggested by the sutures, or gaps between skull bones, intersecting at six large holes called "soft

Human toddlers are dangerously top-heavy.
Courtesy of Corbis.

spots" or fontanels. Normative human development requires much of our extraordinary brain size to be in place before and, especially, soon after birth. But, without those extremely flexible sutures, where adjacent bones can even overlap during human birth (itself much a fetal behavioral initiative, as it turns out), newborns would need to emerge younger, as even more underdeveloped (exterogestate) fetuses, or after the usual 9 months with much smaller and, therefore, even more rapidly growing brains. Over the past 2 million years, however, these options would have increased energetic costs for an infant care pattern already very "expensive," and the third alternative—wider maternal hips—was precluded by the natural selection shaping upright-walking hominid females for 3 million years before we evolved really huge brains. After the neonatal metamorphosis to breathing via lungs, ingesting food orally, and self-regulating core body temperature, our earliest developmental milestones relate principally to the emergence of sophisticated cognitive capacities, including language.

Steep predator pressure on open savannas has caused East Africa's antelopes to evolve the ability to run adroitly at full speed within minutes of birth. This infant impala gazelle (Aepyceros melampus), photographed in Amboseli National Park within 24 hours of birth, for example, shows off legs that constitute a full 60% of neonatal stature, compared with its mother's shape, legs comprising less than 50% of adult stature. Members of the cattle family (Bovidae), gazelles are powerful jumpers, and many are able to maintain running speeds of about 30 mph (48 kph) for long periods of time including bursts of approximately 60 mph (96 kph). Adult impala gazelles are the strongest jumpers of all, able to leap 10 ft (3 m) into the air and travel 30 ft (9 m) in a single bound.
Courtesy of Michael Pereira.

A pair of best friends.

Top: © Anne W. Krause/Corbis;
Bottom:
© Michael Nichols/National Geographic Image Collection.

Sutures accommodate the related tripling of brain volume characterizing our first 2 years outside the womb. Large flexible skulls per se may function additionally to buffer infants' brains effectively against trauma while bipedal locomotion is first attempted. The fontanel atop the skull closes only as most toddlers begin steadying themselves upright, and brain injuries sustained by toddlers are generally far less severe than those observed among adults suffering comparable falls.

Brains are primary foci for developmental considerations because they are core anatomical parts and the ultimate sources of behavior. Across species, relative brain size correlates negatively with rate of physical development and, in mammals, positively with rate of play behavior. Large-brained animals, in other words, generally grow slowly and play frequently. Further, prominent play has recently been linked to specific aspects of neural development. Synapse formation in the cerebellum (which coordinates bodily movement) and parts of the cerebrum (involved in voluntary behavior) is most significantly modified by experience soon after birth, for example, and diverse mammals engage in play most frequently at precisely the ages, in their species, when these windows of opportunity occur. Some kinds of play behavior, then, appear to function as the brain's way to make a better brain. The most recent research on brain development corroborates also for our own species that behavioral initiative and experience sculpt certain regions of neuronal circuitry most dramatically during particular phases of juvenile and teenage development.

Important behavioral factors also relate to more familiar aspects of physical development. The growth spurts and other changes of sexual maturation in primates, for example, constitute metamorphoses nearly as radical as tadpoles' transformations into frogs, although they have not commonly been envisioned in this way. Environmental triggers for puberty remain to be identified, as do behavioral correlates of its particular timing (age of maturation), both across species and among individuals within species.

In this case, consider the baboons of Africa (*Papio* species) and their closest relatives, the macaques of Asia (*Macaca* species). These semiterrestrial monkeys form large social groups within which patterns of dominance and other social dynamics have long been considered extremely similar. Females remain for life in the groups into which they are born, whereas males depart after maturing to join nearby social groups. At puberty, male baboons and macaques generally double in body mass over about 2 years. This rapid growth and correlations among food-intake rates, adult size, and eventual reproductive success make effective feeding competition paramount for maturing males.

In baboons, delay of accelerated growth until fairly large size has already been attained (about 80% of adult female size) enables males to compete effectively with group mates, including all adult females. Pubertal male macaques, by contrast, initiate growth spurts while they are yet less than half the size of adults and, therefore, too small to compete with older females. Consequently, it appears that female baboons, but not female macaques, have evolved a complementary disinclination to engage males in physical fighting. This is suggested by the male social dominance over female agepeers that is invariably observed among juvenile baboons despite that males are no larger than females for 3 full years after weaning and before sexual maturation. Species differences in female perceptions of males as social partners, and male perceptions of females, thus help to differentiate fundamental dynamics within societies that are, in fact, only generally similar across this large tribe of Old World monkeys. Consequences include divergent expression and functioning of "friendships," or special relationships—between males, between females, and between the sexes—among these species.

Using a wide spectrum of perspectives and methods, research of the next several decades will explore how interactions among early behavior, experience, and neuronal development relate to anatomy, life history, and later behavior among individuals of different species, ages, sexes, and personal circumstances. Whereas we already know, for example, that juvenile mammals suffering malnourishment invariably abandon play behavior, we do not yet know what difference, if any, that makes for members of any species. One major class of future investigation will illuminate how patterns of growth and prospects for survival and reproduction—relating to particular perceptions of the environment and approaches to social relationship—differ among individuals whose respective developmental circumstances made possible (or imposed) plenty versus hardly any experience of Type X. When those answers are in, we will much better understand behavior, development, and adaptive evolution.

See also Behavioral Plasticity
Development—*Behavioral Stages*
Development—*Spotted Hyena Development*
Development—*Embryo Behavior*
Development—*Intrauterine Position Effect*

Further Resources

Alcock, J. 2001. *The Triumph of Sociobiology*. New York: Oxford University Press.
Bucknell University Program in Animal Behavior. http://www.bucknell.edu/AnimalBehavior/
    The Bucknell University Program in Animal Behavior was jointly founded in 1968 by the departments of biology and psychology and requires coursework in basic science while encouraging students to study the humanities, relevant social science, languages, and cultures. The availability of animal colonies at Bucknell complements opportunities to conduct fieldwork in a variety of settings off campus.
Hauser, M. D. 2000. *Wild Minds: What Animals Really Think*. New York: Henry Holt.
Hrdy, S. B. 1999. *Mother Nature: A History of Mothers, Infants, and Natural Selection*. New York: Pantheon Books.
Pinker, S. 2002. *The Blank Slate: The Modern Denial of Human Nature*. New York: Viking.
Sherman, P. W. & Alcock, J. (Eds.) 1993, 2001. *Exploring Animal Behavior: Readings from* American Scientist, 1st and 3rd eds. Sunderland, MA: Sinauer Associates.
Strauch, B. 2003. *The Primal Teen: What the New Discoveries about the Teenage Brain Tell Us about Our Kids*. New York: Doubleday.
Wahle, R. A. 1993. *Gimme shelter*. Natural History, 102, 42–49.

Michael E. Pereira

■ Development
Behavioral Stages

Jean Piaget was one of the first psychologists to propose stages of development in children. Since then, a few researchers have written about stages of development in animals. For example, Sue Taylor Parker and Michael McKinney wrote about stages in monkeys and apes. One of the difficulties in comparing animals and humans, is that the traditional tasks used to test human behaviors cannot be directly applied to animals, nor can the tasks for one animal always be applied to another.

The Model of Hierarchical Complexity Allows for Interspecies Comparisons

In recent years, Michael Lamport Commons and his colleagues have proposed the Model of Hierarchical Complexity (MHC) that can be used to determine the stages of animal behavior as well as human behavior. It does so by taking the actions and tasks that animals and humans engage in, and putting them into an *order* based upon how *hierarchically complex* they are. A task action is defined as more hierarchically complex when the higher order action is defined in terms of the actions at the next lower order and organizes these lower-order actions in a nonarbitrary way. Thus, the *order* of hierarchical complexity is obtained by counting the number of coordinations that the action must perform on each lower order action until one reaches a set of elementary order actions. If an action organizes two or more actions from an order before it, that organizing action is by definition one order higher and is therefore more hierarchically complex. *Stage of performance* has the same name and number as the corresponding order of hierarchical complexity of the task it correctly completes.

Behavioral task actions have been described by 15 hierarchical orders of complexity. Theoretically, higher orders are possible. Based on this theory, the table shows stages of animal behaviors. An animal species is characterized by the *highest* stage of performance observed of any member of that species with any amount of training. Animals have been observed to engage in actions up to the concrete stage of development, which is about what 8–10-year-old children do.

The barrier against animals developing abstract stage actions, which is the next stage beyond concrete, is great. At the concrete stage, action involves a small number of specific instances. At the abstract stage, actions involve large-to-indefinite-sized sets. There are no concrete instances for many of these indefinitely large sets and, for this reason, many of the sets are represented by variables. The value of variables can refer to hypothetical things that do not exist. Once one starts using variables it becomes essential to have abstract symbols, such as words, to label those variables. Only humans, thus far, have shown the capacity for using such arbitrary symbols.

See also Behavioral Plasticity
    Cognition—*Grey Parrot Cognition and Communication*

Further Resources

de Waal, F. B. M., & Lanting, F. 1997. *Bonobo: The Forgotten Ape*. Berkeley: University of California Press.
Goodall, J. 1988. *In the Shadow Of Man* (Rev. edn). Boston, MA: Houghton-Mifflin.

Parker, S. T. & McKinney, M. L. 2000. *Origins of Intelligence: The Evolution of Cognitive Development in Monkeys, Apes, and Humans*. Baltimore: Johns Hopkins Press.

Wadsworth, B. J. 1995. *Piaget's Theory of Cognitive and Affective Development*. 5th edn. New York: Longman.

A Comparison of Animals Stages

Order of Hierarchical Complexity	Name (Commons, et al 1998)	Discriminations	Examples of Highest Stage Attained by a Species
0	Calculatory	Exact—no generalization	For computers, only written programmed learning possible. There are no animals that function at this stage.
1	Sensory and motor	Rote, generalized	When water moves, mollusks open shell. Reflexively, if something touches membrane, shell closes. Mobile animals (e.g., Aplysia) habituate, sensitize, and classically and operantly condition. Conditioning produces *generalizations* about which stimuli will elicit the responses of interest.
2	Circular sensory-motor	Open-ended classes	Animals coordinate perception with action, or two or more actions Those whose hunting behavior is controlled by consequences (e.g., most predatory fish, insects) are in this stage. Corrette (1990) observed prey capture in the praying mantis, which coordinated capture and strike movements. Animals, whose hunting behavior is controlled by consequences (e.g., most predatory fish, insects) are also in this stage.
3	Sensory-motor	Concepts	Concepts such as oddity learning in rats (e.g., Bailey & Thomas 1998). Rats discriminated the "odd" one when given two ping pong balls with food odors and one different odor.
4	Nominal	Relations among concepts. Named concepts.	Vaughan (1988) trained pigeons to associate two arbitrary subclasses of slides of trees with different response rates. High response rate was associated with slides in one subclass and low response rate with slides of the other subclass. When slides in the subclass previously associated with high response rate became associated with low response rate (and vice versa), the pigeons changed their associations and correctly responded to each slide after a short reacquisition trial, showing they could attach a virtual label to a subclass.
5	Sentential	Imitates and acquires sequences. Follows short sequential acts.	Pepperberg's (1992) African grey parrot Alex, uttered multiword sentences organizing nominal labels and words. Alex counted two objects, "one, two." To the new question, "What matter [is this] four corner blue [object made of]?" Alex correctly responded, "wood." Dogs and cats run through long arbitrary sequences of actions.
6	Preoperational	Simple deduction without	Brannon & Terrace (1999) trained rhesus monkeys to indicate the larger of two sets of 1 to 4 squares and circles in two rows. Chimpanzees put nuts onto selected flat anvil

(continued)

A Comparison of Animals Stages *(continued)*

Order of Hierarchical Complexity	Name (Commons, et al 1998)	Discriminations	Examples of Highest Stage Attained by a Species
		contradiction excluded. Follows lists of sequential acts	stones and cracked them with selected hammer stones (Inoue-Nakamura & Matsuzawa 1997). Weir, Chappell & Kacelnik (2002) observed that New Caledonian crows make tools by bending a straight piece of wire and then use the wire to pull food out of a tube. Hunt (1996, 2000) observed similar crow behavior in the wild.
7	Primary	Logical deduction and empirical rules. True counting. Simple arithmetic.	Washburn and Rumbaugh (1991) trained rhesus monkeys to select Arabic numerals associated with a number of food pellets. They reliably chose the numeral associated with the larger number of food pellets in a random array of up to 5 numerals. Rumbaugh, Hopkins, et al (1989) showed an adult female chimpanzee removing from a TV display the number of boxes appropriate to the value of a randomly selected Arabic numeral (1, 2 or 3).
8	Concrete		De Waal and Lanting (1997) describe Kanzi, a captive bonobo chimpanzee, using sharp stone flakes and testing the sharpness of each flake with his lips, rejecting nonsharp ones. He then made flakes by throwing a rock against a hard surface, producing many flakes at once. Making simple flake tools is a primary order action. Testing the tools is another primary order action. Coordinating one primary stage action with the another is a concrete stage action. De Waal (1996) describes how a beta male chimpanzee broke up conflicts in an impartial manner. In order to act impartially, the beta male had to consider the perspectives of the other chimps along with his own perspective. Although his awareness of each of these perspectives is a primary action, his ability to integrate all of these perspectives demonstrates that he operates at the concrete.

More Resources

Specific References from the Accompanying Table

Bailey, A. M., & Thomas, R. K. 1998. *An investigation of oddity concept learning by rats.* Psychological Record, 48, 333–344.

Brannon, E. M., & Terrace, H. S. 1998. Ordering of the numerosities 1 to 9 by monkeys. Science, 282, 746–749.

Chappell, J., & Kacelnik, A. 2002. *Tool selectivity in a non-mammal, the New Caledonian crow* (Corvus moneduloides). Animal Cognition, 5, 71–78.

Commons, M. L., Trudeau, E. J., Stein S. A., Richards F. A., & Krause S. R. 1998. *Hierarchical complexity of tasks shows the existence of developmental stages.* Developmental Review, 18(3), 237–278.

Corrette, B. J. 1990. *Prey capture in the praying mantis tenodera-aridifolia-sinensis—coordination of the capture sequence and strike movements.* Journal of Experimental Biology, 148, 34.

de Waal, F. B. M. 1996. *Good natured: The origins of right and wrong in humans and other animals.* Cambridge, MA: Harvard University Press.

de Waal, F. B. M., & Lanting, F. 1997. *Bonobo: The Forgotten Ape.* Berkeley: University of California Press.

Hunt, G. R. 1996. *Manufacture and use of hook-tools by New Caledonian crows.* Nature, 379, 249–251.

Hunt, G. R. 2000. *Human-like, population-level specialization in the manufacture of pandanus tools by New Caledonian crows* Corvus moneduloides. Proceedings of the Royal Society of London B, 267, 403–413.

Inoue-Nakamura, N. & Matsuzawa, T. 1997. *Development of stone tool use by wild chimpanzees* (pan Troglodytes). Journal of Comparative Psychology, 111(2), 159–173.

Lebowitz, B., & Brown, M. F. 1999. *Sex differences in spatial search and pattern learning in the rat.* Psychobiology, 27(3), 364–371.

Parker, S. T. & McKinney, M. L. 2000. *Origins of Intelligence: The Evolution of Cognitive Development in Monkeys, Apes, and Humans.* Baltimore: Johns Hopkins Press.

Pepperberg, I. 1992. *Proficient performance of a conjunctive, recursive task by an African gray parrot* (Psittacus erithacus). Journal of Comparative Psychology, 106(3), 295–305.

Rumbaugh, D. M., Hopkins, W. D., Washburn, D. A., & Savage-Rumbaugh, E. S. 1989. *Lana chimpanzee learns to count by "NUMATH": A summary of a videotaped experimental report.* Psychological Record, 39(4), 459–70.

Vaughan, W. 1988. *Formation of equivalence sets in pigeons.* Journal of Experimental Psychology-Animal Behavior Processes, 14(1), 36–42.

Washburn, D.A., & Rumbaugh, D. M. 1991. *Ordinal judgments of numerical symbols by macaques* (Macaca mulatta). Psychological Science, 2(3), 190–193.

Weir, A. A. S., Chappell, J., & Kacelnik, A. 2002. *Shaping of hooks in New Caledonian crows.* Science, 297:981.

Michael Lamport Commons & Patrice Marie Miller

■ Development
Embryo Behavior

Why Is It Important to Know about Embryo Behavior?

Did you ever wonder what a baby is doing before it is born? Mothers know, of course, that their babies begin to move long before birth occurs. But what does the behavior look like? Are the movements coordinated? Are they influenced by the unique embryonic environment, in which the embryo floats in a pool of amniotic fluid? And, what is the function of these movements? Is there a relationship between prenatal and postnatal behaviors? Does experience before birth influence behavior after birth?

How Can We Study Behavior in Embryos?

If we want to understand embryonic behavior, we first have to find ways of observing it. In mammals, including humans, this is difficult because the embryo develops inside of the mother. The best technique we have available for observing human embryos (called fetuses after 8 weeks of gestation) is ultrasound. This has been very useful in identifying body parts and seeing large movements. Hans Prechtl and his colleagues have pioneered the use of ultrasound images to study human fetal movement. However, the image is relatively

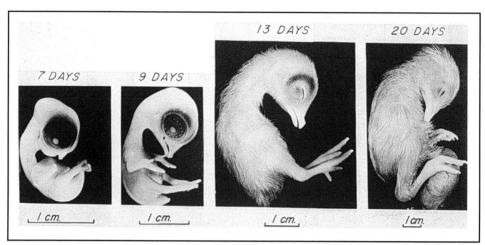

Chick embryos from 7 days, near the time that leg movements first begin, to 20 days, just before hatching.
Courtesy of Anne Bekoff.

"grainy" and the resolution of our current techniques does not allow us to see details of the movements.

For this reason, much of our information about embryonic behavior has come from birds, especially chickens. In birds, the embryos come in neat little packages—eggs. The chick embryo develops inside the egg for about 21 days and then hatches to begin its postnatal life as a baby chick. To see the chick embryo, we can just chip a hole in the shell so that we can see inside. If done carefully, the embryo and the membranes that surround it are not disturbed. The embryo continues to float in the amniotic fluid inside the amnion. Furthermore, we can see that the embryo is moving! Movements begin at about day 3.5 of the 21-day incubation period, as the embryo bends from side to side. By day 6, the embryo's legs and wings begin to move. At day 13, the embryo is nearly continuously active. Every body part moves—trunk, head, beak, legs, wings and tail. These movements, which are called *embryonic motility*, continue up until the time that hatching occurs, on day 21. Beginning at about day 17, movements that are clearly related to hatching begin. The hatching-related behaviors alternate with the embryonic motility during the final days of embryonic development.

What Does Embryonic Motility Look Like?

Viktor Hamburger and Ron Oppenheim and their colleagues made extensive visual observations of chick embryos. Their studies were instrumental in showing that the embryonic behavior was *spontaneous* rather than *reflexogenic*, that is, the embryonic movements were initiated and organized by the central nervous system and did not require sensory stimuli to start or continue. These researchers described the movements as uncoordinated, jerky, and random. Their observations showed that at any given moment, any or all parts of the body might move. They were unable to detect coordination among body parts. In fact, it was not until detailed analysis of videotape recordings were made by Sandra Watson and myself that it became clear that the early movements were, in fact, coordinated. To obtain this result, we put small spots, or *markers*, on the skin of the embryo overlying five major points to delineate the leg: the back, hip, knee, ankle and foot. We then videotaped the embryo as it spontaneously performed its normal repertoire of movements. The videotapes were analyzed on a frame-by-frame basis, and the markers were digitized in each frame. This "kinematic" analysis showed that the hip, knee, and ankle joints generally extended and flexed together, rather than moving independently or in opposite directions. Furthermore, we found an explanation for why the embryonic motility appears uncoordinated. A major factor is that it consists of a mixture of short duration "jerks" and longer, slower movements. The jerky movements fool our eyes and make the movements appear less coordinated than they actually are.

It is also possible to record muscle activity directly via electromyogram (EMG) recordings. I have pioneered the use of flexible suction electrodes to record EMGs from fragile embryonic muscles. This technique has been used to show that the coordinated embryonic movements are due to coordinated activity of the nervous system. For example, extensor muscles of the hip, knee, and ankle are activated together; the hip, knee, and ankle flexors are activated together; and the extensors and flexors alternate. This is similar to the pattern in which these muscles are activated in postnatal behaviors, such as walking. Furthermore, the coordinated muscle activity in embryos begins before any reflexes are functional, and therefore the earliest embryonic movements are centrally organized and do not depend on sensory experience or reflexes.

Is Spontaneous Embryonic Motility an Adaptation to the Unique Embryonic Environment?

Early embryos develop while floating in the amniotic fluid. In this buoyant environment, their limbs can move freely. They don't have to support their weight. Furthermore, their movements don't have to be directed—they don't need to travel or eat or find warmth or safety. Those functions are all provided for them. Furthermore, they are buffered from external stimuli because all sounds, light and mechanical stimuli are damped and attenuated, or completely blocked, by the layers surrounding the embryo. In the case of the bird, these include the shell, various membranes, and the amniotic fluid. Despite the absence or reduction of stimulation, embryos are very active. Near the middle of the incubation period, chick embryos may move almost continuously.

Daniel Drachman and his colleagues performed experimental studies in which chick embryos were prevented from moving by treatment with curare, a drug that stops the muscles from responding to nerve impulses. The embryos developed abnormally. Their joints were fused (*ankylosis*) and the muscles were small and atrophied. These results

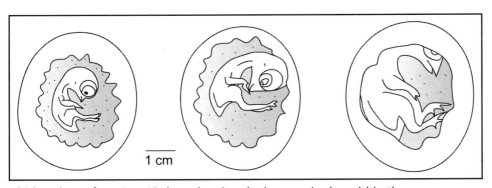

Chick embryos from 9 to 13 days, showing the increase in size within the egg.
Courtesy of Anne Bekoff.

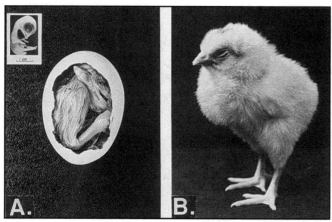

A. Chick embryo folded within the eggshell, ready to hatch at 21 days. Inset: 7-day-old embryo. B. Chick standing up a few hours after hatching.

Courtesy of Anne Bekoff.

showed that the embryonic movements were crucial to the normal development of muscles and joints.

As development proceeds and the embryo grows in size, its body becomes increasingly constrained within the egg. By the time of hatching the embryo will be folded tightly, with legs fully flexed and the head tucked under the right wing so as to fit within the shell. As the embryo gets larger, its movements have an increased potential for damaging itself and the surrounding membranes. For example, if the embryo were to vigorously extend its legs to their full extent, it might not only tear the amnion and other embryonic membranes, but might even damage the foot by hitting the shell. It is probably for this reason that there is a decrease in the amount of movement as the embryo progresses through the second half of embryonic development. Furthermore, the embryonic movements appear to be altered so as to avoid full extension. Therefore, at least in later stages of development, the chick embryo appears to respond to its environment in an adaptive manner.

Is There a Relationship between Prenatal and Postnatal Behaviors?

Substantial evidence suggests that there is continuity between prenatal and postnatal behaviors. For example, as mentioned, my colleagues and I have shown that the earliest leg movements in chick embryos are coordinated. Furthermore, the coordinated pattern of movement seen in embryonic motility appears to represent a generic or *basic* pattern, which is then used and reused in subsequent behaviors, including hatching and postnatal walking. This basic pattern is modulated to produce the specific characteristics that distinguish each behavior and make it distinctive. For example, in embryonic motility, the length extension is approximately equal to the length of flexion: The behavior is symmetrical. However, in walking, the extension phase is typically longer than flexion, and furthermore it varies with speed of walking. This asymmetry is regulated by sensory feedback and helps adapt the steps to speed and terrain. However, removing sensory feedback from a postnatal chick results in the pattern of extension and flexion becoming symmetrical, just as it was in the embryo. Therefore the basic pattern is still there in the postnatal chick and is modulated to produce the asymmetry that we associate with walking.

Does Experience before Birth Influence Motor Behavior after Birth?

There is considerable evidence that early embryonic motility develops before sensory information is available to the embryo. However, it is also clear that, once sensory feedback is available, embryos can respond. The question is whether this prenatal experience affects later behavior. The evidence to date does not fully answer this question. There is some suggestion that at least some aspects of early postnatal motor behaviors, such as walking, can develop normally in chicks even when embryonic motility is abnormal. However, detailed studies using quantitative analysis of electromyograms (EMGs) and kinematic data are needed to answer this question definitively. Recent studies by William Smotherman and Scott Robinson suggest that sensory stimulation before birth in rat fetuses can alter behavior after birth.

See also Behavioral Plasticity
Development—*Intrauterine Position Effect*
Neuroethology

Further Resources

Bekoff, A. 1992. *Neuroethological approaches to the study of motor development in chicks: Achievements and challenges.* Journal of Neurobiology, 23, 1486–1505.
Bekoff, A. 2001. *Spontaneous embryonic motility: An enduring legacy.* International Journal of Developmental Neurobiology, 19, 155–160.
Sharp, A. A. & Bekoff, A. 2001. *Sensory-motor experience during the development of motility in chick embryos.* In: *Motor Neurobiology of the Spinal Cord.* (Ed. by T. C. Cope), pp. 171–191. New York: CRC Press.
Smotherman, W. P. & Robinson, S. R., Eds. 1988. *Behavior of the Fetus.* Caldwell, NJ: Telford.
Von Euler, C., Forssberg, H. & Lagerkrantz, H. (Eds.) 1989. *Neurobiology of Early Infant Behavior.* Wenner-Gren International Symposium Series, vol. 55. London: Macmillan.

Anne Bekoff

■ Development
Intrauterine Position Effect

If you were to look inside a pregnant gerbil, rat, mouse, or most other small rodent you would find her fetuses lined up like peas in a pod, with male and female fetuses randomly distributed within each of the two horns of her uterus. Consequently, you could describe each fetus in terms of its own sex and that of its immediate intrauterine neighbors. A male fetus situated between two male sibs could be described as a two-male fetus (2M male); a female residing between two female fetuses as a two-female fetus (2F female), and so on.

Results of many studies conducted during the last 25 years have shown that the position that a fetus occupies in the uterus, relative to its male and female siblings (its *intrauterine position* or IUP), has life-long effects on its physiology, morphology and behavior. These "intrauterine position effects" are the subject of this article.

To study IUP effects in animals giving birth to more than two young at a time, infants have to be delivered by Caesarian section to determine where they were in relation to one another in the uterus. They then have to be marked permanently so that fetuses from different IUPs can continue to be identified as they mature, and they have to be given to foster parents to rear, because following surgery, mothers may fail to take proper care of their young.

IUP effects exist because male fetuses have higher levels of testosterone in their blood than do female fetuses, and female fetuses have higher levels of estradiol in their blood than do their brothers. These gonadal hormones diffuse from each fetus to its intrauterine neighbors. Consequently, 2M fetuses of both sexes have higher levels of testosterone than do 2F fetuses of the same sex, and 2F fetuses of both sexes have higher circulating levels of estradiol than do 2M fetuses of the same sex. As you might expect, fetuses located adjacent to only one male and/or one female fetus have intermediate levels of estradiol and testosterone, and in adulthood are intermediate between 2M and 2F animals of the same sex in many characteristics.

One important result of studies of IUP effects has been to demonstrate the amazing susceptibility of fetuses during sensitive periods in development to very small differences in exposure to hormones. For instance, the difference in serum testosterone in 2M and 2F male gerbil fetuses is roughly 1.2 nanograms (billionth of a gram) in each milliliter of serum, yet this tiny amount is enough to start a cascade of developmental events that end in profound consequences for reproduction throughout life.

Effects of IUP on males and females of different species are not the same. However, differences between species in effects of IUP should not detract attention from the fact that IUP alters the reproductive life of every species examined to date.

Mongolian Gerbils

Females

The age at which female Mongolian gerbils mature is quite variable. Some show the first signs of puberty before they are 16 days old and while still nursing, others only when 28 days of age or older and already able to live independently. Most unusual, during the period when female gerbils are 22–27 days of age, they very rarely show the first signs of reaching puberty. This strange distribution of rates of maturation is in part an IUP effect. Daughters from 2F IUPs are almost certain to mature early, whereas daughters from 2M IUPs are predominantly late maturing, and age at sexual maturation predicts differences in the lifetime reproductive profiles of female gerbils.

As might be expected, early-maturing gerbil females first reproduce at an earlier age than their late-maturing sisters. Less obviously, early-maturing female gerbils produce more litters during their lifetimes and somewhat larger litters than do late-maturing females, so that early-maturing females produce more than twice as many young during their lifetimes as do late-maturing females. Further, and unexpectedly in view of the strong genetic control of sex of offspring in mammals and the strong effect of IUP on rate of development of female gerbils, litters of early-maturing female gerbils contain more daughters than sons, whereas litters of late-maturing females contain more sons than daughters.

Because there are a greater proportion of female fetuses in the uteri of early-maturing than of late-maturing females, daughters of early-maturing females are more likely to be gestated in 2F IUPs than are daughters of late-maturing females. Thus, mothers and daughters are likely to show similarities in their reproductive life histories because of similarities in their IUPs and, consequently, in their prenatal exposure to gonadal hormones.

Early- and late-maturing female gerbils differ in many aspects of their reproductive behavior. For example, late-maturing females are both better mothers and more aggressive toward unfamiliar males than are early-maturing females. Females that matured early spend less time nursing their young and are less likely to retrieve young that stray from the nest. Late-maturing females are less likely to become pregnant and more likely to attack and injure a male in the weeks following their first encounter with him.

Males

Effects of IUP on male Mongolian gerbils are, if anything, even more profound than the effects of IUP on female gerbils. Fetal male gerbils from 2M IUPs have higher circulating levels of testosterone than do males from 2F IUPs, and this difference in testosterone levels is maintained throughout life.

Perhaps as a consequence of their low testosterone *titers* (concentrations), males from 2F IUPs paired with sexually receptive females are five times more likely to fail to impregnate their partners than are males from 2M IUPs. In fact, a small percent of adult 2F male gerbils have circulating levels of testosterone no different from those of females of their species, and such males never impregnate any females. Perhaps not surprisingly, female gerbils when in breeding condition, but not at other times, prefer 2M males to 2F males.

While 2M male gerbils are more likely to impregnate their mates than are 2F male gerbils, 2F males are considerably better fathers than are 2M males. 2F males generally spend significantly more time in contact with their young, huddled over them and licking them, than do 2M males, and those 2F males that are incapable of impregnating females spend 30 to 50% more time than sexually active 2F males caring for their offspring.

The extra parental effort exerted by 2F males appears to ease the burden on their mates of rearing a litter of young. The average size of second litters reared by females living with 2F males while rearing their first litters is larger than that of females living with 2M males during the same period.

As noted above, males from 2M and 2F IUPs differ throughout life in their circulating levels of testosterone, and differences in parental and sexual activities of male gerbils from different IUPs probably reflect these differences in their testosterone levels. Direct manipulation of the hormone levels of adult male gerbils changes both their sexual and parental behaviors. Males with no testosterone never mate and take very good care of their young, whereas males with artificially elevated testosterone levels are sexually active, but relatively indifferent to their offspring. Thus, testosterone levels in male gerbils, reflecting their IUPs, appear to mediate a trade off between the effort put into directly reproducing and caring for young.

IUP has effects on many behavioral and morphological characteristics of gerbils other than sex and parental behavior. Indeed, everything from whether a gerbil is likely to be right- or left-handed, to the relative size of some structures in its brain and reproductive organs, to its attractiveness to its own parents, is known to be influenced by its IUP. Quite probably, the number of characteristics modified by IUP is greater than we yet know because of the difficulty and expense of studying effects of IUP whether in gerbils or any of the other animals that we consider next.

Mice and Pigs

Intrauterine position has effects on many of the same morphological and behavioral characteristics of other mammals as it does on gerbils, but the direction of IUP effects are sometimes the same and sometimes different. For example, although 2M male mice like 2M male gerbils are more likely to impregnate their partners, and the parental behavior of male mice from different IUPs differ, 2M male mice seem to be more parental and less sexually active than their 2F brethren. In swine, both males and females born in predominately male litters (and therefore more likely to be 2M males) are less likely to conceive as adults than are pigs from litters with relatively few males. Further the relative importance of estrogen and testosterone in mediating IUP effects can differ between species. For example, effects of IUP in gerbils seem to be almost entirely dependent om the testosterone coming from adjacent males, whereas the development of fetal house mice seems to be influenced by the estrogen released by fetal females as well.

Differences in the specific effects of IUP on adult behavior are not unexpected. Although testosterone and estrogen are known to influence the course of development in all vertebrates, their specific effects vary from one species to another. And because IUP effects are a result of differences in prenatal exposure to these gonadal hormones, differences among species in response to hormones should be reflected in differences in the effects of IUP on their development.

Effects of IUP Outside the Laboratory

All of the findings discussed thus far are results of studies carried out in domesticated animals in laboratory settings. Similar effects are known from the wild animals observed in quite natural environments. A particularly ingenious study, in which wild house mice served as subjects, used the oval islands of grass created by the off and on ramps of cloverleaf highway interchanges as enclosures in which to study behaviors of animals from different IUPs under more natural conditions than exist in the laboratory. Wild mice were caught and bred in the laboratory, and their young delivered by Caesarian section, and reared by foster mothers. Once those young had grown to maturity they were released onto highway islands that had been trapped clean of any indigenous inhabitants. The curving highway ramps acted as barriers so that each interchange was a giant cage containing a separate mouse population. Consistent with findings in captivity using female laboratory mice from different IUPs as subjects, 2M females had larger territories than their 2F sisters.

Are There Similar Effects in Humans?

There are surprisingly few studies of effects of sex of uterine neighbors in our own species. Studies that have compared twins of the same and opposite sex have found that women who shared their mother's womb with a brother have more characteristics that some researchers have labeled "male-like" than did twin sisters. However, such an interpretation must be accepted with some caution because "male-like" was often defined in a possibly questionable way, including criteria such as likelihood of using mind-altering drugs and seeking adventure, and susceptibility to boredom. A possibly more convincing, though surely less sensational, study found that sounds produced in the human inner ear, *otoacoustic emissions*, that are more numerous in women than in men, are also more numerous in female members of same-sex twins than in female members of mixed-sex twins. The idea here is that prenatal exposure of female fetuses in mixed-sex twins to testostereone produced by their brothers results in some masculanization of the inner ears of the females, just as in nonhuman animals gestation in IUPs adjacent to males results in masculanization of some body parts of females.

Clearly, much remains to be learned about the role in behavioral development of members of our own species of naturally occurring variation in prenatal exposure to gonadal hormones. IUP effects in nonhuman animals provide important clues as to what sorts of differences in the behavior of male and female humans that experienced different levels of prenatal exposure to gonadal hormones might be looked for.

See also Behavioral Plasticity

Further Resources

Clark, M. M. & Galef, B. G., Jr. 1998. *Perinatal influences on the reproductive behavior of adult rodents.* In: *Maternal Effects as Adaptations* (Ed. by T. A. Mousseau & C. W. Fox), pp. 261–272. Oxford: Oxford University Press.

Vom Saal, F. S. 1989. *Sexual differentiation in litter-bearing mammals: Influence of sex of adjacent fetuses in utero.* Journal of Animal Science, 67, 1824–1840.

Vom Saal, F. S., Clark, M. M., Galef, B. G., Jr., Drickamer, L. C. & Vandenbergh, J. G. 1999. *The intrauterine position phenomenon.* In: *Encyclopedia of Reproduction, Vol. 2.* (Ed. by E. Knobil & J. D. Neil), pp. 893–900. San Diego: Academic Press.

Zielinski, W. J., vom Saal, F. S. & Vandenbergh, J. G. 1992. *The effect of intrauterine position on the survival, reproduction and home range size of female house mice* (Mus musculus). Behavioural Ecology and Sociobiology, 30, 185–191.

Mertice M. Clark & Bennett G. Galef, Jr.

■ Development
Spotted Hyena Development

Spotted hyenas (*Crocuta crocuta*) are cooperative hunters that, unlike other social carnivorans, live in large groups or *clans* that have an unusually complex social organization. Group members team together to bring down large prey and defend their kills against theft, but surprisingly, they also compete fiercely with each other over access to the spoils. Competition is reflected in rank-related priority of access to the carcass, in favor of females, but is primarily manifested in the speed of consumption: Spotted hyenas appear to show the most extreme case of "scramble competition" of any mammal. Thus, the costs and benefits of group living appear in unusually exaggerated form in this species. Moreover, the female spotted hyena displays the most extreme example of masculinization of any female mammal: Her external genitalia are virtually indistinguishable from those of the male, and she is larger and more aggressive than, and unequivocally socially dominant over, the male. Thus, the traditional mammalian pattern of *sexual dimorphism* (i.e., the differences in form between the sexes) is either absent

Spotted hyena mother and cub, Masai Mara, Kenya.
© Joe McDonald / Visuals Unlimited.

or reversed. Understanding how the unusual aspects of social organization and feeding ecology are linked to an even more remarkable suite of male-like morphological and behavioral characteristics in the female has made the spotted hyena a fascinating subject of developmental study.

At a proximate or mechanistic level (a level of analysis aimed at understanding the physiological or developmental events that produce a trait), consideration of the entire array of masculinized female characteristics has focused attention on the role of *androgens* ("male" hormones) in female development. This is because contemporary understanding of sexual differentiation requires that androgens must be present at critical stages to produce male-like patterns. The study of this very unusual species has revealed a largely unexplored set of pathways, potentially functional in other mammals, through which androgens can influence normal female social and sexual development. At an ultimate or causational level (a level of analysis aimed at understanding the evolutionary origins of a trait), consideration of the peculiarities of spotted hyena social organization and feeding behavior have provided insights into the environmental pressures that may have selected for masculinization of the female. Notably, intense feeding competition is hypothesized to be the driving force behind enhanced female aggression and social dominance, traits that presumably are linked to the action of androgens. Thus, through androgen-mediated aggression that enables priority of access to resources, high-ranking mothers and their offspring gain reproductive advantage over lower-ranking families and pass their traits down to future generations. The following sections follow a developmental trajectory to highlight the interplay between ultimate and proximate factors in accounting for some of the unusual features of spotted hyena behavior and morphology.

Reproduction

Although the female spotted hyena's internal reproductive system is similar to that of other female mammals, her external reproductive system is uniquely different. She has no external vagina because the labia majora have fused to form a mock scrotum filled with fatty tissue and having the appearance of testes. Her clitoris is hyperelongated, approximating the size and shape of the male penis, is fully erectile, and is traversed along its entire length by a single urogenital canal (formed by the joining of the reproductive and urinary tracts). Thus, the female has one opening at the tip of her clitoris through which she urinates, receives the male during copulation, and gives birth. Mating through a penis-like clitoris presents unique mechanical challenges for the male, which, owing to his subordinate status and size disadvantage, are initially exacerbated by his extreme approach–avoidance conflict during courtship. Clearly, mating requires female consent and collaboration. A receptive female assumes a seemingly submissive stance, remaining unusually still, with her head down and tail deflected. Using robust retractor muscles, she retracts her otherwise pendulous clitoris so that it is flush with the abdomen and forms a functional vagina. Mating usually involves repeated pairings between the courting couple; however, a female may copulate with different males in a single cycle. Recent DNA evidence shows that multiple mating can result in multipaternity, with littermates being sired by different males.

Gestation and Birth

Clitoral delivery presents its own unique challenges for the female, as the urogenital canal of *primiparous females* (first-time moms) is not large enough to pass a full-term fetus. Therefore, the clitoris must tear to permit delivery of the first cub. Moreover, because gestation is longer in spotted hyenas relative to other carnivores of similar size, the infants are comparatively large (1.5 kg/3.31b) and well developed at birth (e.g., born with their eyes open, incisor and canine teeth fully erupted, and capable of some degree of coordinated movement). The combination of a circuitous birth canal that changes direction in the pelvis, a shorter than necessary umbilical cord that must detach early from the placenta, and a large fetus that must pass through a narrow clitoral passage spells disaster for new mothers and their unborn cubs. Most first-born cubs succumb during the prolonged birthing process, suffocating while trapped in the clitoris before it tears open, or in some cases, the fetus becomes lodged in the pelvis and birthing can claim the mother's life as well. Once the clitoris is torn, however, subsequent births are speedier events that usually occur without incident. In captivity, a female's exposure to androgen blockers during her own period of prenatal development produces a larger and more elastic urogenital opening in adulthood that eliminates the difficulties associated with clitoral birth. Thus, whatever benefits may be associated with prenatal androgen exposure in females are also associated with severe reproductive costs.

The Neonatal Period

In nature, females give birth at a private natal den and initially sequester their young in this protective underground shelter (often an abandoned aardvark burrow or natural crevice). The narrow entrance of the lair prevents access by large predators, including adult hyenas; consequently, maternal contact with her cubs is also constrained. Cubs emerge from the den to nurse, but do not venture far beyond the immediate vicinity of the den entrance—their physical world being of fairly limited scope. Likewise, their social world is also restricted, with partners including their mother and maybe a sibling. Amazingly, the twins that typically constitute a litter are initially intensely aggressive toward each other. Studies of captive animals reveal that their first interactions occur within minutes of birth and involve stereotypic "bite-shake" attacks. One cub latches onto its sibling's upper shoulder region with its teeth and shakes its head vigorously from side to side. Usually within the first day, these interactions establish a dominance relationship between the siblings that, in sisters, can endure a lifetime. Once dominant and subordinate roles are determined, infant aggression rapidly subsides and is replaced in the second week by low-intensity, friendly interactions that gradually become more vigorously playful. However, if the mother's food supply becomes limited, competition between the cubs over access to her nipples escalates. Prolonged or uninterrupted aggressive attacks can produce severe flesh wounds that may ultimately contribute to the death of one sibling, especially if coupled with infection or malnutrition.

Both aggression and play in spotted hyenas emerge on a schedule that is advanced (and in reverse order) by comparison with other carnivores. It is possible that neonatal aggression is an immediate consequence or carry-over effect of fetal exposure to maternal androgens. While in utero, spotted hyena fetuses of both sexes are exposed to androgens derived from their mother's ovaries. As pregnancy progresses, the testosterone titers (concentrations) in maternal blood increase, reaching dramatically high concentrations in the final days preceding birth. Thus, whereas the ultimate driving force behind sibling competition may involve the lifetime benefits accrued to the cub that emerges as the dominant sibling or sole survivor, the proximate mechanism behind neonatal aggression may involve fetal exposure to maternal androgens. In the few cases available for study, cubs exposed prenatally to anti-androgens show a decline in the intensity of neonatal aggression.

Infancy

Once spotted hyena cubs are 2–3 weeks old, their mothers transfer them to the communal den, which similarly includes several underground cavities and a network of tunnels that are inaccessible to adults. The communal den is the center of group activity and is where cubs are introduced to peers and other clan members. The transition from natal to communal den is critical, as cubs experience a dramatic and abrupt change in their social environment. As clans can become quite large (numbering up to 100 members in areas where prey animals are abundant), the sheer number of group members with which young cubs must interact and ultimately forge relationships can be jolting. In addition, cubs remain vulnerable, both to predators and other clan members. They require their mother's protection, yet maternal attendance is constrained by the structure of the den and by her hunting obligations, leaving the youngster often alone and accountable for its social interactions. Thus, to integrate within the clan successfully, infants must arrive at the communal den with a behavioral repertoire that includes appeasement and permits the development of affiliative, as well as competitive, social relationships.

Cubs residing at the communal den must quickly establish their clan membership, learn the rules of appropriate social conduct, and form rank relationships with group members. These activities involve the elaboration and perfection of visual, olfactory, and vocal communication skills. One of the most prominent visual and olfactory signals used by cubs involves erection of their penis or clitoris, and presentation of their erect organ and anal scent glands to clan members, thereby providing the relevant olfactory cues necessary for individual recognition. Although generally unreciprocated, such infantile presentations are the precursors to the ritualized "meeting ceremonies" of older animals, during which two hyenas stand head to tail and lift their legs for mutual inspection and licking. Masculinization of female genitalia allows both sexes to participate equally in these crucially important social displays.

Spotted hyena clans are organized along a linear dominance hierarchy or pecking order, characterized by female social dominance over males. Whereas females typically spend their entire lives within the clan of their birth (*female philopatry*), young males emigrate when they reach puberty (*male dispersal*), ultimately joining new clans. As young females gradually acquire the rank of their mothers, social status within families is extraordinarily stable, passed down across generations of mothers and daughters. Multiple behavioral mechanisms facilitate rank acquisition and maintenance, such as the enhanced aggressiveness of females, maternal intervention, agonistic aiding, and coalition formation. In addition, as an overt and mutual expression of the participants' relative dominance status, the greeting ceremony offers a means of repeatedly advertising and reaffirming social relationships. The same processes of rank acquisition are initially evident in young natal males; however, the slate of their social history is wiped clean upon dispersal. Rank for an adult male is then acquired in his adoptive clan with patience, through years of tenure in residency.

Cubs must also learn survival skills and likely begin to do so through play. During play, rank relationships are momentarily suspended, allowing cubs to gather information about their social environment safely. In addition, play provides a forum for physical development and the practice of motor skills. In keeping with the reverse pattern of sexual dimorphism of behavior, female spotted hyenas engage in more vigorous or "rough-and-tumble" play than do males.

Weaning and the Juvenile Phase

As cubs gradually gain independence through social learning, individual experience, and physical development, the behavioral processes that emerged in infancy continue throughout the juvenile phase. Despite their head start in life, spotted hyena cubs remain dependent on their mother for an unusually long time, residing at the communal den for about 8 months, but being weaned only after about 12 months or more. During this time, the cubs increasingly accompany their mother on her trips away from the communal den, gather information about their physical environment, learn the boundaries of their territory, and further master the skills necessary for their survival. Nevertheless, such prolonged dependency represents an enormous demand and places a premium on the mother's ability to acquire nourishment.

Because spotted hyenas obtain the majority of their diet through active predation, including communal hunting, young cubs must perfect both their solitary and group hunting skills. Although cubs start accompanying clan members on hunting forays at about 6 months of age, when they are still very dependent on mother's milk, they do not kill for themselves until 12–18 months and do not become proficient hunters until adulthood. Likewise, feeding at a social kill (which consists of a frenzied melee of gorging hyenas) also requires a certain degree of social aptitude. Mothers will intervene on behalf of their offspring against lower ranking animals, but eventually youngsters must learn to compete independently.

Puberty and Adulthood

Spotted hyenas grow steadily, lacking the pubertal growth spurt characteristic of many mammals, so they are virtually full grown by 18 months (further emphasizing the energetic drain of nursing). As part of the suite of masculinized features, females are larger and a whopping 12% heavier, on average, than males. The magnitude of this reverse sexual size dimorphism is unusual even among other species that also display size reversals. Males reach puberty at about 2 years of age, but may leave their natal clan at any time between 2 and 5 years, with natal status influencing age at dispersal. Although low ranking males stand to gain little by delaying departure, higher-ranking males reap some benefits from prolonging residence in their natal clan. In accord with their larger size, females also achieve sexual maturity later than do males, at about 3 years of age. At puberty, females experience a rise in estrogen and relaxin, two hormones that increase the elasticity of the urogenital canal and will ultimately allow them to receive a male during mating.

Adulthood is a time for breeding as well as for assuming the social responsibilities that accompany clan membership. For instance, both sexes participate in scent marking to advertise territorial borders, although females take the lead in physical defense of territories during incursions by neighboring clans. A matrilineal social system emerges from the combination of female philopatry and male dispersal, such that subgroups of related females constitute the core of the clan. Perhaps for this reason (ultimately), and as another consequence of masculinization (proximally), females defend their territories more ardently than do males. Unlike many other gregarious carnivores in which only the dominant pair breeds and other group members help rear the young, spotted hyenas have a multimale, multifemale or promiscuous breeding system in which females are the sole providers. Females of all ranks engage equally in breeding, preferentially selecting nonnatal sires, but because of rank-related differences in access to resources, they are differentially successful at raising their young. In addition, males apparently prefer to mate with high-ranking females. Likewise, mating opportunities for the male increase with his social rank, which is a function of residence time in the clan.

Throughout all of these life stages, it is apparent that dominant status is central to survival. In infancy, the dominant cub flourishes over its subordinate sibling, either by eliminating its sibling or by obtaining the majority of the mother's limited resources. In juvenility, cubs acquire their mother's social status, such that cubs of dominant females gain priority of access to the clan's resources. In adulthood, the female's appeal to the male and her ultimate reproductive success are largely foretold by her social status in the clan. Consequently, much of the spotted hyena's social system revolves around aggression-mediated rank acquisition and maintenance, which places a premium on the proximate mechanisms underlying female masculinization.

See also Caregiving—*Parental Care*
Caregiving—*Parental Care and Helping Behavior*
Play—*Social Play Behavior and Social Morality*

Further Resources

Drea, C. M., Hawk, J. E., & Glickman, S. E. 1996. *Aggression decreases as play emerges in infant spotted hyaenas: Preparation for joining the clan.* Animal Behaviour, 51, 1323–1336.

Drea, C. M., Place, N. J., Weldele, M. L., Coscia, E. M., Licht, P. & Glickman, S. E. 2002. *Exposure to naturally circulating androgens in fetal life is prerequisite for male mating but incurs direct reproductive costs in female spotted hyaenas.* Proceedings of the Royal Society of London, B, 269, 1981–1987.

East, M. L., Hofer, H., & Wickler, W. 1993. *The erect 'penis' is a flag of submission in a female-dominated society: greetings in Serengeti spotted hyenas.* Behavioral Ecology & Sociobiology, 33, 355–370.

Frank, L. G., Glickman, S. E., & Licht, P. 1991. *Fatal sibling aggression, precocial development, and androgens in neonatal spotted hyenas.* Science, 252, 702–704.

Glickman, S. E., Frank, L. G., Holekamp, K. E., Smale, L., & Licht, P. 1993. *Costs and benefits of "androgenization" in the female spotted hyena: The natural selection of physiological mechanisms.* In: *Perspectives in Ethology, Vol. 10: Behavior and Evolution* (Ed. By P. P. G. Bateson, et al.), pp. 87–134. New York: Plenum Press.

Holekamp, K. E. & Smale, L. 1998. *Behavioral development in the spotted hyena.* Bioscience, 48, 997–1005.

Kruuk, H. 1972. *The Spotted Hyena: A Study of Predation and Social Behavior.* Chicago: University of Chicago Press.

Mills, M.G.L. 1990. *Kalahari Hyaenas: Comparative Behavioural Ecology of Two Species.* London: Unwin-Hyman.

C. M. Drea

■ Dolphins
Dolphin Behavior and Communication

Despite their grace, dolphins are streamlined aquatic animals with morphological and physiological adaptations shaped over millennia for function. Similar to their terrestrial counterparts, dolphin behavior has been "sculpted" by their environment (in this case, aquatic) with variation exhibited between populations, species, specific habitats, and individuals. The term *dolphin* is often used liberally to refer to all species within the suborder Odontoceti, members of the order Cetacea. However, a generalized categorization is inadequate to represent the incredible diversity in behavior among the roughly 33 species of dolphins in the family Delphinidae that inhabit the world's oceans. For example, in size alone dolphins range from the 1.5 m (5 ft) Hector's dolphin (*Lagenorhynchus hectori*) to the almost 10 m (33 ft) killer whale (*Orcinus orca*). Despite morphological and behavioral differences between

Dolphins

Toni G. Frohoff and Kathleen M. Dudzinski

Dolphins are fascinating mammals, exhibiting unique physiological adaptations to the sea while demonstrating exceptional behavioral similarities with some terrestrial species. Dolphins are members of the taxonomic suborder Odontoceti (derived from the Greek term for sea animals having teeth). Mysticetes (baleen whales) and Odontocetes collectively are known as cetaceans. In addition to having teeth, odontocetes have a single blowhole, a highly specialized echolocation system, and a pronounced forehead, called a "melon." The Odontocetes are comprised of three superfamilies, including the Delphinoidea which include "true" dolphins such as bottlenose dolphins and orcas. The other two superfamilies are the Ziphoidea, (beaked whales) and Physeteroidea (sperm whales).

The evolutionary history of dolphins is the source of much debate. It is generally believed that the earliest known cetacean ancestor (the *Mesonichyd condolarth* from about 65 million years ago) was a hoofed and furred mammal that walked on land, transitioned to a primarily aquatic animal about 45 million years ago, and appeared in the form that we recognize today as dolphins about 5 million years ago. With their anatomy evolved into a streamlined, hydrodynamically efficient shape, dolphins are supremely adapted to life in the sea: Their blowhole is located dorsally and functions as a built-in snorkel. Dolphins are found in all oceans and some rivers around the globe and live from the coastal to pelagic ocean habitats. Dolphins are streamlined for efficient, swift movement through the water and can reach burst swimming speeds of 21 mph (34 kph)—as seen for bottlenose dolphins. Dolphins are able to reach depths exceeding 500 m (1,650 ft) on a single dive because of their flexible, compressible ribcage (to avoid pressure breaks). Added to this anatomical dive adaptation is their abundance of myoglobin (four times that of the hemoglobin in humans) enabling them to store large quantities of oxygen for extended durations. Dolphins can hold their breath for more than 8 minutes.

Despite their different physiologies, some dolphin and terrestrial mammal species (including humans) exhibit striking similarities in behavior including close and complex relationships, sophisticated communication and cognitive skills, highly cooperative foraging strategies, and evidence of culture. Dolphins are typically found in groups of 5-10 individuals or in herds numbering in the hundreds or thousands. Group structure varies from matrilineal groups in which the young remain within access of their mothers for life—to groups of males and groups of females with young. The sensory systems of dolphins have been sculpted by the aquatic environment to aid in their hunting, communication, and navigation. Dolphins exhibit an impressive variety of tactile, gustatory, visual and auditory abilities. Delphinid audition is highly sophisticated with a range of hearing and sound production abilities exponentially greater than that of humans. Dolphins, like bats, are also capable of producing *echolocation* (sonar) that they use to locate prey and for short-range investigation.

Many species of dolphins are facing some of the most serious wildlife conservation challenges today because of worldwide habitat degradation (such as chemical and acoustic pollution), fisheries by-catch, climate change, and various forms of human exploitation.

species, behavioral variation can be highly pronounced *within* species as well; apparently a function of habitat. For example, dusky dolphins (*Lagenorhynchus obscurus*) off Argentina hunt cooperatively while those in the deeper waters off New Zealand do not. And some bottlenose dolphins (*Tursiops truncatus*) in South Carolina and killer whales in various locations (e.g., Indian Ocean, S. Atlantic Ocean shores) temporarily strand themselves on the beach to obtain fish (in the case of the former) or elephant seals and sea lions (the latter).

Research on Dolphin Behavior

Systematic, quantitative research on dolphin group behavior, population dynamics, physiology, distribution, and ecology has only been conducted within the last 35–40 years. Data have been gathered from a variety of different platforms: surface observations from boats (including whaling vessels), land-based stations, aircraft (including airships like blimps), with telemetry devices attached to individual dolphins, and from the captive environment. More recently, detailed observational data have been gathered from underwater in the wild. Detailed behavioral observations of surface actions have been conducted on roughly a dozen dolphin species with virtually all of these studies on coastal or near-coastal populations. For example, coastal bottlenose dolphins, orcas, and short-finned pilot whales (*Globicephala macrorhynchus*) have been studied in many areas. Atlantic spotted dolphins (*Stenella frontalis*) have been observed in the Bahamas and the Azores while spinner dolphins (*Stenella longirostris*) have been studied in Hawaii, Midway, French Polynesia, and in the Eastern Tropical Pacific. Dusky dolphins have been studied intensively around Argentina and New Zealand and humpback dolphins (*Sousa* sp.) off South Africa and Hong Kong. Dolphins have also been studied intensively in captivity.

Nonetheless, research on the behavior and communication of individual dolphins is relatively new. This is likely attributable to the difficulty of studying dolphins. Species that are relatively easy to study (i.e., that live close to shore) typically live in murky water. Dolphins are typically not as accessible to humans as are terrestrial species. Often, dolphins can be difficult to detect and, once spotted, hard to follow, even when they tolerate sustained proximity. The aquatic environment offers no opportunity for researchers to observe dolphins undetected behind "blinds," which limits researchers' abilities to document behavior unaffected by human presence. Studies on captive dolphins circumvent many of the difficulties encountered in the field. Nonetheless, determination of the degree to which these studies can be extrapolated to free-ranging animals has been limited by an absence of comparison with observations in the wild. Thus, it is not surprising that the study of marine mammal behavior lags behind that of some terrestrial mammals such as primates, lions, and elephants. Behavioral studies on dolphins have frequently been conducted from what can be glimpsed from the water's surface. Regardless, extensive similarities between dolphin behavior and that of some terrestrial animals have been reported.

Although most field research has focused on the group activity of dolphins, methods of individual identification (such as photo identification of unique scars, marks, and pigment patterns), improved technology for recording behavior and sounds, and increased tolerance of dolphins to humans in many parts of the world have recently facilitated an increase in research on more subtle and unique behaviors of individual dolphins.

Sensory Systems

Dolphins communicate with one another through an impressive variety of tactile, visual, and auditory abilities. For example, it appears that all dolphin species rely on highly developed acoustical processing for both communication (e.g., whistles) and exploration (e.g., echolocation). Dolphins use echolocation clicks (also called sonar) that are typically high frequency, short and intense pulses of *amplitude-modulated sound* (that is, the frequency essentially remains stable, but the intensity does not) that are used to locate prey and explore the environment. It is generally accepted that all dolphin species echolocate, but this has not been confirmed. Clicks are produced in the *melon*, or forehead, through nasal sacs and are directed at objects. The returning "echoes" are reflections of emitted clicks that result from bouncing off the targeted objects and are received through the lower jaw and transmitted to the brain via the middle and inner ear. Passive listening also seems to play a role in locating food. For example, bottlenose dolphins inhabiting estuarine situations often respond to sounds made by their prey. Many dolphins also produce frequency-modulated, pure-tone sounds, referred to as whistles, that seem to function primarily for communication between individuals.

The majority of dolphins that inhabit water with reasonable visibility have good vision, and thus the ability to make visual signals is an important part of their signal repertoire. Visual communication may be expressed by an extensive variety of postures, gestures, actions, and morphological features (such as dramatic variation in coloration). The use and exchange of visual signals and cues appears to be particularly important in the communication of dolphins inhabiting environments with good visibility. For example, Atlantic spotted dolphins in the Bahamas or spinner dolphins in Hawaii usually frequent water with visibility in excess of 25 m (82 ft). These dolphins seem to rely on sound less than dolphins in murkier waters. For example, Atlantic spotted dolphins in the Bahamas produce one-third the number of whistles of bottlenose dolphins in waters around Mikura Island, Japan, where there is generally less visibility. Further support for this idea that dolphins in clearer water use visual cues more frequently comes from the river dolphins. These fresh water dauphines often inhabit murky, turbid rivers and have small eyes and poor vision as compared with marine species; however, they have exceptional echolocation capabilities. Thus, water visibility may determine, at least to some extent, whether species develop differentially, and rely more heavily upon, visual or auditory cues to exchange information (this is not a comment on the anatomical visual acuity development).

Dolphins have well-developed skin sensitivity which is evidenced by the various ways in which they frequently touch one another. Where and how anatomically dolphins touch one another will convey different meanings ranging from greeting to appeasement to aggression. However, they appear to have only limited gustatory and even less developed olfactory systems, probably as a result of the aquatic environment not being particularly conducive for efficient use of these senses.

Social Behavior

Dolphins are highly gregarious, social animals with both consistent and fluid social relationships. Group size seems to be a function of habitat; nearshore dolphins are often found in smaller groups (such as 2-25 individuals), while offshore dolphins, even of the same species, are often observed in groups of hundreds or even thousands. For many species, the gender, relative ages, and identities of companions often change throughout an individual dolphin's lifetime. Some dolphin communities, such as two populations of bottlenose dolphins studied in Florida and Australia, seem to be *matrilineal*, consisting of females and their accumulated offspring, or sisters and other females. Calves within these groups will often develop stable relationships with each other over a period of years. In these groups, subadult males will usually leave these groups and form *bachelor* groups that often remain together indefinitely. In these two populations, sexually mature males may form partnerships or coalitions with other males and move between the female groups. However, various other social structures have been observed in dolphins. For example, orcas in the Pacific Northwest form stable societies comprised of matrilineal groups that appear to last over the animals' lifetimes. In fact, these are the only known examples of both genders remaining with their maternal natal groups of any mammalian species.

A school of bottlenose dolphins play in the waves of the California coast.
Courtesy of Corbis.

Dolphins live within structurally coordinated social groups where communication between individuals and groups is likely to be important for the maintenance of social life. The behavior of wild dolphins reflects a dynamic interplay of altruism, aggression, social and sexual interactions, exploratory behavior, play, flight, and predator avoidance. Play is an integral part of social relations for dolphins as well as for the development of physical skills required for survival. They are frequently seen riding the bow wave or stern wake of boats and "surfing" on waves. Both juvenile and adult dolphins often chase each other and toss objects, such as seaweed, to one another. Jumping from the water's surface and breaching can also indicate excitement.

Altruistic behavior among dolphins has been frequently reported, and examples include dolphins assisting distressed animals to the surface, physically moving them away from danger, and even attacking the source of danger. Alloparental care has also been observed in some female dolphins that have been observed to "babysit" the offspring of unrelated females. Conversely, aggressive behavior among dolphins is also frequently observed. In fact, bottlenose dolphins who attack porpoises in various parts of the world are one of the few examples of mammals known to direct lethal, nonpredatory aggression toward other mammals. Infanticide has also been documented in Scotland.

Sexual behavior is also a frequently observed form of social interaction among dolphins and often occurs in conjunction with aggressive behavior. In addition to reproduction, sexual behavior may serve a function in social bonding and dominance. Dolphins become sexually mature at variable ages, depending on their species and location. For many coastal dolphin species, females generally reach sexual maturity from about 8-10 years of age, while males sexually mature after 10 years of age. Adults, as well as sexually immature juveniles, are often seen exhibiting sexual behaviors. Dolphins are promiscuous and do not form permanent social bonds with their mates. While courtship behaviors or sexual play may last several hours, intromission is relatively quick, lasting only 10-30 seconds. Some of the sexual behavior of dolphins is similar to that of terrestrial mammals. For example, similar to what Jane Goodall observed in the chimpanzees (*Pan troglodytes*) of Gombe, male bottlenose dolphins in Shark Bay, Monkey Mia, will work in pairs or triplets to herd a female for mating and away from other competing males.

Communication

Physical/Visual Communication. The dolphins' body shape also modifies their methods of communication. For example, modifications for swimming have altered the length and shape of appendages, thus limiting the variety of gestures available to them. In spite of these circumstances, dolphins are adept at using subtle shifts in posture or slight body movements to communicate. For example, aggressive intent or threats by dolphins can be expressed through a direct horizontal or head-to-head approach that is often coupled with jaw claps, head shakes, body hits or slams, and emission of a bubble cloud from the blowhole. The same goes for dolphins exhibiting what is referred to as an "s-shaped" posture (in which the dolphin's tail *flukes* and pectoral fins are lowered and the body appears to be in the shape of an "S") that is sometimes followed by physically aggressive behavior. Dolphins might also flare or extend their pectoral fins (flippers) when posturing aggressively: This action lends the dolphin the illusion of more girth and size and might be intended to "scare off" an opponent. Flipper-flaring by Atlantic spotted and bottlenose dolphins seems to function similarly to ear flapping observed by Cynthia Moss in elephants. Flaring out a single flipper (or even flipping one elephant ear) during nonaggressive interactions tends to send a very different message—maybe one of affection as compared to one of increased size that the double-pectoral flare is likely to convey. However, a slight shift in the angle held by a dolphin's flipper might signify a change in swimming direction to other dolphins in the group. Thus, the same action of the flipper used in varying situations sends a different message to receivers. Such visual displays of posture or gesture are useful for close-range communication among dolphins and provide an alternative or a compliment to acoustic signaling.

The emission of bubbles from dolphins' blowholes likely serves as a visual signal during interactions. For example, some dolphins frequently emit a bubble "burst" beneath the water or at the water's surface immediately prior to approaching a newly-started motor. Some researchers use existence of a bubble stream with a whistle as a method to identify a vocalizing individual dolphin from within a group. Unfortunately, there has been no specific study to examine how dolphins use bubble streams in relation to their sound production. We do know that juvenile dolphins often emit bubble streams from the blowhole when they appear excited. Bubbles are also used during feeding to startle fish or to keep the fish tightly grouped together for easier capture.

Tactile Communication. Dolphins are exceptionally tactile animals; they are often in physical contact with companions. Body contact is an important component in communication, and where and how dolphins are touched conveys different levels of information, or *signal content*, to each individual. Extensive contact and rubbing has been observed for both captive and wild dolphins during play, sexual, maternal, and social (affectionate and aggressive) contexts using the rostrum (beak or "nose"), pectoral fins (flippers), dorsal fin, flukes (tail), belly, as well as the entire body. Touch between dolphins occurs on or with any body part, but in some areas actions appear to have special communicative significance. For example, pectoral fin rubs are sometimes mutually exchanged between two dolphins that have just come together. These rubs may act as a greeting, similar to a handshake or hug between two people—perhaps signal recognition or affection towards another dolphin.

Nonvocal Acoustics. Nonvocal sounds include noise from various body parts striking the water surface as well as the percussive sounds of jaw claps and teeth gnashing. Nonvocal sounds combined with an impressive range of vocalizations make for a highly sophisticated acoustic communications system.

Dusky dolphins breaching. Breaching generally indicates a general excitement.
© Brandon Cole / Visuals Unlimited.

Breaches (i.e., leaping) and slapping chin, pectoral fin, and tail (fluke) against the water produces airborne and underwater sounds upon re-entry that can carry for several miles and may carry communicative messages. Breaching often indicates general excitement deriving from any of several causes including sexual stimulation, location of food, a response to injury or irritation, or parasite removal. Dusky dolphins are well known for three breach types they exhibit in association with three different stages of cooperative feeding: head first re-entry leaps, noisy leaps and social, acrobatic leaps. Noisy leaps may also act as a sound "barrier" to disorient prey and keep them tightly schooled—a signal that is to the dolphin's benefit if not the fishes'. The acrobatic leaps and spins of spinner dolphins seem designed to produce noise since many are common at night, when visual contact is limited.

Tail slapping by dolphins often occurs dozens of times in succession, creates loud, low-frequency underwater and airborne sounds, and usually conveys threat or frustration. The percussive sound of a jaw clap accompanied with a direct approach or aggressive posture is also considered a threat or warning signal. Altogether, the social functions of nonvocal auditory signals produced by dolphins seem limited to long-range communication or to alert others. Coupling nonvocal acoustic cues with other behaviors often exaggerates a signal and its message. Sometimes two or more male bottlenose dolphins will herd or corral a single female for mating purposes. The males create loud jaw pops but also tail slap near the female, push and hit at her body and aggressively approach her. It is not likely that a female pursued with these signals will miss the point of her suitors (aka "attackers").

Vocalizations. Vocal signal exchange is considered the predominant form of communication by dolphins. They live in an aquatic medium where sounds propagate farther and with more reliability than other modalities: Sound travels 4.5 times faster in air than water. Generally speaking, dolphins produce two types of sounds: (1) pulsed, amplitude-modulated sounds (i.e., echolocation clicks) that are broadband, of moderate intensity and high frequency typically associated with nonsocial exploration; and (2) continuous pure tones (i.e., whistles) that are frequency-modulated of mid- to high-frequencies and are typically associated with conspecific socialization.

All dolphins make pulsed sounds and they have been observed to be as high in frequency as 120 kHz. Because not all dolphin species whistle (e.g., Hector's dolphins do not whistle), the use of clicks or echolocation for communication cannot be ruled out. Click trains with very high repetition rates, called burst pulses, possess a social function among dolphins. These clicks are emitted so quickly that they often resemble continuous sounds to the human ear (such as that of a rusty door hinge), rather than a series of discrete clicks. Examples of these sounds include squawks, whines, barks, moans, and more. It appears that burst-pulse sounds are used to communicate among individuals rather than discern information about the environment. We have seen that dolphins at play or while fighting emit squawks, whines and barks. These sounds are harsh and likely convey excitement or frustration. The directional characteristics of many of the pulsed sounds, the relative ease with which they can be localized, their variability, and possibly the power (i.e., intensity of the sound as well as the tactile effects of a loud, powerful sound) with which they can be produced, enhance their value and usefulness as communication signals.

Dolphins produce whistles that overlap with our hearing range (about 2 kHz–20 kHz) but can also reach at least 85 kHz with tones lasting from milliseconds to a few seconds. These sounds often have a rich harmonic content that extends into the ultrasonic frequency range. Whistles vary greatly in contour from simple up or down sweeps to warbles, to U-loops and inverted U-loops. Whistles are thought to function only for communication, but are not produced by all dolphin species.

Why some dolphins whistle and others do not is a very intriguing question. If the whistle appears only under certain ecological or social conditions, one might expect these to be found through careful comparisons of whistling and nonwhistling dolphins. Although a low degree of gregariousness (socializing among group members) is a feature characterizing some non-whistling species, it does not hold for all nonwhistlers. It is fascinating that Hector's dolphins and orcas (the largest of all dolphins) live in social groups, but still do not produce whistles.

With his colleagues, Peter Tyack proposed a leading hypothesis known as the "signature whistle hypothesis" suggesting that dolphins produce a whistle that is individually distinct. More recently other researchers suggested a more limited role for these individually distinct whistles—that of maintaining group cohesion or contact among members of a group. Whistles likely provide a vehicle for maintaining contact and coordination with dolphins while searching for food or traveling out of visual range of peers. Species, regional, or individual specificity in whistles could facilitate identification of schoolmates or familiar associates, aid in the assembly of dispersed animals and in the coordination, spacing and movements of individuals in rapidly-swimming, communally-foraging herds.

Cognition and Culture

The intelligence of dolphins, as evidenced by their behavior in the wild as well as that observed through experiments, has been found to be very advanced—even by human standards. For example, in 2001, Diana Reiss and Lori Marino documented that dolphins—in addition to humans and other great apes—are capable of mirror self-recognition. This is significant, since most scientists previously assumed that this was a uniquely "primate" cognitive attribute.

Also, there is growing evidence that dolphins have culture, similar to that observed in humans and other terrestrial (i.e., chimpanzees and elephants) and avian (i.e., parrots and corvid) species. The delphinid examples are numerous and include free-ranging bottlenose dolphins in Australia, who have demonstrated tool use by using sponges as tools for foraging on the ocean floor, and sperm whales and orcas with vocal dialects that are passed down through generations. In a 2001 article in the journal *Behavioral and Brain Sciences,* Luke Rendell and Hal Whitehead note "The complex and stable vocal and behavioural cultures of sympatric groups of killer whales (*Orcinus orca*) appear to have no parallel outside humans, and represent an independent evolution of cultural faculties."

Killer whales take a look through some holes in the ice.
Courtesy of Corbis.

Cooperative Hunting among Dolphins

Stefanie K. Gazda

Cooperative or group hunting has been reported in a number of mammals and even one bird species. Group hunts that are considered cooperative range from simultaneous chases to hunts that are clearly coordinated.

Harris' hawks, for example, can take different individual roles during hunts. During "flush and ambush" hunts, one to two birds penetrate a bush to flush out a hiding rabbit while others surround the bush and make the kill once the rabbit emerges. The "flush and ambush" strategy of the hawks involves a *division of labor*, which occurs when individuals, working as a team to complete a task, perform different subtasks. Another example of a division of labor in group hunts can be found when individual chimpanzees hunting in the Tai National Forest, Ivory Coast, engage in particular subtasks such as "driving" or "blocking" their Red Colobus monkey prey.

Role specialization is found when individuals take the same subtasks during repeated team tasks. Group hunting with a division of labor and role specialization is extremely rare. The most prominent case is a study of coordinated group hunts in African lionesses (*Panthera leo*). Females in "center" roles wait for prey to move towards them while those in "wing" positions initiate an attack on the prey. Hunting success is higher when lionesses occupy preferred stalking positions.

Cooperative hunting has been described in a number of cetaceans including bottlenose dolphins, killer whales and humpback whales. Accounts of cooperatively feeding bottlenose dolphins include fish being herded into a ball or ahead of dolphins swimming in a crescent formation, or fish trapped against mud banks or between dolphins attacking from either side. Groups of dolphins may even beach themselves to feed on fish they have chased onto mud banks.

In Cedar Key, Florida, group-hunting dolphins engage in a novel behavior while herding fish. One individual in a group of 3-6 dolphins, the "driver," herds the fish in circles, as well as toward the tightly grouped "barrier" dolphins that are closely lined up. The driver may perform fluke-slaps during the drive. Fish being herded leap into the air where some are captured by driver and barrier dolphins. The driver often surfaces alongside the barrier dolphins as the fish begin to leap. Observations of two feeding groups show that individual dolphins herding fish in Cedar Key specialize in the roles of driver and barrier (that is, the same dolphin takes the role of driver in every group hunt), thus meeting the criteria for a division of labor with role specialization.

But are the barrier dolphins just taking advantage of the driver? If so, this behavior would fit a producer-scrounger model, where there is no cooperation between dolphins. This is unlikely, because in such a relationship, one would expect to see attempts by the driver to avoid the barrier dolphins, which was never seen. Furthermore, one group observed is reported to have a stable membership of the same three dolphins, which is not expected in a producer-scrounger relationship. Therefore, this behavior can best be explained as a division of labor with role specialization, possibly the first such case found in bottlenose dolphins or any marine mammal species.

Further Resources

Anderson, C. & Franks, N. R. 2001. *Teams in animal societies*. Behavioral Ecology, 12(5), 534–540.

Caldwell, D. K., Caldwell, M. C. 1972. *The World of the Bottlenose Dolphin*. Philadelphia: Lippincott.

Connor, R. C. 2000. *Group living in whales and dolphins*. In *Cetacean Societies: Field Studies of Dolphins and Whales*, (Ed. by J. Mann, R. C. Connor, P. L. Tyack, & H. Whitehead), pp. 199–218. Chicago: The University of Chicago Press.

Stander, P. E., 1992. *Cooperative hunting in lions: The role of the individual*. Behavioral Ecology and Sociobiology, 29, 445–454.

Wursig, B. 1986. *Delphinid foraging strategies*. In: *Dolphin Cognition and Behavior: A Comparative Approach* (Ed. by R. J. Schusterman, J. A. Thomas, F. G. Wood), pp. 347–359. New Jersey: Lawrence Erlbaum Associates.

Demonstrated intellectual traits in dolphins suggest that they are capable of a highly sophisticated communications system, even if it is very different from that of humans. Controlled experiments have shown that dolphins (specifically, bottlenose dolphins) can understand syntax and semantics. Researchers have presented dolphins with simple sentences—actions, nouns and modifiers—based on simple grammar. Dolphins correctly understood the requests and followed instructions. Currently being examined by these researchers is the capacity of dolphins for logical reasoning, problem solving, and their ability to answer acoustic commands with their own sounds.

Interspecies Interactions

Dolphins of different species interact with one another, and even mate with one another, more often than do many other taxonomic groups. For example, Atlantic spotted dolphins in the Bahamas frequently travel, fight and/or play with bottlenose dolphins. Common and dusky dolphins travel together off Argentina as well as off New Zealand. And mixed delphinid species aggregations are often observed in the Eastern Tropical Pacific. Perhaps delphinid "interspecific sociality" predisposes these animals to having an aptitude for interacting with humans as well.

Dolphins are somewhat unique among wild animals in that it is not uncommon for them to approach humans, and even to initiate "sociable" interactions with them, even without the provisioning of food (although this is certainly a component in some situations). Studies documenting these interactions have recently shed new light on the historical anecdotes of dolphin–human interactions that have accumulated since ancient times. Until recently, scientists largely ignored the subject of "sociable" dolphin–human interactions, largely because of the sensationalization of this topic in the 1960s and '70s. But in the past two decades, researchers have revisited this subject using established ethological techniques that have been useful in managing dolphin interactions with boats and swimmers as well gaining insights into the biology, behavior, and communication of dolphins.

Behavior and Conservation

Many dolphin species are facing some of the most serious of wildlife conservation challenges today, with some species on the verge of extinction. Dolphins are intentionally killed, legally and illegally, in many parts of the world. Dolphins are also unintentionally killed by entanglement in fishing lines and nets. Other formidable threats to dolphins worldwide include prey reduction due to overfishing and bycatch, climate change, and habitat degradation in the form of vessel harassment, acoustic pollution (from boats, oil exploration, and military activities), and chemical and debris pollution. Studies of the behavior of dolphins are critical to understanding how these threats can best be mitigated.

See also Cognition—*Equivalence Relations*
Culture—*Whale Culture and Conservation*
Tools—*Tool Use by Dolphins*
Welfare, Well-Being, and Pain—*Stress in Dolphins*

Further Resources

Cawardine, M. 2002. *Smithsonian Handbooks: Whales, Dolphins, and Porpoises.* New York: Dorling Kindersley.
Mann, J., Connor, R. C., Tyack, P. L. & Whitehead, H. (Eds.) 2000. *Cetacean Societies.* Chicago: The University of Chicago Press.
Norris, K. S., Würsig, B., Wells, R. S. & Würsig, M. (Eds.) 1994. *The Hawaiian Spinner Dolphin.* Berkeley: University of California Press.
Perrin, W. F., Würsig, B. & Thewissen, J. G. M. (Eds.) 2002. *Encyclopedia of Marine Mammals.* San Diego: Academic Press.
Pryor, K. & Norris, K. S. (Eds.) 1991. *Dolphin Societies: Discoveries and Puzzles.* Berkeley: University of California Press.
Reeves, R. S. & Leatherwood, S. P. 1990. *The Bottlenose Dolphin.* San Diego: Academic Press.
Schusterman, R. J., Thomas, J. A. & Wood, F. G. (Eds.) 1986. *Dolphin Cognition and Behavior: A Comparative Approach.* Hillsdale, New Jersey: Lawrence Erlbaum Associates.

Kathleen M. Dudzinski & Toni G. Frohoff

■|Domestication and Behavior

When a free-living animal is taken from its natural habitat and placed in a new and different environment, it frequently exhibits changes in its behavior and physiology as it attempts to adapt to this new environment. In addition to these changes, during the animal's lifetime genetic changes may occur in each succeeding generation of its offspring as they too are exposed to the new environment. *Natural selection* (evolution) will favor those animals that adapt most readily in each generation, and these animals will leave the most offspring (and their genes) for the next generation.

These same processes occur when wild animals are taken from their natural environment and brought into captivity (farms, homes, zoos, laboratories). These newly-captured animals may experience stress as they attempt to adapt to their new environment. The space they live in may look very different and be more limited than in the wild. They may be given unfamiliar food items to eat. They may not have a choice of social companions. Instead, they may be forced to live in close quarters with unfamiliar and sometimes aggressive individuals. Invariably, some of these wild-caught animals will adapt to the captive environment more readily than others. Those which adapt most readily will be most likely to survive and leave offspring (and their genes) for the next generation. This next generation will likely possess those genes that helped their parents adapt to captivity and, being born and reared in captivity, they may have an even greater advantage over their parents in adapting to the captive environment. Behavioral techniques may be learned at a very young age to facilitate the adaptation process.

The generation-by-generation genetic (evolutionary) changes which occur in a population of animals after being taken from the wild and brought into captivity is the same natural selection which occurs in nature. Genetic changes will be greatest in the first few generations in captivity. Then, as the population becomes more adapted to its captive environment, natural selection will be progressively less severe in each generation.

In addition to natural selection in captivity, people who breed animals in captivity will select certain individuals over others to provide offspring for the next generation, a process called *artificial selection*. The animals favored by animal breeders may be the same ones favored by natural selection or they may be different. Artificial selection can be very severe as people attempt to produce animals through genetic change that would be very unlikely to occur by natural selection. The various dog breeds in the world today provide an example of how the appearance and behavior of a single species (wolf) can be dramatically changed by artificial selection. Breeds have been artificially selected for hunting, retrieving, herding, protectiveness, running speed, trainability, and ease of socialization to people. In fact, it is believed that humans have selected dogs which retain their juvenile behaviors into adulthood, a phenomenon referred to as *neoteny*. The argument is that the retention of juvenile characteristics makes adult dogs more responsive to humans and easier to control and, thus, more desirable pets.

Artificial selection has greatly modified the size and appearance of the domestic dog resulting in many distinct breeds.
Courtesy of the American Association for the Advancement of Science.

Certain species naturally possess behaviors and other characteristics that humans desire in animals living under their control (see the table). For example, people favor species which are polygamous (one male will mate with many females) for economic reasons because only a few males are needed to mate with many females. Most all of our common domestic animals are polygamous. Dogs socialize (bond) readily to people just like their wolf ancestors socialize readily to members of the wolf pack. Some species naturally possess behaviors and physiological characteristics that make it easier for them to adapt to a variety of captive environments. Chickens, which have been domesticated from red jungle fowl, eat a wide variety of plant and animal foods and bear offspring which are capable of feeding and moving about with their mother within hours after hatching. Domestic chickens adapt well to living in small or large pens either outdoors or indoors. Such species are said to be preadapted for domestication by humans.

The capacity for being *tamed* (exhibiting a lack of fear of people) is a characteristic shared by nearly all domesticated animals and is considered a preadaptation for domestication. Tameness is normally acquired during an animal's lifetime through rewarding or positive experiences with people. Cattle, sheep, horses and other farm animals reared by humans rather than their mothers are easier to be approached by and to be handled by people. Tameness constitutes an important characteristic of domesticated animals because of the constant and pervasive contact

Behavioral Characteristics Considered Favorable and Unfavorable for the Domestication of Vertebrate Animals

Favorable Characteristics	Unfavorable Characteristics
Social structure of populations	
Social organization—dominance hierarchy	Social organization—territoriality
Large gregarious social groups	Family groups important
Males affiliated with social group	Males typically live in separate groups
Intra- and interspecies aggressive behavior	
Nonaggressive	Naturally aggressive
Sexual behavior	
Promiscuous matings	Form pair bonds prior to mating
Males dominate females	Females dominate males/males appease females
Male initiated	Female initiated
Sexual signals provided by movements or posture	Sexual signals provided by color markings or morphology
Parental behavior	
Precocial young	Altricial young
Young easily separated from parents	Prolonged period of parental care
Response to humans	
Tameable/readily habituated	Difficult to tame
Short flight distance to man	Long flight distance to man
Nonaggressive toward humans	Aggressive toward humans
Readily controlled	Difficult to control
May solicit attention	Independent/avoids attention
Temperament	
Limited sensitivity to changes in environment	Highly sensitive to changes in environment
Locomotor activity and habitat choice	
Limited agility	Highly agile/difficult to contain or restrain
Small home range	Requires large home range
Wide environmental tolerance	Narrow environmental tolerance
Nonshelter seeking	Shelter seeking
Feeding behavior	
Generalist feeder or omnivorous	Specialized dietary preferences/requirements

Modified from Hale, E., 1969.

with humans during their lifetimes. Routine animal management practices used in each generation may insure sufficient tameness for animals to live relatively stressfree in proximity to people. Laboratory rats are routinely reared and maintained in open (to the environment) laboratory cages with little or no opportunity to hide from other animals and people. Reared under these conditions, they are relatively nonaggressive and easy to handle (they make little effort to struggle when held and almost never try to bite the handler). These same animals reared in indoor burrows or outdoor enclosures with ample hiding places are much more aggressive toward humans and difficult to handle. In contrast, when wild Norway rats (the wild ancestors of our common laboratory rat) are reared in open laboratory cages, they retain their aggressiveness toward humans. Their behavior toward people resembles that of domesticated rats reared in burrow systems. Wild rats can be tamed, however, by systematic handling (by humans) early in life. Interestingly, rearing wild rat pups by tame laboratory rat mothers does not make them tamer toward humans. To summarize, wild Norway rats develop a natural fearfulness of humans early in life even when reared in open laboratory cages, but possess the capacity to be tamed. Once they are subjected to artificial selection for ease of handling by their human care-takers, they naturally show less fear of people. In contrast, domesticated Norway rats are genet-ically predisposed to be unafraid of people when reared in open laboratory cages as a result of both conscious and unconscious artificial selection by human caretakers for ease of handling. But domesticated rats have not lost the capacity to behave like wild rats if they are reared in a more natural environment for the species with hiding places. A gene has been discovered that is believed to be at least partially responsible for ease of handling in this species. Untamed wild Norway rats which possess this gene are easier to handle than those without the gene, but the former are still more difficult to handle than their domestic counterparts. Interestingly, this so-called "domestication" gene is present in nearly all domestic Norway rat populations.

Domestic animals that live in proximity to people are not only less fearful of humans than their wild counterparts, but appear to be less fearful and sensitive to other animals and novel objects as well. Studies with rats and foxes have shown that selection for tameness toward people in these species results in changes in brain physiology which not only reduces their fear and sensitivity to humans, but to new and unfamiliar stimuli in general. In nature, fear of strange animals and novel objects can help to protect wild animals from getting in fights or being killed by a predator or from sampling toxic food items. It is not surprising, then, that fear of people is controlled by the same brain centers that control fear of other animals and novel objects. It follows that selection for tameness in captive animal populations not only affects how domesticated animals react toward people but also toward a broad array of novel stimuli. This is made possible by the fact that in captivity humans protect their domestic animals from harmful stimuli and experi-ences in general. Natural selection in captivity (evolution) will also favor animals who are basically nonreactive since they will be less stressed by life in captivity and, thus, may be more likely to reproduce successfully.

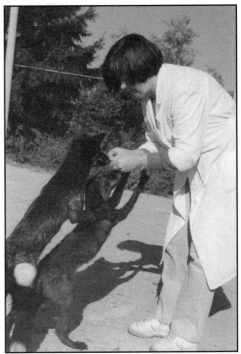

Researcher Irene Plyusnina demonstrates the tameness of silver foxes artificially selected for lack of aggressiveness toward humans.
Courtesy of the Institute of Cytology and Genetics, Novosibirsk, Russia.

These facts point out that animal domestication is a process that involves adaptation to man and the captive environment by some combination of genetic changes oc-curring over generations and environmentally induced de-velopmental events recurring during each generation. Both evolutionary mechanisms as well as adaptations attained during the animal's lifetime contribute to the development of the behavioral, morphological and physiological characteristics we com-monly associate with domesticated animals. With respect to behavior, it is important to point out that domestication has only changed the frequency with which certain behaviors are exhib-ited, not the actual locomotor patterns animals exhibit when engaged in these behaviors. For example, each species has its own characteristic way of fighting. Some species fight with their teeth, some butt their opponent, and still others kick with their hind legs or rear up on their hind legs and attack with their front legs. Domestication does not affect these motor patterns. Domestication only affects the frequency with which animals engage in aggressive behaviors. Domestication typically reduces the frequency of aggression directed toward other animals and humans. Animal caretakers do not like to interact with aggressive animals and, thus, tend to favor nonaggressive individuals when selecting breeding stock for the next generation.

Since the breeding of domestic animals and the environment in which they live are controlled by humans, it is important that we provide for their basic needs and welfare. The ability of our common domestic animals to adapt to the captive environments provided through developmental and evolutionary processes provides for much of their needed welfare. Nondomesticated wild animals can be particularly vulnerable to stress and poor welfare in captivity and should be given special consideration when designing their captive environments and animal management protocols.

See also Cognition—*Domestic Dogs Use Humans as Tools*
Domestication and Behavior—*The Border Collie,
A Wolf in Sheep's Clothing*
Nature and Nurture—*Baldwin Effect*

Further Resources

Coppinger, R. & Coppinger, L. 2001. *Dogs*. New York: Scribner.
Grandin, T. (Ed.). 1998. *Genetics and the Behavior of Domestic Animals*. New York: Academic Press.
Hale, E. 1969. *Domestication and the evolution of behaviour*. In: *The Behaviour of Domestic Animals*, 2nd edn. (Ed. by E. S. E. Hafez), pp. 22–42, Baltimore: Williams & Wilkins.
Jensen, P. (Ed.). 2002. *The Ethology of Domestic Animals*. Wallingford, UK: CABI Publishing.
Price, E. O. 2002. *Animal Domestication and Behavior*. Wallingford, UK: CABI Publishing.

Edward O. Price

■ Domestication and Behavior
The Border Collie, a Wolf in Sheep's Clothing

The gray wolf is the common ancestor of all domestic dogs. This defies credibility when considering for example, a seven-pound Yorkshire terrier-Maltese mix, but when watching a sheepherding trial, it's easy to see the wolf within the border collie, for in many ways, the border collie *is* the proverbial "wolf in sheep's clothing." Through the process of *domestication* (artificial, not natural, selection), humans have converted the predator, the wolf, into the working partner and guardian of the sheep.

In the sheep dog trial, in which a handler and her dog demonstrate their skills at herding sheep on a timed course, it is easy to see show the instincts of the wolf have been preserved—and modified—to meet human ends. The trial consists of six basic components. First, there is the *outrun*, in which the border collie leaves his handler at a post and runs in a large arc, some three-hundred-plus yards, to find and encircle a group of sheep stationed in the distance. In the *lift* the border collie and the sheep make contact. The border collie then *fetches* the sheep through a series of gates to his waiting handler and *drives* the sheep away from his handler. The border collie then forces the sheep into a *pen* and finally, working with his handler, must separate or *shed* off one or more sheep from the rest of the flock.

In the *outrun*, you can see the hunting style of the wolf, who often uses ambush and surprise. A border collie doesn't run head-on at a flock of sheep the way a dog of a another breed would, but circles around the flock and approaches it from behind. A wolf pack may also "herd" livestock when hunting, with some pack members circling widely around a herd and then driving the herd to where other pack members are waiting for the ambush.

In the *lift*, when the border collie and the sheep first make contact, the border collie must assert his authority over the sheep, but he must also calm them: If he frightens them, they may bolt and scatter, but if he doesn't assert his authority, they won't obey him. The most effective border collies can control a flock of sheep from a distance using body language and eye contact. When herding, a border collie will adopt a classic predatory stance with his belly close to the ground, his head held slightly lower than his back, and his tail tucked. In her book *The Versatile Border Collie*, Janet Larson notes that "the trait that sets the border collie apart from most other breeds [of dog] is his use of eye. A crouching, snakelike movement with an intense stare used to hypnotize livestock is what characterizes eye." A border collie uses his eye in the same fashion as a wolf might use eye to assert her dominance and intimidate her prey.

In the *fetch*, *drive*, and *pen*, the border collie demonstrates his ability to move the sheep where he and his handler want them, through a course which involves freestanding gates and into a pen. Finally, in the shed, the border collie and handler combine forces to separate a single animal from the flock, just as a wolf pack might separate a single animal for the kill. The difference, of course, is that the border collie's instinct to kill has been bred out of him and, in this team, a person occupies the role of the alpha wolf.

Dogs have influenced us and changed the course of our history in ways that are equally as profound as the influence we humans have had on dogs. The modern breed of border collie is just over 100 years old, but there are written records about sheepherding dogs that go back before the Romans invaded Britain (A.D. 43), and the significance of the partnership formed between people and herding breeds cannot be underestimated. Without the herding breeds, the history of the Scottish highlands, the development of the open lands of New Zealand and Australia, and the settling of the frontier of the American West would have been radically different because border collies and other herding breeds have enabled people to manage livestock over vast areas of open land. A working sheepdog will run about 10 miles for every mile that sheep move, which means that a border collie must be able to run 60–100 miles (97–161 km) day after day. As Donald McCaig notes in his book, *Eminent Dogs, Dangerous Men*, a trained hill dog in Scotland will finally "be able to gather eight hundred ewes scattered over two or three thousand acres. He will be able to work by himself or to whistled commands at distances of a mile or more. He can run a hundred miles Thursday and get up Friday morning and do it again." (p. 28)

Although the intelligence and work ethic of border collies has long been known to the shepherd, in our modern era those qualities can manifest in unexpected ways. Rico, a border collie, who has been studied by Juliane Kaminski, Joseph Call, and Julia Fischer at the Max-Planck Institute for Evolutionary Anthropology in Leipzig, Germany, has a vocabulary of over 200 words and can learn and retain new words at the same rate as 3-year-old toddlers. In an interview on National Public Radio, Julia Fischer said that Rico was the perfect study subject because "he's a workaholic. He's crazy and he just goes on and on and on and the owner has to stop him. . . . He forgets to drink and eat. I've never seen that before in a dog." Welcome to the world of border collies.

Border collies routinely win in obedience, agility, and other types of competitions and are consistently rated as the most intelligent breed of dogs, something which has led to their popularity as pets. However, border collies are more prone than other breeds to end up at animal shelters because people who adopt them don't understand that the very qualities we admire—their tenacity, intelligence, athleticism, independence, and strong herding instinct—often make them poor pets. Border collies are bred to work, and a border collie who is bored is a dog who sooner or later will get into mischief. In *The Intelligence of Dogs*, Stanley Coren relates the story of a border collie who learned that by running full tilt at the wire mesh in a screen door he could crash through the screen to the outdoors and the freedom that lay beyond. When his owners covered their screen doors within heavy farm wire, the border collie simply started scouting around the home for windows covered in screen mesh and took to jumping his way to freedom. I have a friend who found it impossible to keep her border collie in her backyard because he was able to figure out how to open a variety of handles and hinges that she put on her back gate. When she took to locking her back gate with a padlock, her border collie took to climbing her 6-foot fence.

Max, the author's border collie.
Courtesy of Janette Nystrom.

Before adopting Max, my first border collie, I knew very little about border collies and simply referred to them as farm dogs. Whenever I'd see a historic photograph of the American West, whether it was of farmers or sheepherders, inevitably there would a black-and-white dog sitting on the porch by a group of children, or curled in the back of a wagon, or standing alongside a horse. Along with British farmers, borders collies immigrated to the United States during the mid-1800s, and as early settlers moved west, so too did the border collie. When I asked my mother what kind of dogs were in the old photographs, she told me they were farm dogs, like the kind she had as a child.

In one of my favorite photos of my mother as a child, she is standing alongside just such a farm dog and has her leg cocked around the dog's leg and, my mother tells me, she is trying to trip him. The story of this particular dog is part of family lore, for he showed up on my grandmother's porch one day during the height of the depression and simply refused to leave. At the time my grandmother was working odd jobs and trying to make ends meet and was away for days at a time so this dog became my grandmother's self-appointed helpmate and protector. When he first showed up, my grandmother tried to shoo him away—after all, he was one more hungry mouth to feed—but when he refused to leave and was still on my grandmother's porch the following morning and evening, she relented and gave him a few meager table scraps. "Dog," as my grandmother called him, slept on the front porch and guarded the home from strangers. He once rescued my grandmother from an angry bull by diverting the bull's attention away from her and to himself. He also appointed himself chief babysitter and wouldn't let my mother or her younger sister leave the yard or go anywhere near the ditch that ran along the front of their property. Once, when my mother ignored his barks and proceeded down the street he grabbed her dress and refused to let go until my grandmother, hearing all the commotion, came to investigate.

Border collies can teach us about the story of domestication in many ways: They clearly retain the traits of the wolf, which have been modified to meet our ends, but they also teach us about how, in domesticating animals, our lives have been changed in the bargain. Without the border collie and other working breeds, our agriculture, sheepherding, and hunting success—and consequently, our human history—would have developed differently. We have also been changed in more intimate, personal ways as the friendships we have developed with border collies and other dogs are as much a part of our individual and collective histories as are our relationships with our human friends and family members.

One of my favorite "farm dog" photos from the American West is of a sheepherding camp in Montana at the turn of the century. The photograph shows several men who are relaxing around a small campfire. One man has tipped his cowboy hat upside down and is using it as a watering dish for his small black and white border collie. The other men are looking directly at the camera but this man, who has one hand on his dog's back and another hand balancing his cowboy-hat watering dish, has his eye and his attention clearly focused on his friend.

See also Cognition—*Domestic Dogs Use Human as Tools*
Domestication and Behavior
Wolf Behavior—*Learning to Live in Life or Death Situations*

Further Resources

Coren, S. 1994. *The Intelligence of Dogs: Canine Consciousness and Capabilities*. New York: The Free Press.
Kaminski, J., Call, J. & Fischer, J. 2004. *Word learning in a domestic dog: Evidence for "Fast Mapping."* Science, 304, 1682–1683.
Larson, J. E. 1999. *The Versatile Border Collie*. Loveland, CO: Alpine Blue Ribbon Books.
McCaig, D. 1991. *Eminent Dogs, Dangerous Men: Searching through Scotland for a Border Collie*. New York: The Lyons Press.

United States Border Collie Handlers Association website (http://www.usbcha.com/) includes a calendar of approved open and nursery sheep dog and cattle dog trials in the United States and Canada.

Janette Nystrom

■ Dominance
Development of Dominance Hierarchies

When small groups of animals come together, either in the wild or captivity, they often have a series of aggressive interactions with one another involving actions such as bites, chases, and threats. Eventually these interactions moderate as pairs of animals within the group from dominance relationships—stable arrangements in which one animal is aggressive and the other is not. The overall network or pattern of relationships in the group is known as a dominance hierarchy or pecking order, and once in place, the hierarchy may last weeks or months without change. In many species such as some insects and crustaceans and a large number of birds, fish, and mammals, including human children, these hierarchies usually have the classical linear structure: One animal dominates all the others, a second dominates all but the first, and so on down to the last animal that dominates no one. Higher-ranking animals often produce more viable offspring, suffer less stress, and have greater access to resources that might be in short supply.

How do hierarchies develop their usual linear structures? This is still a mystery, but differences among individual animals are not as important as might be thought. While larger and more aggressive animals do have some advantage in dominance contests, and those that have previously lost, some disadvantage, these factors are not decisive in determining the ranks of animals in dominance hierarchies. For example, in an experiment in which groups of four cichlid fish were assembled to form dominance hierarchies, separated long enough to forget one another, and then reassembled to form second hierarchies, linear structures were extremely common, but the positions of the fish changed, often drastically, from one hierarchy to another. In about three fourths of the groups, two, three, or even all four fish had different ranks in the two hierarchies: The linear structures persisted, even when the fish changed positions in them.

Perhaps surprisingly, especially for animals that might be thought to be not very socially aware, *social dynamics*, interaction in a group context, seems to be a very important factor in causing linear hierarchies to develop. If fish form dominance relationships in pairs, out of sight of the other fish in their groups, linear hierarchies only occur in about half of the groups. But if hierarchies form when the fish are all together, and they can watch each other establish dominance relationships, linear hierarchies occur in almost all groups. What information animals may be gaining when they watch each other interact and how they may act on it to form linear hierarchies is not well understood yet. But one possibility is that fish and other animals may take note of the dominance relationships of other individuals in their groups and, based upon what they see, make inferences on how they should relate to these individuals. For example, if A has already dominated B, and A observes B to dominate C, A may assume that she can easily dominate C, and C may assume the same thing based upon her own observations. Although more research is needed to determine whether a variety of species can make inferences of this sort and apply them to dominance relationships, some recent experiments suggest that primates, crows, and chickens are capable of doing so. Thus, it may be that "social intelligence" is a more important factor in the development of dominance hierarchies than are brawn and force.

See also Aggressive Behavior—*Dominance: Female Chums or Male Hired Guns*
Aggressive Behavior—*Ritualized Fighting*

Further Resources

Bond, A. B., Kamil, A. C. & Balda, R. P. 2003. *Social complexity and transitive inference in corvids.* Animal Behaviour, 65, 479–487.
Brown, J. L. 1975. *The Evolution of Behaviour.* New York: W. W. Norton & Co.
Chase, I. D., Tovey, C., Murch, P. 2003. *Two's company, three's a crowd: Differences in dominance relationships in isolated versus socially embedded pairs of fish.* Behaviour, 140, 1193-1217.
Chase, I. D., Tovey, C., Spangler-Martin, D. & Manfredonia, M. 2002. *Individual differences versus social dynamics in the formation of animal dominance hierarchies.* Proceedings of the National Academy of Sciences, 99, 5744–5749.

Ivan D. Chase & Jeffery L. Beacham

■ Ecopsychology
Human–Nature Interconnections

Ecopsychology surfaced out of much conversation in the early 1990s. The conversations began especially in the San Francisco Bay Area, with the now obvious recognition that environmental problems are essentially due to humanity's lack of meaningful—or soulful—connections with the plants, animals, and elements with which we share the planet. This suggested a deepened appreciation for human responsibility in what some people considered a rapidly growing environmental crisis. It meant becoming responsive to relationships we had seldom considered and reconsidering our anthropocentric behavior and mundane consciousness. Psychology claims to be the science of just that—of human behavior and consciousness—and yet had not seriously considered either behavior or consciousness in relation to the environment, or to nature. Environmental psychology pointed in the relevant direction but remained exceedingly academic in flavor and, more importantly, considered the relationship between humans and the environment solely with reference to human concern; no significant reference was made to the condition of the environment as a function of modern human consciousness. In contrast, ecopsychology's fundamental premise is interdependence. As a developing field, ecopsychology placed itself unabashedly at the heart of the interdependent relationship between humans and nonhumans and, in an effort to clarify human responsibility in this relationship, it has attempted to work out a kind of eco-psycho-spiritual mathematics.

The emotional presence of animals may offer the best medicine for modern malaise.
Courtesy of Corbis.

On the human side of this interdependent equation, ecopsychology suggests that degraded environments yield psychologically unhealthy humans. The theory is based on socio-cultural critique, observations of our collective psychological state, phenomenological narratives, and the etiology and epidemiology of particular psychological disorders. Early thinking about the relationship between psychological states and the environment was largely informed by the work of the human ecologist, Paul Shepard. In short, Shepard suggested that human consciousness coevolved with the natural order while hunting and gathering (while living intimately with the natural order) for some 99% of human history. Given the pace of evolutionary change, the implication is that we are not well adapted to the industrialized world we have recently created. In psychological terms, this misfit looks like variations on the theme of alienation: depression, anxiety, and addiction—all of which are excruciatingly common and on the rise. The further implication is that common psychological disorders are not purely personal matters, but rather, that "the personal is planetary" and, true to psychological interdependence, vice versa.

From the environmental side of the same interdependent equation, it appears that degraded humans, or those suffering from various forms of alienation and dis-ease—what some have called "global malaise"—are the worst offenders against the environment. In other words, fragmented minds fragment, and polluted minds pollute. Turning this around, ecopsychology says that environmentally responsible behavior requires an embodied appreciation for natural beauty, a lived understanding of systems, and a commitment to relationships with integrity. Although this may also appear self-evident, consider this: If integrity (like other life-enhancing qualities) were a coefficient in the complex equation between humans and the natural order, our ecosystems would be, by definition and mathematical fact, increasingly wholesome.

Psycho-spiritual interdependence with the natural world implies the necessity of reciprocating relationships. The fact that common Western behavior seldom includes reciprocity with the nonhuman realm reveals a second basic premise: By virtue of numerous historical events and cultural conditioning, the Western mind is infused with individualism, materialism, and reductionism to such an extent that we tend largely toward divisive, or dualistic, forms of behavior and consciousness. We witness dualistic behavior as fundamentalism and militarism, as our loyalty to scientific fact over lived experience or mystery, and, in the complex calculus of the psyche, as objectification and abuse, including that perpetrated on the environment.

From an ecopsychological point of view, divisive (and ultimately lonely) forms of consciousness are at the root of our environmental catastrophe. The antidote is connectivity in all forms—meaning that relationships are necessarily primary. In a culture within which egocentric behavior and commodification are defining features, this implies no easy task. Ecopsychology asserts that the great diversity of life forms—the magnificent colors and fantastic shapes, the cat calls and bird calls, the wild forms of adaptation and mobility, and perhaps most significantly, the emotional presence of animals—may offer the best medicine for modern malaise, for hearts and souls contracted in response to ideologies that have outgrown their usefulness, and for minds narrowed by anthropocentrism. The fundamental ecopsychology lesson is therefore this: The natural world provides a thousand opportunities to open a heart, deepen a soul, and instruct us in the way of all things organic.

Ultimately, ecopsychology is an urgent and no-bones-about-it call for behaving well within many more relationships then most humans commonly consider, including those with the plants and animals who sustain us. Given everyday modern consciousness, it follows that the essential practice of ecopsychology is mindfulness—with the well-being of all planetary inhabitants held foremost in the mind and firmly in the heart.

Further Resources

Abram, D. 1996. *The Spell of the Sensuous*. New York: Pantheon Books.

Berry, T. 1998. *The Dream of the Earth*. San Francisco: Sierra Club Books.

Fisher, A. 2002. *Radical Ecopsychology: Psychology in the Service of Life*. Albany: State University of New York Press.

Glendinning, C. 1994. *My Name is Chellis and I'm in Recovery from Western Civilization*. Boston: Shambhala Press.

Kidner, D. 2001. *Nature and Psyche*. Albany: State University of New York Press.

Roszak, T. 1992. *The Voice of the Earth*. New York: Simon and Schuster.

Roszak, T., Gomes, M. & Kanner, A. (Eds.) 1995. *Ecopsychology: Restoring the Earth, Healing the Mind.* San Francisco: Sierra Club Books.

Shepard, P. 1982. *Nature and Madness.* San Francisco: Sierra Club Books.

Sewall, L. 1999. *Sight and Sensibility: The Ecopsychology of Perception.* New York: Tarcher/Putnam.

Laura Sewall

■ Education
Classroom Activities in Behavior

You're settling in to watch a nature documentary on TV. A researcher, sitting amidst a group of chimpanzees, jots down notes of the interesting goings-on he or she observes. The pace is frenetic. A fight breaks out between a couple of males. Another male is attempting to mate with a female who is not at all interested. Several youngsters are playing with their mothers close by. Elsewhere, a small hunting party captures a young gazelle. Sound exciting? This is often the perception that people have about the study of behavior, and it is a highly inaccurate one! It probably took a film crew hundreds of hours to compile the exciting footage you see on the screen in a brief 45 minutes. Much of an animal's life is spent resting, casually locomoting, and seeking food. Only a small subset of time is devoted to the thrilling events portrayed on the screen. More important, the researcher you see with a small notebook may be writing down some observations, but he or she is not collecting data. Behavioral data collection is a highly systematic, standardized endeavor. It isn't as easy as you might expect, but it can be conducted by any student with a little training and perseverance. The systematic study of behavior can give you the opportunity to engage in the true process of scientific investigation by developing your own research ideas. Unlike many standard laboratory exercises, the outcome of your investigation may not be known at the outset. Many variables can influence your results, and learning to identify and, where possible, control these factors is critical. *Behavior* is what we observe. It is visible and quantifiable. Developing an understanding of what animals do and why is a starting point for further investigation into ecology, evolution, and population biology.

Practicing some of the basic components of behavioral research projects can help you to develop your skills as an observer and master the methods of behavioral sampling. The sort of casual observation that the researcher you see on TV is doing may provide interesting insights, but it does not provide quantifiable, analyzable data. So, how is behavioral observation really conducted, and how can you acquire these skills?

Formulating a Question

It may seem trivial, but clearly formulating a testable question can be one of the most challenging aspects of behavioral observation research. The kinds of questions one can pose are virtually limitless, but the data that one can readily gather can be used to address only a small subset of these questions. So, how can one form a testable question? First, ask yourself what you can and cannot do in a given situation. Can you manipulate any variables? Knowing what you can control and what you cannot control, and how this will effect your data, is critical. Will you be able to observe the behavior of interest? Rare behaviors, like predation events, may be very difficult to observe. Other behaviors, like reproduction and parental care, may be seasonal. So, a firm understanding of the biology of your organism is an important prerequisite.

When you are observing animals, whether it is in a natural area, a zoo, or your own back-yard, you will undoubtedly come up with many interesting questions that you might want to investigate. Write these down; then assess the feasibility. Outline the type of data that you would collect in order to answer the question. Try to come up with some plans for collecting the specific data, and reassess. This is an iterative (repetitious) process, and should help you narrow the scope of your research question to something manageable and, hopefully, answerable.

Behavioral Sampling Methods

In 1974, a seminal paper was published in the journal *Behaviour*. In this paper, the various systematic methods that researchers used to observe and categorize behavior was formalized and standardized by Jeanne Altmann. Prior to this time, researchers used a variety of methods to *sample* behavior (that is, observe small snippets of behavior in unbiased ways in order to learn about the behavior of an animal), but no clear guidelines existed about how the various methods compared or the advantages and disadvantages of each method. This lack of consistency made it impossible to compare data collected by different researchers at different study sites because it was difficult to evaluate the way in which the information had been gathered. Altmann has spent many years studying baboons in Amboseli National Park in Kenya, and brought this experience to bear in distilling and summarizing observational methodology. The standardization suggested by Altmann's paper facilitated research efforts and allowed for comparisons to be made using data gathered by more than one researcher or at more than one locale.

One of the basic tenets of observational sampling is that behavioral information is observed and recorded (written on a data sheet, entered into a computer, etc.) in an unbiased way. Thus, focusing on the very noticeable behaviors (fights or predation for example) alone isn't appropriate because these are probably rare events. Focusing on them is tempting—as these are electrifying and likely to attract an observer's attention—but this focus could lead to inaccurate conclusions about how animals spend their time.

Another aspect of observational sampling has to do with how observers allocate their time. It is not possible to observe everything an animal does all the time. So, an observer must decide (1) what to observe, (2) whom to observe, (3) when to observe, and (4) how to observe.

What to Observe

You will need to be familiar with the concept of an *ethogram*, which is a catalog or listing with definitions of the species-specific behaviors that make up the repertoire of an animal. Compiling an ethogram is an important prerequisite to any observation study. It can be surprisingly difficult to define behaviors on your ethogram objectively and clearly, but this is critical for any research project. The definition should be clear enough that anyone reading the ethogram would understand it and be able to recognize that behavior when they see it. You can practice writing an ethogram by starting with a familiar animal, like yourself. List the basic activity states that you engage in daily and define these as clearly and concisely as you can. Remember that your ethogram must be exhaustive, so it should include behaviors that cover any situation. The categories may be broadly defined or very specific, depending on the focus of your research question. Based on your extensive experience with your own behavior and schedule, make a prediction about the proportion of your time you spend in each of these activities. This is your predicted activity budget. Given that human beings, in an idea world, should work 8 hours a day and sleep 8 hours a night, you might predict that you spend 33% of your time sleeping and 33% of your time working. How would you determine the accuracy of your prediction? By collecting the data, of course.

There are several ways to do this, based on the observation methods that you've already learned. You might, for example, keep a record of your activities for a day and record your activity at specified time intervals, perhaps every 15 minutes or every hour. Using this *instantaneous sampling technique* could be a grind in the middle of the night, of course. You might instead use a *continuous, or focal sampling technique*, in which you note the time each time you change your behavior. You can then calculate the duration of each behavior from this data by dividing the number of seconds (or minutes) spent in a particular behavior by the total number of seconds (or minutes) spent observing. Try to come up with explanations for any observed deviations between your actual and predicted activity budgets. Perhaps you collected your data on a weekend, and you were up late. You slept for 345 minutes. Thus, you slept for 345 out of 1440 minutes, or 24% of your time. Your results may have been very different had you collected your data on a weeknight. To address this, researchers collect data repeatedly, in order to paint a better picture of activity patterns. If you repeat your study for many more days, you might find that your sleep time does indeed approach 33% of your time. Or, you might not.

Developing an ethogram for a familiar organism may not be too challenging, but try now to compile an ethogram for a less familiar organism. A pet may be a good place to start, or any local wildlife that may be accessible to you. Go to a local park or a backyard birdfeeder. Find an unobtrusive spot and relax. Before beginning any systematic study, it's best to spend some time simply watching, in order to get a good sense of what kinds of activities your chosen subject engages in. As you observe, begin to compile a list of behaviors that you see, and then define them. You're compiling your ethogram. As a first test of your ethogram, go away and come back later in the day, or the following day. Does your ethogram appropriately capture all the behaviors that you see? Do you need to add any behaviors? An even better test is to see how readily someone else can utilize your ethogram. Have a friend or classmate accompany you to observe the organism. How readily can they identify behaviors? Are any of the definitions unclear or confusing? Are there behaviors that are missing from the ethogram?

As you get more comfortable with writing and using ethograms, try writing an ethogram for a species you are even less familiar with, but one that is similar to a more familiar species. Choose a wolf or a tiger (similar to dog and cat, respectively), or a gorilla or baboon (similar to humans). Consider how well an ethogram for the familiar species could be applied to the less familiar one. What kinds of modifications will you need to make? You may want to first do a literature search to see if there is an ethogram already available for the species.

An example of an ethogram follows. This is a portion of an ethogram that will be used to study estrus behavior in gorillas. Some types of behavior are broken down in detail and some are not, because of the research question involved. Can you identify which behaviors are *states* and which are *events*? If the focus of the research changed, so might the specificity of the ethogram. If you were studying foraging behavior, for example, you might lump together certain types of social or reproductive behaviors, but expand foraging to differentiate between searching for food, extracting food, and consuming food. Note the two most important behaviors that will be a part of any ethogram: *Other*, and *Out of View*!

You may have noticed that there are some striking similarities among all the ethograms you have written or reviewed. Behavior represents an organism's way of solving some of the basic necessities of life: finding food and shelter, avoiding predators, reproducing. All animals must engage in these behaviors, but they may not utilize the same strategies to do so. Finding these commonalities across species can be exciting.

Ethogram for Studying Estrus Behavior in Gorillas

- ***Forage:*** Looking for food and/or feeding

- ***Allogroom:*** Allogrooming (running fingers through the fur of another), removing dirt or debris. Mouth may be used, as well.

- ***Locomotion:*** Movement

- ***Alert:*** Inactive, but attending to the environment

- ***Rest:*** Inactive, and not attending to environment (eyes closed)

- ***Other:*** Any behavior not otherwise categorized (you may mention in Comments section if this is a behavior that may be important to our study).

- ***Out of View:*** Subject cannot be seen by observer.

- ***Approach:*** Directed movement toward another individual, from greater than 1.5 m to within 1.5 m (5 ft)

- ***Withdraw:*** Moving away from another individual, from less than 1.5 m to beyond 1.5 m

- ***Chest Beat:*** Pounding of fists on chest.

- ***Affiliative behavior:*** Intentional touching or other positive interaction.

- ***Contact Aggression:*** Aggressive physical contact. Includes behaviors such as hit, slap, push, bite.

- ***Non-Contact Aggression:*** Posture, facial expression, or movement that may or may not be accompanied by vocalization and which results in keeping another at a distance or obtaining desired objects or space or controlling another's behavior. Includes a chest beat directed at another, and chases.

- ***Stare:*** Female stands motionless while watching the male intensely.

- ***Embrace:*** One animal wraps arms and/or legs around another.

- ***Quadrupedal Present:*** Female stands on all fours and orients her anogenital area toward another.

Who to Observe

Your research question and methodology will dictate who you choose to observe. If you are interested in a subset of the group (only females, or only infants, for instance), then this will be the subset you observe. If you are interested in all individuals, then you will need to divide your observation time equally across all subjects. It's important to do this randomly. You can shuffle cards, flip a coin, use a random number generator—just don't pick the first animal you see. Why not? The first animal to catch your attention will likely be the biggest, the loudest, the most boisterous animal in the group. The quiet, subordinate individuals will be less obvious to you. Remember, you don't want to bias your data.

When to Observe

When you collect your data may depend on many factors: what time of day your animals are active, whether the behaviors of interest are distributed evenly throughout the day, the month, or the year, or are clumped, occurring altogether at certain times of the day. Your own schedule may become a factor, too. Are you available to observe at all times of day? These are some of the factors that will go into your observation schedule. If you're only interested in a subset of behaviors (like foraging, for instance) then you might choose to concentrate your observations around the time when your subjects are most likely to be eating. Some animals may feed early in the morning, then rest through the midday period. Many grazing animals feed on and off throughout the day. In a captive setting, feedings may be tightly scheduled. If you don't know what the overall activity pattern is, you may need to scatter your observations throughout the day. Consider the activity budget you constructed on yourself. What would your activity budget have looked like if you only recorded information from 8:00 a.m. until noon, or from 6:00 p.m. until midnight? Because you can't observe all the time, you can "sample" different time periods. You may spread your observations throughout the day in order to get the best picture of how your subject spends its time, or you may concentrate on a particular time of day or particular activity and get a more comprehensive view of that subset of time.

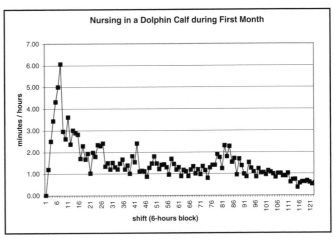

This dolphin graph illustrates changes over time (in this case, nursing).
Courtesy of Sue Margulis.

How to Observe

Perhaps the most important decision you will make is how to collect your data. You've learned about the basic methods of behavioral observation. Different methods are appropriate for different questions. Several important questions to consider are:

1. How many individual animals will you be observing?
2. Can you identify individuals?
3. Are you interested in *stated* behaviors only (i.e., activity budget data)?
4. Will you observe interactions, which are usually events? Answers to these questions will help you to decide on an observation method.

The number of animals you will collect data on, and your ability to identify individuals, may lead you to choose an instantaneous, or scan sampling, approach. It would be very time-consuming to conduct a continuous focal observation on each of 20 or 30 animals in a group, but you could readily scan this group at intervals and record the activity of group members. Even if you cannot identify individuals, you can nevertheless construct a time budget by tallying the number of individuals engaged in each behavior on each of your scans. You may perhaps be able to distinguish certain subsets of your group, such as males and females, or adults and juveniles. This distinction can improve the value of the data you collect.

Classroom Research

Sue Margulis

Often, scientists or students work together to study behavior. To do this accurately, any group of observers must insure that they are all "seeing" and categorizing behavior in the same way. The most appropriate way to evaluate this is to have pairs of observers observe the same animal at the same time and record their observations. This is called an *observer reliability observation*. In an ideal world, these two sets of data should be identical. In the real world though, they seldom are. Two observers watching from different vantage points might not see exactly the same thing; a momentary glance away or even a blink could cause an observer to miss something. As you develop an ethogram or begin a research project, conduct an observer reliability observation with a classmate or colleague. Observe the same animal at the same time and compare your results. How similar are they? Ideally, researchers strive for a high level of agreement between observers (85%–95% is typical). Most important though, is that any deviations be random, and not systematic. If one observer consistently misses a particular behavior, then more training and practice is called for. If an observer can't identify a particular behavior because the definition isn't clear, then the ethogram needs clarification. Usually, 90% agreement is quite attainable.

Studying changes in behavior among individuals in different situations can be fascinating. Because behavior can respond rapidly to an environmental change, one can, in some situations, change the environment and observe any subsequent changes in behavior. If other factors are held constant, the change that you've made will likely explain the observed change in behavior. Using pets or classroom animals, plan some change to an enclosure or routine. This may mean providing a new toy, adding a new food item, or changing a daily schedule. Based on your knowledge of the animals, make some predictions about what effect the change will have. Will it increase activity or decrease activity, and for how long will this effect persist? If it is a new food item, will it be accepted, and if so, how readily? How will you quantify this? If you are to document a change, you will first need to establish a baseline. Conduct observations for some preset period of time, make your environmental modification, and continue to observe. You should be able to quantify changes in behavior. Even if you are observing animals in a natural, or otherwise uncontrolled setting, you can still evaluate the effect of environmental changes on behavior. You may not be able to control those changes, however. A new animal may enter a group, or an individual animal may leave the group. Although such social changes are not in the researcher's control, by conducting behavioral observations over an extended period of time one can evaluate the impact of such social changes by comparing behavior before and after the change.

Instantaneous, or scan sampling is suitable for recording state behaviors, but falls short when events and interactions are of interest. If the latter is the focus of your research, you will likely choose a continuous, focal animal technique.

If you're unsure, you may try different methods and compare the results. The two most common methods of collecting behavioral data are continuous and instantaneous sampling. *Continuous sampling* is best suited for observing small numbers of individuals, and for focusing on interactions. *Instantaneous sampling* is probably the method of choice if you are interested in activity patterns, if your group is large, or if you can't readily distinguish all individuals. Continuous sampling is analogous to videotaping a subject, whereas instantaneous sampling is like taking a series of still or digital photos. If you have access to a video and a still camera, you may use these two techniques to clarify these differences. Select a subject (a person, a pet, a squirrel in the park, a bird), find a partner, and get a camcorder and a polaroid or digital camera. Decide on the ethogram, or behavior categories, that you'll need. Have one person videotape the subject for 10 minutes (a continuous observation), while the other person takes one photo every minute for 10 minutes (an instantaneous observation). Tabulate activity budgets and compare your results. Which method

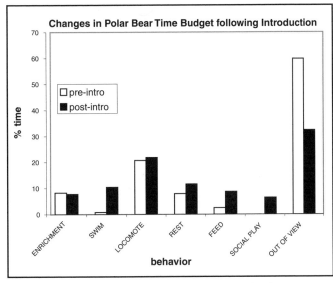

This polar bear graph illustrates a basic time budget comparison before and after a social group change. *Courtesy of Sue Margulis.*

worked better? Why? What would happen if you took a photo every 30 seconds instead, or every 15 seconds?

Selecting an appropriate observational sampling method is critical. This will serve both to facilitate and to limit the kinds of questions you will ultimately be able to answer with your data. It's best to put some time into this choice.

Zoos as Behavioral Research Centers

A zoo can be an ideal place to conduct observations of exotic animals. Typically, the animals will be readily observable, often they will be individually recognizable, and zoo staff may be able to provide you with background information. After first or second grade, classes rarely consider the local zoo to be an appropriate learning destination. And yet, conducting behavioral observations at the zoo, whether it is during a half-day field trip or a semester-long course, can be exciting, rewarding, and beneficial to the zoo. Most zoos welcome students in this capacity. Go to your local zoo, and instead of simply looking at the animals and checking them off on a mental list, select one or a small number of exhibits on which to focus. This is the difference between *looking* and *observing*. Become an observer of animals. Take the time to consider not just what animals may be doing, but how and why. Each zoo exhibit holds a research question waiting to be asked. Go ahead and ask it! Using the methodology you have learned, you can collect data that will allow you to address questions about why and how animals behave the way they do. Often, the answers relate to the ecology of the animal. Consider the various solutions that different animals use to solve the same basic problems of life: finding food, or locomotion, for example. The subtle differences or surprising adaptations can be quite amazing and fascinating.

You may want to spend some time observing a group of primates. Primates utilize a diverse array of strategies for locomoting, foraging, and reproducing. New World primates in the order *Callitrichidae* are largely monogamous; this is a very unusual mating strategy among mammals. Females usually produce twins, and the male, as well as younger siblings, assist in carrying the young. Not only does this relieve the energetic burden on the mother, but it also provides maturing siblings with critical experience in infant care and handling. The lesser apes of the Old World (gibbons and siamangs) are monogamous as well; however, in these species the male's primary role is that of protector. Many other primate species live in large, complex groups with multiple males and multiple females, and may exhibit a *dominance hierarchy*, with dominant animals receiving priority access to resources such as food or mating opportunities. In many species, females remain in their natal groups (the groups in which they were born), and males usually disperse from these groups and attempt to join other groups. Your zoo is likely to have species representing multiple social organizations, and observing the different sorts of interactions in these groups can be exciting. Compare for example, the interaction between males and females in different primate groups. Evaluate how these differences relate to the different mating strategies seen in these diverse species. Grooming is a very important primate behavior. Individuals may groom others in their group. Grooming serves a hygienic function, but its role in maintaining social ties is probably more critical. Make some predictions about whom you might expect to see a male grooming, and whom you might expect to see a female grooming. Does this relate to social organization?

The interesting life histories that you can observe, whether it be at the zoo, a local park, or an aquarium tank in your classroom, are limitless. By utilizing sound scientific principles and objective methodologies, you can investigate a diverse array of research questions in animal behavior. In most cases, your investigation may leave you with more questions than answers, but this is where the excitement lies.

Further Resources

Altmann, J. 1974. *Observational study of behavior: Sampling methods.* Behaviour, 49, 227–265.
Martin, P. & Bateson, P. 1993. *Measuring Behaviour: An Introductory Guide.* New York: Cambridge University Press.
Ploger, B. J. & Yasukawa, K. 2002. *Exploring Animal Behavior in Laboratory and Field.* New York: Academic Press.

Sue Margulis

■ Education—Classroom Activities
All About Chimpanzees!

Chimpanzee Social Life

Chimpanzees live in social groups of up to 80 members. These large social groups are called *communities*. Unlike some other primates who eat, sleep, and travel with the same members of their communities at all times, chimpanzees may "hang out" with someone different from day to day. When food is scarce, individuals will go off alone to find food, but when there is a lot of food in one area, the group may stay relatively close, with all the members feeding near each other. The closest social relationship is between a mother and her infant.

Males like to spend a lot of time with other males, grooming, hunting, and patrolling the boundaries of their range looking for members of other communities. Chimpanzee males probably get along well because they are related. Females spend a lot of time alone because they often do not have relatives in their group. When females reach maturity, they usually leave the community and their mothers in order to find a group with males unrelated to them so that they can mate. Males, however, stay in the group where they were born with their mothers and brothers.

Female chimpanzees lead a more solitary life than do males.
Courtesy of Sue Howell.

Each community has an *alpha male*, a dominant male, to whom all others are submissive. This means if the alpha male wants something, he gets it! Females are usually lower-ranking than the males, but a mother's (or other female's) support is very important for a male who is trying to increase his rank. As long as rankings are stable, life in the group is pretty peaceful. The dominant member handles the others with threats and they usually defer to him to avoid fights. However, fighting does occur, usually when group members try to establish a new rank order. *Intragroup* (within the same community) fighting is rarely fatal and is much less violent than *intergroup* (between different communities) fighting.

Because chimpanzees spend so much time grooming themselves and others, many people think they must have fleas. They don't! Chimpanzees move their fingers through each others hair to clean the skin and keep relations friendly. Sometimes, several chimpanzees will form a line with each chimp grooming the one in front. Chimpanzees help each other out by grooming each other's hard-to-reach places. In particular, adult males spend a lot of time grooming each other—an activity that builds bonds among group members, reduces tension, and prevents fights.

Chimpanzees make and use tools—an activity that separates chimpanzees and humans from most other animals. When people first studied chimpanzees in the wild, they were amazed when they saw chimpanzees "fish" for termites. Chimpanzees shape twigs by pulling off the leaves and breaking the twig to the right length. They then stick the shaped or "modified" twig into a termite hole, wait for the termites to bite, and then quickly pull it out and eat the termites as a spicy snack! Chimpanzees are very inventive in the variety of tools they make. For example, in order to get water, they will at times crumble leaves to use like a sponge, soaking up water and then sucking the water out of the leaf. They also use rocks to break open nuts. Young chimpanzees learn from watching the older ones use tools.

"All About Chimpanzees!"

"All About Chimpanzees!" is a modular classroom education program designed to introduce grade school students to chimpanzees. The program includes five modules that focus on primatology, behavior and communication, morphology, conservation, and animal welfare. Each module includes written lesson plans for teachers, student worksheets and activities, and multimedia materials to bring chimpanzees to life in the classroom. For example, the "Chimpanzee Behavior" module includes classroom activities focused on common student questions: "Why study chimpanzees?," "Where do chimpanzees fit into the animal kingdom?," "What is a primate?," "What is *Pan troglodytes*?," "Where do chimpanzees live?," "What do chimpanzees eat?," and "How do chimpanzees behave and communicate with one another?" Each of these questions is addressed in a separate student workbook activity. Teachers are provided with lecture materials designed to address these questions, and also with a slide presentation, as well as audio and video tape materials.*

The following is an excerpt from the behavior module that focuses on chimpanzee social behavior and facial expressions.

Chimpanzee Communication

Although chimpanzees cannot speak the way humans can, they have a complex set of calls that each mean different things. One such call is the pant-hoot. This is a series of hoots and intakes of breath that can be heard from very far away and functions as a contact call between scattered members of the group. Another call is the wraah bark which means danger! Another chimpanzee vocalization is the hoo. This is a soft and quiet call usually given by an unhappy individual.

Chimpanzees, like humans, also communicate with gestures and body positions. When a person walks into a room with her head down, this signifies that she is sad or shy. When a person smiles, everyone can tell they are happy. Sometimes, when a low-ranking chimpanzee sees a higher-ranking one, he or she bends down, bobbing up and down often with his or her backside facing the dominant one. This is called presenting and it is behavior meant to avoid conflict. When a chimpanzee stands up and sways from side to side, that means he is angry and is trying to threaten others. This is called a display. A display may also include picking up a branch or another available object and dragging it, swinging, or throwing it. Chimpanzees may also be pilo-erect. Pilo-erection is a very interesting form of communication. This is where a chimpanzee's hair stands on end causing them to look bigger and scarier. Chimpanzees also do some very human-like gestures. One behavior is begging where a hand is pushed toward another who has something he or she wants, asking the other to share. Chimpanzees may also hug each other for comfort and kissing is a common form of greeting.

Because chimpanzees really use their vision, they communicate with many different facial expressions—just like humans. In fact, there are many similarities between the facial expressions of humans and chimpanzees because our facial muscles are almost identical. The following are examples of chimpanzee facial expressions. Can you make them all?

Hoot Face

The eyes are wide open and the mouth is closed and relaxed. The lips are pushed forward and curve in the middle. This expression occurs with a low hooting frequency and is usually a response to a sudden change in the environment.

*Requests for modules should be made to the Primate Foundation of Arizona, P.O. Box 20027, Mesa, AZ 85277-0027, jopfa@qwest.net.

Closed Mouth Grin

The chimpanzee's eyes are open wide, and the mouth is closed with the corners withdrawn. The lower lip is pulled back to reveal the teeth and occasionally the gums. This expression is displayed when chimpanzees are fearful or extremely uneasy.

Cry Face

During the cry face, the facial skin is wrinkled. There are wrinkles between the eyebrows, and the eyes are opened wide. The mouth is closed but the lips may be parted. The lips are pushed forward and curl outward. The inside of the lower lip is often exposed. The expression often occurs with a wailing vocalization when the chimpanzees are distressed.

Play Face

The play face is characterized by a retracted lower lip and exposed lower teeth. The mouth is opened and the upper lip is in the normal position or pulled down so that the top teeth are concealed. This expression is often accompanied by pleasure panting (rapid breathing which some people compared to human laughing) and occurs during play.

Play face.
Courtesy of Sue Howell.

Pout Face

All of the facial features except for the lips are in a normal position. The mouth is closed and the lower lip is pushed out so that it sticks out beyond the upper lip while the upper lip is in a normal position. This is a silent expression that often occurs when the chimpanzee is seeking comfort.

Open Mouth Scream Face

The eyes and the mouth are opened wide. The lips and the corners of the mouth are fully opened, exposing the teeth and gums. The cheeks are wrinkled, while the skin above the upper lip is smooth and puffed out. This expression is accompanied by a loud scream that indicates distress.

Scream face.
Courtesy of Sue Howell.

If you enjoyed this unit, you will enjoy the following books:

Barrett, N.S. 1988. *Monkeys and Apes*. London: Franklin Watts.
Dickerson, P. 1989. *Eva*. New York: Delacorte Press.
Goodall, J. 1989. *Chimps*. New York: Atheneum.
Goodall, J. 1989. *The Chimpanzee Family Book*. Saxonville: Picture Book Studio.
Goodall, J. 1986. *The Chimpanzees of Gombe: Patterns or Behavior*. Cambridge, MA: Harvard University Press.
Zihlman, A.L. 2000. *The Human Evolution Coloring Book, 2nd edn*. New York: Barnes & Noble Books.

Sue Howell, Jo Fritz, & Julia Sommerfeld

■ Education—Classroom Actvities
Insects in the Classroom

Easy to care for, small, and fast growing, insects provide excellent models for classroom studies of animal behavior. With some persuasion, students will accept that any behavior of interest can be studied more easily in an insect than in a bird or mammal. While insects lack the "warm and fuzzy" appeal of birds and mammals, their use does not raise the complicated issues of animal care and ethics that stem from using vertebrates in a teaching laboratory. Here, some of the most effective and interesting uses of insects in the classroom are summarized.

A great place to start is with a honeybee observation hive. If the teacher lacks experience with bees, it is best to get the help of a beekeeper in setting up the observation hive. A good observation hive has at least two combs (frames) and includes a queen which has been marked for identification. Honeybee hives require very little maintenance, needing only an exit to the outside so the bees can forage. In the observation hive students can observe the court of workers around the queen, differences between queen and worker behavior, coordination among the workers in caring for the larvae, and dances which communicate food locations.

Ant farms also give a good opportunity to watch queen–worker interactions, coordinated behavior in nest construction, and care of larvae. The best ants for ant farms are in the genus *Formica* because they do not sting. In the spring, queens, workers and larvae will be close to the surface of the soil and can easily be collected (as the summer progresses they move deeper, and collection requires more digging). While commercial ant farms will work well, a quart jar filled with soil is equally acceptable. Hölldobler and Wilson (1990) give more information about identifying and maintaining ants.

Termite workers in the genus *Reticulitermes* can be purchased from biological supply houses. These termites use a trail pheromone to direct the activities of the workers; this pheromone is found in some ballpoint pen inks. Termite workers will follow an ink trail, yielding a fascinating demonstration of how a chemical can direct the behavior of an animal. The teacher should try several brands of ballpoint pens beforehand to ensure that the pens given the students actually contain the pheromone.

The Madagascar hissing cockroach, *Gromphadorhina portentosa*, is often kept in classrooms and can also be obtained from biological supply houses. Males of this species have "horns" which are used in fights, much like deer and elk. They also hiss at each other during fights and when attracting mates. Given the opportunity (a large enough container) males will set up territories, and students can map space use—just as would be done with any territorial animal.

House crickets, *Acheta domestica*, give students an excellent opportunity to study auditory communication. Pet stores sell crickets as food for frogs, making them easily obtainable. Males sing to attract females to their territory; it is a simple matter to record the males' songs and play them back to demonstrate the attractiveness of each male. A slightly more complex experiment involves giving the female a choice between songs. Simple sound-editing computer software can be used to modify songs, changing their attractiveness.

Fruit flies (*Drosophilia melanogasten*), given their small size, are a little more difficult to observe, but provide a wonderful opportunity to observe a complex sequence of mating behavior. Signals used include visual, auditory, and chemical cues; this variety of signaling modes yields a surprisingly complex behavioral sequence prior to mating.

A pint-sized wasp, *Melittobia digitata*, has received considerable publicity as a classroom insect. Termed "wowbugs" by the developers of this species for use in the classroom,

Melittobia offers students the chance to study many aspects of behavior, including host search (this wasp parasitizes other insects) and mating. The only possible drawback is their small size; observations must be made with a magnifying glass or low-power microscope.

Other interesting species that are easy to keep in the classroom include house centipedes (not really an insect, of course, but they fit well in a discussion of insects) whose predatory behavior is easily observed. Various types of mealworms and flour beetles can be used in studies of habitat and feeding preferences. Painted lady butterflies, reared from eggs, will mate in flight cages in the classroom. These species are all easily obtained from commercial suppliers.

Further Resources

For more information about WOWBugs go http://www.wowbugs.com/

Brookes, M. 2001. *Fly: The Unsung Hero of 20th-Century Science*. Collingdale, PA: DIANE Publishing.

Hölldobler, B. & Wilson, E. O. 1990. *The Ants*. Cambridge, MA.: Belknap Press of Harvard University Press.

Matthews, R. W., Koballa, Jr., T. R., Flage, L. R. & Pyle, E. J. 1996. *WOWBugs: New Life for Life Science*. Riverview Press. (706) 369–0931.

Ploger B. J. & Yasukawa, K. (Eds.) 2002. *Exploring Animal Behavior in Laboratory and Field: An Hypothesis-testing Approach to the Development, Causation, Function, and Evolution of Animal Behavior*. New York: Academic Press

Michael D. Breed

■ Education—Classroom Activities
"*Petscope*"

Petscope, a program devised by this author, builds on the fascination that pets hold for many students and transforms community pet stores into comparative animal behavior research centers. Students work in groups or individually and can perform many noninvasive studies of animal behavior. Students can directly observe the similarities and differences among species, develop comparative research skills, and receive practical experience in the study of animal behavior. Petscope projects are ideal for science fair projects, high school biology and psychology courses, and college courses ranging from introductory courses in biology, animal behavior and psychology to more advanced courses such as comparative psychology, research methods, and experimental psychology. Petscope projects can also be readily incorporated into course curriculums as laboratory experiences and/or independent student projects.

Materials Needed

The materials needed are access to a pet store, a working relationship between the pet store manager/owner and instructor, a list of the animals maintained in the pet store, a project plan or worksheet distributed to students, and apparatus required to complete the project such as a data sheet, timer, and writing implement. For projects requiring size measurements, it is helpful to photocopy a ruler onto a worksheet so that students will have a ruler readily available.

Instructions

For the Petscope program to be effective, a good working relationship between the instructor and pet store manager/owner is crucial. This is necessary because the pet store is a privately owned business, and therefore permission must be sought before students begin class projects. Moreover, pet store personnel and the instructor must address issues such as the possibility of students handling some of the animals (i.e., snakes, birds, lizards, rodents, cats, dogs), creating observation stations in front of animal enclosures, establishing a time for student groups to carry out projects that do not interfere with normal business operations, the availability of first aid, the extent to which pet store staff can assist students with their projects, and the possibility of students manipulating the animals' environment either by feeding or adding enrichment items such as "toys" and other stimuli. In addition to the discussion of these issues, pet store personnel should provide the instructor with a list of animals available for study. With this list the instructor is in a better position to direct and plan student research projects, and the student has an opportunity to acquire background information about an animal prior to working with it. We would also recommend that the instructor accompany the students on their first visit so they can be introduced to store personnel.

In addition to developing a good relationship with the pet store, the instructor may want to review behavioral and comparative study methods. Abramson in *A Primer of Invertebrate Learning: The Behavioral Perspective* and Bornstein in *Comparative Methods in Psychology* provide good introductions. Describing basic research methods to the class will provide helpful guidelines for students as they design projects and analyze data. Many projects are suitable for a Petscope program. Three are described below. The first is a project exploring aggressive behavior in male betta fish; the second is a comparative investigation of habitat, physiology, and behavior of pet store animals, and the third is some unobtrusive investigations of fish behavior. These projects are to be considered only as samples. Indeed, one of the more interesting features of the projects is their broad flexibility. In many ways, the only limitation is imagination. For instructors not familiar with the type of projects possible, they may want to look at some of these publications: Abramson (1986, 1990, 1996); Brown and Downhower (1988), Cain (1995), and Kneidel (1993). These articles provide many useful examples for both invertebrate and vertebrate student projects.

Project 1: Aggression in Male Betta Fish

Male betta fish make good subjects for study since they have been bred selectively for bright colors and aggressive behavior. Males courting or confronted with another male become more vivid; many will even react to their mirror image. Students will find stimulating and observing their fish so easy that they can spend time acquiring or improving their basic research and data recording skills.

Students can conduct this exercise individually, in pairs, or in groups depending on the needs of the instructor and pet store staff. The materials needed are a small mirror, a picture of a betta fish (pictures are easily found at the library or in on-line encyclopedias or web sites), and writing materials. Most pet stores keep male betta fish separately in small glass or plastic containers; such containers are ideal for this project.

After arriving at the pet store, each research group will follow this procedure:

1. Make sure that each betta container to be observed is at least two feet from other betta bowls.
2. Observe betta 2 to 3 minutes. Record color, size, movement behavior, and breathing rate (breathing can be measured by watching gill movement).

3. Move the container next to another betta container so that the containers are paired up. Students may want to record how far apart containers are when bettas start to react to each other.

4. Observe bettas in paired bowls and record color, movement behavior, and breathing rate. If the bettas behave differently, how could behavior differences be explained? Are the two bettas different colors? Different sizes?

5. Return the bowls to their original separated positions. Allow bettas to return to their original color and behavior.

6. Hold a mirror up to the side of the betta container and record color, movement behavior, and breathing rate. Does the betta react? Is the reaction similar to its reaction to the real betta?

7. Remove the mirror. Allow betta to return to its original color and behavior.

8. Hold the betta picture up to the side of the container and record color, movement behavior, and breathing rate. Compare any reactions with reactions to the mirror and real betta.

When the experiment is concluded, students as a class can compare reactions to the various stimuli, looking for similarities and differences among the fish. A wide variety of discussion or writing topics can arise from these comparisons: Why would fish of the same species react to the same stimuli differently? Does fish color relate to type of reaction? Why do only male betta fishes display aggression and bright colors? How is the data to be analyzed and graphed? Why is there a need for careful measurement? What is the importance of replication of data from experimenter to experimenter? What are the advantages and disadvantages of working with a research team?

Project 2: Creating Petscope Data Cards

The project was inspired by the old Time-Life animal cards. In the Time-Life version, each card contained interesting facts about animals such as alligators and jaguars. Our version contains both a library component and an observational/research component. To create Petscope cards students visit the pet store and select animals in which they have an interest. Alternatively, the instructor may decide to focus the students' attention on a few related species such as goldfish, angel fish, and fighting fish. Once the animals are selected, a decision must be made to have the students work as part of a research team or individually. Having made this decision students are assigned to the pet store and begin to create Petscope cards.

The library component will require the student to gather information on classification (class, order, family, genus, species), behavior, social reproductive learning, mate selection, related species, range, physiology and anatomy of the animals they select. Students with access to scanners may wish to incorporate visual images into their cards. A useful addition is to include information describing scientists who conduct research with that species. This information would include selected references, a summary of research findings, and address. A benefit of this addition is that students learn about the work of, and perhaps network with, professors in their home institution. Moreover, the library component encourages library use, Internet searches, and discussions with experts such as pet store personnel. We have found that students enjoy helping to decide what information to include in the Petscope cards.

Sample Petscope Card

Scientific Name _____

Common Name(s) _____

_____

Description (Size, Coloration, etc.) _____

_____

_____

Picture

Notes on Observed Behavior _____

_____

_____

_____

_____

Notes on Library Research
(References, Latest Findings, Current Projects and Researchers)

_____

_____

_____

_____

_____

_____

The observation/research component of the Petscope card relies on information gathered directly by students at the pet store and can include information such as descriptive anatomy (e.g., color, length, weight), feeding and mating strategies, growth rate, diet, habitat, size, and popularity.

Once completed, comparisons among the cards can be made on any of the dimensions listed above. Subsequent class discussion topics include the importance and difficulty of classification, the role of evolution and ecology in shaping biological, anatomical, and behavioral processes, the influence of pollution and human encroachment on the development of species, and the importance of pets in mental health. When the assignment is completed the cards can be donated to the pet store and displayed, or donated to local Head Start programs and other educational institutions. The following figure provides an example of a Petscope card worksheet.

Student Worksheet: Comparative Anatomy of Fish

Species Studied (Scientific and Common Names)

_____

In the space below, draw your species. Note important characteristics like size, color, any range of colors within species, differences between make and female fish, etc.

Observe and describe visible fin and tail differences and similarities between your species and two other species. Are fins and tails the same size and shape? Do all 3 species seem to use fins and tails in the same way?

Species 1 (Name) _____

_____

_____

Species 2 (Name) _____

_____

_____

Species 3 (Name) _____

_____

_____

_____

Sample Petscope Worksheet

Project 3: The Study of Fish

Various species of fish are available in pet stores. Students can be divided into small groups and assigned a tank. Suggested areas of investigation are listed below. See the sample worksheet for an example of a project.

1. _Ethological investigations_: Students can observe, describe and compare environments, coloration of animals, ratio of male to female, food gathering strategies, defensive strategies, time allocated to various activities.

2. _Learning investigations_: Many people have noted that their pet fish swim to the top of the tank when the aquarium lid is lifted or the aquarium light is turned on. Students can use this response to study whether fish respond to stimuli paired with feeding and

to see whether time of day or the introduction of novel stimuli inhibit feeding. Students can also observe responses to new situations or novel stimuli.

3. *Comparative anatomy*: Students can compare development, shape, size, and function of anatomical features such as fins and tails.

4. *Social behavior*: Students can observe and describe mating behavior, courtship behavior, group interaction, and species recognition. How do the fish react when new fish are placed into the tank? Do they behave differently if the new fish are of a different species?

5. *Comparative physiology*: Basic observations could include variations in breathing rates based on species, size of fish, or male vs. female fish.

6. *Research methods*: Depending on the research question asked and the abilities of the students, research methods can range from recording numbers of animals to the creation of behavioral profiles for various species.

Discussion

After conducting a project, students should have a general understanding of the comparative approach and the importance of animal research. Moreover, students will have learned observational and data recording techniques. If Petscope cards are created, students will have exposure to CD-ROM based databases and other library resources. Moreover, they will have developed writing skills and graphic design skills. In sharing information with the class, a student's communication and public speaking skills will also be utilized.

It has been our experience that pet store owners are happy to participate in the Petscope program because of the program's educational value. There is also the perception that the program increases business and expands the customer base. A limitation, however, is that some pet stores have a narrow range of species. In such cases the instructor can restrict the number of species or solicit participation from two pet stores that differ in the type of animal displayed.

Writing Component

In addition to creating Petscope cards, other writing assignments could include further research and description of animals from the perspective of home pets. Students can also use their observational skills to chronicle the behavior of family pets and those of their friends.

See also Social Organization—*Social Order and Communication
in Dogs*

Further Resources

Abramson, C. I. 1986. *Invertebrates in the classroom*. Teaching of Psychology, 13, 24–29.
Abramson, C. I. 1994. *A Primer of Invertebrate Learning: The Behavioral Perspective*. Washington, DC: American Psychological Association.
Abramson, C. I. 1990. *Invertebrate Learning: A Laboratory Manual and Source Book*. Washington, D.C.: American Psychological Association.
Abramson, C. I., Onstott, T., Edwards, S., & Bowe, K. 1996. *Classical-conditioning demonstrations for elementary and advanced courses*. Teaching of Psychology, 23, 26–30.
Bornstein, M. 1980 (Ed.). *Comparative methods in psychology*. Hillsdale, N.J.: Lawrence Erlbaum Associates.
Brown, L., & Downhower, J. F. 1988. *Analyses in Behavioral Ecology: A Manual for Lab and Field*. Sunderland, MA: Sinauer Associates.

Cain, N. W. 1995. *Animal Behavior Science Projects*. New York: John Wiley and Sons.

Kneidel, S. S. 1993. *Creepy Crawlies and the Scientific Method*. Golden, CO: Fulcrum Publishing.

Web Sites

Fields of Knowledge houses the "Infography." The Infography contains citations to books, internet sites, journal articles and other information related to many subjects including the behavior of animals. The web site is: http://www.fieldsofknowledge.com/

Charles I. Abramson

■ Education—Classroom Activities
Planarians

Concept

Planarians as subjects for psychological demonstrations in animal behavior offer many advantages over more traditional vertebrate animals for laboratory course work and independent study. They are inexpensive to procure and maintain; planarians suitable for a class of 25 can be obtained for about $10.00. Planarians can be trained in a variety of mazes, runways and conditioning "troughs" that cost dollars rather than hundreds of dollars and that can be constructed from everyday materials such as plastic tubes and tubing connectors. They can be used to demonstrate principles of the comparative analysis of behavior, ethology, biochemistry of learning, physiological psychology, and behavioral ecology. Also, the nervous system of planarians is well suited to explore questions about the underlying physiology of behavior and the transmission of learned information from one animal to another. High school and college students can reenact some of the more controversial studies in the history of animal learning such as the transfer of information between animals by cannibalism and by regeneration.

Planarians are in the phylum Platyhelminthes. This phylum contains planarians, flukes, and tapeworms. In these animals we find the first appearance of characteristics critical for the development of complex behavior. These characteristics include mirror images of the left and right side of the body (bilateral symmetry), the appearance of a "brain," polarized neurons, and a definitive front (anterior) and back (posterior) orientation—with the anterior end containing a head and "eyes." These new advances make habituation and modifying behavior based on consequences (instrumental learning) possible.

In this exercise, students use planarians to learn about principles of habituation and instrumental conditioning. *Habituation* refers to the reduction in responding to a stimulus as it is repeated. In habituation, the animal does not learn to do something new or different—only an existing behavior is modified. For a decline in responsiveness to be considered habituation, it must be determined that the decline is not related to what behavioral scientists have called sensory adaptation and motor fatigue. For example, the planarian may stop contracting to an airpuff, not because it has learned to do so, but because of a temporary adaptation of its touch receptors (known as *sensory adaptation*). Alternatively, the planarian may still sense the airpuff, but can no longer make the effort necessary for it to contract (known as *effector fatigue*).

Studies of habituation show that it has several characteristics, including the following:

1. The more rapid the rate of stimulation is, the faster the habituation is.
2. The weaker the stimulus is, the faster the habituation is.
3. Habituation to one stimulus will produce habituation to similar stimuli.
4. Withholding the stimulus for a long period of time will lead to the recovery of the response.

Materials Needed

Each research team should have one planarian, one 20 cc plastic syringe (without needle), one stop watch, one data sheet, one conditioning tray, one small paint brush, and enough spring or pond water to fill the conditioning tray(s). The instructor should also have two containers: one to hold the stock colony and the other to place conditioned planarians. In addition, the instructor should have extra planarians to replace any that fail to respond.

Conditioning Tray

A plastic cheese cutting board makes an excellent low cost conditioning tray. The tray costs about $2.00 and is available from most supermarkets. A .8 cm deep by 1.4 cm wide trough lies along the perimeter of the cutting board and serves as the conditioning trough. To prepare the trough, fill it with 15 cc of spring water. Planarian pond water can also be used. This produces a depth of approximately 5 mm which is deep enough for the animal to swim yet still react to the airpuff. If the cheese cutting board is not available, a plastic butter dish or petri dish will substitute. To administer the habituating stimulus, or in the case of instrumental conditioning, the punishment (e.g., the airpuff), a plastic 20 cc syringe, available at any drug store or biological supply house, is used. An alternative to the syringe is to use a plastic soda straw. If a straw is used, the airpuff is presented by blowing into one end of the straw while directing the opposite end to the animals' head.

Planarians

Planarians are easy to obtain and care for and are available from any number of biological supply houses. Any species can be used. For the more adventurous, planarians can be captured in ponds, streams, lakes and rivers. When planarians are ordered commercially, instructions for feeding and maintenance are included. Planarians can be housed in any container that can hold water. The most important factor in maintaining a good planarian culture is clean water. The water should be changed once a week and be chlorine free. A good time to change the water is after a feeding. Planarians are primarily carnivorous (and cannibalistic) and show a preference for fresh liver, although they will eat frozen liver that has been defrosted. A weekly feeding is quite sufficient to maintain healthy planarians. When the experiment is completed the instructor can decide to keep a stock colony for future demonstrations. If this is not desirable, the stock colony and conditioned colony can be donated to a biology/zoology department.

Data Sheet

A data sheet suitable for both demonstrations is illustrated in the figure on p. 541. The data sheet contains room for student information, observations, and data.

Sample Data Sheet

Name: _____   Experiment: Habituation, Instrumental, Other

Date: _____   Time: _____   Subject #: _____

Number of trials: _____   ISI / ITI: _____   Subject size: _____

Stimulus intensity: _____   Stimulus duration: _____   Other: _____

Trial/Response	Trial/Response	Trial/Response
1	21	41
2	22	42
3	23	43
4	24	44
5	25	45
6	26	46
7	27	47
8	28	48
9	29	49
10	30	50
11	31	51
12	32	52
13	33	53
14	34	54
15	35	55
16	36	56
17	37	57
18	38	58
19	39	59
20	40	60

Notes and Observations:

Instructions

Identification of Contraction Response

For habituation experiments, the response to an airpuff is a pronounced contraction of the planarian away from the direction of the airpuff.

Identification of a Turning Response

For instrumental conditioning, the turning response is a contraction followed by an extension of the animal. Following several contractions and extensions, the animal begins to turn away from the direction of the airpuff and swims in the opposite direction. For example, if the animal is swimming in a clockwise direction, repeated presentations of an airpuff elicits a number of contractions and extensions. These behaviors are followed by the animal turning and swimming in a counterclockwise direction. In our demonstration the dependent variable of interest is the number of airpuffs required to make the animal swim in the opposite direction. As training progresses the number of airpuffs required to produce a turning response steadily declines.

Transferring the Planarian from the Home Container to the Conditioning Trough

Use the paintbrush to gently pick up the planarian. Once the planarian is on the paintbrush, move the paintbrush to the conditioning trough and place it in the water. The planarian will quickly crawl off the paintbrush and into the water. Alternatively, you can use an eyedropper. A limitation of the eyedropper is that a planarian will often get stuck to the sides of the dropper when the water is expelled.

Conducting the Experiments

Habituation

In our demonstration, the time between airpuff presentations (i.e., interstimulus interval) is 60 seconds, the intensity of the airpuff is 20 cc of air, the duration is approximately 1 second (the time to depress the syringe), and the number of trials (experiences) is 50. Having selected the experimental parameters, fill the conditioning trough with water and remove a planarian from the colony with a paint brush and gently place it in the conditioning trough. Following a 5-minute adaptation period, administer the airpuff by pulling back on the plunger of the syringe until the plunger reaches the 20 cc mark. Next, direct the tip of the syringe to the head of the planarian and depress the plunger. Upon contact with the airpuff, a planarian will typically contract to about one half of its length. After the syringe is depressed, the experimenter must do two things. *First*, start the timer to begin the interstimulus interval (e.g., the time between airpuffs), and *second*, record the presence or absence of a contraction response on a data sheet. It is often convenient to record a "1" if the animal contracted and a "0" if it did not. If graph paper is placed underneath the conditioning trough (assuming the trough you use is transparent) the contraction response may be quantified and expressed as the number of boxes exposed—be sure to use graph paper with many gridlines. If the planarian does not contract, for example, no boxes will be exposed and you will enter a 0 on the data sheet. With repeated applications of the airpuff, the number of contractions decrease. When the experiment is completed, the planarian should be returned to a home tank reserved for conditioned animals.

The experimental procedure for demonstrating habituation does not contain control procedures. It may be desirable for some members of the class to perform baseline experiments to determine the base rate of contraction in the absence of airpuff. Students also find it informative to modify the procedure to include sensitization trials to rule out sensory adaptation and motor fatigue as alternative explanations for the decrease in contraction they observe. One way to provide such stimulation is to gently touch the animal with a bristle from the paint brush. Alternatively, a small drop of water can be splashed near the animal using an eyedropper. The animal should contract to the introduction of the new stimulus and contract again when the airpuff is reintroduced. Such a response rules out sensory adaptation and effector fatigue because the planarian is showing you that it is still responsive to stimulation. If class time permits, students will find it interesting to vary the intensity of the airpuff by increasing or decreasing the volume of air in the syringe. The duration of the airpuff can also be easily altered by varying the speed in which the plunger of the syringe is depressed.

Instrumental Conditioning

In this demonstration, students use the same airpuff and cutting board to explore principles of instrumental conditioning. Instrumental conditioning is behavior controlled by its consequences. In contrast to habituation, instrumental conditioning does require the animal to learn something new or different. Studies of instrumental conditioning show that it has several characteristics, including the following:

1. The shorter the delay between response and consequence, the faster the learning.
2. The greater the consequence, the stronger the learning.
3. The more pairings between a response and consequence, the stronger the learning.
4. Applying the consequence in the absence of the response reduces learning.
5. When the consequence no longer follows the response, the response gradually becomes weaker and eventually stops occurring.
6. When a learned response has been established to a particular situation, similar situations may elicit the response.

When filled with water, the trough in the cutting board makes an excellent rectangular planarian runway. A runway can be conceptualized as a maze without choice points. To demonstrate that even in an animal as simple as the planarian, behavior can be controlled by its consequences. The planarian is given a series of airpuffs until they turn in the opposite direction. As the training session continues, the number of airpuffs required to elicit turning behavior decreases.

In our demonstration, the intensity of the airpuff is 20 cc of air, the duration of the airpuff is as short as possible (1 second or less), the intertrial interval is fixed at 1 minute, and the number of trials needed to conclude the experiment is 60. When the parameters of the experiment are decided, gently remove the planarian from its home container (with the paint brush) and give it a 5-minute adaptation period in the conditioning trough.

When the planarian is in the trough, it will begin to swim in one direction (e.g., clockwise or counterclockwise). Following the 5-minute adaptation period, the experimenter will try to reverse this movement by giving the planarian a series of airpuffs directed to its head. When the planarian changes direction, depress the stop watch to time the next intertrial interval, and record the number of airpuffs on the data sheet. Typically it will take 12–20 airpuffs before the animal changes direction. This will soon decrease to about 4–5. The number of airpuffs administered before the planarian changes direction is called the *dependent variable*. When the experiment is completed, return the subject to a home container reserved for conditioned animals.

As in the habituation experiment, students enjoy creating their own experimental designs. Some of the more common variations include manipulating the intensity of the airpuff, presenting the airpuff on a partial schedule of reward, delaying the time between the response and presentation of the airpuff, no longer presenting the airpuff following a turn (known technically as extinction), varying the intertrial interval, pretreating the animal with drugs such as caffeine, and investigating whether instrumental behavior survives when the animal is cut in half and allowed to grow into two new animals (known as regeneration). A particularly interesting variation is to show students the effect of pollution on learning. This can be accomplished by placing planarians in polluted water obtained from, for example, a local stream or water tap.

Unresponsive Animals

Occasionally a student may encounter an unresponsive planarian. Therefore, whenever possible extra planarians should be brought to the class to replace those not contracting to the airpuff or, once contracted, not re-extending. Such behavior is likely if the animals are dropping their tails, are over stimulated, or come from a polluted environment. If such behavior is observed, the animals should be immediately removed from the conditioning chamber and gently returned to their home container.

Discussion

The advantage of the demonstrations reported here are their extremely low cost and versatility. Moreover, students enjoy the hands-on approach. An entire research station for each team containing a planarian and apparatus will cost under $6.00. In a small class each student can condition his or her own planarian. If the demonstration is presented to a large lecture class, the instructor can divide students into research teams where one member of the team, for example, presents the stimuli and the other member records data. An effective alternative to creating research teams is to place the apparatus on top of an overhead projector so the entire class can view the performance of a single animal.

Although not required, our experience suggests that students enjoy the planarian demonstrations more if they are integrated into a component on learning and memory. We also recommend that the component include information on the natural history and biology of these interesting organisms. Students also enjoy hearing about the unique place planarian learning holds in the history of psychology (Rilling 1996).

The use of planarians to demonstrate principles of habituation and instrumental learning is also useful in generating classroom discussion about the importance of learning. Points for discussion include the usefulness of cognitive explanations of behavior, what are suitable control groups, what is a suitable statistical analysis, and what is the biological significance of conditioning.

Additional Reading Material

The material in this reading list was carefully selected to provide an instructor or interested student with information on various aspects of using planarians in the classroom. The material can be used to create a reading list on a specific topic or range of topics, offer tips and insights into various aspects of planarian behavior, stimulate classroom discussion and provide ideas for future student projects and instructor demonstrations.

Other Classroom Demonstrations Using Planarians

Abramson, C. I. 1990. *Invertebrate Learning: A Laboratory Manual and Source Book.* Washington, DC: American Psychological Association.

Katz, A. N. 1978. *Inexpensive animal learning exercises for huge introductory laboratory classes.* Teaching of Psychology, 5, 91–93.

McConnell, J. V. 1967. *A Manual of Psychological Experiments on Planarians.* 2nd edn. Ann Arbor, MI: Journal of Biological Psychology.

Owren, M. J. & Scheuneman, D. L. 1993. *An inexpensive habituation and sensitization learning laboratory exercise using planarians.* Teaching of Psychology, 20, 226–228.

Review Articles

Corning, W. C. & Kelly, S. 1973. *Platyhelminthes: The turbellarians.* In: *Invertebrate Learning: Protozoans through Annelids Vol. 1* (Ed. by W. C. Corning, J. A. Dyal & A. O. D. Willows), pp. 171–224. New York: Plenum Press.

Corning, W. C. & Ratner, S. C. 1967 (Eds.). *Chemistry of Learning.* New York: Plenum.

Corning, W. C. & Riccio, D. 1970. *The planarian controversy.* In: *Molecular Approaches to Learning and Memory* (Ed. by W. Byrne), pp. 107–150. New York: Academic Press.

McConnell, J. V. & Shelby, J. M. 1970. *Memory transfer experiments in invertebrates.* In: *Molecular Mechanisms in Memory and Learning* (Ed. by G. Unger), pp. 71–101. New York: Plenum Press.

Sarnat, H. B. & Netsky, M. G. 1985. *The brain of the planarian as the ancestor of the human brain.* Canadian Journal of Neurological Sciences, 12, 296–302.

Sheiman, I. M. & Tiras, K. L. 1996. *Memory and morphogenesis in planaria and beetle.* In: *Russian Contributions to Invertebrate Behavior* (Ed. by C. I. Abramson, Z. P. Shuranova & Y. M. Burmistrov), pp. 43–76. Westport, CT: Praeger.

Habituation and Sensitization Experiments

Westerman, R. A. 1963a. *Somatic inheritance of habituation to light in planarians.* Science, 140, 676–677.

Westerman, R. A. 1963b. *A study of habituation of responses to light in the planarian Dugesia dorotocephala.* Worm Runner's Digest, 5, 6–11.

Classical Conditioning Experiments

Crawford, F. T. 1967. *Behavioral modification of planarians.* In: *Chemistry of Learning* (Ed. by W. C. Corning & S. C. Ratner), pp. 234–250. New York: Plenum Press.

Jacobson, A. L., Fried, C. & Horowitz, S. D. 1967. *Classical conditioning, pseudo-conditioning, or sensitization in the planarian.* Journal of Comparative and Physiological Psychology, 64, 73–79.

McConnell, J. V. 1962. *Memory transfer through cannibalism in planarians.* Journal of Neuropsychiatry, 3, 42–48.

McConnell, J. V., Jacobson, A. L. & Kimble, D. P. 1959. *The effects of regeneration upon retention of a conditioned response in the planarian.* Journal of Comparative and Physiological Psychology, 52, 1–5.

Thompson, R. & McConnell, J. V. 1955. *Classical conditioning in the planarian,* Dugesia dorotocephala. Journal of Comparative and Physiological Psychology, 48, 65–68.

Vattano, F. J. & Hullett, J. H. 1964. *Learning in planarians as a function of interstimulus interval.* Psychonomic Science, 1, 331–332.

Instrumental/Operant Conditioning Experiments

Best, J. B. 1965. *Behavior of planaria in instrumental learning paradigms.* Animal Behaviour, 13, Supplement 1, 69–75.

Corning, W. C. 1964. *Evidence of right–left discrimination in planarians.* Journal of Psychology, 58, 131–139.

Crawford, F. T. & Skeen, L. C. 1967. *Operant responding in the planarian: A replication study.* Psychological Reports, 20, 1023–1027.

Humpheries, B. & McConnell, J. V. 1964. *Factors affecting maze learning in planarians.* Worm Runner's Digest, 6, 52–59.

Krebs, E. K. 1975. *Factors in conditioning and orientation of planarians: Learning in a polar field.* Journal of Biological Psychology, 17, 21–23.

Lee, R. M. 1963. *Conditioning of a free operant response in planaria.* Science, 139, 1048–1049.

Wells, P. H. 1967. *Training flatworms in a Van Oye maze.* In: *Chemistry of Learning* (Ed. by W. C. Corning & S. C. Ratner), pp. 251–254. New York: Plenum.

Influence of Pollutants and Drugs on Behavior

Bonner, J. C. & Wells, M. R. 1987. *Comparative acute toxicity of DDT metabolites among American and European species of planarians.* Comparative Biochemistry and Physiology (C): Comparative Pharmacology and Toxicology, 87, 437–438.

Congiu, A. M., Casu, S. & Ugazio, G. 1989. *Toxicity of selenium and mercury on the planarian* Dugesia gonocephala. Research Communications in Chemical Pathology and Pharmacology, 66, 87–96.

Johnson, L. R., Davenport, R., Balbach, H. & Schaeffer, D. J. 1994. *Phototoxicology: 3. Comparative toxicity of trinitrototoluene and aminodinitrotoluenes to* Daphnia magna, Dugesia dorotocephala, *and sheep erythrocytes.* Ecotoxicology and Environmental Safety, 27, 34–49.

Kessler, C. C. 1973. *The effect of magnesium pemoline on learning in the planarian.* Journal of Biological Psychology, 15, 31–33.

Comparative/Ethological Investigations

Dolci Palma, I. A. 1987. *The hunting and eating of prey by planarians of the species* Dugesia tigrina *kept in the laboratory.* Ciencia e Cultura, 39, 557–560.

Loh, P. Y., Yap, H. H., Chong, N. L. & Ho, S. C. 1992. *Laboratory studies on the predatory activity of a turbellarian,* Dugesia sp. (Penang) on Aedes aegypti, Anopheles maculatus, Culex quinquefasciatus *and* Mansonia uniformis. Mosquito Borne Diseases Bulletin, 9, 55–59.

Mason, P. R. 1973. *Size and other factors determining planarian behaviour.* Journal of Biological Psychology, 15, 8–13.

Mason, P. R. 1975. *Chemo-klino-kinesis in planarian food location.* Animal Behaviour, 23, 460–469.

Nixon, S. E. 1974. *Some behavioral observations on a cave dwelling planarian.* Journal of Biological Psychology, 16, 32–33.

Peters, A., Streng, A. & Michiels, N. K. 1996. *Mating behaviour in a hermaphroditic flatworm with reciprocal insemination: Do they assess their mates during copulation?* Ethology, 102, 236–251.

Reynierse, J. H. & Gleason, K. 1974. *Determinants of planarian aggregation behavior.* Animal Learning and Behavior, 3, 343–346.

Rivera, V. R. & Perich, M. J. 1994. *Effects of water quality on survival and reproduction of four species of planaria* (turbellaria: tricladida). Invertebrate Reproduction and Development, 25, 1–7.

Zoological/Natural History Information

Barnes, R. D. 1974. *Invertebrate Zoology* (3rd edn). Philadelphia: W. B. Saunders.

Pearse, V., Pearse, J., Buschbaum, M. & Buschbaum, R. 1987. *Living Invertebrates*. Pacific Grove, CA: The Boxwood Press.

Web Sites

Web Source: Fields of Knowledge houses the "Infography." The Infography contains citations to books, internet sites, journal articles and other information related to many subjects including the behavior of animals. The Web site is: http://www.fieldsofknowledge.com/

Planarium Research Equipment Sources

The following are some sources for the equipment described.

Conditioning Tray: A plastic cheese cutting board manufactured by Arrow Plastic Manufacturing, 701 E. Devon Ave. Elk Grove, IL 60007, part number 00711 makes a good tray. They can be contacted at (847) 595-9000 or http://www.arrowplastic.com

Planarium pond water is available from Ward's Biology, Rochester, NY, part number 88 W 7010. Web address: http://www.wardsci.com.

Planarians are available from a number of biological supply houses. One is Ward's Biology, Web address: http://www.wardsci.com.

Writing Component

One of the strengths of this exercise is that it can be easily incorporated into a writing assignment. Students will formulate a hypothesis, conduct the literature review, design and carry out the experiment, analyze, graph, and discuss the results. An alternative writing assignment is to have students conduct a literature review with the goal of comparing their planarian results with those obtained with other species (Abramson 1986, 1990, 1994; Abramson, et al 1996; Sheiman & Tiras 1996). Students can also be asked to provide written answers to the discussion questions mentioned above and to create a poster presentation based on their library and classroom research.

Further Resources

Abramson, C. I. 1986. *Invertebrates in the classroom*. Teaching of Psychology, 13, 24–29.

Abramson, C. I. 1990. *Invertebrate Learning: A Laboratory Manual and Source Book*. Washington, DC: American Psychological Association.

Abramson, C. I. 1994. *A Primer of Invertebrate Learning: The Behavioral Perspective*. Washington, DC: American Psychological Association.

Abramson, C. I., Onstott, T., Edwards, S. & Bowe, K. 1996. *Classical-conditioning demonstrations for elementary and advanced courses*. Teaching of Psychology, 23, 26–30.

Rilling, M. 1996. *James McConnell's forgotten 1960s quest for planarian learning: A biochemical engram, and celebrity*. American Psychologist, 51, 589–598.

Sheiman, I. M, & Tiras, K. L. 1996. *Memory and morphogenesis in planaria and beetle*. In: *Russian Contributions to Invertebrate Behavior* (Ed. by C. I. Abramson, Z. P. Shuranova & Y. M. Burmistrov), pp. 43–76. Westport, CT: Praeger.

Charles I. Abramson

■ Emotions
Emotions and Affective Experiences

What exactly is an emotion? There is no generally accepted scientific answer to this question. Although practically everyone has an intuitive grasp of the concept, it has been difficult for psychologists and animal behaviorists to agree on a scientific definition of this multifaceted phenomena. Some take the position that an adequate definition can only be generated at the end of relevant scientific inquiries rather than at the beginning, especially after adequate study of the underlying brain mechanisms. From that perspective, a list of provisional criteria has been offered for the types of neural systems that should be sought (Panksepp 1998, p. 48–49). The resulting brain-based definition is quite similar to the psychological definition described below, and unlike many previous definitions, it accepts the essential importance of the internally felt, subjectively experienced, aspects of emotional arousal, even in other animals. Without the criterion of emotional feelings, it would be difficult to distinguish emotional processes from other psychological functions.

After discussing hundreds of past definitions, Kleinginna & Kleinginna (1981) argued that "a formal definition of emotion should be broad enough to include all traditionally significant aspects of emotion, while attempting to differentiate it from other psychological processes," they suggest the following working definition:

> Emotion is a complex set of interactions among subjective and objective factors, mediated by neural/hormonal systems, which can (a) give rise to affective experiences such as feelings of arousal, pleasure/displeasure; (b) generate cognitive processes such as emotionally relevant perceptual effects, appraisals, labeling processes; (c) activate widespread physiological adjustments to the arousing conditions; and (d) lead to behavior that is often, but not always, expressive, goal-directed, and adaptive. (p. 355)

These criteria are in general agreement with a neural definition noted above. However, in animal research we must employ the last two criteria even more than we do in humans, since we can only measure subjective aspects of animal emotions by the characteristics that are evident in their physiological responses and outward behaviors.

The feeling aspect of emotions must always be studied indirectly through the study of instinctual actions and the use of various learned approach and avoidance tasks. Among the best measures are *conditioned place preference* and *avoidance* measures (Bardo & Bevins 2000). Animals do not go back into environments where they have had aversive experiences, but they readily return to those where they had positive feelings. More direct vocal reports of emotional experiences are becoming available (Knutson, Burgdorf & Panksepp 2002). For instance, laboratory rats exhibit rapid,

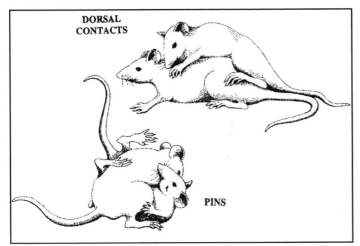

DORSAL CONTACTS

PINS

"Chirping" in rats occurs most frequently in the midst of play.
Courtesy of Jaak Panksepp.

high-frequency 50 kHz chirps (which humans cannot hear without special equipment) in environments where they have previously received rewards, and they exhibit lower frequency 22 kHz squeals in environments where they have experienced dangers or have been punished. Indeed, 50 kHz chirps can be aroused simply by tickling the animals, which has led to the suggestion that such sounds may be a primitive laughter response (Panksepp & Burgdorf 2003). Such an interpretation challenges all behavioral scientists to contemplate the potential feelings of their animal subjects more seriously.

Across species, it is probably easier to study the affective-feeling aspects of emotions than the accompanying cognitive contents because the feeling aspects are more genetically ingrained than the cognitive aspects. There is much evidence that emotional affects emerge substantially from the instinctual action systems that generate spontaneous emotional behaviors. Thus, angry feelings arise to some extent from brain systems that spontaneously generate angry behaviors along with the resulting sensory changes. Unfortunately, there has been a long-standing historical bias to consider instinctual behaviors as being largely unconscious motor outputs of the brain. In fact, such instinctual tools for living that evolution built into the neurodynamics of our brains long ago may be the fundamental sources of our emotional feelings. In this context, it is worth noting what Sigmund Freud said in *Beyond the Pleasure Principle* (1920/1959):

> No knowledge would have been more valuable as a foundation for true psychological science than an approximate grasp of the common characteristics and possible distinctive features of the instincts.

In any event, one critical aspect we must focus on in studying animal emotions are the objective instinctual emotional expressions which arise from evolutionarily-based neural systems that govern how animals spontaneously respond to various life-challenging situations. Now we know that the instinctual brain systems that generate emotionality in animals are situated subcortically—deep below the massive external canopy of the brain's neocortex. These ancient brain areas "light up" when human brain activities during intense emotional feelings are imaged with sensitive tools such as Positron Emission Tomography or PET scans (Liotti & Panksepp 2003). These are also the brain areas from which affective changes and instinctual emotional behaviors can be generated in both animals and humans (Panksepp 1998).

However, the category of affective feelings is probably larger than the number of emotional instincts and basic emotional feelings. Thus, before we discuss the basic emotional feelings of animals, let us briefly consider the broader question: *What is affective experience in humans?* Emotional feelings are not the only category of affective experiences we can have. Others are linked more closely to our sensory systems (e.g., the pleasure of taste) and other to bodily imbalances (e.g., the discomfort of hunger).

The importance of affective experiences in human affairs has long been related to the concept of *utility*. As Jeremy Bentham, in his 1789 *Introduction to the Principles of Morals and Legislation* famously said: "Utility is . . . that property in any object, whereby it tends to produce benefit, advantage, pleasure, good, or happiness. . . or. . . to prevent the happening of mischief, pain, evil, or unhappiness." Many sensorially-linked affects reflect brain cost–benefit "calculations" that signal the survival utility of objects.

Affective feelings often appear to reside within the objects of the world, but they are in fact reflections of evolved brain processes. However, our brains were evolutionarily designed to project feelings onto temporally related perceptions of the world, which allows learning to help guide our behavioral choices, maximizing those that produce positive feelings and minimizing those that produce negative ones.

The remarkable cognitive skills of humans permit such behavioral adjustments to be done across very long spans of time, helping explain why one might forego short-term pleasures for long-term satisfactions. Presumably animals also project their feelings onto the objects of the world, allowing them to make shorter-term self-centered behavioral choices that also facilitate their survival. This is not to deny the existence of unconscious reinforcement processes in the molding of behavior, but to suggest that other mammals are also *active-agents* who can look after their own affective welfare.

Let us now focus on the *emotional* affects—those feelings that arise from major instinctual action systems of the brain. As already noted, one of the most productive neuroscience viewpoints is that the internally experienced feelings of emotions emerge from the same "instinctual" brain systems from which the spontaneous expressive behaviors and accompanying bodily changes arise. This viewpoint is based on the fact that when localized electrical stimulation of specific, subcortical brain regions is used to generate coherent emotional responses (such as fear, anger, separation distress, maternal, sexual, and exploratory behaviors) animals readily learn to approach or avoid the perceived sources of such stimulation. This strongly suggests that other animals are also internally experiencing emotional states, and it provides a coherent scientific strategy to unravel how affective states are generated in the human brain (Panksepp 1998).

Evidence indicates that other mammals have affective experiences that resemble our own.
© *Michael Lewis/National Geographic Image Collection.*

The idea that subjective emotional experiences constitute scientifically meaningful and solvable issues was resisted in many areas of behavioral science for most of the twentieth century, presumably because such hidden processes of the animal mind could never be studied scientifically. As a result, many scientists still assert that the only rational path is to remain agnostic, and hence silent, about such supposedly "unobservable" matters. Of course, the down side of such exceedingly critical thinking is the potential rejection of a great deal of the diversity of mental life on the face of the earth, which can promote cruelty.

This classic philosophical problem—the conundrum of subjectivity—was incorporated into a famous essay by Thomas Nagel (1974) entitled "What Is It Like to Be a Bat?" Most agree it is impossible to fathom, with any precision, the thoughts of other animals. Despite that self-evident difficulty, it may actually be easier to decipher the feelings of other animals, because we share evolutionarily homologous mind–brain functions. If so, anthropomorphism, the attribution of human psychological capacities to other animals, becomes less of a sin (e.g., Panksepp 2003) as we recognize the types of emotional continuities among mammals that Charles Darwin (1872) emphasized in *The Expression of the Emotions in Man and Animals.*

Many behavioristically trained investigators still deny the utility of such working hypotheses, and continue to assert (as did René Descartes, 1596–1650 who brought us the

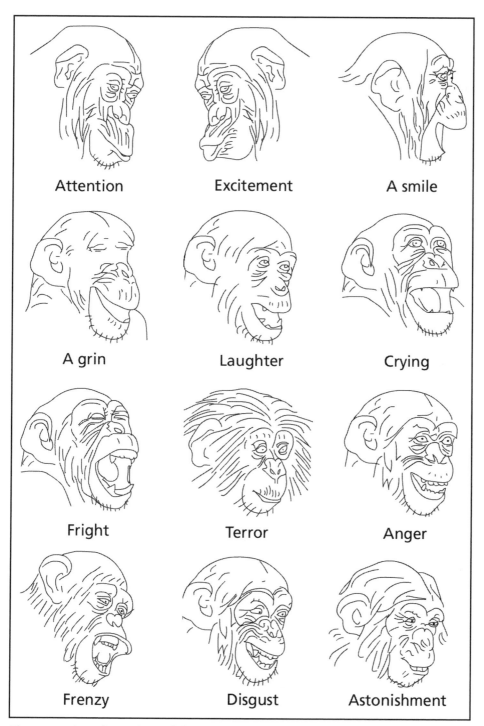

Attention   Excitement   A smile

A grin   Laughter   Crying

Fright   Terror   Anger

Frenzy   Disgust   Astonishment

The expression of emotions of chimpanzees.
Courtesy of the National Library of Medicine.

mischief of mind–brain dualism) that other animals are unconscious "zombies." This untenable belief is often based on the assumption that all forms of consciousness arises largely from the cortical "thinking cap" which is, of course, more modest in most animals than in most humans. However, a cortical locus of control seems unlikely for "affective consciousness" (Liotti & Panksepp 2003). It is impossible to generate discrete emotional behaviors or intense affective states by stimulating specific regions of the neocortex, while it is easy to obtain such effects from essentially the same subcortical brain regions in all mammals (Panksepp 1998). Thus, Descartes' many intellectual descendants are probably wrong concerning the cortical sources of emotional feelings.

Let us now consider the instinctual emotional systems we share with other animals (CAPS are used in labels to highlight that we are discussing distinct emotional circuits). All mammals have brain systems for:

1. investigating the world and foraging for resources (a SEEKING system);
2. becoming angry if access to resources are thwarted (RAGE);
3. becoming scared when one's bodily well-being is threatened (FEAR);
4. various sexual desires that are somewhat different in males and females (LUST);
5. urges to exhibit loving attention to one's offspring and friends (CARE);
6. feelings of precipitous distress when one has lost contact with loved ones (PANIC); and
7. the boisterous joyousness of rough and tumble playfulness (PLAY).

Each of these systems has some specific neurochemical codes (e.g., neuropeptides) that can serve as excellent targets for new drug discoveries. All these systems are also under the control of very general neurochemical circuits, such as those that convey information by glutamate, norepinephrine and serotonin, that regulate the overall arousal of practically all brain systems, and that can lead to dramatic shifts in mood. Indeed, our ability to manipulate such systems pharmacologically led to the revolution in biological psychiatry that started 50 years ago.

What do these systems help animals do? During SEEKING urges, animals move forward exploring and manipulating interesting new aspects of their worlds; with learning, this system allows animals to anticipate rewards effectively. The RAGE system becomes aroused when animals are frustrated, yielding intense anger-type of attack patterns. The FEAR system induces animals to hide and freeze during mild threats (as when predators are far away) and to flee when danger is close at hand (as when predators are nearby). The LUST systems, somewhat different in males and females, promote characteristic forms of courting leading to copulation. The CARE system, typically stronger in females than males, promotes nurturant behaviors that help assure the survival of the young (e.g., in rats, retrieving and gathering pups together, licking them, feeding them, and protecting them). The PANIC system leads young animals that are separated from their parents to cry pitifully until they are found, usually by the mother, and returned home. The PLAY system coaxes young animals to frolic with each other in characteristic ways that vary in detail from species to species, but which are sufficiently similar to be grouped under the same general category. For instance, animals pounce on each other and get into dynamic chasing and wrestling matches that often end up with one animal briefly "winning," before the sequences of behavior start over again.

In sum, we humans have descended from ancestral creatures that already had such brain operating systems, and the positive and negative feelings created in our brains are probably linked to the arousal of such systems. Of course, there are also many additional "socially-derived" feelings when various basic emotions are thwarted and blended with ongoing cognitions (yielding frustrations and feelings such as pride, shame, jealousy, guilt, embarrassment, courage, and so on, many of which may exist only in certain species and some are uniquely human). One compelling line of evidence for shared basic feelings is that other mammals are attracted to the drugs that humans commonly enjoy and crave, and they avoid drugs that we humans do not like.

So what, in a deep neural sense, are emotional feelings? They reflect the various types of neurodynamics that establish characteristic, mentally experienced "forces" that regulate and reflect action readiness within the nervous system. The distinct dynamics are evident in the pounding force of anger, the shivery feelings of fear, the caress of love, the urgent thrusting of sexuality, the painful pangs of grief, the exuberance of joy, and the persistent "nosy" poking about of organisms seeking resources.

And how do the material events of the brain actually get converted into the mystery of subjective experience? No one is certain, but some have suggested that the core of being in all vertebrate animals is organized around brain mechanisms that generate instinctual action tendencies (i.e., primal neural mechanisms for self-representation). Although it is commonly believed that motor systems of the brain do not have the capacity to generate conscious states, that idea may be flawed. All organisms may have primitive action representations in their brains which are essential for the generation of emotional feelings that may be elaborated at the very core of mental life (Panksepp 1998). The basic types of neurodynamics that can inundate such a core "self" are evident in the instinctual emotional action of animals.

Although a fuller understanding of how the transformation of brain activities into emotional experiences actually transpires deserves more scientific attention, from what we already know, it seems that affective values were constructed early in mammalian brain evolution. Such ancient brain mechanisms provide a solid affective grounding to animal existence. This view of brain–mind organization has more potential to contribute to a more admirable scientific image of mental life, and hopefully to a better treatment of other animals with whom we share the world, than does any form of denial or agnosticism on the issue.

Considering all the evidence, it now seems wiser to accept animal emotions and associated feelings as part of the natural order. This does not mean that the other animals think about their feelings as deeply as we humans do. The evidence simply indicates that other mammals do have affective experiences that resemble our own. The nature of these feelings can be scientifically clarified by studying the ancient emotional operating systems of their brains. The most important test of the validity of such concepts will be at the neurochemical level, with the development of new pharmaceutical agents, based on animal brain research, that can modify human feelings in predictable ways.

See also Cognition—*Fairness in Monkeys*
Cognition—*Social Cognition in Primates and Other Animals*
Cognition—*Theory of Mind*
Communication—*Modal Action Patterns*
Cooperation
Emotions—*Emotions and Cognition*
Empathy
Play—*Dog Minds and Dog Play*
Play—*Social Play Behavior and Social Morality*
Welfare, Well-Being, and Pain—*Behavior and Animal Suffering*

Further Resources

Bardo, M. T. & Bevins, R. A. 2000. *Conditioned place preference: What does it add to our preclinical understanding of drug reward.* Psychopharmacology, 153, 31–43.

Kleinginna, P. R. & Kleinginna, A. M. 1981. *A categorized list of emotion definitions with suggestions for a consensual definition.* Motivation and Emotion, 5, 345–355.

Knutson, B., Burgdorf, J. & Panksepp, J. 2002. *Ultrasonic vocalizations as indices of affective states in rat.* Psychological Bulletin, 128, 961–977.

Liotti, M. & Panksepp, J. 2003. *Imaging human emotions and affective feelings: Implications for biological psychiatry.* In: *Textbook of Biological Psychiatry* (Ed. by Panksepp, J.), pp. 33–74. Wiley: New York

Nagel, T. 1974. *What is it like to be a bat?* Philosophical Review. 83, 435–450.

Panksepp, J. 1998. *Affective Neuroscience: The Foundations of Human and Animal Emotions.* New York: Oxford University Press.

Panksepp, J. 2003. *Can anthropomorphic analyses of "separation cries" in other animals inform us about the emotional nature of social loss in humans? Comment on Blumberg and Sokoloff 2001.* Psychological Review, 110, 376–388.

Panksepp, J. & Burgdorf, J. 2003. *Laughing rats and the evolutionary antecedents of human joy?* Physiology & Behavior, 79, 533–547.

Jaak Panksepp

■ Emotions
Emotions and Cognition

Free-living animals seem to be faced with an almost impossible task to perform: When and where to find the most valuable food items at the lowest cost while at the same time avoiding the risk of predation in an environment which may change over short periods of time. So, they have to build up a cognitive map of the environment (exploration) while maintaining their energetic balance and avoiding unnecessary risks (assessing costs and benefits). They have to find the most profitable long-term scenarios, integrating different pieces of information. How do they do that? Let us look at an experiment in humans run by scientists (Antonio Damasio and colleagues and recently by myself), which illustrates the problem at hand and sheds light upon some of the underlying mechanisms. Imagine that in front of you there are four decks of yellow cards indicated by the capitals A through D. The experimenter tells you that you are allowed to turn the cards one by one from any deck you like without any time constraint for your choices. Furthermore, the experimenter tells you that on the cards it says that you either receive money from the experimenter, indicated by a plus sign, or that you have to hand money over to the experimenter, indicated by a minus sign. Of course, you should try to earn as much money as you can, and to start with you get $2,000 as a loan. Finally, the experimenter says that he will tell you when to stop. What happens? This problem is basically the same problem a foraging animal is faced with: Extract information from the environment to find the best long-term scenario.

You need to know the ins and outs of the game, so you start turning cards and begin to build up a knowledge database of the experiment. The most efficient way, of course, is to do this in a systematic way, turning the cards from the different decks in an orderly sequence, for instance A-B-C-D-D-C-B-A. Indeed, that is what some people do. For instance, after 40 trials they have taken 10 cards from each deck. What may you have learned along the way? First, you learn that decks A and B contain cards which have a "+$100" all the time, whereas decks C and D contain cards which have a "+$50" all the time. Thus, in decks A and B the benefits are twice as high as in decks C and D. Second, you find that in decks A and B you have taken great losses as well. Deck A losses vary from $150 to $350, and deck B has one big loss of $1250. In decks C and D the losses are smaller, varying between $25 and $75 in deck C, and a one-time big loss of $250 in deck D. So, superficially the costs in decks A and B are higher than in decks C and D. If you would have been able to keep track of all the benefits and costs, you would have been able to calculate precisely that decks A and B have a net negative balance of $250 whereas decks C and D have a net positive balance of $250. In practice, few if any people make such a calculation or are able to do so, but they still roughly estimate that decks A and B are less profitable than decks C and D, which will be explained below.

Now how can you proceed? You can decide, based on this information, either to completely avoid A and B and take cards only from decks C and D, or to balance your choices between C and D, and A and B such that the majority of choices comes from decks C and D. The first strategy will earn you a lot of money (+$1500 if we stop at trial 100), but at the expense of having little knowledge of the game as a whole, because you have no knowledge that the next 10 cards of decks A and B are programmed in the same way: They may or they may not. The fact that decks C and D remain profitable in the same way says little of decks A and B. In the second strategy, you will gain more information visiting the decks A and B, but at the expense of earning less money in the end because you may run into losses. One thing which you will discover is that, as in decks C and D, the different cost cards are at random positions in decks A and B. I have run this experiment on my colleagues and students in our lab using 22 participants. Two groups emerged: people (15) who won money after 100 trials (on average $692) and people (7) who lost money (on average $686). The former had a less even distribution of the cards over the decks than the latter after 100 trials. So those with a final positive cost–benefit outcome "decided," on the basis of less overall information, what to do in the long run than those with a final negative cost–benefit outcome. Thus, this experiment shows how to maintain a balance between obtaining knowledge of the decks in order to be able to predict what might happen in future cards, and keeping a net gain while continuing the game even though the costs cards are not regularly ordered and without knowing when the game might end.

How are we able to perform such a task? Insight into this comes from patients with brain lesions studied by Antonio Damasio and his colleagues. They have performed this experiment with patients that had lesions in either the amygdale (AMY) or the ventromedial prefrontal cortex (VMF). Both groups of patients lost money in this game. When I analyzed the data of the VMF patients in the same way as I did above, it turned out that these patients appeared to be very much like the "losers" I observed, only much more so. Damasio and colleagues measured the emotional responsiveness of the control group and his patients through the skin conductance response (SCR). When emotionally aroused, we sweat a little, and therefore this arousal may be measured by measuring the sweat induced change of the conductance capacity of the skin by sending a little current through electrodes which are near one another.

They found the following interesting findings. First, in controls an increase in SCR was observed after turning over the benefit (+) and the cost (−) cards. As expected we are aroused when we run into stimuli signalling rewards or punishments (it should be noted that money is not the real reward, it serves as secondary reinforcer for primary rewards such as food). Second, over the course of the experiment, an increase of SCR emerged prior to making a choice for decks of cards. The SCRs for the "bad" decks were stronger than for the "good" decks. It seems as if an internal signal is generated that "tells" us what is good or bad. Indeed, when asked during the course of the experiments whether they knew the differences between the different decks, it turned out that long before people could tell exactly the difference between the different decks these signals were already present and the performance already well underway. Indeed, like Damasio, I have observed that not all people may tell exactly the differences between the different decks, but reply in a more general sense that the A and B decks are no good, while the C and D are all right, and do perform well. Third, when the AMY patients were measured in exactly the same way, they had no emotional response after turning the cards and, accordingly, also not before turning the cards. In contrast VMF patients were emotionally aroused when they turned the benefit and cost cards, but did not develop the anticipatory SCR increases. In fact, some VMF patients correctly identified the A and B decks as bad, but did not behave accordingly.

What do these data tell us? First, that we behave efficiently long before we are able to articulate why we do so. Our linguistic or reflective capacity runs behind in this sense. Second, we have a system in the brain, the connected structures of the amygdala and ventromedial prefrontal cortex, that plays a crucial role in enabling us to estimate roughly "good" from "bad" stimuli and/or scenarios. We are aware of the activity of this system as "gut feeling" or intuition. This system may be labelled the *emotional system*. Third, we have a system in the brain which deals with building up a cognitive map of the "test environment," such as the spatial arrangement of the different "good" and "bad" decks and retaining information of the different events within the decks. It deals with the probability that, based upon present knowledge, the next event will be the same or not, in a given location. This is what people express in this experiment when they refer to trying to detect rules about the way the cards in the decks are organized (expressed as statistics of running into cost cards), and in the way they point out in which location they have seen high or low cost cards. This system may be labelled the *cognitive system*. This system comprises the connected structures of the hippocampus (HICC) and dorso-lateral prefrontal cortex (DLF). Damasio tested patients with lesions in the DLF and HICC and observed that their scores were within the lower (DLF) or just below the (HICC) normal range of selecting cards from the "good" decks, suggesting that the cognitive load of this task is not extremely high. Fourth, both systems (emotional and cognitive) are needed to guide behavior in an adaptive way for the future. Already in the early 70s, the Canadian psychophysiologist M. Cabanac expressed that, for the emotional system, pleasant equals useful; that is, what "feels good" is good in the long run. These systems do not appear to be very precise, but rather make rough estimates about the situation to "decide" for the future in an integrated way.

What may these studies tell us about animals' behavior in similar situations, such as finding the best long-term scenarios for obtaining food? For one thing, vertebrate nonhuman animals possess the basics of the neuroarchitecture for carrying out such activities, and numerous studies tend to converge to the point that animals organize their behavior on their awareness of relationships in their environment and their internal states. Many studies exist which have dealt with the separate aspects of the activities of both the emotional and the cognitive. For instance, researchers have looked for parameters which indicate whether animals "like" what they consume, and whether they are able to look forward to items which they "like." Furthermore others have analyzed how animals may build up spatial and temporal maps of events in their environment. Few studies, however, exist which have tried to model this interplay in a changing environment and push the animals to the limit. Together with my colleagues, I have recently tried to model this in animals.

The experiments we are performing in rats are aimed at creating the following situation. In a test box, which has several arms radiating out of a start box, rats find variable amounts of food items to which variable costs are attached. Each of the arms has its own pattern of changing costs and benefits, trial by trial, but some arms are, in the long term, more profitable than others and some are absolutely not. The animals cannot see beforehand how costs and benefits vary trial by trial. Our initial experiments thus far show that rats are able to gauge costs and benefits this way, but individual differences do exist in their ability to do so.

To come back the question of how animals perform the near impossible task of doing the right thing in a changing environment—we see that they are equipped with systems in their brains to perform this job. They have systems that are good at creating spatio-temporal maps of the environment in which they are living, which tell them what happens when and where. They also are equipped with systems that are good at creating emotional maps of their own performance, at gauging costs and benefits for different scenarios so that the best ones are performed since they "feel the best." Do we understand how this whole machinery works as yet? The answer is no. And this is, of course, science's future challenge.

See also Cognition—*Fairness in Monkeys*
      Cognition—*Social Cognition in Primates and Other*
        Animals
      Cognition—*Theory of Mind*
      Emotions—*Emotions and Affective Experiences*
      Emotions—*Pleasure*
      Empathy
      Play—*Social Play Behavior and Social Morality*

Further Resources

Cabanac, M., 1992. *Pleasure: The common currency*. Journal of Theoretical Biology, 155, 173–200.

Spruijt, B. M, van den Bos, R. & Pijlman, F. T. A. 2001. *A concept of welfare based on reward evaluating mechanisms in the brain: Anticipatory behaviour as an indicator for the state of reward systems*. Applied Animal Behaviour Science, 72, 145–171.

Tranel, D., Bechara, A. & Damasio, A. R. 2000. *Decision making and the somatic marker hypothesis*. In: *The New Cognitive Neurosciences*, 2nd edn. (Ed. by M. S. Gazzaniga), pp. 1047–1061. Cambridge, MA: The MIT Press.

Van den Bos, R., 2000. *General organizational principles of the brain as key to the study of animal consciousness*. Psyche, 6(5), February 2000, http://psyche.cs.monash.edu.au/v6/psyche-6-05-vandenbos.html

Van den Bos, R., Houx B. B. & Spruijt, B. M. (2002). *Cognition and emotion in concert in human and nonhuman animals*. In: *The Cognitive Animal: Empirical and Theoretical Perspectives on Animal Cognition* (Ed. by M. Bekoff, C. Allen & G. Burghardt), pp. 97–103. Cambridge, MA: The MIT Press.

Ruud van den Bos

■ Emotions
How Do We Know Animals Can Feel?

In veterinary school, students find that one of the greatest and most common sins is to be anthropomorphic. *Anthropomorphism* is the attribution of human behavior or characteristics to animals or any nonhuman items. Students naive enough to express concerns ranging from the housing of laboratory animals to postoperative pain, were metaphorically clucked under the chin and advised that they were being anthropomorphic and/or ridiculous.

Oddly enough, this never occurs now that I am a practicing veterinarian. My clients never call and say, "Spot is acting in such a way as to imply that he is feeling pain, anxiety or pleasure." They seem to accept at face value that expressions of pleasure or pain are just that. But when discussing the behavior of nonhuman animals, how do we know their actions are not just instinct? Are we sure that they are self-aware and conscious? How do *we* know that *they* know what they are doing?

First, animals are not the first group about which these questions have been raised. Before it was recognized that women were sentient creatures, men thought women "had a mind no more than a goose." People of color and others not part of the socioeconomic majority have also been singled out as being mindless, unable to feel pain and suffer, and essentially considered walking automatons. The roots of this type of thinking are based in Cartesian philosophy. The seventeenth-century mathematician and philosopher René Descartes developed a philosophy, which among other things, claimed nonhuman animals did not feel pain; that animals were machines or automatons. That is how dogs, cats, birds, rabbits, guinea pigs, horses, ferrets, and other animals were perceived by many for hundreds of years. Fortunately, society has advanced, and today many of those historically denied consciousness have been granted, based on *sentience* (the ability to feel things), the rights they deserve; except nonhuman animals. Today, because of science, we have the technology and understanding to refute these accusations and misunderstandings, even about nonhuman animals.

Before science and the scientific method, opinions, like those of Descartes, ruled the day. Opinion leaders drew their authority from heredity, the church, or some such basis, and took the lead in declaring for the rest of the people what was true. Clearly, society needed a better way to settle disputes about the material world than just relying on opinion. Science filled that gap. Science has given society most of what is today considered essential, for example, computers, air travel, antibiotics, vaccines, safe surgery, indoor climate control, fresh water in abundance, space travel complete with the spinoff technology, satellites, indoor plumbing systems, cell phones, microwave ovens, and global positioning satellites. There is no question that science explains the universe like no other method of thought.

Can science decide whether a cat feels things? We can start by looking at ourselves. Solipsism is a very similar belief to the one that says nonhuman animals are unconscious of their actions. *Solipsism* is the belief that you are the only sentient, real thing in the universe. Everything else is illusion. The proof that you are not alone in the universe is similar to the proof that others—men, women, dogs, cats, chimpanzees, and so forth—are conscious creatures capable of feeling pain, living a life of interest to themselves, and are conscious of, and have control over, their behavior.

The way to prove this starts with science. The scientific method uses observation and experimentation to find truth. We can observe that when dogs are injured, they behave as we do. We can observe that when cats are afraid, they also act as we do: They hide or fight. When animals are hungry, they eat; when they are thirsty, they drink. When you return home and your dog is happy to see you, she expresses it. Maybe it is not exactly as a human does, but the dog expresses it nonetheless. Our observation of such behavior indicates there is a high probability that they since they act like us they feel like us. Or, such behavior at least gives us sufficient reason to see if we can find more data to support our position.

Today, our knowledge of neuroanatomy and neurophysiology has advanced to the point that we know animals and humans have brains that are similar enough to expect that when a dog acts as if he is excited to see you, he really is. He feels things. Maybe not exactly like we do, but then I probably do not feel things like pain and pleasure exactly like you do.

But we both feel, and we both feel more or less the same things: pain, pleasure, hunger, thirst, joy, depression, anxiety, and so forth. And the part of the body that allows us to feel these things, the brain and specifically certain sections of the brain, have been shown to exist in animals and humans. So we have observations that indicate animals know what they are doing and experimental data that supports the observation.

Let's look in depth at just one example of what we feel. The mechanism of pain has been studied extensively in human and nonhuman animals. By studying the anatomy and physiology of human and nonhuman animals, it has been found that the hardwiring for pain recognition and transmission is essentially the same in all vertebrates. If the nervous system is thought of as a computer, the hardware in all vertebrates is essentially the same, whereas the software makes us humans, dogs dogs, cats cats, and chimpanzees chimpanzees. A pain impulse can originate, for example, in the skin. Various receptors signal the pain process to begin, and it is possible to study single nerve fibers in human and nonhuman animals to ascertain which stimuli result in triggering these receptors. From the skin, the pain signal is transmitted to the spinal cord. In the spinal cord, the pain impulse leaves the nerves that were stimulated by the pain and continues on up the spinal cord by different nerve fibers. In order to go from one nerve to another, a chemical must be released between the two, thus signaling the new nerve to be active. Numerous chemicals have been isolated in nonhuman animals and humans that stimulate the new nerves. From the spinal cord the nerve impulse travels to the brain.

Once the pain stimulus reaches the brain, and here our knowledge of the anatomy and physiology of what happens decreases. We know that multiple areas of the brain are involved in pain recognition and control, but exactly where and how remains unclear. Humans, cats, rats, and monkeys have been studied more extensively than other species. Brain imaging techniques such as PET and MRI have helped us locate where these nerve tracts end, and thus where pain occurs in the brain. One definitive thing that can be said about the transmission of pain signals in humans and animals is that once they reach the brain, they appear to be transmitted to many parts of the brain. It is clear that pain is not perceived at one "center" or only one location in the brain. Indeed, numerous nerve tracts exist to transmit painful stimuli to numerous areas in the brain in human and nonhuman animals. The same is true of cognitive functions. So, human and nonhuman animals have in common the hardwiring necessary for pain and other cognitive functions, and demonstrate the behavior expected from such hardwiring. Humans, besides the human we know as self, have also been shown to have this same hardwiring. For this reason solipsism is not an option, and for the same reason, neither is Cartesianism.

In summary, we have the same scientific evidence that nonhuman animals feel pain that we have that humans other than ourselves feel pain. I cannot prove that *you* feel pain with the same degree of assurance that I know *I* feel pain. I cannot measure pain with an x-ray or blood test, either in humans or nonhuman animals. The ability to use language to express pain does not prove that pain exists. Humans, deaf and mute from birth, dogs, cats, rabbits, and other animals experience pain and express vocalizations, but those vocalizations are not language. There is no magical genetic tie between the ability to use human language and the ability to feel pain or feel other things. The neural mechanisms are different. Based on the evidence we have available, nonhuman animals, infants, mute humans, and others are capable of feeling things like pain. To the best of our knowledge, plants, stones, and inanimate objects lack, among other things, the necessary hard wiring to experience pain and indeed, do not experience pain, and indeed are not conscious. We have no way of "getting inside the other's skin," but I do know that I feel pain and other things, and science tells me the reason for my feelings lies in my brain and nervous system. You, and nonhuman animals, have a brain and nervous system capable of receiving painful stimuli, and I can observe behavior that suggests pain, such as withdrawal from the stimuli and vocalizations, such as "ouch" or "that hurts" or a bark or growl.

I cannot prove either your existence or your ability to feel pain as a mathematical certainty. Only if we are willing to accept the reasoning behind solipsism, can we disregard the evidence for feelings such as pain in animals. If we have the same hardwiring of our nervous system for pain as a rat, cat, or monkey, and if the behavioral response to the stimulation of that portion of the nervous system responsible for pain is the same, then it would be illogical to assume that the rats, cats, monkeys, and so forth do not feel pain; likewise for other cognitive functions. It would be similar to assuming that the hearts of humans pump blood but the hearts of dogs do not. Granted the proof of animal pain offered in this article does not fulfill the criteria for a proof in geometry, which is why some use the fallacious argument know as *argumentum ad ignorantium* (an argument from ignorance) to defend their position that animals are not conscious. Arguments from ignorance go something like this: Q. How do we know that ghosts don't exist? A. Because no one has ever seen them. Or Q. How do we know dogs are behaving based on instinct and are really not conscious of their surroundings? A. Because they have never expressed verbally, in a language we can understand that they have feelings and an interest in life. Arguing that something does not exist because its existence has not been proven to the satisfaction of some arbitrary critique is fallacious.

The arguments presented here and elsewhere for the sentience of nonhuman animals *do* fulfill the requirements of science. Aristotle once said, "It is the mark of an educated man not to demand proof that is inappropriate for a given subject." Readers of this encyclopedia should remember nonhuman animals know what they're doing. If someone says they don't, the burden of proof is on him or her to prove it, not on us. I believe that science has already proven our position. I know, based on science and my experience, that the dogs and cats, among other animals, and other people featured in this collection, have a life that is of interest to themselves, know what they are doing, and feel things much as you and I do.

See also Emotions—*Emotions and Affective Experience*
　　　　Emotions—*Emotions and Cognition*
　　　　Emotions—*Pleasure*

Further Resources

Books Examining the Mental Life of Animals

Allen, C. & Bekoff, M. 1999. *Species of Mind: The Philosophy and Biology of Cognitive Ethology.* Cambridge, MA: MIT Press.
Bekoff, M. 2002. *Minding Animals: Awareness, Emotions, and Heart.* Oxford: Oxford University Press.
Shanks, N. 2002. *Animals and Science.* Santa Barbara, CA: ABC–Clio.
Griffin, D. 1992. *Animal Minds.* Chicago: University of Chicago Press.

Books Examining the Role of Science in Modern Life

Curd, M. & Cover, J. A. 1988. *Philosophy of Science.* New York: W.W. Norton.
Gross, P. R. & Levitt, N. 1997. *Higher Superstition.* Baltimore, MD: Johns Hopkins University Press.
Gross, P. (Ed.) 1997. *The Flight from Science and Reason.* New York: New York Academy of Sciences.
Valiela, I. 2001. *Doing Science.* Oxford: Oxford University Press.

Books Examining the Role of Animals in Society

Dombroski, D. 1997. *Babies and Beasts: The Argument from Marginal Cases.* Champaign, IL: University of Illinois Press.

Books on Pain in Humans and Animals

Bonica, J. J. 1990. *Management of Pain.* Philadelphia, PA: Lea & Febiger.
Rollin, B. E. 1991. *The Unheeded Cry: Animal Consciousness, Animal Pain and Science.* Oxford: Oxford University Press.

Jean Swingle Greek

■ Emotions
Laughter in Animals

Viewpoints are changing. Recent research on animal emotion has garnered a polite reception from scientists, rather than being ridiculed. In this atmosphere of academic freedom, research on play in a variety of animals, including dogs, is on the rise. Recent DNA research conducted by Carles Vilá and colleagues suggests that dogs and humans have been cohabitating for well over 100,000 years. Common communication signals between closely allied species almost seem a necessity. In fact, there are dozens of examples and observations of animals using the signals of other species to flee from danger or find a rich cache of food (Belding's ground squirrels).

Marc Bekoff of the University of Colorado postulates, "experiences with play promote learning about the intentions of others" and is "training for the unexpected." So a dog, through social play and the accompanying signals, might learn that the play bow she presents when she wants to play is the same as a play bow presented by another dog and so means the same thing.

While the bow is probably the most often mentioned play signal, the gamut includes lowering the head, making a quick side step, and vocalizations. Another signal widely used by humans, chimpanzees, and rats during play is laughter. Jane Goodall noted chimpanzees making a breathy exhalation through the mouth when engaged in play. Allen and Trixie Gardner also noted the same sound when playing with chimps in their language studies. Burgdorf and Panksepp (2001) found that rats when tickled would emit high frequency chirps. Patricia Simonet in 1997 reported that Asian elephants, during play, emitted quiet breathy sounds. Although, she was not reporting these sounds as laughter, she noted that they produced these sounds during solitary play as well as social play. Dogs appear to have a similar vocalization.

Nicola Rooney of the University of Southampton investigated play signals between species, to see if humans could signal play to their dogs. She found that when owners whispered to their dogs, the dogs were more likely to engage in play. Simonet chose to explore a similar sounding (to a human whisper) vocalization in dogs. Using a parabolic microphone and computer spectrogram analysis, Simonet identified barks, growls, whines, and a forced breathy (mouth) exhalation that she came to identify as a dog-laugh. While the barks, growls, and whines were used in various types of encounters, the laugh was only heard during play or friendly greetings. Simonet found it in both social and solitary play—a dog playing alone and animating an object by tossing it into the air often made the dog-laugh, in effect laughing at his own antics.

The dog-laugh and dog pant have very different properties; the sound waves and the frequency are quite different. Simonet conducted further experiments to test whether the laugh actually had meaning.

Recordings of growling mixed with whining (play growling), the dog-laugh, and a human imitation of the dog-laugh were played to shelter dogs, one at a time, in a room

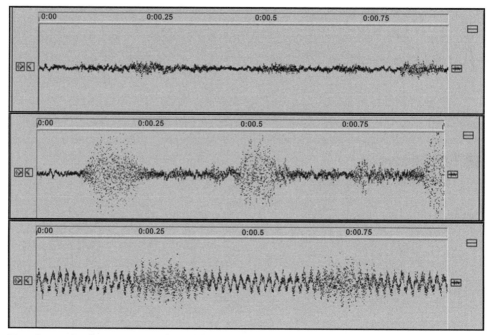

Top: Dog pant for one second; middle: dog-laugh for one second; bottom: human imitation of dog-laugh for one second.
Courtesy of Patricia Simonet.

with either the human experimenters or the humans and a "neutral" dog. When presented with the recorded play-growl, the dogs performed a variety of submissive behaviors, varying by age. When the dog-laugh was played with the neutral dog in the room, all the dogs but one tried to initiate play with the neutral dog. The one different dog engaged in solitary play with a toy.

The human imitation of a dog-laugh seemed to confuse the dogs for a moment. They went first to the computer generating the sound, then to one of the humans in the room, then finally to the neutral dog to offer to play. Dogs hearing a dog-laugh made a play face and then offered a play bow or chase to the individual in the room, even if that individual was a human rather than a dog. So powerful is this stimulus, that humans can initiate play with dogs by using the dog-laugh without any other play signal, such as a play bow.

Further Resources

Dog Laughter

For information, including spectrographs and recordings go to www.laughing-dog.org

Laughter in Rats

Burgdorf, J., & Panksepp, J. 2001. *Tickling induces reward in adolescent rats.* Physiology and Behavior, 72, 167–173.

Chimpanzee Laughter

Gardner, R. A. & Gardner, B. T. 1989. *Teaching Sign Language to Chimpanzees.* New York: State University Press of New York.

Goodall, J. 1986. *The Chimpanzees of Gombe: Patterns of Behavior.* Cambridge: Belknap Press.

Elephant Laughter

Simonet, P. R. 1997. *Environmental enrichment in captive Asian elephants: Importance of novelty.* In: The Published Proceedings of the Third International Conference on Environmental Enrichment in Orlando, FL, Oct. 1997.

Pets

PeTalk. www.petalk.org
> *PeTalk, sponsored by Patricia Simonet, provides information on different types of pets and offers services for pets with behavior problems. PeTalk also conducts noninvasive behavioral research in cognition and animal welfare for both domestic and captive wild animals.*

Patricia Simonet

■ Emotions
Pleasure

During a recent trip to Assateague, Virginia, I watched two fish crows (*Corvus ossifragus*) land on an old wooden billboard that protruded incongruously from a cattail marsh. Hoping they would stay awhile, I swiveled my telescope and focused on them. They first engaged in flight play, then over the next 10 minutes, one bird (always the same one) repeatedly sidled up to the other, leaned over, and pointed his/her beak down, exposing the nape. The other bird responded by gently sweeping his/her bill through the feathers as though searching for parasites. There was every indication that they were mates or good buddies, and that their contact was as pleasurable for both giver and receiver, as a massage or caress between two humans.

Were these two birds interacting purely for the pleasure of it? A traditional view of animals would deny this, based on the prejudiced notion that animals' lives are a constant struggle for survival, with pleasure a luxury reserved only for humans. But adaptive behaviors can be pleasurable, and conscious, thinking creatures are almost certainly motivated to act according to what feels right at that moment, and not according to some awareness of survival value. Pleasure probably evolved in part as a positive motivator to engage in behaviors that promote survival and reproductive success. Should you doubt this, ask yourself if you enjoy any of the following: foraging (e.g., berry picking), eating, resting and/or sleeping, competing, eliminating bodily wastes, playing, or being among friends or with a lover.

Two sea lions play with each other off the coast of California. Animals can also experience pleasure while engaged in activities such as exploring their environment and being with their companions.
Courtesy of Corbis.

Here are some examples that suggest pleasure in other animals. I exclude play because it is well documented elsewhere, including in this volume.

Careful scientific studies find that rats, especially young ones, utter short, ultrasonic chirps during rough-and-tumble play and when tickled on the nape. They also chirp when anticipating various treats. When tickling is withheld, the rats will nip one's hand gently to solicit more. This solicitation behavior resembles that of domestic dogs, who may nuzzle one's hand after being petted, stroked, or rubbed on the belly or nape. Grooming causes the release of opiate chemicals in an animal's body, suggesting that the activity is pleasurable. Specialized marine fish have evolved a cooperative relationship with other species, from whose bodies they remove old sloughing skin and scales, and parasites, by gentle nibbling. At these fish cleaning stations, large predators line up patiently, then open their mouths and gills to allow access to these cleaner fishes (whom they could eat, but never do). Similarly, crocodiles allow birds to clean their teeth, a symbiosis that is probably mutually pleasurable. Hippopotamuses deliberately splay their toes and spread their legs to provide easy access or to solicit cleanings from fish species that share their African springs.

The domestic cat's response to catnip suggests that this herb is intrinsically rewarding, or pleasurable. Under its spell, the cat romps, rolls about, and generally behaves friskily—by all indications having a good time. There is no food reward involved, nor anything else to be gained from catnip that suggests survivorship value to the cat.

Sexual pleasure should be adaptive because it promotes and reinforces procreative behavior, and though poorly studied in other animals, we are almost certainly not the only species to experience it. A captive dolphin turned a hurdle apparatus into a sex toy when he began deliberately failing to clear the bar, instead rubbing his genitals against it on each pass. Chimpanzees, and bonobos especially, have sex more than do humans. Bonobos use the pleasure of sexual contact in many ways, for example, to calm tensions following a squabble, or to facilitate sharing of a large food source. This is not irrefutable evidence that these animals derive pleasure from sexual contact, but it is suggestive. And given all the similarities between humans and other vertebrate animals, the burden of proof should be placed on those who would deny such pleasure in animals.

There is evidence in some animals of a sense of esthetics. The displays of peacocks, pheasants, and birds of paradise, and the courtship songs of other birds, strongly suggest that females appreciate beauty. Male bowerbirds decorate their elaborate nests with flowers and colorful, shiny objects that suggest an esthetic sense akin to ours. While we can easily relate to these examples, we should be mindful that other species may have disparate esthetic preferences. I doubt that vultures find their diet of rotting flesh repulsive, and I would pity a warthog who did not find other warthogs attractive.

The notion that other creatures have a sense of humor is controversial, but great apes, at least, seem to. When Koko, a captive mountain gorilla who has learned to communicate in American Sign Language, was asked to identify the color of a white towel held up by a trainer, she signed "red." When asked again, she repeated this three times. Then, grinning, she plucked off a bit of red lint clinging to the towel, held it up to the trainer's face and signed "red" again. Koko also makes plays on words, signing Koko + nut = 3D coconut, and putting a straw to her nose and calling herself a "thirsty elephant." These anecdotes from a single animal are not enough to conclude that some animals possess a sense of humor, but they do suggest that the capability may exist beyond humans.

A crane amuses itself with a husk of corn.
© *National Geographic.*

Is our human capacity for sensory experience the "gold standard" by which all other species should be measured? Such a view would falsely assume that we share the same sensory capabilities as other creatures. Many animals can see light spectra that we cannot, hear lower and/or higher sounds, produce and detect electrical fields, navigate by the earth's magnetic field, and perceive time far more acutely than we can. These sensory abilities do not necessarily have relevance to experiencing pleasure, but they do suggest realms of pleasure for animals that we as humans can only imagine.

For too long scientists have denied the existence of positive sensory experiences in other species because we cannot know for certain what another being feels. But in the absence of compelling evidence to the contrary, it is more reasonable to assume that other creatures, who share so much in common with us through our shared evolutionary origins, do, in fact, experience pleasure. We cannot feel the hummingbird's response to a trumpet flower's nectar, the dog's anticipation of chasing a ball, or the turtle's experience of basking in the sun, but we can imagine those feelings based on our own experiences of similar situations. What we can observe in animals, combined with our capacity to empathize from our own experience, leaves little doubt that the animal kingdom is a rich repository of pleasure. And as we grow to accept and acknowledge the pleasure that attends animals' lives, evidence for it will proliferate, for we are more likely to find something when we are looking for it.

See also Emotions—*Emotions and Affective Experiences*
Emotions—*Emotions and Cognition*
Play—*Social Play Behavior and Social Morality*

Further Resources

Barber, T. X. 1993. *The Human Nature of Birds*. New York: St. Martin's Press.
Bekoff, M. (Ed.). 2000. *The Smile of a Dolphin: Remarkable Accounts of Animal Emotions*. New York: Discovery Books.
Burger, J. 2001. *The Parrot Who Owns Me: The Story of a Relationship*. New York: Villard.
Griffin, E. R. 1992. *Animal Minds*. Chicago: University of Chicago Press.
Knutson B., Burgdorf, J. & Panksepp, J. 1998. *Anticipation of play elicits high-frequency ultrasonic vocalizations in young rats*. Journal of Comparative Psychology, 112, 65–73.
Masson, J. M. & McCarthy, S. 1995. *When Elephants Weep: The Emotional Lives of Animals*. New York: Delacorte.

Jonathan Balcombe

■|Empathy

The term *empathy* means that you *feel* the emotion of another individual as a result of witnessing their emotional state, or thinking about their emotional state. Because most of the literature on empathy focuses on examples of distress, we will talk about empathy with respect to distress, keeping in mind that, in theory, empathy can apply to any emotion, including joy and happiness. When you have empathy for another individual in distress, you feel the distress of the other person and you may have a desire to help them.

Empathy is very similar to *sympathy* and sometimes people use the term sympathy for such cases. However, colloquially, and in the modern scientific literature, empathy differs from sympathy because with sympathy you feel *sorry for* the other individual, but you do not feel the same emotion as they feel. There are opposing cases where you feel the other individual's emotion so strongly that you yourself become distressed and need to be soothed, and thus you cannot help the other individual. When this happens, it is called *emotional contagion*. In empathy, you feel the emotion of the other, but you remain aware that the other individual is the one in distress, and you can focus on their state.

Many people believe that animals other than humans do not experience empathy and sympathy. This is primarily because people believe that empathy results from a high-level thinking process where you imagine what it must be like to be in the situation of the other individual, and then you make inferences about how the other individual must feel. If this is how empathy is achieved, then it would be a high-level cognitive process that requires a large information-processing capacity, and many animals may not have such a capacity. When one individual has conscious thoughts about the state of another and makes mental models of the other's feelings and needs, it is called *cognitive empathy*. Research suggests that cognitive empathy may not exist in creatures other than humans and possibly apes. But most research indicates that the ability to feel the state of another and to try to help the other is an innate and automatic process that does not require conscious thought and exists across species.

Charles Darwin himself expressed in *The Descent of Man* such a view, partly inspired by moral philosophers such as Adam Smith and David Hume. Darwin saw morality, including one of its pillars—sympathy—as a natural tendency. In his *Introduction to Social Psychology* (1908/1923), William McDougall also anticipated current evolutionary ideas of empathy when he stated that empathy must exist in all group-living animals, or those with the "gregariousness instinct," because these animals are innately affected by the emotions of others. Empirical research supports their view, revealing empathy to be a phenomenon that exists to varying degrees in nonhuman species including rodents, dogs, monkeys, dolphins, and apes. For example, in experiments with human children by Carolyn Zahn-Waxler and colleagues, a mother feigns distress and a researcher observes the reaction of the children. Oftentimes, the family dog also displays consolatory behaviors toward the feigning mother. Both rats and pigeons in the laboratory display a profound emotional response to the suffering of a conspecific (an animal of the same species) and act to terminate the stress. Monkeys react similarly in experimental distress situations, even starving themselves to prevent a conspecific from being shocked in their presence. There are many striking examples of empathy in apes. Much research has empirically demonstrated the existence of consolation in chimpanzees, whereby one animal will act to soothe the distress of another.

To whet the appetite for cases of empathy in animals, consider the following anecdotal example from Frans de Waal's book *Bonobo: The Forgotten Ape.*

> Kidogo, a 21 year old bonobo (*Pan paniscus*) at the Milwaukee County Zoo suffers from a serious heart condition. He is feeble, lacking the normal stamina and self-confidence of a grown male. When first moved to the Milwaukee Zoo, the keepers' shifting commands in the unfamiliar building thoroughly confused him. He failed to understand where to go when people urged him to move from one place to another.
>
> Other apes in the group would step in, however. They would approach Kidogo, take him by the hand, and lead him in the right direction. Caretaker and animal trainer Barbara Bell observed many instances of spontaneous assistance, and learned to call

upon other bonobos to move Kidogo. If lost, Kidogo would utter distress calls, whereupon others would calm him down, or act as his guide. One of his main helpers was the highest-ranking male, Lody. These observations of bonobo males walking hand-in-hand dispel the notion that they are unsupportive of each other.

Only one bonobo tried to take advantage of Kidogo's condition. Murph, a five-year-old male, often teased Kidogo, who lacked the assertiveness to stop the youngster. Lody, however, sometimes interfered by grabbing the juvenile by an ankle when he was about to start his annoying games, or by going over to Kidogo to put a protective arm around him. (p. 157)

Where Does Empathy Come From?

How can animals experience empathy to varying degrees? What does it mean to have some degree of empathy? At the root, empathy results from the way that the nervous system is designed. When you perceive an action or an emotion, you activate the part of your brain that you yourself use to generate that action or emotion. This is called the *perception–action* design of the nervous system. You can see that this is true if you think of an extreme case. Think about when you watch a very intense sports event or when you see someone in a movie who is very upset. You can become so involved in the action that you actually make their gestures for them. You jerk your own arm as the goalie stretches to catch the ball, and you frown and start to cry as the main character experiences a tragedy. The sight of their actions stimulated the action in your own brain so strongly that you actually generated an action potential that traveled down your spinal column to stimulate the muscles. These are called *ideo-motor actions*. In less extreme cases, when you attend to the actions and emotions of another, you still activate these parts of your brain, but to a lesser degree. Thus, you feel the emotion of the other, and you understand what it is like to be them, but you do not necessarily show any outward sign of this.

The perception–action design of the nervous system is not a recent development in evolution reserved solely for humans; on the contrary, it seems that all chordates have aspects of a perception–action design, and that this is extremely important for survival. For example, if an animal that lives in a group sees something dangerous, usually a predator, an alarm call is given and in most cases the group collectively moves away from the source of danger. Thus, the alarm of one individual alarms others. This phenomenon is well-documented for many species, including ground squirrels, birds and monkeys. Given this mechanism, danger is more likely to be detected even though each individual spends less time on vigilance. This "more eyes" phenomenon allows greater investment in activities that promote reproductive success, such as feeding and finding mates.

The spread of positive emotions, such as excitement, is also representative of this perception–action design of the nervous system. Wild dogs, for example, are described as nosing, licking, squeaking, and jumping at each other before the onset of a hunting expedition. Flocks of geese flap their wings and hop up and down on the ground before flying off. Sled dogs similarly jump up and down, barking and whining, before the beginning of a mushing drive. In these situations, the energy is concentrated in time and intensity, but spreads to reach all individuals in the area, thereby maximizing the success of the effort. These examples demonstrate the importance of the perception–action design of the nervous system for coordinating group activities that are crucial for escape from predation, for foraging, hunting, and mass migrations—all of which directly affect the reproductive success of the individual.

Alarm and vicarious excitement are very basic forms of empathy, more akin to emotional contagion. These basic forms of empathy were the first to evolve and are the first to appear in development in primates. For example, Frans de Waal describes a scene from his observations of rhesus macaques (*Macaca mulatta*) where a severely distressed infant will often cause other infants to approach, embrace, mount, or even pile on top of the victim. These macaque scenes seem to result from the spread of distress to the other infants who then seek contact to soothe their own emotional arousal. Emotional contagion is also thought to be the first stage of empathic response in humans, exemplified when infants in a nursery cry in response to other infants' cries, and one-year-old children seek comfort after witnessing the injury of another.

Extensive research from primates indicates that the mother–offspring relationship is crucial for proper development of empathy. In the mother–offspring relationship, the fact that the infant can feel the state of the mother allows the infant to learn about the world simply by watching how the mother reacts to things. In experiments by Susan Mineka and colleagues, infant monkeys learn to fear snakes just from one instance of seeing the mother react fearfully to a snake. In a similar experiment designed by Joseph Campos and colleagues, infant humans look to the reaction of their mother before deciding whether or not to crawl off of a potentially dangerous cliff.

It is also adaptive that the mother can feel the state of the infant because it allows the mother to provide proper care for the infant. The fact that the infant receives immediate and appropriate responses to its needs does two things.

1. The infant does not need to throw a loud, disruptive, and physiologically stressful tantrum (that may draw unwanted attention from group members and predators) to communicate its needs.
2. The infant develops the ability to regulate its own emotions so that it can eventually soothe itself and change moods as necessary without help from the mother.

Data from animals and humans indicate that this emotional learning is necessary for the development of empathy since an individual that cannot regulate its own emotions will become overly distressed from perceiving distress in another, and will not be able to help the other.

Thus, the perception–action design of the nervous system is very adaptive. It allows group-living animals to coordinate their activities, and in the mother–offspring bond it stimulates mothers to take appropriate action for their infants, and it allows infants to develop emotional regulation abilities and to learn about the environment by watching others.

How is it that empathy is extended beyond the mother–infant relationship? Caregiving in the parent and offspring relationship results when the infant requests care using distress signs, and the mother soothes the infant by satisfying the infant's needs. These same cues, once set up in infancy, can also be used to generate empathy outside of this relationship. A distressed chimpanzee, for example, who has just lost a major battle will "pout, whimper, yelp, beg with outstretched hand, or impatiently shake both hands" in order to solicit the consolatory contact of others (de Waal & Aureli, 1996). In his book, *Love and Hate* (1971/1974) Irenäus Eibl-Eibesfeldt argues that the infantile releasers of caregiving are used throughout adult human life, such as the use of a high-pitched voice or "baby names" between lovers. The following data from nonhuman animals attest to the fact that displays of distress in one individual evoke distress and helping in unrelated adult subjects.

Evidence for Empathy in Nonprimates

Russell Church, in 1959, first established that rats were affected by the emotional state of conspecifics while testing to see if the pain reaction of a conspecific could be used as a conditioned stimulus for a subject. According to his conditioning model, if the distress of the object is followed by a painful stimulus to the subject, then the subject will be conditioned to fear the pain reaction of the object. This was thought to be a possible mechanism for altruism, as subjects would learn to help others in the absence of any obvious reward. In the beginning of Church's experiment, all subjects showed fear when they observed an adjacent rat being shocked (the fear was measured with a decrease in bar pressing, thought to be a behavioral indication of fear). Although this response was much greater if the observing animal previously experienced a shock paired to that of the other animal, even subjects who had previously experienced shock without the conditioned pairing showed fear when the adjacent rat was shocked. This experiment was replicated with pigeons. The fact that animals showed distress without having a shock paired to that of the other animal indicates that rats and pigeons are sensitive to the state of the other. This is further supported by the following study, which investigated the potential for altruism in albino rats.

In 1962, George Rice and Priscilla Gainer presented a rat with the sight of another rat being suspended just off the floor by a hoist. Bar pressing by the subject lowered the stimulus animal onto the floor and thus terminated its distress responses (wriggling of the body, distress vocalizations). Subjects in this experiment increased their bar pressing to the sight of the suspended animal, thus displaying what the authors referred to as an "altruistic response," operationally defined as a behavior that reduces the distress of another. The interpretation was bolstered by behavioral data as the subjects spent the duration of each trial in a location close to and oriented toward the suspended rat. Notice that the behavioral response of fear is traditionally represented by a *decrease* in bar pressing. Thus, the subjects in this experiment were likely aroused by the sight of the object (in the form of emotional contagion), but were not pressing the bar out of fear for their own safety.

Further discounting the role of fear in producing the "altruistic" response, in a subsequent experiment by George Rice in 1964, the rat subjects witnessed delivery of electric shocks to the object. Rather than pressing the bar to eliminate the shock of the object, the subjects "typically retreated to the corner of their box farthest from the distressed, squeaking, and dancing animal and crouched there, motionless, for the greater part of this condition." Noting the interference of fear with bar pressing in this study, Rice concluded that the increase in bar presses in the original "suspended distress" study was not the result of distress in the subject. The behavioral descriptions indicate that only the subjects in the shock experiment were overly stressed by the sight of the other and were thus unable to surpass their own distress to act altruistically. This is consistent with many findings in the human literature, which show that an overly distressed subject is less likely to respond with empathy or sympathy. It also provides further support for the idea that the development of emotion regulation abilities is important for empathic responding, even in rats.

These data provide compelling nonprimate evidence for the perception–action link between individuals and its role in producing empathic or altruistic behavior. Faced with the mild distress of a live animal, rats and pigeons react as if the object affects them emotionally and take measures to eliminate the distress of the object. This is not to say that these reactions necessarily involve an intention to help the other, because the extent to which rats and pigeons understand the impact of their own behavior on others remains unknown. Also, learning played a crucial role in the duration of the subjects' responses, illustrating that emotional contagion is not simply an innate response, but is affected by past experience. There exists vast support for the effect of previous experience with the distress situation or distressed individual.

Evidence for Empathy in Monkeys

Macaque monkeys show a high degree of tolerance and helping toward handicapped group members, attesting to the fact that they perceive the different abilities of other individuals, and take these into account when interacting with them. Azalea, a rhesus macaque who lived in a socially-housed group, had a genetic disorder, autosomal trisomy. As a result of her disorder, Azalea had motor defects and delays in developing social behavior. She had a high dependency on her mother and kin and poorly defined dominance relationships. Azalea was tolerated and accepted in the group despite her physical and social defects, even though rhesus macaques are typically highly structured and aggressive. Up until her death at age 32 months, Azalea was not peripheralized from the group, and there were no signs of aggressive rejection by other group members. A wild female Japanese macaque (*M. fuscata*) named Mozu had congenitally deformed lower portions of her legs and arms. Despite her related difficulties with locomotion, foraging and care of young, Mozu was an integrated member of her group and had five offspring who lived to reproduce.

Few experimental studies have been conducted on empathy in nonhuman primates, and most of those occurred in the 1950s and 60s. The most extensive inquiry comes from the Department of Clinical Science in Pittsburgh. John Murphy, Robert Miller and Arthur Mirsky were the first to show that the emotion of one monkey could act as the conditioned stimulus for another. This interanimal conditioning provided a springboard for a long and successful investigation into the communication of affect between rhesus monkeys.

In one experiment in 1958, subjects were trained until they understood that they could terminate a shock to themselves by pressing a bar in their chamber. After this, the experimenters instead delivered the shock to another monkey, in the presence of the subject, but it was still the subject that could terminate the shock by pressing the bar. Seventy-one percent of the time on the first day the subjects would press the bar to terminate the shock to the other animal. The other monkey quickly learned that the shock came after conditioned stimulus of a change in lighting and "began to leap and run around whenever its compartment was illuminated." Seventy-three percent of the bar presses by the subject occurred to this distress in the other monkey, before the shock was even administered. In addition, the subjects displayed "piloerection, urination, defecation and excited behavior" at the sight of the distressed other monkey. While this shows that the monkey subjects did not want to observe the distress of a live conspecific, in subsequent experiments, the monkey subjects did not respond to the sight of an albino rat being shocked or to a monkey-like puppet thrashing around. The response could be reinstated using pictures of monkeys, especially pictures of familiar monkeys showing fear, but this response was less strong and less clear than to the live stimulus animals. The familiarity response is expected given that it would be most adaptive to help individuals that are close to you or that you are related to. There is also empirical support for this bias toward familiarity in the human literature from subjects of all ages, infants to undergraduates.

Thus, similar to the findings with rats and pigeons, after learning the consequences of shock, these monkeys were aroused by the sight of a conspecific in distress, acted to eliminate the suffering of the stimulus animal, but were less responsive to artificial or unfamiliar stimuli.

A similar task tried to condition subjects to respond to the positive emotion of other monkeys by associating the emotion with the delivery of food. This version of the task was less successful than the distressing version. Heart rate data from the same experiment revealed that when the task was successful, the subject monkey and the other monkey had similar trends in their heart rates. These results confirm the proposed basis for empathy in emotional contagion and the perception–action link. They further attest to the fact that circumstances involving risk and distress are more emotionally salient to the subjects. The emotion literature in general has been much more successful in finding the biological basis of negative emotions than positive. This makes sense given that the failure to detect danger can be deadly, which has a much greater impact on reproductive success.

In a similar series of studies done at Northwestern University Medical School, monkeys were found to refrain from bar pressing to obtain a reward if it caused another monkey to receive a shock. In these experiments, the animals were first trained to pull one of two chains for the delivery of food, depending on the color of the light stimulus. Subsequently, one of the two chains was reprogrammed to also deliver a shock to the other monkey in view of the subject. Thus, the subject would have to witness the shock of the other monkey in order to receive the food reward for that chain. In the second set of experiments, 10 out of 15 subjects displayed a preference for the nonshock chain in testing even though this resulted in half as many food rewards. Of the remaining 5 subjects, one stopped pulling the chains altogether for 5 days and another for 12 days after witnessing the shock of the other monkey. These subjects were literally starving themselves to keep a conspecific from being shocked. In agreement with the rat findings, starvation was induced more by visual than auditory communication, was more likely to appear in animals that had previously experienced shock themselves, and was enhanced by familiarity with the shocked individual.

Familiarity effects are also seen in cognitive empathy experiments with monkeys. Although the monkeys in these experiments were unable use perspective-taking information to switch roles with their partners, subjects who were housed together performed better than unfamiliar individuals, and familiar subjects shared more food with their partners after the experiment. In a later experiment, pairs of macaque monkeys that were trained to cooperate for food showed a dramatic increase in their tendency to get along. These data support the role of familiarity in facilitating communication and cooperation, further supporting theoretical models and empirical evidence for empathy and altruism.

The above experiments suggest that monkeys will act to avoid witnessing the distress of a conspecific. Subjects in some experiments accepted reductions in food, sometimes to the point of starvation, to avoid participating in or witnessing the distress of the object. Across experiments, familiarity with the stimulus animal affected the ability of animals to communicate emotion and intention. Further, subject's responses were facilitated by their familiarity with the particular distress situation. Given that these monkey species are often characterized as aggressive and of inferior intelligence, they showed remarkable empathic and altruistic responses to the distress of their conspecifics. This is especially notable given the unnatural laboratory conditions in which the studies were done, and the lack of traditional social bonding opportunities for the animals in most cases.

Evidence for Empathy in Apes

In a content analysis of over 2,000 anecdotal reports of nonhuman primate empathy, Sanjida O'Connell counted the frequency of three types of empathy: *emotional* (understanding another's emotion—closest to the definition used here), *concordance* (understanding nonemotional states—similar to cognitive empathy) and *extended* (acts of helping tailored to the other's needs). Chimpanzees exhibited all three types of empathy. Examples where one chimpanzee displayed an understanding of the emotion of another (excitement, grief/sadness/frustration, and fear) were extremely common, with most outcomes resulting in the subject comforting the chimpanzee in distress. Chimpanzees appeared to comprehend the emotions, attitudes, and situations of another individual and even endangered their own lives to save conspecifics in danger. In one case, reported by Jane Goodall, an adult male chimpanzee died trying to rescue an infant who had fallen over the electric fence into a moat on the other side. Monkey displays of empathy, by contrast, were restricted to mediation of fights, adoption of orphans, and reactions to illness and wounding (as seen in the tolerance toward handicapped individuals mentioned above).

Consolation is a primary example of empathy in chimpanzees. Consolation, as first defined by Frans de Waal & Angeline van Roosmalen, occurs when a bystander approaches a recipient of aggression, shortly after a fight. De Waal describes in *Chimpanzee Politics: Power and Sex among Apes* (1982) how two adult female chimpanzees in the Arnhem Zoo colony used to console each other after fights: "Not only do they often act together against attackers, they also seek comfort and reassurance from each other. When one of them has been involved in a painful conflict, she goes to the other to be embraced. They then literally scream in each other's arms."

Consolation involves contact initiation by a previously uninvolved bystander who is assumed to be less distressed, and directs consolatory efforts to the victim rather than to itself. Thus, in consolation, there is no direct benefit for the consoler. One can say that in consolation, the consoling individual has become distressed from the sight of the victim and seeks comfort for his or her own feelings (which would be emotional contagion, but not empathy). However, the consoler often does not show signs of distress, such as facial expressions or vocalizations, and may wait until after the most intense displays of distress have disappeared to approach the other.

The tendency to console seems to be unique to apes and humans: It has not been found in any monkey species despite intensive efforts to find it. The reports on chimpanzees are far from anecdotal. De Waal & van Roosmalen based their conclusion on an analysis of hundreds observations, and a recent study by de Waal & Filippo Aureli includes an even larger sample.

There is also sporadic anecdotal evidence for cognitive empathy in chimpanzees. There is the example of Kidogo from the beginning of this entry. In another case from de Waal's *Good Natured*, a male chimpanzee saw a female struggling with a technical problem and waited until she had left the scene to solve it and bring her the item she was after. In another case, Kuni, a bonobo female at the Twycross Zoo in England, tried to "help" a bird:

> One day, Kuni captured a starling. Out of fear that she might molest the stunned bird, which appeared undamaged, the keeper urged the ape to let it go. Perhaps because of this encouragement, Kuni took the bird outside and gently set it onto its feet, the right way up, where it stayed looking petrified. When it didn't move, she threw it a little, but it just fluttered. Not satisfied, Kuni picked up the starling with one hand and climbed to the highest point of the highest tree where she wrapped her legs around the trunk so that she had both hands free to hold the bird. She then carefully unfolded its wings and spread them wide open, one wing in each hand, before throwing the bird as hard she could towards the barrier of the enclosure. Unfortunately, it fell short and landed onto the bank of the moat where Kuni guarded it for a long time against a curious juvenile. By the end of the day, the bird was gone without a trace or feather. It is assumed that, recovered from its shock, it had flown away.

Such anecdotes hint at underlying cognitive capacities rarely acknowledged in animals other than ourselves. Familiarity with their imaginative understanding and well-developed caring capacities explains why experts of apes acted with little surprise to the most famous case of nonhuman empathy, the rescue of a 3-year-old boy at the Brookfield Zoo, on August 16, 1996. The child, who had fallen 6 meters (20 ft) into the primate exhibit was scooped up and carried to safety by Binti Jua, an 8-year-old female western lowland gorilla. The gorilla sat down on a log in a stream, cradling the boy in her lap, giving him a gentle back pat before she continued on her way. This act of sympathy touched many hearts, making Binti a celebrity overnight. Whereas some commentators have tried to explain Binti's behavior as the product of training or a confused maternal instinct, her behavior fits entirely with everything else we know about apes, which is that they respond comfortingly to individuals in distress. The only significant difference was that in this case the behavior was directed at a member of a different species.

In sum, data currently exist for cases of subjects appearing distressed, but consoling the object (emotional contagion), and for cases where the subject consoles the object, but does not appear to be distressed (empathy or sympathy). Anecdotes such as the ones above, with Kidogo the bonobo or Kumi the chimpanzee, point to the existence of cognitive empathy, but without being able to know what these animals were feeling or thinking at the time, it is hard to make a firm conclusion. The data on consolation, the striking anecdotes of cognitive empathy, and other data on cognitive perspective-taking abilities in apes suggest that apes evaluate the emotions and situations of others with a greater understanding than is found in most other animals apart from ourselves. More research is needed, however, to determine the extent to which ape consolation is actually similar to human consolation.

Cognitive Empathy

Only humans and the great apes, together classified as the Hominoids, have been cited as evincing cognitive empathy (also known as *perspective-taking*). The above reports on consolation behavior provide the only systematic data indicating a substantial, perhaps radical difference between the way chimpanzees respond to distress in others caused by aggressive conflict. This difference seems to fit with the overall higher cognitive level and tendency to take another's perspective of apes relative to monkeys. In a cognitive empathy experiment developed by Daniel Povinelli and colleagues, two animals cooperated to manipulate a lever device to obtain food from opposing sides of an apparatus. Each of the subjects had a different task requirement. If one subject could successfully do the other's task just from watching, he was considered to have empathy. Chimpanzees but not monkeys succeeded in the task, further suggesting that apes can use perspective-taking to gain knowledge. In a similar experiment by William Mason & J. Hollis, monkeys were also unsuccessful in learning the role of their partner in a communication experiment.

Experiments from human development and from evolution suggest that cognitive empathy evolved at the same time or in the same species as many other social–cognitive capabilities. Some suggest that cognitive empathy was made possible by the evolution of other abilities such as perspective-taking. Others suggest the opposite, that empathy is a prerequisite for perspective-taking! While this debate is a long way from being resolved, it is true that the quality of empathy has changed along with other cognitive abilities in recent evolutionary history, starting well before the appearance of our species. This change correlates with a disproportionate increase in the prefrontal cortex in recent primate evolutionary history.

The same nervous system link between perception and action that helps us navigate the physical environment helps us to navigate the social environment. Thus, the perception–action link allows individuals to acquire motor skills and social skills easily. It is hoped that once we fully understand how empathy evolved and how the nervous system accomplishes empathy, it will be easier for people to see how empathy is rooted in more basic processes such as emotional contagion and that we share this ability with many other species.

Further Resources

Darwin, C. 1871/1982. *The Descent of Man, and Selection in Relation to Sex.* Princeton: Princeton University Press.
Darwin, C. 1998/1872. *The Expression of the Emotions in Man and Animals,* 3rd edn. New York: Oxford University Press.
Eisenberg, N. & Strayer, J. 1987. *Empathy and Its Development.* New York: Cambridge University Press.
de Waal, F. B. M. 1996. *Good Natured: The Origins of Right and Wrong in Humans and Other Animals.* Cambridge, MA: Harvard University Press.
de Waal, F. B. M. & Aureli, F. 1996. *Consolation, Reconciliation, and a Possible Cognitive Difference between Macaques and Chimpanzees.* New York: Cambridge University Press.
de Waal, F. B. M. 1997. *Bonobo: The Forgotten Ape.* Berkeley: University of California Press.
Preston, S. D., & de Waal, F. B. M. 2002. *The communication of emotions and the possibility of empathy in animals.* In: *Altruism & Altruistic Love: Science, Philosophy, and Religion in Dialogue* (Ed. by S. Post, L. G. Underwood, J. P. Schloss, & W. B. Hurlburt), pp. 284–308. Oxford: Oxford University Press.
Preston, S. D. & de Waal, F. B. M. 2002. *Empathy: Its ultimate and proximate bases.* Behavioral and Brain Sciences, 25(1), 1–72.

Stephanie D. Preston

■ Exploratory Behavior
Inquisitiveness in Animals

Peter Marler and William J. Hamilton III in *Mechanisms of Animal Behavior* started their chapter on exploration and play with the following statement: "Animals spend much of their time in motor activity, the function of which is often difficult to identify." (1966, p. 159). Twenty-eight years later, H. Keller, K. Schneider & B. Henderson (1994) wrote in the foreword to their collective book: "... we have not attempted to arrive at a definition of curiosity and exploratory behavior upon which every contributor to this volume would agree. Given the state of the art in research and theory on curiosity and exploratory behavior, we thought to attempt to do so would be counterproductive." (1994, p. 3).

Some present-day textbooks on animal behavior even fail to include that form of behavior in the index. Does that mean that the subject itself is unimportant and uninteresting? Apparently not. It is commonly agreed that, among the so-called higher animals, exploration is the principal form of behavior and may be analyzed together with other behavioral forms (Keller, Schneider & Henderson, 1994). However, in comparison with such activities as feeding, caring for young, nesting or attracting mates, exploratory behaviors prove to be much more difficult to analyze in terms of their adaptive value. This is because exploratory behaviors:

- include elements of other forms of behavior;
- are triggered by a variety of stimuli (it is impossible to pinpoint key stimuli or specific releasers);

- are noncyclic in contrast to, for example, feeding; and
- have unclear consequences; that is, we do not know exactly what a given animal gains as a result of exploration.

The history of scientific interest in the research on exploration and curiosity dates back, as most of the comparative psychology, to Charles Darwin. He was the very first person to emphasize the significance of curiosity in the higher psychical faculties formation. H. Jennings (1906) included in his immense book on the behavior of lower organisms extensive examples of activities that he described as trial movements, and which today we could call exploratory. For instance: "When stimulated, the earthworm frequently responds by moving the head first in one direction, then in another, often repeating these movements several times. It then finally follows up those movements which decrease stimulation" (Jennings 1906, p. 247). Jennings also quoted S. J. Holmes (1905), who stated, "The lives of most insects, crustaceans, worms ... show an amount of exploration that in many cases exceeds that made by any higher animals. Throughout the animal kingdom there is obedience to the Pauline injunction, 'Prove all things, hold fast to that which is good'." Presumably, these authors understated that the intensity of trial movement correlates with the complexity of animal's sensory systems. It is their undeniable commitment however, to emphasize the role of exploration even in simple organisms. Research on exploration in animals reached its peak at the beginning of twenty-first century, and this heightened interest is clearly associated with cognitive approaches to the study of animal behavior.

The Stimulation Relevant for Exploratory Behavior

The analysis of the stimulus properties that evoke exploratory activity does not provide a clear picture. In other behavioral categories, such as feeding or fighting, it is usually possible to assign behavior specific stimulus properties that are crucial for releasing given actions. As far as the exploratory actions are concerned, one could say that it is not the content of the stimulus, but its formal characteristic, such as intensity, complexity, and novelty, that makes the organism explore it. To be even more specific, the releasing properties of the stimulation are shaped, not by an absolute value of a given property, but by changes in the values. It seems that the concept of discrepancy between the organism's expectations and the actual stimulation inflow may be a useful tool for analyzing the relationship between the stimulation properties and the exploratory acts.

There is undeniable evidence of the existence of the relationship between the intensity of exploration and the magnitude of the novelty. It takes the form of the reversed "U." Within the range of low and moderate novelty magnitude, the stimulus is likely to evoke exploratory acts, whereas stimulation with high novel magitude is likely to elicit withdrawal, freezing, or other forms of defense response.

The Adaptive Value of Exploration

One of the first hypotheses regarding the process of exploration was advanced by Tolman (1948). He described the spatial map constructed by the animal in the process of exploring a new area. At the initial stage of research on the adaptive value of exploration, it was assumed that this form of behavior was related to the organism's cognitive functioning. A number of authors followed that approach in their studies of the effects of exploration.

There is no doubt that exploratory activity enables the animal to gather information about its environment. This process involves certain costs, the principal one being the risk of encountering a predator. We could attempt to analyze it through simulation involving games theory. This requires the assignment of arbitrary point values to individual events. Let us assume that in our analysis the initial exploratory act guarantees (probability = 1) 10 points of information ($I_1 = 10$). Will every subsequent act yield the same information value? Probably not. There are a number of arguments for the thesis that information about immediate surroundings is more valuable to the animal than information about events more distant in the animal's "space–time." Therefore, the value of subsequent acts (I) of exploration must be defined as relative to previously acquired information, for example, as in the formula $Ii = 10/i$. Therefore, the amount of information gathered as a result of exploratory actions will be the total of information values of subsequent exploratory acts, CI:

$$CI = \sum_{i=1}^{n} Ii$$

At the same time, each exploratory act carries the risk of encountering a predator. Let us assign -100 points to that event and assess its probability at 0.02. If we take the expected value, i.e., the product of multiplying the value of an event by its probability, as our point of departure, we can calculate that in our example every exploratory act will entail the cost of -2 points. Therefore, each exploratory act will result in gathering information of the value I and bearing cost $PR = 2$. The value of a single exploratory act can be expressed as a sum of gain Ii and cost PR, i.e., $Vi = i + PR$. It can be calculated that if we adhere to the parameters of the above-mentioned example, with the number of individual explorations $I = 5$ the value shall be 0, and that it shall be negative for subsequent explorations.

Obviously, the animal cannot regulate its exploratory activity on the basis of the value of a single exploratory act Vi, as it can only be known *post factum*. Therefore, the mechanism of regulating that activity must be based on the data about sequences of events gathered in the course of a species' evolution. In our example, it will be the assessment of the total value of an exploratory sequence CV that shall constitute that measure:

$$CV = \sum_{i=1}^{n} Vi$$

Since, after reaching a certain value, Vi becomes negative, the total value of exploration begins to drop. This relationship is shown in the figure on p. 577.

Consequently, despite the fact that exploration increases knowledge about the environment, the animal does not explore continuously. The level of exploration (equated in our analysis with its usefulness) will be specified by the sum of the value (negative) of risk associated with the next act of exploration and the increase in knowledge resulting from that act.

Data significant for the evaluation of this model derive from research involving a comparison between species living in habitats with differing levels of predatory pressure.

Konrad Lorenz claimed that exploration as a form of behavior is characteristic of animals capable of functioning in diverse environments, feeding on various kinds of food, and relocating in space. It is illustrated by the animal's ability to initiate a sequence of exploratory actions in response to a wide variety of stimuli and, subsequently, to choose the appropriate behavior by selecting (learning) a motor pattern. Its function is to gather information stored *ad acta*. The result of that process is the creation of latent knowledge store, which seems a particularly relevant assertion.

Survival (together with a chance for further exploratory behaviors) requires the animal to continue exploration at such a rate that the balance between the expected gains and costs is as favorable as possible. The fact that among some species there exists a mechanism of forcing low status members of the group to engage in exploration illustrates the complex nature of the relationship between behavior intensity and adaptive success.

Levels of Exploratory Behavior, Environment Variability, and Individual Differences

Taking as our point of departure the analysis of the adaptive function of exploration outlined above, we should ask the question: Which part of the behavior regulation system is affected by

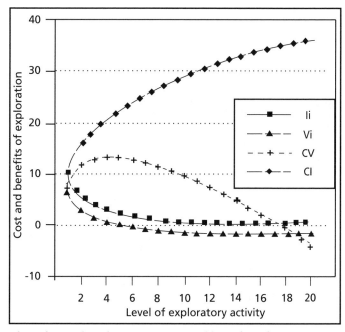

The relationships between costs and benefits of exploration. See text for symbol descriptions.
Courtesy of Wojciech Pisula.

the above-mentioned situational parameters? The information value of exploration depends on the characteristics of a given environment. Let us assume that the organism functions in a highly structured environment, for example, a flat surface consisting of black and white squares arranged in a chessboard pattern. Subsequent exploratory acts will add little to the spatial representation formed by the animal. If, however, the environment consists of elements with different characteristics, changeable and unpredictable, each exploratory act will result in a significant enrichment of the spatial representation. Therefore, variability and predictability of a given environment can affect the levels of exploration by changing the value of an exploratory act (I). This parameter depends strictly on the characteristics of the environment and not of the individual. However, it does affect the value of sensory reinforcement resulting from the change in the stimulus field.

Exploration *intensity* is one of the parameters of individual characteristics. An animal exploring its environment "does not know" the actual probability of encountering a predator. The essence of adaptive behavior is to act in such a way so as not to verify that probability "empirically." Hence, the animals need to be equipped with mechanisms for risk assessment, and the factor causing higher vertebrates to withdraw from a new environment is undoubtedly the negative emotional arousal. The facility of triggering that response is one of the basic dimensions characterizing the individual's temperamental profile. Individual differences affecting the levels of exploration will therefore be related to the parameter specified in the above discussion as *PR*. While analyzing the relationships between the postulated mechanisms of exploratory behavior regulation, it is useful to consider the hypothetical comparison of the results of exploratory actions performed by individuals with differing *PR* values, in environments with different levels of variability, and, consequently, with different values of *I*. They are shown in the figure on p. 578.

The model proposed here is a simplification. For example, it assumes that environment variability does not influence the value of subjective *PR*. This is one of the testable hypotheses

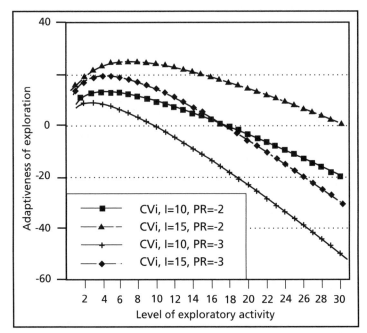

Changes of the exploration adaptive values, depending on the given parameter changes.
Courtesy of Wojciech Pisula.

that can be formulated on the basis of this model. It is worth noting that the most significant differences occur in the values of the total amount of information gathered in the course of exploration. Direct effects, resulting from the exploration levels, are relatively weak.

Exploration and Play

Play behavior has been considered by ethologists and comparative psychologists as closely related to exploration. The view has been widespread although there is little empirical evidence for it. Theoretical analyses of the phenomenon also fail to give ground to final conclusions. There is no doubt that both play and exploratory acts provide an animal with some sort of information. In certain situations this information may be basically the same. For instance, both object play and object-directed investigatory acts may produce the same experience. On the other hand, locomotor exploration and locomotor play generate definitely different feedback. To test the general hypothesis about the close relationship of exploration and play, two sets of data are needed. One source of evidence comes from between-species comparisons. Animals of species that play, mostly avian and mammals, show the most intensive and sophisticated exploratory behavior, and therefore, they seem to be most suited for such analyses. It is important, however, to mention here that the time scale of play evolution has been underestimated, and we are now just beginning to understand the role of play in animal behavior.

The other source of data comes from studies on individual differences. In rats, the primary form of play is play fighting. It starts at weaning, reaches its peak at the age of 30–40 days and then declines most rapidly when animals reach sexual maturity. The development of exploration presents a slightly different pattern since the peak of exploratory activities is reached between 70–90 days of age. However those individuals that show a high level of juvenile play fighting manifest also a higher level of adult investigatory behavior than their counterparts. Play and exploration share the mediation of similar, direct, and physical contact with an external source of stimulation. Moreover, both forms of behavior involve motor activity that triggers feedback from the playmate or investigated inanimate object. These facts support the postulate to see both play and exploration within a shared behavioral category.

The Integrative Levels of Exploration

Exploration is manifested in various behavioral forms. Usually, it can be observed that an animal that is placed in a novel environment or in proximity to a novel object, is moving throughout the space and performing various actions, such as: sniffing, touching, manipulating, and many others. There are various forms of behavior that we tend to call "exploration," "exploratory acts," or "novelty seeking." W. Pisula (1998) proposed to arrange them within a hierarchical integrative system, reflecting their phylogenesis, ontogenesis, and temporal occurrence, as demonstrated in the figure below.

The figure shows the forms of exploratory behavior, arranged hierarchically from the most elementary (at the bottom) to the most complex. The column labelled "Mechanism of behavior" includes elements of mechanisms underlying a given form of exploratory behavior, which seem to be specific and necessary for that level. A similar rule applies to a purpose or function of behavior. A given behavioral form may serve functions other than those listed in the figure. The lines with arrowheads show a postulated pathway from the mechanism at a given level through behavior and its function to the mechanisms operating at a higher level.

More specifically, *orienting response* (or reflex) describes any turning of the body with respect to the position of the specific stimulus. Actually this meaning is synonymous with taxis, though this term is most often used with reference to lower organisms. In the most general meaning, orienting response means any attentional response to a stimulus, such as head turning or ear raising. *Locomotor exploration* refers to the situation in which an

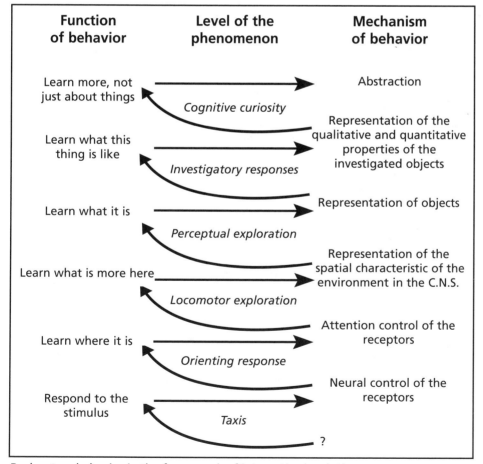

Exploratory behavior in the framework of integrative levels theory.
Courtesy of Wojciech Pisula.

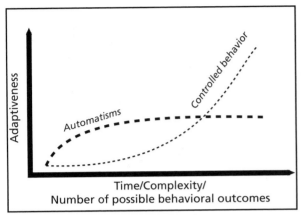

The general relationship between the number of possible behavioral outcomes, complexity of the environment, time and the adaptiveness of automatic and controlled components of behavior.
Courtesy of Wojciech Pisula.

organism is moving within, or approaching a novel environment. The term *patrolling*, on the other hand, is often used to describe movement through a well-known area, usually the animal's own territory. *Perceptual exploration* refers to a prolonged perception (by any sensory system) of a specific stimulation. This is a rather direct extension of the orienting response. *Investigatory behavior* includes various behavioral activities, such as interaction with the investigated object, manipulation of the object, investigation of a particular area (prolonged sniffing), and so on. *Cognitive curiosity* covers novelty, information and stimulation seeking at the cognitive level. In analyzing these forms of exploration, one could raise the questions about the possible roles of automatic and controlled components of behavior.

The theoretical analysis of the phenomenon leads to a conclusion that *time, complexity* and the *number of possible behavioral outcomes* are the crucial variables mediating the process of automatic vs. controlled behavior regulations. The

Controlled components of behavior	Level of the phenomenon	Automatic components of behavior
"Asking" questions not just about things; seeking information, not just stimulation	*Cognitive curiosity*	Processes underlying concept formation
"Asking" questions about properties of objects; testing involving feedback	*Investigatory responses*	Absorbing qualitative aspects of object stimulation and combining them with representation of objects
"Asking" questions about occurring objects	*Perceptual exploration*	Absorbing stimulation from specific sources (objects), as clusters forming representation of objects
Emergence of spontaneous locomotor activity, initiated both by external stimulation and internal factors	*Locomotor exploration*	Absorbing incoming sensory stimulation into existing representation of the spatial characteristic of the environment
Response duration is affected by properties of stimulation, such as intensity and novelty	*Orienting response*	Arousal mechanisms, "taxis" like directing receptors toward stimulus source, specific movements (head turning)
Almost none; habituation is the only sign of emerging control	*Taxis*	Neural circuits working as pure automation

Automatic and controlled components of exploration.
Courtesy of Wojciech Pisula.

ethologists were the first to find strong evidence that *automatisms* provide animals with abilities to behave adaptively without spending time on behavior modification. In some circumstances these abilities form the main behavioral basis for survival and reproduction. Time may be analyzed on three different scales pertinent for psychological phenomena. These are: duration of a stimulus situation, individual longevity, and the duration of a given species evolution. In all three cases one may predict that the more time available, the greater role of controlled processes emergence. Moreover, the complexity of a stimulus situation and the number of possible behavioral outcomes also increase with time. This relationship is summarized in the graph on p. 580.

Identifying the elements of exploratory behavior that belong to automatic vs. controlled behavioral activity, we find elements specific for a given level of integration of exploration. The simplified summarization of the components of exploration is shown in the next figure.

Animal and human exploratory behavior continues to be an elusive area of research However, the significance of that research is growing. In many cases, studies involving psychoactive drug testing are based on the exploratory activity measurement. On the other hand, our knowledge about animal exploratory acts may give us a chance to get insight into the evolution of cognition, at least within the area of novelty related behavior.

See also Curiosity

Further Resources

Jennings, H. S. 1906. *The Behavior of Lower Organisms*. New York: Columbia University Press.
Keller, H., Schneider, K. & Henderson, B. (Eds.) 1994. *Curiosity and Exploration*. Berlin: Springer Verlag.
Marler, P., & Hamilton, W. J. III. 1966. *Mechanisms of Animal Behavior*. New York: John Wiley & Sons, Inc.
Pisula, W. 1998. *Integrative levels in comparative psychology—the example of exploratory behavior*. European Psychologist, 3, 62–69.
Tolman, E. C. 1948. *Cognitive maps in rats and men*. Psychological Review, 55, 189–208.

Wojciech Pisula

Feeding Behavior
Grizzly Foraging

A mature, mid-sized Alaskan brown bear in his teens, Waldo gained 100 pounds in one month when he fished at McNeil Falls. In July/August of each year, the salmon swim up the McNeil River to spawn, providing a feast for grizzlies from miles around.

In comparison, what sort of diet would a teenaged wrestler need to gain over 3 pounds a day? If the wrestler were airlifted into coastal Alaska, where Waldo roamed the vast sedge meadows, where would he find that many calories? Would he be able to survive 6 months of cold winter without eating or drinking, as Waldo did? Questions about "what to eat when" and "where to go to find it" are basic to the study of foraging behavior in animals. Many scientists have been studying their behavior. Their studies are listed at the end of this entry.

Waldo is one of over 200 grizzlies who gather annually at the McNeil River State Game Sanctuary, where Larry Aumiller and his colleagues have documented their feeding behavior for several decades. In 2002, Dan Zatz set up online real-time viewing of the foraging bears, using a remote video camera (*www.seemorewildlife.com*). As documented in the video "Showdown at Grizzly River," this long-term study is an example of how noninvasive technology permits researchers to answer questions about foraging behavior, without disturbing long-lived mammals like the grizzlies.

Scientists seek answers about the *what* and *where* questions of foraging behavior by comparing different species as well as the same species at different times and locations. By analogy to human behavior, the "what" questions are about items on a grocery list and the "where" questions are about shopping decisions.

Jennifer Owen has explained that the *what* questions deal with tastiness (*palatability*) and nutritional gain (*profitability*). For example, protein tastes good to grizzlies when they first wake up from their winter sleep, and the calories in fish eggs are most appealing when bears fatten up before going to sleep for the winter. In contrast, bamboo sprouts would not attract grizzlies, although they are the preferred diet of pandas.

Decisions about which food items to include in the diet differ between "generalists" like grizzlies and "specialists," such as pandas. Specialists have relatively fewer items in their diet compared to generalists. For example, pandas specialize on bamboo, and grizzlies eat whatever is available. An analogy might be that specialists have a short shopping list compared to generalists.

As specialists, pandas have evolved digestive mechanisms that detoxify the bamboo's chemical defenses, poisons that make the bamboo distasteful for other species. Stephen J. Gould explains why both giant pandas and red pandas evolved a wrist pad that grasps bamboo like a thumb in a catcher's mitt, a structure found in none of the bear species. The benefit for pandas is less competition with other species. The cost is that they cannot switch foods as easily as the grizzlies.

Although specialists have less competition, they are more vulnerable when their food disappears because they have few choices. For example, if the bamboo forests were cut where the pandas live in China, they would have no other source of food. In contrast, the grizzlies readily switch to feeding on other sources and may be lured to garbage and livestock. Such attractive food sources lure grizzlies into conflict with humans throughout the northern portion of the globe, as Robert Cheek described for the sad case of Chocolate Legs. She was unusually comfortable around people and after enchanting (and frightening) many visitors to Glacier National Park, Chocolate Legs was relocated to a remote area where she threatened no one. However, after reverting to her old ways with two large cubs, she was blamed for the death of a hiker Craig Dahl, who reportedly would have been most upset if he had known her family would have to be killed, in the name of protecting public safety.

For scientists studying foraging behavior, Jennifer Owen explained that the "where" questions have to do with availability and accessibility of food. In a grocery store, some fruits are available all year, and others come into season only at one time of year. Likewise, when a grizzly like Waldo wakes up in the spring, the only available food may be the carcasses of other animals that died during the winter or the freeze-dried berries left from the fall. After spring thaw, the flowering plants that stored sugars and starches in thick root bulbs are the first to grow sprouts eaten by the bears. Tom McNamee describes how later the bears dig rodents out of burrows, find insect grubs under rocks or decaying logs and catch the vulnerable calves of elk, caribou or moose. Spawning fish run the rivers at predictable times in midsummer. Berries ripen toward the end of the summer. In contrast, bamboo is available year-round for the specialist pandas.

How do grizzly cubs learn where to go to find each of the foods that come into season? They follow their mothers. Many of the foods are not accessible to cubs. However, the long claws of their mothers are handy tools for digging and catching food items that would otherwise be inaccessible to cubs. By analogy, an unopened jar of baby food is useless to a toddler. Some of the cubs that have visited McNeil Falls with their mothers have taken up the same distinctive fishing techniques that their mothers demonstrated, reports Larry Aumiller. In Yellowstone National Park, Frank Craighead discovered that some bears, who had learned from their mothers to feed at garbage dumps, were unsuccessful at finding natural sources of food when the dumps were closed.

Adult grizzles are smart about learning how to get at food that is inaccessible. In the video "Grizzly Diaries," Tim Treadwell shows footage of a grizzly who had learned her own special technique for opening clams stranded at low tide out on the coastal mudflats. After digging up the clam, she held it down with one paw and used another claw like a bottle opener to pry open the clam. Other bears had learned different individualized techniques for opening the clams. It's no wonder bears learn easily how to open campers' ice coolers.

Tom Bledsoe has discovered that rainfall can influence the fishing access for the bears at McNeil Falls. Bears can catch the salmon better in shallow water at the edge of the river. In a year of high rainfall, fewer bears were successful at fishing because the water was too deep.

Competition among the bears also influenced their access to salmon at McNeil Falls. When the water was high and the fishing was poor, fights broke out more frequently among the grizzlies. Those bears that were most successful at intimidating others were more likely to get the choice fishing spots. Mothers with cubs were more likely to give up fishing when a male approached, fearing for the safety of their vulnerable offspring.

Larry Aumiller has observed that the most successful bears became more picky about what they ate during their fishing binges at McNeil Falls. For example, Groucho's record catch was 96 salmon in one day. He did not eat the whole fish, stripping only the fatty skin and eggs for himself, leaving the lean protein for other bears to scavenge. One observer noted that it was as if an athlete ordered a pile of cheeseburgers, then ate just the cheese and fries, throwing the rest away. This condition of insatiable appetite is called *hyperphagia*.

The function of hyperphagia is rapid weight gain prior to winter, when no food or water is available in the frozen homeland of grizzlies, as Sue Healy and Michael Bright describe. Instead of storing food in the freezer, grizzlies store calories in their bodies. Those skinny bears that did not store fat, did not live through the winter. In this manner, the genotypes for slim bears would have been "edited out" of the population in the past history of the species.

Bears are able to survive the winter without eating because their metabolism slows during *hibernation*, a sleep-like state lasting for months. During hibernation, heartbeat and breathing are so slow that a grizzly burns a bare minimum of

A grizzly bear waits for salmon at the Katmai National Park, 2003.
Courtesy of Gillian Bowser.

calories from its stored body fat, which is used up by the spring. It's like balancing a checkbook: To keep money in the bank, one has to cut back expenses when there is no paycheck.

Grizzlies are a good example of how genes may influence appetite to change with the seasons. When bears wake up from hibernation in the spring, they have little appetite, a condition called *anorexia*. With the spring thaw, plants begin to sprout and grizzlies choose the most nutritious. For example, in the Rocky Mountains grizzlies dig up the nutritious little root bulbs of plants like spring beauty, glacier lilly, biscuitroot, and chives. Later they feed on the growing buds and plant shoots of sedges, which provide protein in easily digestible liquid form. After the spring growth spurt, many of these plants produce defenses that make them taste bad or hard to digest, so grizzlies switch to other foods like the fish as their appetites increase.

How does Waldo know to go on an eating binge at the end of the summer, then switch off his appetite to make it through lean times in winter and spring? David Stephens hypothesized that animals don't really make a conscious decision about the future; rather they follow short-term rules that happened to have worked out over the long-term, which means during the past history of the species. Judging from the fossil record, the history of grizzlies goes back a quarter of a million years, before the ice ages and the glaciers that repeatedly carved the northern landscapes where grizzlies live.

For grizzlies, the shorter days and colder nights of July trigger a change in their metabolism, the body system that controls how much fat is stored in the body or used for exercise. During the autumn eating binge, bears choose food items that are highest in calories. During the spring recovery from winter starvation, grizzlies choose food items high in proteins, starches, and sugars. So Waldo's choice of food would depend not only on what was seasonally available, but also on his body condition. He eats enough in 6 months to last the whole year. It is as if his body says, "turn on the appetite when the days grow short," and, "turn off the appetite when the fat bulges are thick enough."

In summary, grizzly bears such as Waldo help illustrate some of the scientific hypotheses about foraging behavior. Some of these hypotheses are about *what* animals choose to include in their diets, and others are about decisions of *where* to go to find the food. This is analogous to grocery lists and decisions of where to shop when food availability changes with the seasons. Generalists such as grizzlies have long shopping lists and learn from trial and error, and from their families where to find food. For northern species like grizzlies, their appetite increases when fatty foods are readily available and turns off during the winter when no food is available. During stressfully long winters, their anorexia is matched by inactivity, an adaptation that allows them to survive from fat stored in their bodies. Due to the genetic history of this species, changes in physiology match changes in the environment.

See also Feeding Behavior—*Scrounging*
Feeding Behavior—*Social Foraging*

Further Resources

Bledsoe, T. 1987. *Brown Bear Summer*. New York: E. P. Dutton.
Bright, M. 1997. *Intelligence in Animals*. New York: The Reader's Digest Association, Limited.
Cheek, R. 2001. *Chocolate Legs*. Columbia Falls, MT: Skyline Publishing.
Craighead, F. C. Jr. 1979. *Track of the Grizzly*. San Francisco: Sierra Club Books.
Gould, S. J. 1980. *The Panda's Thumb: More Reflections in Natural History*. New York: W.W. Norton & Company.
Healy, S. 1994. *Foraging and storing*. In: *Animal Behavior*. (Ed. by T. Halliday), pp. 57–63. Norman: University of Oklahoma Press.
McNamee, T. 1984. *The Grizzly Bear*. New York: Alfred A. Knopf.
Owen, J. 1982. *Feeding Strategy*. Chicago: The University of Chicago Press.
SeeMore Wildlife Systems, Inc. www.seemorewildlife.com. Accessed Sept. 10, 2003.
Stephens, D. W. & Krebs, J. R. 1986. *Foraging Theory*. Princeton: Princeton University Press.
Walker, T. & Aumiller, L. 1993. *The Way of the Grizzly*. Stillwater, MN: Voyageur Press.

Videos

Showdown at Grizzly River. 2000. Thirteen/WNET New York (Shown on PBS Television Series, *Nature*.)
Treadwell, T. 1999. *Grizzly Diaries*. Discovery Channel Video.

Jane M. Packard

■ Feeding Behavior
Scrounging

Living in groups has advantages, such as predator avoidance. The group may be a partner market and may provide the stage for courtship and mate choice. Also, individuals in groups can benefit from the experience accumulated in the group, often over generations (by social learning). However, there are also disadvantages of group living, such as competition for food, mates, and so on. In general, group members may either compete via their efficiency in finding and processing food (*scramble competition*), or they may start defending food patches against other group members (*interference competition*). Group members may exploit others and may themselves be exploited. For example, individuals may *scrounge* food others have produced. Scrounging may be parasitic, which is the case if the producer enjoys no fitness benefit from sharing food with a scrounger either via inclusive fitness or tit-for-tat relationships. If food is not shared voluntarily, scrounging may be considered to be a special case of interference competition.

A *scrounger* profits from the prior investment by a *producer*. The time spent in the search for and handling of food items are the major costs of foraging because this determines the time lag between successive items and hence, *profitability* (energy intake per time) of individual foraging. Also, time matters because feeding often means exposure to predators, and there are other indispensable obligations to be accommodated in individual time budgets, such as social interactions and ultimately, reproduction. Also, during many times of the year, individuals employ an energy maximizing tactic, for example, when feeding young. Therefore, scrounging will develop if it increases individual foraging efficiency.

Various Contexts of Scrounging

Opportunistic Scrounging

As an example, a probing bird, such as a redshank or a northern bald ibis first needs to locate worms or arthropods in the soil via its mechanoreceptors in the beak. Once found, the prey item needs to be extracted before it can be swallowed. A scrounger snatching away the food item from a producer just after it was produced, therefore saves search and handling time and will either specialize in scrounging or employ scrounging as an opportunistic or alternative tactic, besides producing.

The northern bald ibis (*Geronticus eremita*), for example, use pastures and short-vegetation steppe habitats to forage for small invertebrates and vertebrates. After 400 years of absence from Europe, a free-flying, reproducing group of these highly threatened birds has been established at the Konrad Lorenz Research Station in Grünau, Austria. Natural foraging of these birds at local pastures is closely monitored. Within the group, scrounging pairs may form. The potential scrounger, usually a male, forages closely behind a producer, usually a female. When the producer extracts a worm or a beetle larva from the soil, the scrounger tries to snatch the item from the probing hole or the even beak of the producer. In this case, scrounging is an opportunistic tactic, potentially accounting for more than 10% of the food volume ingested by a scrounger. Scrounging males were dominant over the producing females and pairing was random; that is, a male rarely scrounged at his reproductive partner. In this case, scrounging may be seen as an opportunistic tactic of dominants to increase their foraging success.

Scrounging for Safety

If producing food is acutely dangerous, scrounging may be employed as a safe alternative. Ravens, for example, often cofeed at carcasses produced by predators, such as wolves, bears or human hunters. However, cofeeding with wolves is dangerous. In our setting (Bugnyar and Kotrschal 2002a), for example, up to 10 wild ravens per year (mainly inexperienced, young males) are killed by the wolves. Hence, not all ravens in the group take the risk of producing meat directly from the carcass, but scrounge from their daring conspecifics (members of the same species) via a number of mechanisms.

As a competitive tactic, ravens fly off with pieces of meat, cache it out of sight of other ravens, return to the carcass and start another round. These producing–caching cycles explain why a relatively small group of ravens may strip even large carcasses of all the meat within short time. These producing and caching ravens also give other group members the possibility to avoid direct exposure to the wolves at the carcass and to obtain meat from producers in a number of different ways by *displacing* (a dominant scrounger forces a producer to give up his piece of meat), by *stealing* (a nearby scrounger quickly snatches meat from a producer and retreats), by *sharing* (the scrounger being a social partner of the producer and hence, being allowed to cofeed with, or even being fed by, the producer), by *chase flying* (the producer flying off with food is harassed by a scrounger and finally drops its item) and by *cache raiding* (a scrounger watches caching from a distance raids a cache when the producer has left).

The raven example shows that scrounging is not always a matter of brute force, and that sophisticated games may develop over scrounging, such as the one between cachers (producers) and cache raiders (scroungers). The cacher tries to hide out of the sight of conspecifics, and the potential raider watches as inconspicuously as possible. This game is probably not necessarily played according to rules, but may involve bluffing and some insight into the mind of the opponent.

Scrounging and Foraging Skills

There may be differences in foraging skills within the group. Innovators, for example, may find novel ways to produce food, thereby opening to other group members an opportunity to scrounge and/or to learn the relevant skills. In this case, producers are at the same time models for potential observers, and hence, may initiate a social tradition within the group. Innovators do not seem to be a random sample of a population, but are probably predisposed by how they cope with challenge. Whether scroungers/observers will finally acquire the novel skill, seems mainly a matter of profitability. If scrounging is sufficiently profitable, this will inhibit the cultural transmission of food finding skills, as, for example found in pigeons. If however, producing is more profitable than scrounging, a novel skill may spread quickly in a population, particularly if facilitated by a tolerant relationship between producer and scrounger.

The previous examples show that often, but not always, scroungers are dominant over producers, and that complex social interactions may develop over scrounging, which may even lead to the formation of cultural traditions.

See also Feeding Behavior—*Grizzly Foraging*
Feeding Behavior—*Social Foraging*
Social Organization—*Herd Dynamics: Why Aggregations Vary in Size and Complexity*

Further Resources

Bugnyar, T. & Kotrschal, K. 2002a. *Scrounging tactics in free-ranging ravens*, Corvus corax. Ethology, 108, 993–1009.
Bugnyar, T. & Kotrschal, K. 2002b. *Observational learning and the raiding of food caches in ravens, Corvus corax: is it "tactical" deception?* Animal Behaviour, 64, 185–195.
Fritz, J. & Kotrschal, K. 2002. *On avian imitation: Cognitive and ethological perspectives.* In: *Imitation in Animals and Artefacts.* (Ed. by K. Dauterhahn & C. L. Nehaniv), pp. 133–156. Cambridge, MA: MIT Press.
Giraldeau, L.-A. & Caraco, T. 2000. *Social Foraging Theory.* Princeton, NJ: Princeton University Press.
Krause, J. & Ruxton, G. D. 2002. *Living in Groups.* Oxford Series in Ecology and Evolution. Oxford: Oxford University Press.
Pfeffer, K., Fritz, J. & Kotrschal, K. 2002. *Hormonal correlates of being an innovative greylag goose.* Animal Behaviour, 63, 687–695.

Kurt Kotrschal

■ Feeding Behavior
Social Foraging

Introduction

Animals often gather together when searching for food or feeding. Flocks of pigeons, herds of wildebeests and schools of bluefin tunas illustrate the wide-ranging nature of social foraging in animals. Nevertheless, several species, such as squirrels or hawks, rarely forage in groups. Research on social foraging reveals the advantages and disadvantages of foraging in a group for individuals and why some species forage in groups while others forage alone.

Definition

Foragers may gather together in the same part of the environment because each individual is attracted independently to a specific feature such as a large fruiting tree. Despite the congregation in space, such fortuitous groups, which dismantle as quickly as they form, are not considered cases of social foraging. Social foraging would occur, how-

Bluefin tuna schooling near Baja California.
© Richard Herrmann / Visuals Unlimited.

ever, if individuals were actively attracted to the tree by the presence of companions. In this case, the attraction is beneficial since individuals can obtain food more quickly than by searching on their own. In general, social foraging occurs when the success of foragers depends on the behavior of companions within communication range. The behavior of companions can influence success at each step in the food procurement process, from the initial search to the actual consumption of resources. Success can be enhanced by the presence of companions at certain steps of the foraging process but may remain unchanged or even decrease at other steps as long as social foraging proves beneficial overall. Success is defined in terms of survival chances. Individuals that obtain food more efficiently or evade predation attempts more often are considered more successful.

Advantages of Social Foraging

Social foraging can increase success by increasing feeding efficiency and/or decreasing predation risk. In terms of feeding efficiency, the presence of more eyes in a group enables group members to locate food patches more quickly when they are scattered widely in the environment, thereby reducing search time considerably. In many species, such as gulls and vultures, foragers search for food over a wide area. When one bird locates food, all other foragers within sight can rapidly converge on the food source.

Other factors can also contribute to an increase in feeding efficiency. Groups can tackle different types of prey that are not accessible to solitary foragers. For instance, although solitary lions usually eat small prey, groups of lions can attack larger and potentially more rewarding prey such as buffaloes. In addition, individual group members can focus on different tasks, thus increasing the likelihood of capturing the prey. In lion groups, some individuals stalk the prey while others lay in wait to bring down the fleeing prey.

The presence of several foragers in the same area at the same time can allow individuals in groups to herd prey more effectively. Herons walking systematically in line in a field can flush more prey than by adopting an independent search pattern. Systematic foraging also occurs in groups of dolphins circling schools of fish until all prey are forced to a restricted area where attacks will be more successful.

In terms of predation avoidance, more eyes in a group mean that predation threats can be detected more often and more quickly. In addition, should the group be under attack, the mere presence of several companions nearby decreases the risk that any one individual in the group will fall prey to the predator. Generally, the perception of predation risk is much reduced in a group, enabling individuals to allocate more time to feeding and less to looking for predators.

In addition to greater detection of predation threats, groups of prey can provide a better defense against predators. This is true for musk oxen, which form phalanxes when attacked by wolves. In addition, the presence of companions may buffer more centrally-located foragers from side attacks, sending individuals jostling to central positions during a predation attempt.

Disadvantages of Social Foraging

Social foraging can decrease success when foraging in groups reduces prey availability or when group members interact negatively with one another. In terms of prey availability, the presence of several foragers in a group is more likely to disturb mobile prey leading to lower foraging success. For instance, stalking predators such as large cats, which rely on surprise attacks, usually forage alone to minimize prey disturbance. Groups also deplete resources more quickly than solitary foragers, and if food is scattered widely in the environment, groups may be forced to travel longer distances each day to satisfy their energy needs. Thus, compared to small groups, large groups of red colobus monkeys spend more time searching for food each day. The cost of increased travel may be prohibitive if moving from one place to another is energetically demanding.

Negative relationships among group members are quite common. For example, some individuals in groups steal prey obtained by their companions instead of searching for food. In groups of oystercatchers foraging for mussels, the foraging success of all but the most dominant birds decreases sharply as the number of nearby companions increases since more and more food gets stolen. In addition, potential victims spend an increasing amount of time trying to avoid dominant companions at the expense of foraging. Fighting is often more prevalent in large groups as individuals struggle to displace companions from food or protect their own resources. Time spent in aggressive interactions detracts from foraging time and decreases success.

In several species that forage in groups, individuals pool their search effort and all share resources located by the group. This sharing of information ensures that food patches are encountered more regularly, thereby compensating for the fact that resources, once located, must be divided among all group members including the finder. Individual search effort, however, is sometimes parasitized by other group members, leading to scrounging. When scrounging, individuals invest no time in searching for new food sources and wait instead to share the food discoveries of others. Scrounging may be used occasionally by all group members or all the time by a few foragers. Specialization is more likely to occur when individuals greatly differ in their ability to locate food or to displace others from food. In many seed-eating birds, such as dark-eyed juncos, dominant group members scrounge the food discoveries of more subordinate companions, while in other species, such as pigeons, scrounging is not related to dominance status. When scrounging occurs, success is expected to decrease since the group becomes less productive when fewer individuals spend time looking for food.

In many seed-eating birds, such as dark-eyed juncos, dominant group members scrounge the food discoveries of more subordinate companions. © Steve Maslowski / Visuals Unlimited.

As demonstrated by food theft and scrounging, the larger number of resources that a group can potentially uncover creates opportunities for individuals to parasitize the efforts of companions. In addition, these two factors illustrate how foraging success may differ quite considerably among members of the same group. Inequalities can be even more pronounced, for example, when some group members are forced to occupy parts of the habitat that contain less food or that are more exposed to predators.

Group Size Choices

The advantages and disadvantages of social foraging will vary with the number of foragers present in the group. In small groups, individuals may experience less competition for food, but may be more vulnerable to predators. In large groups, predation risk may be reduced, but increased scrounging or fighting may limit success. If individuals behave in ways that make them as successful as possible, the choice of group size should be based on a compromise between advantages and disadvantages.

In many species, individual success shows a complex relationship with group size. For instance, success can increase with group size when groups are small, but then decrease when groups are large. Thus there is an optimal group size somewhere in the middle, one that should be the most prevalent for a given species, because individual success cannot be improved by an alternative choice of group size. However, it has been argued that such groups may not be stable, and that larger groups should in fact be expected. Consider a solitary forager who has the option either of remaining alone or of joining a nearby group containing the optimal number of foragers. Solitary foraging is generally not a very successful option, both in terms of feeding efficiency and predation avoidance. Hence, joining the group represents the best choice for the solitary forager. Joining such a group, however, will reduce the success of all members already present. Should other solitary foragers in the area also join, this would lower success even more. The process of joining should stop when joining the now much larger group becomes less successful than remaining alone. At this point, ironically, social foraging would not provide any extra advantages over solitary foraging. This conundrum could be avoided if individuals could control access to the group and restrict joining opportunities.

Active manipulation of group size could also be helpful at the other extreme when group size is too small. For instance, group members could actively recruit companions when an increase in group size is needed to obtain more benefits. House sparrows in small flocks use chirrup calls to attract companions to food sources that can be shared by all. Hooting calls emitted by many primate species at rich food sources have also been considered recruitment signals to build up group size.

In species where success decreases steadily with group size, foraging alone probably represents the best option. The question is why social foraging occurs at all in many of these species. One possibility is that feeding opportunities may be so limited in time and space that foragers have no other choice than to feed together despite the reduction in success. It addition, being in a group can provide advantages in nonforaging contexts, perhaps compensating for any loss in foraging success. In lions, for example, the ability to repel intruder males trying to overcome the pride and kill the cubs is thought to be a major factor in the evolution of group living in this species.

Comparative Analyses

Studying variation in group size and its relationship to individual success is a powerful way to determine how social foraging benefits individuals. Another way to examine the adaptive value of social foraging is to relate variation in mean group size across different species to differences in ecological factors experienced by these different species. Comparative analyses examine variation in the expression of sociality across species as a function of ecological factors such as food distribution and predation risk.

In birds, for example, results indicate that species foraging in flocks are more likely than solitary species to exploit resources that occur in clumps. Social foraging is probably helpful in locating widely scattered clumps of food that can be shared by several foragers. If social foraging were mostly related to a reduction in predation risk, social foraging should be more prevalent in more vulnerable species, such as small species, because of their relative lack of weaponry, and species living in more open habitats, which are potentially more exposed to predators. In birds, however, social foraging was not related to habitat openness and was less, not more, prevalent in smaller species. Resource characteristics appear the best predictor of social foraging tendencies across species in birds.

Future Directions

Many issues require more scrutiny in the study of social foraging. It is not clear in many species whether individuals forage in groups that actually maximize success. This stems in part from a lack of knowledge of the advantages and disadvantages that foragers experience in groups of different sizes, but also in part because it is not clear how and if group foragers can manipulate access to their groups. The ecological factors involved in the evolution of social foraging are also not fully understood. The use of comparative analyses in a wide variety of taxa will be useful to examine how the balance between the various advantages and disadvantages of social foraging is struck in different species.

See also Behavioral Plasticity
        Cognition—*Cache Robbing*
        Cognition—*Caching Behavior*
        Feeding Behavior—*Grizzly Foraging*
        Feeding Behavior—*Scrounging*
        Social Evolution—*Optimization and Evolutionary*
          Game Theory
        Social Evolution—*Social Evolution in Bonnet Macaques*
        Social Organization—*Herd Dynamics: Why Aggregations*
          Vary in Size and Complexity

Further Resources

Beauchamp, G. 2002. *Higher-level evolution of intraspecific flock-feeding in birds.* Behavioral Ecology and Sociobiology, 51, 480–487.

Clark, C. W. & Mangel, M. 1986. *The evolutionary advantages of group foraging.* Theoretical Population Biology, 30, 45–75.

Elgar, M. A. 1989. *Predator vigilance and group size in birds and mammals.* Biological Reviews, 64, 13–33.

Giraldeau, L.-A. & Caraco, T. 2000. *Social Foraging Theory.* Princeton: Princeton University Press.

Packer, C., Scheel, D. & Pusey, A. E. 1990. *Why lions forage in groups: Food is not enough.* American Naturalist, 136, 1–19.

Pulliam, H. R. & Caraco, T. 1984. *Living in groups: Is there an optimal group size?* In: *Behavioural Ecology* (Ed. by J. R. Krebs & N. B. Davies), pp. 122–147. Oxford: Oxford University Press.

Guy Beauchamp

■ Feeding Behavior
Social Learning: Food

Feeding is essential for living. Specialist animals have diets consisting of one type of food, whereas generalists (or omnivorous) animals have flexible diets which include a wide variety of foods. Whereas in the former species, learning plays almost no role in food choice, for the latter experience is important, since the individual has to recognize and feed on certain foods among those potentially available. The individual can learn by itself what to eat, or it can rely on what its group members do.

Individuals living in social groups can learn about foods from others. This is particularly relevant for omnivorous animals that need to include novel foods in their diet. When encountering a novel food, they are faced with the so-called "omnivore's dilemma": Eating the novel food might result in being poisoned (some plants and insects defend themselves from herbivorous animals and predators with poisonous secondary metabolites, such as glycosides and alkaloids), or avoiding the food might result in loosing the opportunity to discover something nutritious.

Our interest is to describe how animals of different species acquire their diets and balance the need to enlarge them with the caution necessary to avoid ingesting deleterious substances. In particular, the focus will be on the interplay between learning by individual experience and social influences on individual learning. Although these two processes are difficult to separate, for clarity's sake, we will first discuss the contribution of individual behavior and physiology (taste, food neophobia, and food aversion learning) to the acquisition of the diet, and then examine how social influences can affect individual experience about foods.

The Contribution of Individual Experience

Taste

The information conveyed by the senses, and especially by taste, may help animals to sample a potential food and decide whether to eat it or to avoid it. Substances high in calories, such as sugars, are perceived as sweet by humans and are readily accepted by both humans and nonhuman animals. Conversely, substances that humans find bitter (for example,

The information, conveyed by the senses and especially by taste, may help animals such as this capuchin monkey sample a potential food and decide whether to eat it or avoid it.
Courtesy of Elisabetta Visalberghi.

glycosides and alkaloids) and that are usually dangerous to ingest, induce rejection responses in animals. In humans, acceptance and rejection responses are evident in early infancy, before experiencing the consequences of the ingestion of sweet or bitter substances, and these responses enable them to avoid toxic substances and to maximize energy intake. Moreover, since odor often signals the edibility or toxicity of a food, it is often used as a cue by macrosmatic (capable of smelling at a distance) animals who rely on olfaction.

Although many toxic compounds have a bitter taste that allows their detection, some very poisonous substances (e.g., dioscin, a lethal alkaloid present in the plant *Dioscorea dumetorum*) are almost tasteless. In these cases, relying on taste alone might be dangerous. In addition to using odor and immediate taste feedback as cues for evaluating a food, the individual has behavioral and physiological features that allow a further evaluation after food ingestion. We will illustrate how cautious exploration and ingestion of novel foods and food aversion learning decrease the risk of dying by eating an excessive amount of poisonous food.

Food Neophobia

Animals begin life with little information about their environment, so exploration may help them to adapt. *Novelty* (defined as introduction of new features in a otherwise familiar context) elicits an ambivalent response, characterized by interest, or *neophilia*, and avoidance, or *neophobia*. The predominance of one over the other, or vice versa, is affected by the degree of the stimulus novelty, the species' ecology and the individual's characteristics (age, sex, and so on). Surprisingly, in a completely novel situation, neophilia may overcome neophobia. For example, when a troop of macaques from Japan was transplanted to Texas, at first they did not show any neophobic response toward the many new plants they encountered for the first time. Afterwards, when the macaques acquired a staple diet, they became more neophobic.

Animals of many species, from birds to nonhuman and human primates, carefully approach, explore, and taste novel food.
Courtesy of Elisabetta Visalberghi.

Studies of food neophobia involve the introduction of a novel food item, or of a novel object in association with a familiar food item, and the assessment of the animal's behavioral responses (latency and frequency of approach, duration of exploration and/or manipulation, and amount of food ingested). Although food neophobia has been studied mostly in rats with the aim of controlling the growth of wild populations, animals of many other species, from birds to nonhuman and human primates, watchfully approach a novel food by exploring and tasting a little amount of it.

Caution toward unknown food items reduces the risks of ingesting a lethal quantity of poisonous substances. However, omnivorous species should overcome neophobia because their success depends also on their ability to exploit new food sources, especially when staple food is scarce. In children, as well as in capuchins, neophobia decreases after an individual has repeatedly encountered and consumed a novel food without gastrointestinal illness, and that food becomes part of the individual's diet.

Food Aversion Learning

If ingestion of a novel food has noxious consequences, an individual will associate these consequences with its taste, and it will not eat that food anymore. This phenomenon, also called *food aversion learning* or *Garcia effect*, was first discovered in rats and later demonstrated in many other animal species, including humans. In this robust learning mechanism, gastrointestinal illness, following the ingestion of a food, leads to complete avoidance of the food. Moreover, the avoidance persists even if that food is no longer noxious, and the avoidance learning process is quicker when the food is novel than when the food is familiar. Finally, the aversion is often established after only a single taste experience with the noxious food. Although olfactory cues are less effective in the establishment of conditioned aversions, an odor that has been presented in combination with a taste is highly effective.

A brown bear shows her cub how to catch salmon in the Brooks River, Katmai National Park and Preserve, Alaska.
© Kevin Fleming / Corbis.

Taste aversion learning has traditionally been considered a variant of classical, or Pavlovian, conditioning (when a dog salivates to the sound of a bell which it has experienced to precede its feeding time). However, taste aversion learning has some unusual features, which has set it apart from classical conditioning. First, it is rather resistant to extinction. Second, it occurs despite of long intervals (hours!) between food ingestion and its negative consequences, whereas in classical conditioning the interval is of a few seconds.

Taste aversion learning is not a laboratory artifact and occurs in the wild where the situation is often more intricate. In fact, when foraging, omnivorous animals switch from one food to another, and the encounter with a novel food occurs while consuming foods belonging to the usual diet. How is it then possible that animals are able to associate illness with the ingestion of the novel food and not with that of foods that are already part of their diet? An experiment carried out with Japanese macaques has shown that after having experienced illness, monkeys at first stop consuming the most novel food they have ingested before their illness and, only if they continue to be ill do they stop eating other foods (usually starting to eliminate the most novel ones). This stepwise strategy reduces the risks of suffering illness and starving to death.

Knowledge on food aversion learning has been used to develop techniques to prevent pests (rats, coyotes, monkeys) from raiding cultivated or urbanized areas and killing livestock. These kinds of interventions are successful to a limited degree, mainly because usually they are started when the target food is already familiar to the animals. Moreover, if the aversive substance is hidden inside it, the animals are able to detect the difference and avoid the baited food. Finally, there is no evidence that animals learn to avoid food from aversively conditioned group members, whereas they are inclined to follow others in their raiding attitude. This means that, since every single individual should be "discouraged" by giving it baited food, the whole process becomes extremely time-consuming.

Social Influences on Diet Acquisition

The combination of food neophobia and food aversion learning allows the individual to explore novel (potential) food items without risking too much and, in the future, exploit them if palatable. The next point to be considered is whether by being with other more experienced group members, the individual can learn more quickly and more safely. In particular, do animals learn from others about what to consume and what to avoid? In general animals are thought to have this capacity, and it is current thought that we, as humans, learn a lot from others in this domain.

In describing salient cases of social influences affecting the acceptance of novel food, the focus will be on how the mother and group members (mother or any other individual) influence the learning processes of the naive individual.

Learning through Mother's Cues

In many mammals (e.g., rats, rabbits) youngsters' food choices are strongly affected by their mothers' diet during pregnancy and lactation. In rats, chemical cues coming from the food eaten by the dam reach the pups while still in uterus via the blood stream; therefore fetal rats become familiar with the flavors of at least some of the foods that their mother is eating. Also later, during lactation, flavors of foods ingested by the mother are conveyed to her young in the milk; at weaning, these rat pups prefer foods with flavors that they experienced while nursing. Finally, the mere sight of the mother and group members eating is very attractive for weaning rats; this attraction makes it very likely that they will eat their first solid food in the immediate vicinity of more experienced group members.

Also rabbit pups prefer the diet eaten by their mother. Despite the limited amount of time spent by mother rabbits in the nest, information is transferred to rabbit pups in at least three ways: prenatally through the bloodstream, via the fecal pellets deposited by the mother in the nest, or during nursing through the milk. Also, human infants are exposed to a large variety of flavors present in their mother's milk. Because of this early experience, later in life breast-fed infants tend to accept novel foods more readily than formula-fed ones.

Influences from Group Members

In many species (from birds to rats and primates including humans) acceptance of a novel food is higher when group members are nearby. Social tolerance among group members and interest for the other's activities increase the probabilities for individuals to encounter, taste, and eat the same foods. It follows that, for a naive individual, simply being with more knowledgeable group members and taking food scraps from them fosters opportunities for experiencing the foods group members eat. Since members of the same species have similar taste and similar physiological responses to food, being together with knowledgeable individuals strongly increases the chance of eventually eating the same foods.

Social influences on food neophobia have been thoroughly studied in rats. In this species, both behavioral and physiological factors are responsible for the increased acceptance of novel food. Intimate relations result in the transmission of information from demonstrators (i.e., more experienced group members) to observers. Olfactory cues from the digestive tract of demonstrators and bits of food clinging to their fur allows observers to prefer the foods that demonstrators have eaten. Simple exposure of observers to the

Two chimpanzees share fruit. Fruit collected in the Tai Forest is unselfishly shared by individuals, increasing social bonds in the community.
© Michael Nichols/National Geographic Image Collection.

smell or taste of a food is not sufficient to enhance an observer's preference for that food. Observers' preferences for food are strongly enhanced by the combination of the smell of a food and odors produced by the demonstrator rat (in particular, the carbonyl disulfide and carbon disulfide present in rat breath). Although wild rats are likely to experience a larger variety of cues related to food than rats living in a laboratory, the above studies demonstrate the many subtle ways in which individual learning is biased by social cues.

In many avian species, visual cues mediate socially-induced food selection. Red-winged blackbirds (*Agelaius phoeniceus*) repeatedly exposed to group members eating only one of two kinds of colored pastries (green or orange), and then presented with a preference test between green pastries and orange pastries, chose the pastries of the same color as those eaten by their group members. Similarly, adolescent Burmese red jungle fowls (*Gallus gallus spadiceus*) exposed to live group members feeding from visually distinct food dishes (or to videotapes showing this), preferred feeding from food dishes of the same type as those the group members were eating from.

Children's food preferences can be modified by exposure to peer models that have different preferences, or to adults eating a novel food.

Finally, food sharing (i.e., actively giving food by one animal to another)—a common behavior in many species (nidicolous birds, carnivores, several monogamous South American primates, and humans)—is possibly a way to gain information about food. For example, in golden-lion tamarins (*Leonthopithecus rosalia*)—a South American primate—infants beg for food and try to procure it from adults. Food exchange in family groups often occurs. Golden-lion tamarins use an offering posture, eye contact, and a food call to invite a group member to take food from them. The food call seems to be the most unambiguous and significant signal in predicting food transfer. In fact, whereas both eye contact and offering position occur in contexts other than food interactions, the food call was only heard in food-related contexts.

In golden-lion tamarins, adults are more likely to share familiar foods than novel ones, and are also more likely to share rare foods than abundant ones. Consequently, food sharing seems to be more a way to ensure that infants receive adequate amounts of food (difficult to locate or to acquire) than a way of informing them about which novel foods to eat. This interpretation is in agreement with the lack of evidence of active teaching in animals, in which the teacher conveys information and influences the behavior of the learner by taking into account what the learner knows and what it does not know. At present, humans are the only species in which a cognitively demanding behavior commonly occurs in a variety of circumstances.

Purposefulness of Social Influences

Social influences can lead to safe feeding only if the target food encountered by naive individuals and the food eaten by the knowledgeable group members are the same. In the case of novel foods, if social facilitation of eating is specific (i.e., oriented toward the same food as that which the group members are eating), the naive animal ends up eating a food that is likely to be palatable. Vice versa, if social facilitation is generic (i.e., directed to food in general), the naive individual does not match the appearance of its own food with the appearance of a group member's food, and merely learns to accept the food more readily than it would have done alone. In capuchin monkeys, the eating activities of group members increased the frequency of eating behavior and the amount of food eaten by an observer, even though they were eating a food of a different color. Therefore, the hypothesis that social facilitation is a generic process seems supported.

Animals Do Not Learn from Others to Avoid Harmful Food

There is no doubt that animals are capable of learning individually that a food is noxious. The question now is whether they are capable of learning it from others (i.e., by watching them avoiding it, or in other words, not eating it). This would seem a quicker and less troublesome way to learn that a food is unpalatable.

So far, there is no evidence that this is the case in animals other than humans. Observer rats exposed to demonstrators experiencing an acute gastrointestinal distress, exhibited preferences for, rather than aversions to, whatever food their ill demonstrators had eaten. However, social influences on rats' food choices are extremely powerful in reversing a learned aversion. In fact, observer rats aversively conditioned to a food, often abandon their aversion after interacting with a rat that has eaten that food.

There is no clear evidence that primates are capable of learning to avoid a food from others. For example, in Japanese macaques, when a dominant male (i.e., the individual that all group members are interested in and watch attentively) was aversively conditioned not to eat a given food, group members did not significantly decrease their consumption of that food.

At present, the only positive experimental evidence comes from birds. In red-winged blackbirds, the location of the food (and not the food itself, as in the studies cited above) was the salient cue the observers had to rely on. The observer saw a demonstrator eating on an orange box and showing signs of illness immediately afterwards. When the observer itself was given a choice between food in an orange box or in a red box, it chose to eat from the latter.

Finally, because of the similarity between nonhuman and human primates, the role of social influences on feeding is usually overinterpreted in monkeys and apes more than in other mammals. For example, it has often been assumed that experienced individuals actively intervene to prevent naive group members from tasting toxic food. However, at present there is only anecdotal evidence of this kind of behavior in chimpanzees and gorillas (the primate species more genetically related to humans). Anecdotes are open to different interpretations; for example, the instance, described by Watts in 1985, in which a mother gorilla pushed away the stem of an unidentified plant (not eaten by other gorillas) that her young daughter had pulled toward herself could be either a case of maternal intervention to prevent food ingestion, or more parsimoniously one of the many instances in which a mother takes food (no matter if bad or good) away from its offspring. These kinds of anecdotal observations need to be validated in controlled experiments.

In conclusion, the strong preferences for sweet foods and the dislike for bitter substances which are often associated with toxic compounds, and the neophobic response coupled with food aversion learning, efficiently reduce the risk of making a fatal mistake. In this scenario, the individual does not need to observe others to learn what to ingest and what to avoid. The individual is equipped with the behavioral and physiological tools necessary to select energy-rich foods and to avoid deleterious ones. Nevertheless, individuals do not learn in a void context: Group members' presence and behavior influence the individual's learning opportunities.

See also Behavioral Plasticity
    Cognition—*Caching Behavior*
    Feeding Behavior—*Social Foraging*

Further Resources

Barnett, S. A. 1963. *The Rat. A Study in Behaviour*. Chicago: Aldine Publishing Company.
Capaldi, E. D. (Ed.) 1996. *Why We Eat What We Eat: The Psychology of Eating*. Washington, DC: American Psychological Association.
Dominy, N. J., Lucas, P. W., Osorio, D. & Yamashita, N. 2001. *The sensory ecology of primate food perception*. Evolutionary Anthropology, 10, 171–186.
Greenberg, R. & Mettke-Hoffmann, C. 2001. *Ecological aspects of neophobia and exploration in birds*. Ornithology, 16, 119–178.
Heyes, C. & Galef B. G. (Eds.) 1996. *Social Learning in Animals: The Roots of Culture*. San Diego, CA: Academic Press.
Hladik, C. M. & Simmen, B. 1996. *Taste perception and feeding behavior in nonhuman primates and human populations*. Evolutionary Anthropology, 5, 58–71.
Visalberghi, E. & Addessi, E. 2003. *Food for thoughts: Social learning and the feeding behavior in capuchin monkeys. Insights from the laboratory*. In: *Traditions in Non-human Animals: Models and Evidence*, (Ed. by D. Fragaszy & S. Perry). Cambridge: Cambridge University Press.

Elsa Addessi & Elisabetta Visalberghi

■ |Friendship in Animals

Almost every human being has some friends, or at least wishes he did. For many people close friends are as important as family. Friends are so essential that it is hard to imagine life without them. What about other animals? Do they have friends, too?

The answer, at least for some species, is a resounding "yes." Individuals form friendships in many social mammals, including wolves, bottlenose dolphins, chimpanzees, and elephants. Different kinds of friendships exist. In many social species, juveniles play over and over with certain partners, and it seems reasonable to think of these play partners as friends.

A male and female yellow baboon stand together in the green savanna of Amboseli National Park, Kenya.
© Corbis.

In bottlenose dolphins, immature males associate with one or two males of about the same age. These early friendships often last into adulthood and develop into long-term alliances. In bonobos (pygmy chimpanzees), each adolescent female leaves the community she was born in and moves to another group. Knowing no one, she picks out one adult female and is especially nice to her. She will groom the older female and even offer to have sex with her (female–female sex is common in this species). If she succeeds in developing this friendship, it will help her to become an integrated member of the new community.

In order to study friendship, scientists must first be able to recognize all of the individuals in their study group. Once they know everyone, they can collect information about who "hangs out" with whom. There are two main ways to do this. First, the observer can scan the whole group at regular time intervals (such as every 15 minutes) and record the nearest neighbor of every animal at that instant. After doing this hundreds of times, the researcher may discover that some animals tend to be in proximity to one another much of the time. A second way to find out who associates with whom is to observe one animal closely for a predetermined period of time (such as 30 minutes) and record who this "focal" animal interacts with. These two methods complement one another, and scientists often use both.

Friendships between adult male and adult female savanna baboons are probably the best-studied animal friendships. We can use them as a case study to illustrate how we investigate animal friendship and to better understand what friendship may mean to the animals.

Baboons are large, ground-dwelling African monkeys that live in troops ranging from 30–150 members. I studied one troop of about 150 that included 18 adult males and 36 adult females. I conducted focal samples on all of the females over a period of 2 years. When I analyzed the data, I found that each adult female associated mainly with just one or two of the 18 adult males, and different females associated with different males. Females were very choosy as to which males they groomed, and these were nearly always the same males they spent time near. For each female, I thus defined a male "friend" as any male that had an exceptionally high score for both proximity and grooming.

Once I had defined friendship, I could ask who formed friendships with whom? There was no evidence that females preferred high-ranking males as friends, but they did tend to favor males who had been in their troop for at least 6 months (male baboons, like female bonobos, transfer into a new troop as adolescents). Adult females of all ages formed friendships with adult males of all ages, but adolescent males and females tended to form friendships mainly with each other. These youthful friendships tended to be unstable, often lasting only a few months, whereas some of the friendships between fully adult males and females lasted for many years.

Such friendships between adult males and females have been documented for every savanna baboon population that has been studied. This suggests that friendships have been favored by natural selection. How might they benefit females and males? When a male becomes friends with a female, he also develops a close bond with her youngest offspring. He will sit with, groom, and play with her infant. If another adult male draws near, the friend will often scoop up the infant and carry her off. Such protection may be very important, because adult male baboons sometimes kill young infants. Males that have just moved into a troop, who have not yet fathered any infants of their own, are particularly murderous. In a series of ingenious experiments, scientists studying savanna baboons in southern Africa used a tape recorder to play the screams of infant baboons to various males. They found that most males ignored these screams, but the male friends of the mother stared at the hidden tape recorder, suggesting that they were especially attuned to the cries of their female friends' infants. It thus appears that female baboons form friendships with males at least in part to gain protection from aggression by other males. Male friends also protect the females themselves when other males, or even females, attack them.

Protecting a female or her infant against other males can be risky, so why would an adult male do this? In some baboon populations, males direct their protection toward infants they are likely to have fathered; they may be protecting their own genetic contribution to future generations. But in the troop that I studied, males often formed friendships with females they had never mated with, so they could not be the fathers of those infants. Female baboons don't mate when they are pregnant (6 months) or nursing (up to 24 months), but eventually they resume estrus and become sexually appealing to males. I found that most females, although they mated with multiple males, preferred to mate with their male friends. Thus, by sticking with a female and protecting her infant, a male may increase the chances that he will father her offspring in the future.

These hypotheses help us to understand why, in evolutionary terms, baboons might form friendships, but it is also important to ask how the baboons themselves experience friendship. I compared the kinds of interactions that females had with their males friends *versus* males in general. Striking differences emerged. Females approached their friends but moved away from all other males, indicating that they really wanted to be near their friends. Friends sometimes snuggled, suggesting they enjoyed close physical (nonsexual) contact (I never observed nonfriend couples snuggling). When a female approached a male and presented her hindquarters to him in greeting, he was much more likely to respond if he was a friend, and greetings between friends were slower and more relaxed than greetings between nonfriends, which were often jerky and tense. If we had comparable data on human friendships, we would not hesitate to conclude that friends were emotionally attached to one another. We might hypothesize, then, that baboons make friends because they enjoy having them, just as humans do!

Unlike bonds among kin or allies, friendship is not based on a single, measurable factor, such as geneological relatedness or the ability to support another in a fight. Instead, friendships seem to derive from affectionate feelings based on idiosyncratic preferences that differ from one individual to the next. This means that animals who form friendships have the capacity to distinguish the personal traits of different individuals and use those traits as a basis for choosing whom they want to be with. What could be smarter than that?

See also Cooperation
　　　　Play—*Dog Minds and Dog Play*
　　　　Play—*Social Play Behavior and Social Morality*
　　　　Social Organization—*Alliances*

Further Resources

Altmann, Jeanne. 1980. *Baboon Mothers and Infants*. Cambridge: Harvard University Press.
Smuts, Barbara. 1999. *Sex and Friendship in Baboons*. 2nd edn. Cambridge: Harvard University Press.

Barbara Smuts

■ Frisch, Karl von
(1886–1982)

Karl von Frisch was born in Vienna in 1886, and, as a boy, was a lover of pets. His research emphasized both sensory physiology and behavior. He studied color vision in fish and provided the first conclusive demonstration of color vision in insects. He is best known for his discovery of the dance language with which honeybees communicate the distance and direction of food sources to their hive mates. He also showed that, like other arthropods, bees respond to polarized light, which won him the Nobel Prize in 1973 for Medicine or Physiology, which he shared with Konrad Lorenz and Nikolaas Tingbergen.

Donald A. Dewsbury

■ Frisch, Karl von
Decoding the Language of the Bee

Nobel Lecture, December 12, 1973

Some 60 years ago, many biologists thought that bees and other insects were totally color-blind animals. I was unable to believe it. For the bright colors of flowers can be understood only as an adaptation to color-sensitive visitors. This was the beginning of experiments on the color sense of the bee (1). On a table outdoors I placed a colored paper between papers of different shades of gray and on it I laid a small glass dish filled with sugar syrup. Bees from a nearby hive could be trained to recognize this color and demonstrated their ability to distinguish it from shades of gray. To prevent too great a gathering of bees, I instituted breaks between feedings. After these breaks, only sporadic scout bees came to the empty bowl and flew back home; the feeding table remained deserted. If a scout bee, however, found the bowl filled and returned home successfully, within a few minutes the entire forager group was back. Had she reported her findings to the hive? This question subsequently became the starting point for further investigations. In order that the behavior of foragers could be seen after their return to the hive, a small colony was placed in an observation hive with glass windows, and a feeding bowl was placed next to it. The individual foragers were marked with colored dots, that is, numbered according to a certain system. Now an astonishing picture could be seen in the observation hive: Even before the returning bees turned over the contents of their honey sack to other bees, they ran over the comb in close circles, alternately to the right and the left. This round dance caused the numbered bees moving behind them to undertake a new excursion to the feeding place.

But foragers from one hive do not always fly to the same feeding source. Foraging groups form: One may collect from dandelions, another from clover, and a third from forget-me-nots. Even in flowering plants the food supply often becomes scarce, and a "feeding break" ensues. Were the bees in the experiment able to alert those very same foragers who were at the bowl with them? Did they know each other individually?

To settle the question, I installed two feeding places at which two groups from the same observation hive collected separately. During a feeding break, both groups stayed on the honeycomb and mingled with each other. Then one of the bowls was refilled. The bees coming from the filled bowl alerted by their dances not only their own group but also bees of the second group, which responded by flying to their customary feeding place where they investigated the empty bowl.

However, the natural stopping places of bees are not glass bowls but flowers. Therefore, the experiment was modified; one of two groups of bees collected food from linden blossoms, the other one from robinias. Now the picture changed. After the feeding break, the bees returning from the linden blossoms caused only the linden bees to fly out again; the robinia collectors paid no attention to their dances. On the other hand, when bees returned successfully from robinia blossoms, the linden bees showed no interest in their dances, while members of the robinia group immediately ran to a dancer in their vicinity, following along behind her and then flying out. Some clever bees also learned to use both sources of food, depending on the occasion. They would then send out the linden gatherers after returning from the linden source, and the robinia gatherers after visiting the robinias. Thus, the bees did not know each other individually. It appeared that the fragrance of the specific blossom attached to their bodies was decisive. This was confirmed when essential oils or synthetic scents at the feeding place produced the same effect.

When feeding was continuous, new recruits showed up at the food source next to the old foragers. They, too, were alerted by the dance. But how did they find their goal?

Peppermint oil was added to the feeding place next to the hive. In addition, bowls with sugar syrup were put on small cardboard sheets at various places in the nearby meadow; some of the sheets were scented with peppermint oil and the others with other essential oils. The result was unequivocal: A few minutes after the start of feeding, recruits from the observation hive appeared not only at the feeding place next to the foraging bees but also at the other peppermint bowls posted at some distance in the meadow. The other scented bowls, however, remained undisturbed. The smell of lavender, fennel, thyme oil, and so forth had no attraction. When the scent at the feeding place was replaced by a different one, the goal of the swarming recruits changed accordingly. They let themselves be guided by the scent on the dancers. Scent is a very simple but effective means of communication. It attains full significance, however, only in combination with another condition. If the sugar syrup becomes scarce or is offered in weaker concentrations, after a certain point the dancing becomes slower and finally stops even though the collecting may continue. On the other hand, the sweeter the sugar syrup, the more lively and lengthier the various dances. The effect of advertising is thereby enhanced, and it is increased further by the scent gland in the forager's abdomen which is activated upon arrival at a good source of supply. Thus it signals "Come hither!" to recruits searching in the vicinity. Many female insects have scent glands to attract the male. In worker bees, which are mere workhorses devoid of any sexual interest, the scent organ is put to the service of the community.

Karl von Frisch
Courtesy of the National Library of Medicine.

Let us now imagine a meadow in the spring. Various types of plants blossom simultaneously, producing nectar of differing concentrations. The richer and sweeter its flow, the livelier the dance of the bees that discover and visit one type of flower. The flowers with the best nectar transmit a specific fragrance which ensures that they are most sought after. Thus, in this simple fashion, traffic is regulated according to the law of supply and demand not only to benefit the bees but also to promote pollination and seed yield of plant varieties rich in nectar. A new and hitherto unknown side of the biological significance of flower fragrance is thus revealed. Its great diversity and strict species specificity communicate a truly charming scent language.

This was how things stood in 1923 (2), and I believed I knew the language of the bees. On resuming the experiments 20 years later, I noticed that the most beautiful aspect had escaped me. Then, for the first time, I installed the feeding place several hundred meters away instead of next to the hive, and saw to my astonishment that the recruits immediately started foraging at that great distance while paying hardly any attention to bowls near the hive. The opposite occurred when the foragers located the sugar syrup, as before, near the hive. Could they possess a signal for distance?

Two foraging groups were formed from one observation hive. One feeding place was located 12 m from the hive, the other at a distance of 300 m. On opening the observation hive, I was astonished to see that all foragers from nearby performed round dances, while long-distance foragers did tail-wagging dances. Moving the nearby feeding place step by step to greater distances resulted in the round dances changing to tail-wagging dances at a distance of about 50 m. The second feeding place was brought back step by step, past the first feeding place close to the hive. At the same critical distance of about 50 m, the tail-wagging dances became round dances (3, 4). I had been aware of the tail-wagging dance for a long time, but considered it to be typical of pollen collectors. My mistake was due to the fact that, at that time, bees with pollen baskets always arrived from a greater distance than my sugar syrup collectors.

Thus it became evident, and subsequent experiments confirmed (5), that the round dance is a signal that symbolically invites the hive members to search the immediate vicinity of the hive. The tail-wagging dance sends them to greater distances, not infrequently several kilometers. The signal "closer than 50 m" or "farther than 50 m" alone would not be of much help. In fact, however, the pace of the tail-wagging dance changes in a regular manner with increasing distance: its rhythm decreases. According to the present state of our knowledge, information on flight distance is given by the length of time required to go through the straight part of the figure—eight dance in each repeat. This straight stretch is sharply marked by tail-wagging dance movements and simultaneously toned (in the true meaning of the word) by a buzzing sound (6, 7). Longer distances are expressed symbolically by longer tail-wagging times. For distances of 200 to 4500 m, they increase from about 0.5 second to about 4 seconds (6, 8). The tail-wagging dance not only indicates distance but also gives the direction to the goal. In the observation hive, the bees that come from the same feeding place make their tail-wagging runs in the same direction, whereas these runs are oriented differently for bees coming from other directions. However, the direction of the tail-wagging runs of bees coming from one feeding place does not remain constant. As the day advances the direction changes by the same angle as that traversed by the sun in the meantime, but in the opposite rotation. Thus, the recruiting dancer shows the other bees the direction to the goal in relation to the position of the sun (5, 6). Those hours at the observation hive when the bees revealed this secret to me remain unforgettable. The fascinating thing is that the angle between the position of the sun and the dancer's path to the goal is expressed by the dancer in the darkness of the hive, on the vertical surface of the comb, as an angular deflection from the vertical. The bee thus transposes the angle to a different area of sense perception. If the goal lies in the direction of the sun, the tail-wagging dance points upward. If the goal is located 40° to the left of the sun's position, the dancer shifts the straight run 40° to the left of the vertical, and so forth (5, 6). On the comb, members of the hive move after the dancer and maintain close contact with her, especially during the tail-wagging runs, and take in the information offered. Can they follow it and with what accuracy?

The indication of direction was tested by us using the following method (9). At a certain distance from the hive, a feeding place was installed at which numbered bees were fed on an unscented platform with a sugar solution so dilute that they did not dance in the hive and therefore did not alert forager recruits. Only at the start of the experiment did they receive concentrated sugar solutions slightly scented with (for example) lavender oil. At 50 m closer to the hive, plates baited with the same scent but without food were placed in a fan-shaped arrangement. The number of forager recruits arriving at the plates was an indication of the intensity with which they searched in various directions.

Since such fan experiments proved that indication of direction was successful, we made a step-by-step test of distance-indicating procedures. Here, all scented plates were located in the same direction as the feeding place, from the hive area to a distance well beyond the feeding place. Incoming flights of forager recruits to the feeding site itself were of course not evaluated because here an additional attractant was created by the food and the visiting bees (6).

To sum up, this and preceding experiments taught us that the information on the direction and distance of the goal was adhered to with astonishing accuracy—and not only in gathering nectar and pollen. The same dances are observed on a swarm. Here the scout bees indicate to the waiting bees the location of the domicile they have discovered. Of greatest interest here is that the intensity of the promotional message depends on the quality of the domicile discovered, that the various groups of scouting bees compete with each other, and that therefore the decision is finally made in favor of the best domicile (10).

While not doubting that direction and distance of the goal can be discerned from the tail-wagging dances, a group of American biologists led by A. M. Wenner does not agree that the forager recruits make use of this information. According to them, these bees find the goal by using their olfactory sense only (11). This view is incompatible with many of our results (6, 12). It is refuted by the following experiment, to cite only one.

Numbered bees from an observation hive collected at a feeding place 230 m from the hive. The hive was turned on its side so that the comb surface was horizontal; the sky was screened. Under these conditions, the dancers could orient themselves neither by gravity nor by the sky, and danced confusedly in all directions. Plates with the same scent as that at the feeding place were located at various distances in the direction of the feeding place and in three other directions. They were visited in all directions and in great numbers by forager recruits, with no preferences being given to the direction of the feeding place. The observation hive was now turned back 90° to its normal position so that the dancers could indicate the direction of the goal on the vertical comb surface. Within a few minutes, the stream of newly alerted bees flew out in the direction of the feeding place; the scented plates in this direction were increasingly frequented, and in a short time no forager recruits at all appeared at the scented plates in the three other directions. No change had occurred at the sources of scent in the open field or in the other external conditions. The change in the behavior of the forager recruits could be attributed only to the directional dances.

It is conceivable that some people will not believe such a thing. Personally, I also harbored doubts in the beginning and desired to find out whether the intelligent bees of my observation hive had not perhaps manifested a special behavior. I opened an ordinary hive, lifted up one of the combs and watched the expected dances. Curious as to what would happen, I turned the comb in such a way that the dancing area became horizontal. Gravity as a means of orientation was thus eliminated. However, without any signs of perplexity, the bees continued to dance and by the direction of their tail-wagging runs pointed directly to the feeding place, just as we show the way by raising an arm. When the comb was turned like a record on a turntable, they continued to adjust themselves to their new direction, like the needle of a compass (13). This behavior can be studied at leisure at a horizontal observation hive. It is basically very easy if we recall that the direction of the tail-wagging run relates to the **sun's** position. During the tail-wagging run on the comb, the bee has only to set itself at the same angle to the sun as it maintained during its flight to the feeding place. Afterward, when the recruits set their line of flight at the same angle to the sun, they are flying in the direction of the goal.

This type of discretional indication is nothing unusual. Incoming foragers not infrequently begin to dance facing the sun on the horizontal alighting board of the hive if they are met here by nonworking comrades. The transmission of information through horizontal dancing is easier to understand than that when the angle is transposed to the vertical comb surface. We also seem to have here the original, phylogenetically older type of directional indication. In India there still exist several strains of the species *Apis.* My student and co-worker, Martin Lindauer, went there to use them for "comparative language studies." The small honeybee, *Apis florea,* is on a more primitive level than our honeybee and other Indian strains. The colony builds a single comb out in the open on a branch; the comb has a horizontally extended top edge that serves exclusively as a dancing floor. When these bees are forced onto the vertical comb surface of the side, they cannot render the sun's angle by dancing and their tail-wagging dances become disoriented (14).

Let us now return to our own bees and the observation of dances on a horizontal hive. There can be no doubt that the sun's position is decisive for the direction of their dancing. The sun may be replaced by a lamp in a dark tent. By changing its position, the bees are made to dance in any desired direction. But there was one big puzzle. To prevent excessive heating during most of the experiments, a protective roof was installed over the observation hive. The dancers were unable to see the sun. Nevertheless their dance was usually correct. Orientation by heat rays, by penetrating radiation, as well as other explanations seemed possible and had to be discarded—until I noticed that a view of the blue sky is the same as a view of the sun. When clouds passed over the section of the sky visible to the bees, disoriented dances immediately resulted. Therefore they must have been able to read the sun's position from the blue sky. The direction of vibration of polarized blue light differs in relation to the sun's position across the entire vault of the sky. Thus, to one that is able to perceive the direction of vibration, even a spot of blue sky can disclose the sun's position by its polarization pattern. Are bees endowed with this capacity?

The following test furnished an answer. The observation hive was set horizontally in a dark tent from which the dancers had a lateral view of a small area of blue sky. They danced correctly toward the west where their feeding place was located 200 m away. When a round, rotatable polarizing foil was placed over the comb in a way as not to change the direction of the vibration of the polarized light from that part of the sky, they continued to dance correctly. If, however, I turned the foil right or left, the direction of the bees' dance changed to the right or the left by corresponding angle values. Thus, bees are able to perceive polarized light. The sky, which to our eyes is a uniform blue, is distinctly patterned to them (13,15). They use this extensively and, in their orientation, guide themselves not only by the sun's position but also by the resulting polarization patterns of the blue sky. They also continue to recognize the sun's position after it has set or when it is obscured by a mountain. Once again the bees appear to us miraculous. But it is now clear that ants and other insects, crayfish, spiders, and even octopuses perceive polarized light and use it for orientation, and that among all these animals the human being is the unendowed one, together with many other vertebrates. In one respect, however, bees remain singular: Only they use polarized light not only for their own orientation but also to communicate to their colonies the direction to a distant goal (6).

Thus the language of the bee, which was initially brought to our attention by the physiology of sense perception, has now led us back to it. It also had already led to general questions of orientation in time and space. When bees use the sun as a compass during their own flights as well as to inform their comrades, one difficulty arises: With the advancing hour of the day, the sun's position changes, and one would imagine that it can serve as a geographic marker for a short time only.

I had long contemplated an experiment whose execution was postponed from one year to the next by the feeling that it would not amount to much. However, in the early morning of a fall day in 1949, we sealed the entrance of our observation hive standing in Brunnwinkl on the shore of the Wolfgangsee, transported it across the lake, and placed it 5 km away in a completely different area unknown to the bees (15). Numbered bees from this colony had visited a feeding place 200 m to the west on previous days. From the familiar lakeshore and steep wooded hills they now found themselves in flat meadows; none of the known landmarks could be seen. Four feeding bowls with the same scent as at the former feeding place were placed 200 m from the hive toward the west, east, north, and south, and the entrance was then opened. Of the 29 marked bees that had visited in the west during the previous afternoon in Brunnwinkl, 27 found the bowls within 3 hours: 5 in the south, 1 each in the east and north, but 20 in the west. Each was captured on arrival and was thus unable to send others out by dancing in the hive. Only the sun could have guided those who arrived. It, however, was southeast of the hive, while on the preceding day during the last foraging flights it had been close to the western horizon. Bees possess excellent timing, an inner clock, so to speak. During earlier experiments, by feeding at certain hours only they trained themselves to arrive promptly at the table at that time—even if the table was not set. The above trial, repeated in many modifications (6, 15, 16), has now taught us that they are also familiar with the sun's daily movement and can, by calculating the hour of the day, use this star as a true compass. The same discovery was simultaneously and independently made by Gustav Kramer using birds (6).

During the past few years, an old and persistent question has opened a new field of work for bee researchers. In discussing the direction indication, I initially kept something from you. The dancers did not always point correctly to the food sources. At certain hours they were markedly off to the left or the right. However, no inaccuracies or accidental deviation were involved; the errors were consistent and, when recorded under the same conditions, time and again gave the same curves for a typical daily routine. Thus they could correct, for example, for a different spatial position of the comb. Errors arose only with transposition of the dancing angle; in horizontal dances there is no "incorrect indication of direction." Observations over many years, made jointly with my co-worker Lindauer, finally led us to a conclusion which seemed acceptable (6). However, it was disproved by Lindauer, who persisted in his experiments together with his student H. Martin. They recognized the magnetic field of the earth as a cause for incorrect indication of direction. If this is artificially screened out, the error disappears; and by artificially altering the course of the lines of flux, the incorrect indication of direction was changed correspondingly (17). The idea that the magnetic field might play a role in the puzzling orientation performance of animals was rejected for a long time. During the past years it has been confirmed by new observations, especially in birds and insects (18). Nothing so far points to the possibility that bees, in their purposeful flights cross-country, are making use of the earth's magnetic field. Unexpectedly, however, it proved equally significant biologically but in a different way. When a swarm of bees builds its combs in a hive furnished to them by the beekeeper, their position in space is prescribed by the small suspended wooden frames. In the natural habitat of the bee, perhaps in the hollow of a tree, there are no wooden frames present. Nevertheless, thousands of bees labor together and in the course of one night achieve an orderly structure of parallel combs; the individual animal works here and there without getting instructions from a superintendent. They orient themselves by the earth's magnetic field and uniformly have in mind the comb position which they knew from the parent colony (20). However, these are problems whose solution is fully underway, and we may expect quite a few surprises. By this I do not mean that problems such as the perception of polarized light have been conclusively solved. On the contrary: A question answered usually raises new problems, and it would be presumptuous to assume that an end is ever achieved. It was not possible to present more than just a sketchy illustration in this lecture and to point out a few important steps in the development of our knowledge. To corroborate and extend them requires more time and work than the outsider can imagine. The effort of one individual is not sufficient for this. Helpers presented themselves, and I must express my appreciation to them at this time. If one is fortunate in finding capable students of whom many become permanent co-workers and friends, this is one of the most beautiful fruits of scientific work.

References and Notes

1. K. Von Frisch, *Zool. Jahrb. Abt. Allg. Zool. Physiol. Tiere* 35, 1 (1914–1915).
2. ———, *ibid.* 40, 1 (1923).
3. ———, *Experientia* 2 No. 10 (1946).
4. The threshhold of transition from the round dance to the tail-wagging dance varies with each race of honeybees; according to R. Boch [*Z. Vergl. Physiol. 40, 289 (1957)*], it is about 50 m for *Apis mellifica carnica,* about 30 m for *A. mellifica mellifica* and *A. mellifica intermissa,* about 20 m for *A. mellifica caucasia* and *A. mellifica ligustica,* and 7 m for *A. mellifica fasciala.* The fact that the strain we used mostly in our experiments, the Carniolan bee, has the largest round dance circumference was of benefit in our experiments.
5. K. von Frisch, *Österreich. Zool. Z.* 1, I (1946).
6. ———. *Tanzsprache und Orientienung der Bienen,* Springer-Verlag. Berlin, translation: *The Dance Language Orientation of Bees,* Belknap, Cambridge Further references are found in this book.
7. H. Esch, *Z. Vergl. Physiol.* 45, 1 (1961); A. M. Wenner, *Anim. Behav.* 10, 79.
8. K. von Frisch and R. Jander, *Z Vergl. Physiol.* 40, 239 (1957). 1965) (English, Mass., 1967)]. (1962).
9. I use the word "us," since the open-field experiments had assumed such proportions that they could no longer be carried out without trained assistants.
10. M. Lindauer, *Z Vergl. Physiol.* 37, 263 (1955).

11. A. M. Wenner, *The Bee Language Controversy: An Experience in Science* (Educational Programs Improvement Corp., Boulder, Colo., 1971).

12. K. von Frisch, *Anim. Behav.* 21, 628 (1973).

13. *Naturwissenschaften* 35, 12 (1948): *ibid.,* p. 38.

14. M. Lindauer, *Z. Vergl. Physiol,* 38, 521 (1956).

15. K. von Frisch, *Experientia 6,* 210 (1950).

16. M. Renner, *Z. Vergl. Physiol. 40, 85* (1957); ibid. 42, 449 (1959).

17. M. Lindauer and H. Martin, *ibid.* 60, 219 (1968); M. Lindauer, *Rhein. Westjäl. Akad. Wiss. Rep. No. 218* (1971).

18. H. Martin and M. Lindauer, *Fortschr. Zool.* 21, Nos. 2 and 3 (1973).

See also Communication—*Honeybee Dance Language*

History—*History of Animal Behavior Studies*

Lorenz, Konrad Z.—*Analogy as a Source of Knowledge*

Nobel Prize—*The 1973 Nobel Prize for Medicine or Physiology*

Tinbergen, Nikolaas (1907–1988)

◼ Gorillas
Gorillas/Koko

Studying the behavior of free-living gorillas has its limitations. Access to the gorillas is constrained and the ability of the gorillas to communicate and assist researchers in understanding their behavior is obviously limited.

Therefore the kind of extended research done with the gorilla Koko is uniquely valuable in better understanding the behavior of gorillas and in better understanding the behavioral differences among the great apes which include gorillas, chimpanzees, orangutans, bonobos and humans.

Some approaches to studying and understanding the behavioral differences between humans and nonhuman apes involve looking at DNA or anatomical differences. But one of the most valuable tools for studying these behavioral distinctions is an analysis of the cognitive differences among the great apes. One of the best ways to do that is to work with apes who have the ability to communicate, like Koko.

Koko is a female western lowland gorilla who has learned to communicate with humans using a form of sign language. She knows over a 1,000 signs and often combines them into statements that are like English language sentences. She also understands over 2,000 words of spoken English. Like other great apes who have been raised and educated by humans, Koko's accomplishments challenge the idea that only humans can use language. Koko's interactions with people and animals have also revealed her sensitive, caring personality and have helped to change society's perceptions of gorillas.

Koko was born on July 4, 1971, at the San Francisco Zoo. When she was one year old, Koko was placed in the care of Dr. Francine "Penny" Patterson, a psychologist at Stanford University. Inspired by earlier research done with chimpanzees, Patterson wanted to determine whether a gorilla, also, could learn language. She taught Koko American Sign Language by signing and speaking to Koko during everyday activities. Within just a few weeks, Koko was correctly using the signs for the words *food*, *drink*, and *more*.

Encouraged by these results, Patterson added several hours of more structured lessons to Koko's daily routine. By the time Koko was four-and-a-half years old, she was using more than 75 signs. Additional words in her vocabulary included *toothbrush*, *dog*, *that*, *out*, *necklace*,

Koko the gorilla signs the word "sad" for psychologist Penny Patterson.
© 2003 Ronald H. Cohn/Gorilla Foundation www.koko.org.

open, hurry, pour, and *there*. She also began to put signs together into longer expressions such as *open hurry* or *pour there drink*.

Koko now uses language the same way people do—to make requests, ask questions, or describe what she sees and feels. In terms of understanding gorilla behavior, Koko's ability to express herself has revealed that she has other complex capabilities such as self-awareness, imagination, memories, and emotions. As a youngster, Koko referred to herself as a "fine animal gorilla"; as an adult she has said she is a "fine gorilla person." Koko plays pretend, just like a child, and forms her dolls' hands into signs. Koko grieved over the death of her kitten and regularly tells her friends that she loves them. Koko can also express her feelings and emotions artistically. She paints pictures using appropriate color and form and even titles her paintings using relevant language related to the subject matter.

Despite more than 30 years of experimental work with Koko, some scientists doubt that Koko is actually using language. Instead, they contend that she has been conditioned to perform behaviors that only resemble language. Other researchers believe that while great apes can learn individual words, they cannot use grammar or syntax the way humans do.

Patterson responds that Koko's signing cannot be explained by conditioning because she invents new signs and regularly starts conversations without any prompting or rewards from her teachers. Koko will even "talk" to herself in sign language when she is alone. Koko also creates new combinations of signs to describe new objects, for example, calling a ring a *finger bracelet*.

With regard to grammar, Patterson states that it is misleading to compare the written translations of Koko's signing with English sentences because the structure of sign language is very different from English. More study is needed to determine how apes use grammatical rules.

Koko's conversations with people have given us a unique window into the mind of a gorilla, enabled us to learn more about gorilla behavior, and to conclude that gorillas are much more like us than we once believed.

See also Cognition—*Consciousness*
Cognition—*Talking Chimpanzees*
Cognition—*Theory of Mind*

Further Resources

Gardner, R. A. & Gardner, B. T. 1969. *Teaching Language to a Chimpanzee*. Science, 165, 644–672.

Patterson, F. 1978a. *Conversations with a Gorilla*. National Geographic, (154)4, 438–465.

Patterson, F. 1978b. *The Gestures of a Gorilla: Language Acquisition in Another Pongid*. Brain and Language, 5, 72–97.

Patterson, F. & Linden, E. 1981. *The Education of Koko*. New York: Holt, Rinehart and Winston.

Patterson, F. 1986. *The mind of the gorilla: Conversation and conservation*. In: *Primates: The Road to Self-Sustaining Populations* (Ed. by K. Benirschke), 933–947. New York: Springer-Verlag.

Patterson, F. & Gordon, W. 2001. *Twenty-seven years of project Koko and Michael*. In: *All Apes Great and Small*, Volume I (Ed. by N. E. Briggs, B. M. F. Galdikas, J. Goodall, G. L. Shapiro, & L. K. Sheeran), pp. 165–176. New York: Kluwer.

Vessels, J. 1985. *Koko's Kitten*. National Geographic, (167)1, 110–113.

http://www.koko.org
A Web site sponsored by the Gorilla Foundation with information about Koko and the Maui Ape preserve.

Penny Patterson

■ Griffin, Donald Redfield (1915–2003)
Windows on the World

Born in Southampton, New York, Donald Griffin grew up in Massachusetts, where he honed his powers of observation on the birds and small mammals of his Cape Cod home. He prepped at Phillips Andover, and earned both his undergraduate and Ph.D. degrees at Harvard. He was on the faculty at Cornell, then Harvard, and finally at Rockefeller University. After his nominal retirement, Griffin was a visiting professor at Princeton, and then again at Harvard. He died in Lexington, MA, in November 2003.

While still an undergraduate at Harvard, Griffin discovered that two common species of bats emit ultrasonic pulses of sound. Then in 1939 he and a fellow student, Robert Galambos, showed that these bats use the returning echoes to navigate. (The delay between emission of the cry and the return of its echo reveals the distance to the target.) Griffin proceeded to demonstrate that these species of bats also hunt flying insect prey with same sonar system, and coined the term "echolocation" to describe the strategy. We know now that at least shrews, oilbirds, and some marine mammals also employ echolocation in navigation and/or prey detection, often with pulses audible to human ears.

These unexpected findings profoundly changed Griffin's perspective on animal sensory experience—changed it in ways that led him in time to incite a revolution in the study of animal behavior. When Griffin first discovered the ultrasonic vocalizations of insectivorous bats, he thought they must be some sort of communication system. While he had toyed with the idea that echoes from the pulses could be useful in navigation, the notion was so new and beyond the scope of known behavior that he found it impossible to stir up any interest among the biologists on the Harvard faculty. Griffin's rueful remark in his classic *Listening in the Dark* (1958) says it all: "Excessive caution can sometimes lead one as far astray as rash enthusiasm." The rest of the career of this modest and unassuming man was in large part devoted to demolishing the conventional strictures that fettered the study of animal behavior. He argued for such radical ideas as transcending parsimony when dealing with behavior and imagining the impossible. Then he challenged us to put the resulting theories to rigorous testing.

Cognitive ethologist Donald Griffin at the Concord Field Station of Harvard University in 2001.
© *Ed Quinn / Corbis.*

Other scientists took up the torch. Over the next few decades, researchers inspired by Griffin's discoveries of a new sense—a "sensory window" as he called it—found species that could detect patterns of polarized light, infrared wavelengths, the ultraviolet, infrasonic sounds, electric fields, the earth's magnetic field, atmospheric pressure, relative humidity, carbon dioxide concentrations, and so on. His emphasis on the physiological bases of behavior led to groundbreaking work on the neural pathways and specialized processing areas that systematically refine raw sensory experience into a mental "reality" animals can use—processing that produces localized activity in the bat brain indistinguishable from that generated by their visual experience. Bats, it would seem, do "see" in the dark. The importance of highly evolved circuits for data analysis was borne in upon him when he set about developing echolocation devices for blind humans. Despite a brain volume hundreds of times larger than that of bats, people are barely able to learn to use echoes.

Griffin's work on bats was put on hold after Pearl Harbor. During World War II, Griffin worked on a variety of projects for the Department of the Army. One involved devising better microphones and headsets for use in the incredibly noisy environment of tanks. The technology had hardly changed in 25 years, so that the existing equipment produced highly distorted sounds limited to a narrow range of frequencies. Like so many scientists involved in war research, Griffin learned much about the stubborn inertia of the military and their disdain for civilian experts: What had been good enough for General Pershing's troops was good enough now—period. Many harrowing hours in maneuvering tanks gave Griffin enough data to get his point across, and the Army installed the higher-fidelity equipment he had designed.

His next project for the Army encountered fewer obstacles, probably because there was nothing in stock to replace. Griffin worked on designing clothing for extreme cold weather, including electrically heated flight suits for air crews. This took him to Alaska, where he developed a life-long interest in arctic birds, most of which are long-distance migrants that fly thousands of miles each spring to feast on the abundant insect life flourishing during the few summer months of very long days of the Arctic. He performed a classic (and presumably unauthorized) study on the energetics of summering robins in Umiat, Alaska, which he then compared to similar data from Ohio. This kind of comparison—the essence of the field later to be called behavioral ecology—showed how natural selection has acted to alter the natural history (especially the clutch size) of a species to optimize its reproductive success. Griffin also participated in developing a night vision device designed to see in the infrared.

After the war, Griffin divided his time between studying bats and investigating homing and migration in birds. He found two general types of echolocation pulses: Some species emit long steady ultrasonic tones, whereas others produce click-like sounds which sweep down from higher to lower frequencies. Other species use a hybrid form of these calls. The *constant-frequency* signals are common among species that navigate at night but do not catch insects; this group includes fruit bats, pollinating bats, and vampire bats. The *frequency-modulated* cries are characteristic of insect-catching bats, and are now known to provide information not only on the distance and direction of the target, but on its rate of movement (indicated by the Doppler shift of the echo), wingbeat frequency, size, and texture. Griffin even found fishing bats that use the echoes off the surface ripples produced by shallow-swimming fish: Targeting the center of the disturbance, they hook the prey animal with claws dropped below the surface at just the right moment.

Griffin's war work brought him into contact with the engineers involved in the research and development of radar. The secrecy surrounding radar persisted well into the Cold War, and thus the problems with "false" echoes (often called "angels") were known only to a few biologists. Griffin's wide understanding of bird behavior suggested to him that many of the most puzzling angels were, in fact, flocks of starlings, while others were migrating birds. Although dismissed as absurd at first, this conjecture proved exactly correct. Thus Griffin began a field now called *radar ornithology*. Radar allows researchers to track birds at night (when most species migrate), and yields the kind of exact data on altitude, air speed, and direction that had previously been unavailable. Later in his career, Griffin was able to show that night-migrating birds can continue to fly accurately in or between layers of clouds—an observation that implicated the earth's magnetic field as a "backup" system by migrants.

In his lucid summary of the field—*Bird Migration* (1964)—Griffin describes how he put the first reality into the mystery of homing. Researchers had tried displacing a variety of birds hundreds or thousands of miles from home, and had reported that the birds returned to their nests. When a skeptical Griffin examined the data, however, he discovered that most of the kidnapped birds never returned at all, and those that did, got back only very slowly. He calculated that these data were consistent with a relentless bird searching at random. Focussing on two key species, Griffin repeated the tests with much more rigorous controls. The herring gulls he took from Cape Cod to locations as remote as Lake Michigan and the coast of Georgia homed determinedly, as did the Manx shearwaters he moved as much as several thousand miles. This implied a "map" sense wholly alien to human experience.

To discover if homing (outside of the relatively fast and reliable performance of pigeons) was real, Griffin decided that he would have to actually track the birds in flight—so he learned to fly. In his most famous tests, he moved gannets (large white birds easily tracked from the air) from their nests near the Gaspé Peninsula on the Atlantic coast of Quebec. The majority were released at the airport in Caribou, Maine. The birds were slow to return home, mainly because they seemed to enjoy hours soaring in updrafts. Quite incidentally, therefore, Griffin (equipped with a motion picture camera in his plane) was enabled to make some of the first major discoveries about soaring, a strikingly effective trick for reducing flight costs.

The flight paths of the gannets and gulls Griffin tracked were nothing like the beeline routes most researchers imagined. Instead, though a high proportion of his birds returned, they spent the first day wandering about Maine and Quebec more or less aimlessly—at least until Griffin's plane ran low on fuel. When he followed homing pigeons from the air, on the other hand, he found a much reduced tendency to wander, and more initial orientation toward home; nevertheless, there was still a surprising amount of "noise" in a species bred to this one task. Through these tracking studies, however, he was able to demonstrate that homing pigeons must have a map sense.

Returning to the shearwaters, Griffin found that their initial orientation after movement over long distances was as accurate as that of homing pigeons. And the initial orientation of the gulls, gannets, and the terns he later tested was, upon further inspection, also fairly accurate. Though these birds have a sense of where they are relative to home after being displaced (often in darkness), and set off in roughly the right direction, they seem rather poorly motivated to return, and easily distracted from the task researchers want to study. During migration season, on the other hand, they fly fast, far, and accurately. Thanks to Griffin's groundwork, we now have a nearly complete picture of the variety of sensory cues used in bird migration, and their constant recalibration as the animals move across different latitudes.

After the War, Griffin learned that the famous Austrian zoologist, Karl von Frisch, had made an astonishing discovery. While under house arrest at his family home, von Frisch had found that forager honeybees have an abstract language for specifying the distance, direction, and quality of food sources they have discovered. At the time the idea that an insect might have the second-most complex language on the planet (a distant second to our own, to be sure) was literally incredible. But to the discoverer of the equally incredible phenomenon of echolocation—and a naturalist with a long-standing interest in bees—Griffin took it upon himself to bring von Frisch from a shattered Europe to the United States for an extended series of lectures. It is fair to say that without Griffin and William Thorpe (at Cambridge, who translated the first bee-language paper into English), it would have been years before the English-speaking world knew of von Frisch's discoveries.

This trip was instrumental in inducing the Rockefeller Foundation to provide long-term funding for von Frisch's experiments in postwar Germany, where there was essentially no money for science. It was also on this trip that Griffin introduced von Frisch to Edmund Land, who had found in one set of tests described in the lecture a strong hint that bees might orient to polarized light in the sky. Land provided von Frisch with his latest invention: a UV-transparent polarizing filter. With this filter von Frisch proved Land's surmise correct: Like the bats, bees were navigating with a sense humans are blind to. When Griffin tried to arrange for von Frisch's lectures to be published as a short book, he encountered the kind of blinding skepticism he had been fighting for a decade. Experts in bee behavior labeled the work as the "wild ravings" of a shell-shocked professor. (The book was eventually published by Harvard Press.)

Griffin's interest in honeybees deepened. Beyond the wealth of new sensory channels von Frisch used the dance to demonstrate and study, Griffin wondered whether the language could provide a look into the mind of the insect itself. He encouraged and sponsored work that led to a nearly complete understanding of how honeybees use navigation cues, learn to recognize specific species of flowers, make cost/benefit decisions about foraging and choosing a new nest cavity, manipulate pictures in their mind, and plan novel routes. Thanks to his inspiration, the behavior of the honeybee is better understood than that of any other insect. One of the major insights Griffin's prompting led to was the confirmation of Konrad Lorenz's intriguing, but largely unsupported, speculation that most learning is innately directed, the product of natural selection. This idea, so readily demonstrated in the experimentally convenient honeybee, has proven the best guide to understanding most learning in birds and mammals as well. Like most great ideas, in retrospect it seems obvious.

Honeybees, then, are astonishingly complex. Their complexity is evident to us only because they are willing to be trained to a food source, can be induced to live in a transparent hive, and often dance upon their return; presumably other insects needing similar abilities have an equally remarkable set of mental abilities. But this view of bees contradicts the dogma that the simplest explanation for behavior should be taken as the most likely. A reaction to the wildly speculative accounts of early naturalists, the Law of Parsimony had hit the study of behavior particularly hard. When a horse named Clever Hans, whose reading, computational, and philosophical skills drew researchers from around the world, was proved to be responding to cues provided by his unwitting questioners, the stage was set for a reductionistic backlash.

That the best scientists of Europe had been convinced of Hans' abilities led to an extreme response: Animals henceforth were only to be studied under highly controlled conditions (laboratories) and in tests without humans present to cue the responses. No higher mental processes were to be inferred if any alternative explanation was possible; in time, researchers generally stopped asking such questions at all. Behaviorists (the dominant psychological school) considered animals to be instinct-free learning machines with no cognitive capacity; ethologists (the main biological approach) thought of animals basically as instinct-driven robots, again with no cognitive capacity. Behaviorists and ethologists disagreed about nearly everything *except* that animals had no true mental faculties.

In 1976 Griffin stepped into this intellectual minefield with an astonishing book, *The Question of Animal Awareness*. In this thin, but deeply subversive volume, he suggested that the principle of evolutionary continuity actually made explanations based on simple kinds of thinking, less strained and complex than the elaborate mind-free accounts of psychology and animal behavior. Not surprisingly, the reaction of Griffin's colleagues was hostile. Griffin responded with thought-provoking articles suggesting that animals were capable of creating mental maps, planning, forming concepts, recognizing their images in mirrors, making and using tools. He illustrated his arguments with examples drawn from an increasing body of data from well-designed and well-controlled experiments. Within a few years, the tide of opinion had turned to a comparative trickle of skepticism.

Experiments inspired by Griffin's daring and relentless attack on established wisdom added more and more evidence for mental activity in animals. While there remain numerous doubters, the majority of biologists and psychologists are now convinced that at least some animals—perhaps as phylogenetically remote as honeybees and hunting spiders—process information in ways we describe in terms of thinking when they occur in our own minds. (Whether we are correct in this introspective analysis is another question.) Another result has been a major effort at redesigning zoos so that the animals do not become bored and lapse into psychotic repetitive behaviors; instead, they are given toys and games and other challenges to deal with.

Griffin's four pioneering books on animal mentality established him as the father of the new and intensely active field of *cognitive ethology*. Like much of his previous work, they demonstrate how one courageous, energetic, and well-informed scientist can change his field, or start a new one. Impatient with sloppy thinking or inelegant experimental design, Griffin endeavored to strip the blinders from dogmatic researchers across disciplines. He took on biologists, psychologists, linguists, statisticians, and philosophers, challenging them to look at the evidence, think outside the box, and test their theories fearlessly. His personal and professional contributions have opened the eyes and expanded the horizons of all his fellow animals.

See also Cognition—*Cognitive Ethology: The Comparative Study of Animal Minds*
Communication—Auditory—*Bat Sonar*
History—*History of Animal Behavior Studies*

James L. Gould

Hamilton, William D., III
(1936-2000)

William David Hamilton, the last of six children, born in 1936 to Archibald and Bettina Hamilton, grew up in a small cottage in little English town called Sevenoaks. From early childhood on, Hamilton was an avid naturalist and insect collector, spending the vast majority of his youth out in the woods surrounding the family cottage. Bill first learned about evolutionary biology from his mother Bettina, as they walked the four short miles between their house and Darwin's "Down House."

Once Hamilton understood the way that natural selection worked, he was hooked, and he saw everything—including behavior—through this lens. This is nicely illustrated in a wonderful story that Bill's sister, Mary Bliss, shared with me. When he was a teenager, Hamilton lined up his siblings and snapped a photo. From the photo he measured the height of each sibling, and then he examined whether the distribution of heights matched the normal distribution that Charles Darwin's cousin, Francis Galton, had predicted for complex traits like height.

Hamilton spent 1956–1957 serving mandatory national service in the British military, where he acted as a recruiter for the British Army's Royal Corp of Engineers. During this period, Hamilton kept a journal of his daily activities, and what strikes the reader of these journals most is the virtual absence of a discussion of anything military. Instead, Hamilton's journals are filled with wonderful observations of natural history—particularly insect natural history. Even when he was in national service, Hamilton used every free minute he had to pop off to the woods and study nature.

After his military service, Bill Hamilton entered Cambridge University as an undergraduate. It was during his days at Cambridge that Hamilton came across Sir Ronald Fisher's book, *The Genetical Theory of Natural Selection*. Fisher championed Darwin's ideas and spoke at length both about genetics and about how natural selection can operate on behavioral traits. Hamilton became a self-admitted "Fisher freak," and immersed himself in Fisher's book. Partly as a result of Fisher's work, Hamilton became obsessed with understanding the evolution of altruistic behavior.

Altruistic behaviors are defined as those that are costly to the individuals undertaking them, but provide a benefit to others. For example, when a honeybee stings an intruder in defense of its nest and dies as a result of her action, we are looking at altruism. But how could natural selection have ever favored such behaviors? Surely, any behavior that helps others, but harms self should be eliminated—and fast—by the process of natural selection. Yet, Bill Hamilton, as well as many other naturalists, knew that altruism was a real phenomenon, as they had seen over and over again in nature.

Darwin first raised the riddle of the evolution of altruism in *The Origin of Species*, but it was not until Bill Hamilton's Ph.D. dissertation work (done jointly at University College London and The London School of Economics) that a true theory for the evolution of altruism emerged. What Hamilton demonstrated, in a series of eloquent mathematical models, was that altruism can evolve *if* altruists help their blood kin. The reason is that blood kin are, by definition, genetically similar to the altruist, so by saving their kin, altruists are in effect saving copies of their own genes—copies that just so happen to reside in the bodies of their blood relatives. Translated in the cold language of natural selection, relatives are worth helping in direct proportion to their genetic (blood) relatedness.

Hamilton's work demonstrated that altruism evolves via blood kinship when the following condition is met: $rb \geq c$. In this little equation, r represents the genetic relatedness of two individuals, c stands for the cost of an altruistic act that one of them might undertake to aid the other, and b equals the benefit that a recipient of such altruism receives. The actual mathematics in Hamilton's paper are far more complicated than this but, remarkably, it all simplifies to *relatedness* times *benefit* must be equal to or greater than *the cost of an altruistic act*. The equation $rb \geq c$ is known as "Hamilton's Rule," and it has revolutionized the study of evolution and behavior. After Hamilton's work, hundreds of behaviors that seemed to be evolutionary paradoxes suddenly fit nicely into an evolutionary framework. If an evolutionary biologist today observes cooperation, altruism, caregiving, protection from predators, babysitting and so on, his gut response, based on a knowledge of Hamilton's Rule, is to try to learn if the individuals involved are blood kin—and they often are.

If Bill Hamilton had done nothing but explain the evolution of altruism between relatives, he would have been regarded as one of the most important evolutionary biologists of his generation, but he refused to sit on his laurels. After moving to the University of Michigan in 1977, he proceeded to examine the evolution of altruism among *unrelated* individuals, and, in collaboration with political scientist Robert Axelrod, came up with another path-breaking discovery. If unrelated individuals *reciprocated* acts of goodness, than altruism could evolve without any link to genetic relatedness.

In 1980, after being awarded a prestigious Royal Society Professorship at Oxford University, Hamilton returned to England. This was a no-strings-attached research position reserved for the very best minds in science, and Hamilton took advantage of the opportunity by publishing new ideas on everything from the evolution of sex and the role of parasites in evolution, to the reason some trees become more colorful than others each fall. With respect to autumn color patterns, Hamilton and his colleague, Steve Brown, hypothesized that bright coloration served as a signal to warn insect pests that the tree they are considering for attack is in fact quiet healthy and not worth the effort. This hypothesis—and the data that support it—nicely illustrate Bill Hamilton's ever-engaged mind, his extraordinary creativity, and his passion for explaining the natural world.

On March 7, 2000, 63-year-old William D. Hamilton died. Hamilton had caught malaria while in the Congo trying to determine once and for all whether HIV had initially spread from other primates to humans. Following his usual modus operandi, Hamilton wanted to study the question first hand, in nature, before applying his mathematical skills—even if it meant, as it did in this case, collecting chimpanzee feces. He felt that the issue of HIV's origin needed resolution, and that this would require the keen eye of an evolutionary biologist. Unfortunately, before he was able to make much progress, he was struck with malaria and forced to return to England, where, after weeks in a coma, he passed away.

Above and beyond his intellect, one of the most endearing features that Bill Hamilton possessed was his gentle nature. Everyone knew that his work stood alone, high above that which others have accomplished, and yet Hamilton was the antithesis of arrogance. I was lucky enough to have breakfast with Bill on one of my trips up to Harvard University. To say that I was nervous before we met is the understatement of the year. Yet once we sat down over coffee, I realized that I was chatting with one of the nicest, most sincere and humble people I could ever hope to meet.

Further Resources

Axelrod, R. & Hamilton, W. D. 1981. *The evolution of cooperation*. Science, 211, 1390–1396.

Hamilton, W. D. 1963. *The evolution of altruistic behavior*. The American Naturalist, 97, 354–356.

Hamilton, W. D. 1964. *The genetical evolution of social behaviour. I and II*. Journal of Theoretical Biology. 7, 1–52.

Hamilton, W. D. 1996. *Narrow Roads of Gene Land: The Collected Papers of W. D. Hamilton*. Volume 1. Oxford: W.H. Freeman.

Hamilton, W. D. 2001. *Narrow Roads of Gene Land: The Collected Papers of W. D. Hamilton*. Volume 2. Oxford: Oxford University Press.

Hamilton, W. D. & Brown, S. 2001. *Autumn tree colors as a handicap signal*. Proceeding of the Royal Society of London, 268, 1489–1993.

Lee Alan Dugatkin

■ | Hibernation

Animals that live in cold climates must solve the problem of staying alive during the winter when air temperatures are below freezing and food supplies are limited. Staying warm and finding food is a particular problem for birds and mammals because, like humans, they are homeothermic endotherms that maintain a high body temperature through metabolic generation of heat. Birds and mammals must eat frequently to meet the energy costs of heat generation, but finding adequate food is often difficult in winter. Some animals solve the problem of staying warm and finding food by migrating to a milder climate for the winter, but this option is limited to animals that can fly. In northern temperate zones, most birds and some bats migrate south, often traveling thousands of miles—as far as South America, Africa, and southern Asia—to avoid winter. Another solution used by birds and mammals is growth of a thick covering of insulation that keeps the animal warm while it searches for food. Arctic animals like snowy owls and snowshoe hares that remain active through the winter depend on feathers and fur for insulation. Mammals that neither migrate nor remain active use a third strategy to survive the winter; they hibernate.

The types of mammals that are most likely to hibernate are bears, some bats, and some rodents. Among rodents, none of the tree-dwelling squirrels hibernate, but most of the ground-dwelling squirrels do hibernate. In North America, over 30 species of chipmunks (*Tamias*), ground squirrels (*Spermophilus*), prairie dogs (*Cynomys*), and marmots (*Marmota*) spend many months in hibernation each year. Some other rodents that hibernate are jumping mice (*Zapus*) and pocket mice (*Perognathus*) in North America and dormice (*Glis*) and hamsters (*Cricetus*) in Europe. Hibernation is very rare in birds, but it has been reported for poorwills (*Phalaenoptilus*).

Hibernation is an ecological term that describes the prolonged period of time when the animal is inactive in a safe place such as a den, a cave, or an underground burrow. Mammals spend most of hibernation in the physiological state of *torpor*, a dormant condition in which body temperature is cool and body functions such as metabolism, heart rate, and breathing are slowed down. Mammals prepare for hibernation by storing a large amount of energy either as fat on the body or as seeds cached in the hibernation site. Some rodents eat so much food before they enter hibernation that 40–50% of their body weight is fat. Most of this fat is white, but hibernating mammals also store small amounts of brown fat, also known as brown adipose tissue (BAT).

In small hibernating mammals like bats and ground squirrels, body temperature during each torpor bout approximates that of the air in the cave or the soil surrounding the burrow. At the start of the hibernation season, body temperature may only decline to 15°C (59°F), but by the end of winter, when the cave or burrow is very cold, body temperature can go as low as –1°C (30.2°F) before the animal is in danger of freezing. If the environmental temperature drops even lower, the hibernating mammal maintains its body temperature near –1°C so that the tissues never freeze. The ability of dormant mammals to keep the body at a constant low temperature when the environmental temperature is well below freezing indicates that body temperature continues to be regulated by physiological means during torpor.

Small mammals do not remain cold continuously throughout hibernation. Instead, hibernation consists of a series of torpor bouts that last for days or weeks, interrupted by brief periods of less than a day when the animal rewarms back to normal mammalian body temperature of 37°C (98.6°F). During these brief periods of warmth, the animal usually remains in the cave or burrow. Most hibernators do not eat between torpor bouts, but a few small species like chipmunks store most of their energy as seeds rather than fat and they eat seeds during the intertorpor periods. Small mammals that spend up to 8–9 months of every year in hibernation experience dozens of torpor bouts and rewarming events in a single hibernation season. Hibernation in large mammals like bears differs from hibernation in bats and rodents. Large hibernators become inactive in winter but their body temperature drops only slightly, to about 32°C (89.6°F), and they do not go through alternating bouts of torpor and rewarming.

When small hibernators rewarm between torpor bouts, a large amount of body heat must be produced in a period of just 1–2 hours. Brown fat has special properties that allow animals to rewarm rapidly. Whereas white fat is stored all over the body, brown fat is found only in a few locations near the heart and between the shoulder blades. Brown fat looks dark because it contains many cellular organelles called mitochondria, and because it has a rich supply of blood vessels. When an animal is ready to rewarm from a torpor bout, the mitochondria in brown fat cells produce heat that is then carried by the blood to warm up the brain and other body organs. Heat production by metabolic means in cells is called *nonshivering thermogenesis*. Mammals also shiver during the rewarming process to produce additional heat in the muscles from shivering thermogenesis.

Many reasons have been proposed for why small hibernating mammals warm up between torpor bouts during hibernation. Some, such as the need to urinate or the need to exercise the muscles, are not supported by watching what captive animals do. In fact, ground squirrels mostly sleep during the intertorpor periods and often do not leave the nest. The observation that the longest torpor bouts occur late in winter when body temperature is coolest has led to the hypothesis that the colder the animal, the lower its metabolic rate and the less frequently it needs to rewarm. Experiments with ground squirrels support this relationship between body temperature, metabolism, and rewarming, perhaps indicating that some metabolic processes necessary for detoxification of metabolic wastes work best when the mammal is warm.

Although hibernation is primarily a strategy to avoid the cold winter, many hibernating mammals enter hibernation long before the cold weather arrives. For example, adult Richardson's ground squirrels (*Spermophilus richardsonii*) living on the prairies in Canada and the northern United States hibernate for 8–9 months, from July to March. Because underground soil temperature even in July is less than 15°C (59°F), ground squirrels conserve a lot of energy by starting hibernation in summer. Young Richardson's ground squirrels do not begin hibernation until August or September because first they must grow before they can store the fat necessary to survive hibernation.

When an animal is in the physiological state of torpor it is very vulnerable to predation because it cannot escape or defend itself. Hibernators minimize the risk of predation by seeking out remote locations, such as deep within caves for flying mammals like bats or in underground dens. Rodents create a safe place to hibernate by excavating a chamber known as a *hibernaculum* that is usually 50–100 cm (19–39 in) underground. When Richardson's ground squirrels excavate the hibernaculum, the chamber initially connects to the rest of the burrow system. When the animal is ready to begin hibernating, it blocks off all the interconnections by plugging up tunnels with soil so predators such as weasels cannot find the hibernaculum. Once the ground squirrel has enclosed itself in the hibernation system, it does not leave until spring, so it must provision the hibernaculum with everything needed for the next 8 months. For female ground squirrels, all they need is a large amount of dry grass to form a spherical nest in which to hibernate. Males need grass bedding too, but they also store seeds underneath the grass nest. Male Richardson's ground squirrels gather those seeds 4–6 weeks before they enter hibernation. They do not eat the seeds in the intervals between torpor bouts, but save them up to eat after the last torpor bout. The stored seeds then enable males to regain some of the weight they lost during hibernation so that they are better prepared for the intense fighting associated with finding mates in the upcoming mating season.

Because spring conditions vary from year to year, hibernators need to be able to adjust the end of hibernation to coincide with improving conditions. The method by which animals that hibernate underground know when aboveground conditions are appropriate requires further study, but one theory is that ground squirrels investigate conditions in a special tunnel that dead ends near the surface. Intertorpor rewarming occurs more frequently as spring approaches, increasing the number of opportunities for animals to investigate conditions near the surface. When ready to resume activity in spring, the ground squirrel digs a chimney-like tunnel up to the surface.

Ground squirrels hibernate alone. Although each animal excavates its own hibernaculum, family members often construct this hibernaculum as an offshoot of the same burrow system. Some marmots are thought to hibernate communally, with all family members sharing the same hibernaculum. Mother bears hibernate with their unweaned offspring. Bats often huddle together in caves, sometimes forming large aggregations when hibernating.

Timing of reproduction relative to timing of hibernation varies among mammals. Bears mate before hibernation and the infant is born and suckled during the period of winter dormancy. Some bats mate before or during the hibernation season, but embryo development does not begin until spring. Ground-dwelling squirrels mate as soon as hibernation ends in spring. In all mammalian species, infants must then grow and fatten in time to be ready for the next hibernation season.

Hibernating mammals are not the only homeotherms capable of torpor. Many small mammals and birds that weigh less than 50 g (1.75 oz) use torpor on a daily basis to conserve energy during the part of the day when they are not feeding. Nocturnal mammals like mice and bats go into torpor during the daytime, whereas diurnal birds like hummingbirds and chickadees go into torpor at night. Torpor in these animals can occur whenever conditions are chilly, not just in wintertime and not just in extreme latitudes. Tiny hummingbirds that live in the tropics go into torpor overnight at any time of year when conditions are too cool for them to maintain a high body temperature. Torpor bouts last only a few hours during the normal rest period; the animal then warms up in time to begin foraging again the next day. The major difference between mammals and birds that go into daily torpor and hibernating mammals is the duration of the torpor bouts. Hibernators stay continuously in torpor for many days, sometimes as long as 3–4 weeks, so they practice seasonal torpor rather than daily torpor.

All animals that live in cold climates must cope with winter, but only birds and mammals face the problem that their normal body temperature is much higher than winter temperatures. Reptiles and amphibians also hibernate in a safe place as a means to escape harsh winter conditions, but these animals are ectotherms with a very limited ability to generate body heat internally. Consequently, hibernating reptiles and amphibians stay continuously cool for as long as environmental conditions stay cool. In extreme cases, some reptiles and amphibians actually freeze in winter, but the freezing process is highly regulated and confined just to the water that is outside the cells, such as the fluids in the body cavity. Some insects that withstand exposure to incredibly cold conditions likewise survive by regulating the freezing process in their bodies, whereas others survive by generating antifreeze chemicals in their bodies. Collectively, animals have solved the problem of coping with cold conditions in many different ways, but only certain small mammals and one bird species use a form of hibernation that is characterized by torpor bouts interspersed with brief periods of rewarming to normal body temperature.

See also Behavioral Physiology—*Turtle Behavior and Physiology*

Further Resources

Barnes, B. M. 1989. *Freeze avoidance in a mammal: Body temperatures below 0°C in an arctic hibernator.* Science, 244, 1593–1595.
Davenport, J. 1992. *Animal Life at Low Temperature.* New York: Chapman & Hall.
French, A. R. 1988. *The patterns of mammalian hibernation.* American Scientist, 76, 569–575.
McKechnie, A. E. & Lovegrove, B. G. 2002. *Avian facultative hypothermic responses: A review.* Condor, 104, 705–724.
Michener, G. R. 1992. *Sexual differences in over-winter torpor patterns of Richardson's ground squirrels in natural hibernacula.* Oecologia, 89, 397–406.
Michener, G. R. 2002. *Seasonal use of subterranean sleep and hibernation sites by adult female Richardson's ground squirrels.* Journal of Mammalogy, 83, 999–1012.
Storey, K. B. & Storey, J. M. 1990. *Frozen and alive.* Scientific American, 263, 92–97.

Gail R. Michener

■ History
History of Animal Behavior Studies

The study of animal behavior has a rich history reaching from the drawings of the early cave dwellers to the sophisticated methods and theories of the twenty-first century. About 200 decorated caves from the Upper Paleolithic period survive and reveal an interest of early humans in animals, such as horses, bison, and oxen, and their habits. The history moved through the early uses of animals in hunting, as pets, as food, and in folk tales.

A number of ancient Greek and Roman thinkers, including Herodotus, Aristotle, and Pliny the Elder had well-reasoned views concerning animal behavior. Aristotle believed the study of all species to be of value and provided a wealth of information concerning sensory function, and social, reproductive, and other behavioral patterns. He was a strong believer in the value of careful observation. Although he believed species to be fixed, not evolving, he thought that the various species lay along a continuous scale—that there were no gaps such as that often suggested between humans and other species. He suggested the existence of a *Scala naturae*, or great chain of being, along which species could be ranked along a single scale according to the degree of their perfection and the complexity of their structure and function—an idea no longer widely accepted by scientists today.

Although many animal behaviorists would date the modern study of animal behavior as beginning with the work of Charles Darwin in the nineteenth century, there was much relevant interest in animal behavior before him. For one thing, there were many European naturalists, including Frederick II of Hohenstaufen, John Ray, Gilbert White, Charles-Georges Le Roy, Georges Louis Leclerc Comte de Buffon, and Hermann Samuel Reimarus, who made careful observations on such topics as bird song, nesting habits, migration, and feeding. The emphasis was upon natural behavior that could be observed in the field; topics included animal feelings, consciousness, and choice, often framed within a religious context. Humans also were interested in animals for practical reasons, as in controlling the behavior of domesticated species and in falconry. The control over behavior developed by the early falconers who trained their birds to hunt for them was extraordinary. Finally, philosophers had much to say that was relevant to the study of animal behavior. René Descartes became known for advocating a great divide between humans and other species with only humans possessing the rational soul and other species as automata able to carry on simple mental functions, but unable to think or have language. The precursors of the nature–nurture debate that would separate English- and German-speaking scientists in the twentieth century can be seen in the writings of philosophers such as John Locke, who believed the mind to be a blank slate, and Immanuel Kant, who believed in an active mind innately possessed of certain propensities.

It was Charles Darwin who set the occasion for the development of the field of animal behavior studies as we know it today. Although the notion that species were continuous was not new with Darwin, it was his work that made the idea more acceptable and laid the foundation for a comparative science of behavior. Darwin suggested that evolution works through a process of natural selection favoring organisms able to reproduce competitively within the environment in which they lived and not according to some predetermined plan. He proposed notions of sexual selection—the

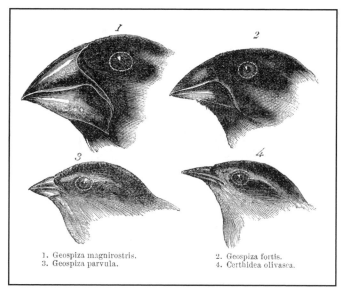

1. Geospiza magnirostris. 2. Geospiza fortis.
3. Geospiza parvula. 4. Certhidea olivasca.

Four of the species of finch observed by Darwin on the Galapagos Islands, showing variation in beak size and shape. The study of the flora of the Islands contributed to Darwin's theory of evolution.
HIP/Scala / Art Resource, NY.

competition among individuals for reproductive opportunities. Further, Darwin himself studied animal behavior in a comparative framework as with his 1872 book *The Expression of the Emotions in Man and Animals*.

Darwin was followed in England by several scientists who developed the study of animal behavior in significant ways. Douglas A. Spalding emphasized the development of behavior and the role of genetic and environmental influences—nature and nurture. He found birds reared in small enclosures able to fly when first released and young chicks reared in isolation to respond to the call of a hen they had never seen or heard. Spalding studied imprinting, the tendency of young *precocial birds* (birds who are capable of a high degree of independent activity from birth) to follow the first moving object they encounter and to become preferentially attached to that object. Darwin's neighbor, Sir John Lubbock made important methodological contributions as he studied maze learning and problem solving in insects. George John Romanes was Darwin's protégé who studied a wide range of species. Although Romanes was a competent scientist, as exemplified in his 1885 book on *Jelly-Fish, Star-Fish, and Sea Urchins*, he is often caricatured for his uncritical use of anecdotes and his anthropomorphic writing style in his 1882 book, *Animal Intelligence*—perhaps the first textbook in comparative psychology.

C. Lloyd Morgan, both a friend and a critic of Romanes, conducted a series of simple experiments on learning, instinct, and imitation in a variety of species. He is best know for authoring what has become known as "Morgan's canon" in his 1894 *Introduction to Comparative Psychology*: "In no case may we interpret an action as the outcome of the exercise of a higher psychical faculty, if it can be interpreted as the outcome of the exercise of one which stands lower in the psychological scale" (p. 53). Animal behaviorists are still debating the meaning of this statement. No meaningful "psychological scale" has ever been widely accepted. It is clear, however, that Morgan did not intend to exclude the possibility that animals might be conscious or possess some of the complex processes known in humans. Rather, he wanted to insist that alternative explanations, in which processes that are somehow psychologically less advanced or complex, must be excluded. Another British scientist who studied animal behavior was Leonard T. Hobhouse, who studied problem solving in a variety of species of pets and zoo animals and concluded that complex processes were indeed operative. For the early part of the twentieth century, English studies of behavior would again come into the hands of naturalists, such as Edmund Selous, Eliot Howard, and Julian Huxley.

This flurry of activity among British scientists took place near the end of the nineteenth and beginning of the twentieth century. At this point, the main action shifted to North America and Germany. Great Britain lacked institutional support for research in animal behavior; such work was conducted by the landed gentry or others who found a living outside of universities. As the American university system developed, animal behaviorists were able to gain positions and status and the stage was set for the next phase. It should not be forgotten that some American naturalists, such as John James Audubon, George and Elizabeth Peckham, and L. H. Morgan made important observations of animals in the field. It was in the universities, however, where the field of animal behavior began to flourish among both zoologists and psychologists.

Among prominent zoologist/animal behaviorists of the early twentieth century were Charles O. Whitman, Wallace Craig, Jacques Loeb, and Herbert S. Jennings. Whitman's most substantial behavioral research was on the behavior of pigeons. His most influential proposal was that the comparative method could be used to study the evolutionary history of instincts just as it could be to analyze structure. Thus, we might be able to construct a *phylogeny* (or evolutionary history) of behavior in a manner similar to that of limbs or wings. Craig, a student of Whitman, also studied pigeons and doves. He later became known for introducing the distinction between appetitive behavior and consummatory acts. The *appetitive behavior* is the variable, searching behavior often associated with drives and appetites. This behavior ends with a more stereotyped chain of responses when the appropriate stimuli are encountered. Thus, a bird may be "searching" for a mate and exhibit a stereotyped courtship display when a potential mate is found. Whitman and Craig would later be seen as important anticipators of ethology. Loeb believed that much behavior could be explained as relatively simple, forced movements or tropisms and sought to control behavior. Jennings, by contrast, believed that even the behavior of invertebrates has more complex determinants and emphasized influences internal to the animal's body.

Meanwhile, the proposals that mind had evolved and could be studied from a comparative perspective encouraged development of American comparative psychology around the turn of the twentieth century. The leading players were such men as Linus Kline, and Willard Small at Clark University, Robert Yerkes and his students at Harvard University, Edward L. Thorndike at Columbia University, and others at Cornell University, the University of Chicago, and the Johns Hopkins University. This body of research was characterized by the study of a range of species of invertebrates and vertebrates, study of the naturally occurring behavior of species before experimentation was undertaken, and the relation of behavior in the laboratory to that in nature. Early in his career, Yerkes, for example, studied such species as medusae, turtles, green crabs, crayfish, crows, mice, and cats. Thorndike studied a variety of problems but was best known for his studies of cats in puzzle boxes. A cat was placed inside of a handmade crate of one of several designs, all of which required that the animal operate some kind of device in order to escape and get to the food. Thorndike concluded that the cats, and indeed all species, learn by trial and error and were incapable of higher thought processes. This was Morgan's canon grossly over interpreted. Thorndike's critics, such as Wesley Mills and Hobhouse, by contrast, suggested that Thorndike had so rigged the situation that there was nothing else that the animal could do but to make the requisite response, and any display of insight was precluded by the test situation.

A prominent woman in the field, rare at this time, was Margaret Floy Washburn, a graduate of Cornell, who wrote the 1909 textbook, *The Animal Mind*, that would be the standard for the next quarter century. African–American zoologist–psychologist Charles Henry Turner conducted a remarkable serious of studies of such topics as feeding and mating in parasitic bees, learning in cockroaches, hearing in moths, and prey removal in gallery spiders—all in the first quarter of the twentieth century.

Significant events were also transpiring in Germany. German biologist Oskar Heinroth made proposals regarding the evolution of behavior that were similar to those of Whitman, although in a more functional context. Jakob von Uexküll proposed the notion of the *Umwelt*, the unique, subjective perceptual world of each animal. Psychologist Wolfgang Köhler conducted his classical research on the mentality of apes, including such tasks as stacking boxes to reach a banana suspended above their enclosure and joining two short sticks to make a long one to reach a banana placed outside their cage. Köhler, unlike Thorndike, believed that animals displayed insight in problem solving.

The foundation had been laid for the development of a broadly-based comparative psychology. Several events led to a detour. In psychology departments, with strong pressures for application of psychological knowledge in educational and other applied situations, animal research became less highly valued, and the faculty members trained in animal behavior research were pressured into more applied fields. That and the outbreak of World War I decimated the group of young comparative psychologists who had been educated in the new laboratories.

Another factor was the development of behaviorism by John B. Watson. Early in his career, Watson worked within the broad comparative framework of the time. His most notable research consisted of field studies of two species of birds, noddy terns and sooty terns, off the coast of Florida. Watson studied their postures and displays as well as their orientation and migration. Much of this work anticipated that of later European ethology, as will be discussed shortly. Thus, he can be regarded as a "proto-ethologist." However, later in his career he developed the system of behaviorism that came to dominate his thinking and that of many colleagues. The original core ideas of behaviorism were that behavior should be studied objectively and experimentally without reference to introspection or appeals to consciousness. The behavior of humans and animals should be studied on the same plane as essential to a true science of behavior. What Watson proposed was not completely new, but he presented it an a coherent format and publicized it so that it became an important approach. As behaviorism matured during the 1920s, however, it became associated with an extreme environmentalism—a view that nearly all behavior is learned and that genetic influences are minimal. In reaction to proposals from the pens of authors such as William James and William McDougall that humans possess a long laundry list of instincts, there was an anti-instinct revolt in psychology led by Zing-Yang Kuo and Knight Dunlap. Under the influence of Thorndike, Watson, and others, much of animal psychology became devoted to the study of learning in domesticated species such as laboratory rats. Although this is a worthy endeavor in and of itself, it was often conducted without reference to the comparative and evolutionary roots of the field and thus directed at a different set of goals. Nevertheless, a thread of a true comparative psychology flourished again during the 1930s.

John Watson, the "father of Behaviorism" ca. 1920.
© *Underwood—Underwood / Corbis.*

During the period from the 1930s into the 1960s the most prominent activity in animal behavior studies lay in the development of comparative psychology in the United States and ethology in Europe.

Ethology was developed primarily by the Austrian Konrad Lorenz and the Dutchman Niko Tinbergen. The pair shared the 1973 Nobel Prize for Physiology or Medicine with Karl von Frisch, noted for his studies of honeybee vision and communication. It is difficult to define ethology. It is often identified as the study of animal behavior under natural conditions although its primary founder, Lorenz, did virtually no such research. Rather, he was a master of keeping captive animals and observing their naturally occurring behavior. It has been written that ethologists emphasized studies of instinct and evolution with careful observations and field experiments on birds, fish, and insects. This is contrasted with animal psychologists who studied learning and development in controlled laboratory experiments with statistical analysis on mammals, especially laboratory rats. There is a grain of truth to these characterizations, although there are notable exceptions within both groups for virtually all of these characteristics.

Several behavioral patterns were analyzed carefully by ethologists and provide an indication of the flavor of their work. During the breeding season male sticklebacks, small fish, develop a red belly, build a small nest, and defend a territory against intruding males. If a female enters the territory, the male and female perform a wonderfully choreographed set of displays that ends with the female laying her eggs in the nest and the male fertilizing them. Ethologists carefully studied the stimuli generated by each partner and responded to by the other in the integration of this complex sequence. The behavior of each becomes a signal for the next behavior in the chain for the partner. Another favorite was the pattern of egg rolling in greylag geese. When an egg rolls from the nest of an incubating goose, she retrieves it using a very stereotyped pattern to roll it back into the nest. These and many other behavioral patterns provided a basis for the studies and theories of the ethologists.

In Lorenz's basic system, the stereotyped behavioral patterns, or *fixed action patterns (FAP)*, were viewed as rigid sequences that were unlearned, triggered by simple environmental stimuli, and characteristic of all appropriate members of the species. The stimuli that triggered the patterns were the *sign stimuli*. Sign stimuli were thought to act on a specific site in the nervous system, the *innate releasing mechanism*, to allow expression of the FAP. The FAP did not have to be stimulated; rather, the nervous system was active and the stimulus served to remove inhibition and thus allow the behavior to be expressed. The probability that this would occur was jointly determined by the strength of the stimulus and the amount of *action-specific energy* in the nervous system. The hypothetical energy source associated with each FAP was viewed as increasing spontaneously during the time that the FAP is not released and depleted by its expression. Indeed, after a long time without release, the FAP could occur in the absence of the sign stimulus, a *vacuum activity*. Lorenz's formulation served early ethology well. As the field developed, however, many problems were encountered. The FAPs were found to be more variable than previously thought, the innate releasing mechanism came to be thought of as the entire sensory and nervous system, and many behavioral patterns did not function as expected from the notion of action-specific energy. New models were developed as the field matured.

In discussing the parallel development of comparative psychology the true comparative thread that was closest to its biological roots should be emphasized. In 1930 Robert Yerkes founded an institution that became known as the Yerkes Laboratories of Primate Biology and was the largest collection of chimpanzees for research in the world. Yerkes and his associates studied sensory function, naturally occurring behavioral patterns, learning, and cognition. In other developments begun during this period, Harry F. Harlow developed a colony of rhesus monkeys at the University of Wisconsin with which he studied patterns of development and learning. He emphasized the importance of the effects of the mother and siblings on the development of normal behavior. C. Ray Carpenter undertook a series of landmark field studies of the social behavior of nonhuman primates. He is often regarded as the "father" of such research that later became associated with scientists such as Jane Goodall, Dian Fossey, and George Schaller. Frank A. Beach undertook a series of studies of the physiological and environmental influences on social and sexual behavior in laboratory mammals. T. C. Schneirla developed an extensive program of laboratory and field research on the behavior of ants. As should be apparent, this part of animal psychology remained close to its biological foundations.

Although there were some interactions between ethologists and comparative psychologists before World War II, it was after the war that contacts became more frequent. In the early days, there was much conflict. The two groups had developed very different theoretical viewpoints. Where the ethologists emphasized evolutionary analyses of instinctive behavior, the psychologists emphasized the importance of the environment and learning in the development of behavior. Whereas ethologists emphasized species-characteristic behavior, psychologists stressed its plasticity. With repeated contacts, however, both sides were modified. Ethologists increased their attention to the complexities of behavioral development and came to appreciate the experimental control of the psychologists' research. Many psychologists came to increase the attention paid to naturally occurring behavioral patterns in relation to the environment and evolutionary history. Although the English-speaking

Some Leaders in Animal Behavior Studies

Frank A. Beach (1911–1988)

Frank A. Beach was the son of a professor of music and grew up loving both music and writing. Beach later became the foremost comparative psychologist of his time. His research was focused on the roles of experience, the environment, and physiology on animal behavior. He was especially noted for research on hormones and behavior and his insightful integrative writings. An outgoing and fun-loving man, he played an important role in bringing ethology and psychology together.

Robert M. Yerkes (1876–1956)

Robert M. Yerkes grew up on a Pennsylvania farm and earned a Ph.D. in psychology at Harvard University. He was known as the dean of comparative psychology for his long-term influence on the field. Early in his career, he studied a great variety of species. During World War I he spearheaded a program of mental testing for the U. S. Army. Later, he conducted important early studies of the learning and behavior of chimpanzees. A shy, formal man, he nevertheless was influential as both a scientist and an administrator.

The Yerkes National Primate Research Center at Emory University in Atlanta, Georgia, is named for Yerkes, "whose early studies of chimpanzees and other apes during the 1920s persuaded Yale University, the Rockefeller Foundation and the Carnegie Foundation to fund the establishment of the first laboratory in the United States for the study of nonhuman primates" (Yerkes National Primate Research Center Web site, http://www.emory.edu/WHSC/YERKES/about_history.html.) The center was started in Florida and eventually was moved to Emory University. Today, it is one of the foremost institutions in the study of primates.

ethologists tended to be more accommodating than many German-speaking ethologists, there was a general sense of harmony and cooperation between the groups as they exchanged students and had many cooperative interactions at professional meetings. Comparative psychologist Daniel Lehrman, founder of the Institute of Animal Behavior of Rutgers University and a student of reproductive behavior in doves, played an important role in bringing the groups together.

Tinbergen developed a formulation according to which a complete analysis of behavior had to include *immediate causation* (the stimuli and internal factors controlling behavior), *development* (the interacting effects of genes and environment in the ontogeny of behavior), *evolutionary history* (the past phylogenetic sequence of behavioral change across species), and *adaptive significance* (the role of behavior in promoting the survival and reproduction of the animals displaying the behavior). This became a useful framework for studies in both ethology and comparative psychology.

Several other developments occurred between World War II and the present. These include the advent of sociobiology, comparative cognition, and evolutionary psychology. During the 1960s and 1970s a new approach to the study of animal behavior was developed and came to dominate the field. The overlapping fields of sociobiology and behavioral ecology sprang from a few key ideas. One was the reemphasis of the notion that, in most cases, natural selection works at the level of the individual or gene, not that of the group or species. It is individuals or genes that are favored through natural selection; behavior that benefits a group, to the disadvantage of the actor, should be rare. Coupled with that was the notion that stemmed from the work of William D. Hamilton of indirect, or kin selection. Because individuals share genes with their close relative, their fitness, or representation in future generations, might be enhanced by aiding those relatives. The fitness benefits of aid would have to outweigh any fitness costs. Related concepts included inclusive fitness, altruism, reciprocal altruism, and evolutionary stable strategies as are discussed elsewhere in this encyclopedia. These concepts were collected and organized in Edward O. Wilson's 1975 book *Sociobiology: The New Synthesis*. Although Wilson's last chapter, dealing with human behavior, proved highly controversial, the core ideas of this approach came to dominate the field of animal behavior studies. Some believed that Wilson's contribution in this book, analogous to Watson's before him, was not so much the development of new ideas as in packaging them in a single book with a catchy title and publicizing them. Richard Dawkins would soon follow in both developing these ideas and bringing them to a broad audience. The emphasis in animal behavior studies came to be focused on Tinbergen's questions of the adaptiveness of behavior. Sexual selection, foraging, and cooperation became hot topics.

At about the same time, the so-called cognitive revolution occurred in psychology and related fields. The study of thought processes had continued during the dominance of behaviorism but had been forced underground. During the 1960s cognitive approaches returned to favor as behaviorism, now in the form of B. F. Skinner's very different version, flourished but in a less dominant role. Cognitive approaches became increasingly important in animal behavior studies as well in the mainstream psychology of human behavior. Much of the research on animal learning was reformulated in a more cognitive context. Remarkable studies of language-like behavior and number competency in apes were a part of this emphasis. A subfield of cognitive ecology developed. The most extreme development was a cognitive ethology originated by American ethologist Donald R. Griffin. Griffin urged a move away from the cautions of both behaviorism and much of traditional ethology with a consideration of the possibility that animals have significant mental experiences and that a window on the minds of animals can be opened through the study of communication processes. He suggested that discussions of awareness and imagery that had been banished a century ago could be now revisited in the light of the progress of a century of research on animal behavior.

Sociobiologist Edward O. Wilson with microscope.
© *Steve Liss / Time Life Pictures / Getty Images.*

In the recently developed field of evolutionary psychology, scientists analyze human behavior in the light of evolutionary principles building upon developments in sociobiology and cognitive psychology. Whereas earlier evolutionary interpretations of human behavior had been focused on the current adaptive value of existing behavioral patterns, evolutionary psychologists try to assess innate psychological mechanisms as they evolved and were adaptive in the context of the hunter–gatherer societies of the environments of the Pleistocene era. Humans are viewed as possessing specialized modular learning mechanisms that function, not generally, but only in specific domains. Thus, the mind is composed of a set of modules adapted to solve specific problems of survival and reproduction as faced by our ancestors.

Early in the twenty-first century, activity in the field of animal behavior studies seems concentrated on work in behavioral ecology and animal cognition. A significant trend is apparent in the interest in integrating research on Tinbergen's questions of adaptive significance and immediate causation. In this work, analyses of physiological and cognitive determinants of behavior are combined with, and related to, those of their function in the natural lives of animals.

And, like behavior itself, the field of animal behavior studies is always evolving and changing in relation both to new research and thought and to the broader intellectual and social context within which animal behaviorists function. Like the behavioral patterns themselves, it is likely that the field will continue to show patterns of change in response to these pressures.

See also Craig, Wallace (1876–1954)
Darwin, Charles (1809–1882)
Frisch, Karl von (1886–1982)
Frisch, Karl von—*Decoding the Language of the Bee*
Griffin, Donald Redfield (1915–2003)
Hamilton, William D. (1936–2000)
History—*Niko Tinbergen and the "Four Questions" of Ethology*
Lorenz, Konrad Z. (1903–1989)
Lorenz, Konrad Z.—*Analogy as a source of Knowledge*
Nobel Prize—*The 1973 Nobel Prize for Medicine on Physiology*
Tinbergen, Nikolaas (1907–1988)
Tinbergen, Nikolaas—*Ethology and Stress Diseases*
Whitman, Charles Otis (1842–1910)

Further Resources

Boakes, R. 1984. *From Darwin to Behaviourism*. Cambridge: Cambridge University Press.

Burkhardt, R. W., Jr. 1981. *On the emergence of ethology as a scientific discipline*. Conspectus of History, 1, 62–81.

Dewsbury, D. A. 1984. *Comparative Psychology in the Twentieth Century*. Stroudsburg, PA: Hutchinson Ross.

Dewsbury, D. A. (Ed.) 1985/1989. *Studying Animal Behavior: Autobiographies of the Founders*. Chicago: University of Chicago Press.

Glickman, S. E. 1985. *Some thoughts on the evolution of comparative psychology*. In: *A Century of Psychology as Science* (Ed. by S. Koch & D. E. Leary), pp. 738–782. New York: McGraw-Hill.

Hess, E. H. 1962. *Ethology: An approach toward the complete analysis of behavior*. In: *New Directions in Psychology* (Ed. by R. Brown, E. Galanter, E. H. Hess, & G. Mandler), pp. 157–266. New York: Holt, Rinehart and Winston.

Richards, R. J. 1987. *Darwin and the Emergence of Evolutionary Theories of Mind and Behavior*. Chicago: University of Chicago Press.

Singer, B. 1981. *History of the study of animal behaviour*. In: *The Oxford Companion to Animal Behaviour* (Ed. by D, McFarland), pp. 255–272. Oxford: Oxford University Press.

Thorpe, W. H. 1979. *The Origins and Rise of Ethology: The Science of the Natural Behaviour of Animals*. London: Praeger.

Warden, C. J. 1927. *A Short Outline of Comparative Psychology*. New York: Norton.

Donald A. Dewsbury

■ History
Human Views of the Great Apes
from Ancient Greece to the Present

For those curious about how our views about animals change, how the West formed its views about great ape intelligence may be one of the most telling tales. The West has traditionally considered high intelligence as the quality that divides human from animal—in particular, the ability to reason. Animal intelligence is held to operate only at concrete levels, bound tightly to immediate senses, so animals lack the capacity to solve problems mentally, the essence of reasoning. They may be able to respond in complex ways to events around them, but can never see beyond the here and now. Animal minds are held to differ so greatly from human minds that they represent a distinct *kind* of intelligence, "animal" intelligence, which operates by a more primitive set of rules and processes.

We know for sure where Western views on great apes began: with nothing. Great apes are the primates we know as chimpanzees, bonobos, gorillas, and orangutans. They do not inhabit any part of Europe, the source of many Western views, and, until a few centuries ago, their homelands were beyond Europe's reach. So our views about them must owe to a mix of borrowed ideas, ignorance, and current worldviews, with knowledge contributing only recently. Tracing these views through time then promises interesting insights into the forces at play.

Great apes have been square pegs that refuse to fit into round holes. They seem to sit right on the bridge between humans and animals, refusing to move to either side. They are obviously animals, in the sense of not human: They are hairy, walk mostly on all fours, and cannot speak. But even casual exposure makes it equally obvious that they resemble humans more than typical Western animals do. Zoo-goers are often overheard making comments like, "that gorilla looks just like your uncle," or "see how the mommy chimpanzee tickles her baby to make him laugh, just like us." Anatomically and physiologically, great apes are so like us that we select them for research on AIDS and travel into space. People who work with them also realize, rapidly, that great apes do not have typical animal minds—or typical human minds. Even the media knows of both messages. It has portrayed gorillas as *King Kong* style beasts, driven by crude desires, but also as "gentle giants" capable of empathy and understanding.

Western thinkers have struggled for centuries with the ambiguities the great apes create, especially the problems they create for a tidy human–animal divide marked by reason. That struggle is not yet entirely resolved, and how it has played out offers clues into what makes us change our views of animals.

Aristotle's Divide

The seeds of western views about great ape intelligence are understood to come from Ancient Greece, primarily from Aristotle in the third century BCE.

Aristotle is credited with originating the Western view that reason divides animals from humans. The greatest philosopher of his time on the nature of living beings, he collected extensive information about animals from all over the world. Based on this evidence, he concluded that animals and humans as very similar, even in their intellect, but animals have very concrete minds and cannot reason as humans do. He even considered apes, although Greece has none, because the ancient Greeks had already contacted parts of Africa and India that apes inhabit.

A second notion we owe to Aristotle is *The Great Chain of Being*, the theory that all things in the universe have their proper station, unchanging and unchangeable, ordered from lowest to highest. This theory strongly influenced views of animals during the Middle Ages, when all reality was seen as organized from least to most real, in successively higher stages: elements to plants, to animals, to humans, to angels, then to God.

A third Ancient Greek notion that influenced Western ideas is that "monstrous races" lived: troglodytes, who were cave dwellers, pygmies, little people who lived with them, cynocephali, a dog-headed race, and satyrs, a four-legged race with human figures, goat feet, horns, and hairy bodies. Satyrs were shadowy creatures of the woods, lascivious, shameless, two-formed, wantonly inclined. Pan was their leader, a great personage based on the Greek god of flocks, shepherds and woods. Like them, Pan was human-bodied, goat-legged, hairy, and horned; he also walked erect with a staff and indulged, goat-like, in playful lechery and nymph chasing.

All three notions affected great apes. Because of Aristotle's view, the West has assumed, even insisted, that great apes cannot reason. Because of the Great Chain of Being, Europeans interpreted legitimate great ape talents as ersatz copies of human abilities. And because of beliefs in monstrous races, great apes were assumed to have savage traits attributed to mythical satyrs, troglodytes, and Pan.

With the benefits of hindsight and modern science, it is clear that the seeds of Western views about animals are flawed. Aristotle's view, neat though it is, has a fatal error: Aristotle had no knowledge of great apes. His world didn't extend to the equatorial regions of Africa or Southeast Asia that great apes inhabit. The apes he knew of were probably Barbary macaques, tail-less monkeys of Gibraltar, not great apes at all. So Aristotle was ignorant of great apes, among *the* most important animals to consider, when he developed his view of the human–animal divide. As for the Chain of Being and the monstrous races, the first turns out to be wrong and the second mythical, so both sent Western views of great apes in unrealistic directions.

You Can't Get There from Here

There was little way for Western thinkers to develop realistic views of great apes without better evidence. As it turned out, that had to wait for centuries, until Europeans reached equatorial Africa and Southeast Asia. Until then, when and where it seemed useful, European thinkers reshaped existing views this way and that to suit the times. Often, they simply dredged up old Greek ideas, dusted them off, added their own twist, and forwarded them as established truth.

The ancient Romans could have changed things because they apparently thought Greek philosophy was about as useful as drilling holes in millet seeds. They nonetheless lifted much of their "information" from Greek philosophers, reviving beliefs about monstrous races and rejuvenating Aristotle's view of the human–animal divide. Early Christianity promoted Aristotle's view but adopted only the parts it liked (e.g., only humans have reason), exaggerated some others (e.g., made the human–animal boundary unbreachable, apparently to distance Christian from pagan views), and simply dropped the rest (e.g., human–animal similarities).

Through the Middle Ages, European views on great apes remained about as valid as views on whether goblins are human, based on material in fairy tales. The hardened version of Aristotle's view held, although attitudes to primates softened a bit. By the twelfth century thinkers acknowledged human–animal similarities again, even putting nonhuman primates in a special category bridging the human–animal divide because of their human-like abilities to imitate and solve problems. But they still twisted facts to fit their view of the order of things, The Great Chain of Being. They believed all animals have tails, for instance; only humans do not, because God removed Adam's tail. Apes also lack tails, as it turns out. But instead of accepting that "tail-less-ness" is not uniquely human, medieval thinkers asserted that apes' tail-less-ness betrayed them as disgraceful imposters, even devils, with presumptuous desires to rise above their proper station. They also took primates' human-like abilities as signs of the devil or foolish caricatures of humans. Many of these ideas came straight from the ancients—animal tails from Leviticus, images of the devil perhaps from ancient Greece, and the Chain of Being of course from Aristotle.

Outsiders

Windows of opportunity were opening for Europeans to obtain real evidence of great apes by the ninth century. Islamic Arabs had reached the islands of Sumatra and Borneo and on returning, described creatures that are recognizably orangutans. Qazwînî, in 1383, reported remarks of two ninth century scholars about these creatures. On Sumatra, he wrote, there lived "human beings" four spans in height who dwell in the treetops, eat fruit, and shun man. They went naked and barefoot, had hair on their bodies that covered their pudenda, and faces covered with russet-colored down. Their language, a sort of whistling, was incomprehensible.

Later, local peoples who shared the forests with great apes could also have offered rich information about them. The Dayak peoples of Borneo, for instance, tell many tales about orangutans. Some of their tales have an inventive cast, like one that ponders how orangutans came to be. Two birdlike beings created all life on earth, it says, and one day they created humans. To celebrate this great accomplishment, they feasted late into the night. The next day they tried to create more humans, but not feeling so well after the night's revels, they messed up the recipe and created orangutans instead.

Other Dayak tales describe orangutans convincingly. Dayaks who have met orangutans directly know they are fruit lovers, appearing where fruit is abundant and fading away when it is not. They describe how orangutans seek honey, reaching into a hive with one hand while covering their eyes with the other against the swarm of angry bees. They recount with admiration how an orangutan can bend a tree by pulling a single leaf, playing the forces so subtly that the leaf does not snap, or recall with pleasure the antics of orangutans who were village "pets." Dayak lore consistently portrays orangutans as powerful, intelligent persons very close to humans and worthy of respect. Iban Dayaks of Sarawak called them "forest cousins" and tabooed eating them. Head-hunting Dayaks saw their heads as powerful symbols, perhaps for use in animistic ceremonies. As for not speaking, Dayaks of northern Borneo believed in their high intelligence and held that they *chose* not to, to avoid having to work.

These reports were from other cultures and religions, however, and found few European ears. About the only item that took root is the name by which we know one of the great apes, *orang utan*, a local name meaning "person of the forest." So an important avenue of understanding great apes was essentially ignored. Quick to believe their own myths, it seems, Europeans were loath to credit others'.

Europe Meets the Great Apes

Europeans gained the wherewithal to create informed views of great apes in the seventeenth century, when seafaring took them to unfamiliar worlds and their reports started to reach home. The first credible reports of great apes came in the 1600s from Battell, an

English privateer who had been imprisoned in Africa, and Bontius, a Dutch physician who had worked in Batavia (Jakarta, today). Bontius described great apes he had seen himself in Batavia. Battell told of two dangerous monsters in Africa, *M'Geko* and *M'Pungu*, that we now know were chimpanzees and gorillas.

Europeans also began to capture great apes and send them home. The first ones we know of went to a Dutch menagerie in 1630 and to London in 1697. Anatomists took the greatest interest in them, probably because they didn't survive long. Those who paid interest—a Dutch physician, Tulp, and England's foremost anatomist, Tyson—were highly respected, and their anatomical studies began to dispel ignorance. But in 1641 they flipped out of scientific mode and back to familiar beliefs. Tulp declared his ape was Pliny's Indian satyr, a man of the woods, *Homo sylvestris*. Bontius, in 1658, then Tyson, in 1699, followed suit and proclaimed that their apes were *Homo sylvestris*, too, though Tyson said his was the pygmie of the ancients, *Simia troglodytes*. Claiming these apes were incarnations of the monstrous races entrenched old misunderstandings even more firmly.

Realistic appreciation of great apes' nature, intelligence included, was even slower in coming. Undoubtedly, this was partly because the earliest great apes to reach Europe died quickly, and while they survived, must have wasted under poorly informed care. More orangutans and chimpanzees were surviving the voyage by the mid-eighteenth century, but even with living evidence, European thinkers twisted what they saw to fit the ideas of the day.

Late in the eighteenth century, during the European Age of Enlightenment, thinkers like Rousseau thought great apes could help humans return to a more harmonious state. He saw them as the natural, pristine men of the woods that humans were before they fell from grace. Orangutans briefly enjoyed great favor because they were believed to be the "real" link between humans and other animals. If fairly accurate portrayals were a positive result, other consequences were less so. The fame meant large numbers of orangutans were captured and shipped to European and American zoos, for instance, where they still died quickly.

The nineteenth century brought the Industrial Revolution, so work, commerce, and competition became the values of the day, and Nature became frightening, virtually worthless, and no longer the original state of harmony. Great apes were once again portrayed as the wild men of the ancients, uncivilized, ferocious, and lewd like Pan and his satyrs. Their names—*Simia satyrus, Simia troglodytes*, and *Pan troglodytes*—no doubt fuelled the prejudice. Edgar Allan Poe, for instance, made the orangutan, in his 1841 *Murders in the Rue Morgue*, a smart but wildly ferocious brute.

Great apes were reshaped again in the mid- to late nineteenth century, this time by scientists, to fit the new ideas about evolution of great minds like Darwin and Lamarck. Both thought great apes were the "ancestors" that "fathered" humans, ergo an evolutionary "missing link" must logically exist between the two—a man–ape or ape-man. At the end of the century, in quest of the imagined missing link, Eugene Dubois set sail for Southeast Asia. On his very first stop, Sumatra, he found a wealth of orangutan fossils. Little interested in this evolutionary issue, however, he abandoned them to pursue rumors of more human-like fossils in Java.

Thoroughly Modern Apes

The twentieth century saw renewed attention to great ape minds from a newly powerful scientific community interested in examining human minds from the perspective of evolution. Chimpanzees took premier status in this research as humans' closest living relatives, dramatically extroverted, with minds like quicksilver. Slow-moving orangutans were back-benched as the most obtuse, stupid, and boring of the great apes. Even interest in chimpanzees quickly faded, however, as psychologists flocked to study rats and anthropologists to study cultures and fossils.

Very few early twentieth century scholars looked beyond old beliefs. Those that looked carefully at orangutans found reflective, insightful beings. Yerkes, a prominent great ape specialist, described them as nearly mechanical geniuses. Köhler noted their capacity to solve problems by insight. Furness and Haggerty both reported that orangutans could understand human language and learn by imitation. Furness ranked their minds within the human grade, perhaps only a step below the level of what were considered the most primitive human beings. These findings had little effect on mainstream twentieth century views of great apes, however, which opted to disregard them rather than give up Aristotle's view of the human–animal divide.

Two events reawakened interest in great ape minds in the 1960s, the end of World War II and the cognitive revolution. The story since then is short in time but long in substance because science contributed more new material to views on great apes and their minds in the last 50 years than all other sources did in the previous 2500.

After WWII, international cooperation flourished, helping primate specialists around the world to share their views. Physical anthropologists, notably Sherwood Washburn and Louis Leakey, showed the value of studying living primates for insights into human evolution. Japanese primatologists introduced their views to Western researchers: interest in primate social structures, the value of long-term field studies, and the importance of the individual in primate groups. They even named individual monkeys, a practice dismissed as "not objective" in America. Europeans offered the ethological view, which emphasizes the biological significance of behavior and studying species-typical behavior in its natural environment. American comparative psychology brought the opposite, tightly controlled experimental methods for studying behavior in laboratories. To this, the cognitive revolution added renewed interest in studying the mind, even in nonhuman species, after half a century of opinion that science could not study such intangibles.

The first successful long-term field studies of wild chimpanzees, gorillas, and orangutans were launched by the 1960s. Researchers did not aim to study great ape mentality, but the intelligent behavior they saw was so striking that most reported it. And now the West was finally ready to listen. Jane Goodall rocked a major view about what makes human minds unique, "man the toolmaker," when she reported wild chimpanzees use and manufacture tools in the wild. Others reported that orangutans use traditional travel routes, navigate by "mental maps," and create highly sophisticated maneuvers to negotiate the complexities of travel in the trees.

Experiments on captive great apes flourished, too. Thanks to the cognitive revolution, researchers began developing ways to test for reason in animals. They needed to. The many tests already developed to study animal intelligence couldn't even detect reason because they assessed only the simpler abilities that animals were assumed to use. Researchers also borrowed ideas about human intelligence to guide studies of nonhuman primate intelligence. An important idea they borrowed was Piaget's: Human intelligence develops, or changes with age. Development advances with brain growth plus life experiences, and complex abilities, like reason, take years to build. So human infants, children, and adults don't just look different, they think differently too. Studies taking this view found striking similarities between great apes and humans in their intellectual development. They also showed that scientists had been systematically underestimating great ape intelligence because they had typically tested subjects who were too young or too deprived to have developed their intellectual abilities to their species' best potential.

Nonexperimental projects on captive great apes also looked beyond established views and found impressive mental abilities. Ape language studies are one example. People like the Kelloggs and the Hayes raised great apes in their homes in the 1950s, dreaming of talking to the apes. Although they found that it was an impossible dream—physically, great apes can't make human speech sounds—they sparked interest in ape language. Later projects used gestural signs or visual symbols instead, plastic or printed, and great apes tutored these ways typically acquired about 130 words and made and understood short sentences. They were not simply parroting. Several invented new signs and sentences. Chantek, an orangutan, invented "eye–drink" for contact lens solution. Washoe, a chimpanzee, signed "baby in my drink" on seeing a doll in her cup. Their progress was much slower than humans', however, and never progressed beyond 2–3 word sentences.

Ape language is important on its own, but it is also about intelligence because intelligence delimits what an individual has to say. Evidence now shows that great apes attain similar levels in their language and intelligence, roughly comparable to 2–3-year-old humans. Chantek, for instance, could learn by imitation, an ability that requires simple reason. He could even imitate on demand. Lyn Miles, who worked with him, used this ability to teach Chantek new signs: She demonstrated them, then simply asked him to imitate. Chantek's language also showed off his intellect—that he could lie, for instance, and plan ahead. He'd sign "dirty," meaning he needed the toilet, but once in the bathroom ignored the toilet and played. He signed "in milk raisin" to himself, *before* heading inside and asking a caregiver for milk.

Experiments on problem solving produced similar results. In one, an orangutan was shown five boxes. The first box held keys to the other four, which were locked. One locked box held food, one was empty, and two held keys—*one* of which opened the food box. To get the food, the orangutan had to pick the right key from the first box (the key to the box with the food box key), unlock that box, take the key, then unlock the food box. He whizzed through the mental maze. Other studies showed that great apes can draw simple pictures (circles, crosses), tie simple knots, untie complex knots, use "maps" (e.g., scale models to find items, television, photos), form simple concepts (e.g., food, other species), classify items by multiple features (e.g., color and shape), use part–whole relations (e.g., one quarter of an apple), sort items into a graded series, count, add, apply analogies, and judge quantities (e.g., which of two bowls has more candies). All great ape species have also mastered complex tool skills. Some even mastered meta-tools, tools to make or modify other tools: An orangutan and a bonobo each used two stones to make a stone flake tool, then cut with the flake. Great apes have shown signs of *insight* in these tests, that is, using reason instead of trial-and-error fiddling, but it is devilishly difficult to prove.

All this sparked interest in studying intelligent behavior in the field, especially since the 1980s. Wild chimpanzees showed even more complex tool abilities, including *tool kits* (individuals master a whole set of tool skills), *tool sets* (several tools used for one task), *composite tools* (tools made of many interrelated parts), and *meta-tools*. The most complex are stone hammer–anvil sets that West African chimpanzees make for cracking nuts. A few have even added a meta-tool, wedging a third rock under their anvil to level it. Other complex problem solving also came under study. Mountain gorillas showed intricate manipulative skills for transforming nasty plants into neat edible packages. Orangutans showed tool and nontool skills of similar complexity—combined with elaborate acrobatic maneuvers to enable using them in the trees. Orangutans and chimpanzees both make tools in advance and may carry them considerable distances to their work site, so they also plan and make tools from a mental design. These skills, too, all point to the capacity to reason at simple levels.

The idea that primate intelligence might be "social," adapted primarily for negotiating complex social life, arose in the 1950s. It really captured interest in the 1980s, when books like *Chimpanzee Politics* and *Machiavellian Intelligence* showed how nonhuman primates use their wits, and even calculated deceit, to best competitors. Many other faces of social intelligence soon came under study, from competition (e.g., reading another's visual perspective or intentions) to cooperation (e.g., joint problem solving, food sharing, social learning, exchanging favors) and conflict management (reconciliation, conflict mediation, coalitions and alliances).

Great apes at first seemed similar to monkeys in social intelligence, but newer work shows that their maneuvers are more complex. Their deceptive capers are more elaborate, conniving, and calculating. Menzel, for example, engineered a conflict between Belle and Rock in a captive chimpanzee group. He showed Belle where he had hidden food in a field then let the whole group into the field. Belle went straight to the food and uncovered it—but all the others had followed, and Rock stole it. Belle began to approach the food slowly, even sitting on it without uncovering it, until Rock left. So Rock started pushing her aside to search beneath, or moving off without looking back—to about 5 m away, when he could often whirl around to catch her uncovering the food. Belle began staying farther from the food or walking away from it while Rock watched. But Rock used her "indirect" cues, enlarging his search radius around her and monitoring her fidgeting to detect when he was getting "hot." Belle could not control her fidgeting, so that ended the contest. Great apes also show more complex social learning and cooperation than monkeys. They can use imitation in acquiring skills and experts can demonstrate to learners how best to handle difficult tasks. As for cooperation, West African chimpanzees' team hunting of colobus monkeys is especially impressive. Team members divide the labor, take complementary roles, and even switch roles mid-hunt if necessary. Great apes' social intelligence limit seems to be "mind reading," understanding others as mental beings. If unable to master that, they are nonetheless highly astute at reading others' perspectives and attention.

Primate social intelligence has a final face, culture. Japanese primatologists began studying Japanese macaque social life in the 1950s as a way of exploring the evolutionary roots of human cultures. For species as closely related as human and nonhuman primates, cultural similarities today could point to shared cultural origins in the past. Great apes became prominent in this arena in the 1990s, when enough evidence had accumulated on many wild communities to examine whether cultural traditions exist. In chimpanzees and orangutans, evidence for traditions is now so compelling that it gained publication in two of the world's premier scientific journals, *Science* and *Nature*. Cultures are about intelligence, just as language is, because the traditions that cultures can support depend on the abilities to learn, teach, and communicate. The more powerful these abilities, the richer and more complex the traditions that can be shared. Great apes' cultural processes appear to support much of the complex tool use observed in wild chimpanzees and orangutans, stone hammer and anvil use included. Among nonhuman primates, the great apes alone show abilities to learn by imitation, teach by demonstration, and communicate at near symbolic levels, so it is no surprise that their cultures are more complex. Yet again, all these key abilities entail rudimentary reasoning.

Back to Aristotle's view, this scientific evidence suggests that reason is not what divides humans from great apes, but it may be what divides great apes from other nonhuman primates. Great apes differ less from other nonhuman primates in what they do than in how they do it. Monkeys also make and use tools, deceive, and learn socially, but at simpler intellectual levels than great apes. In their tool use, deception, social learning, food sharing, sense of self, division of labor, symbolic communication, and culture the great apes show intellectual abilities traditionally viewed as uniquely human. These intellectual abilities fall short of humans', but they reach reason—simple reason, but reason nonetheless. Finding a better way to describe the great ape–human intellectual divide is proving difficult. Premack's rule of thumb may capture it as well as any at this point: Great apes have the intellectual potential of human children 3.5 years of age, but not beyond.

That great apes can reason is not gaining easy acceptance. Some remain unconvinced by their best performances, emphasizing their failures instead. Most important, however, may be the mixed pattern—smart one time, dumb the next. It might simply reflect varied testing methods, but it might also be a clue to how great ape intelligence operates. We can check whether this might be so by asking how intelligence operates in human children under 3.5 years old. Over about 1.5 years old, such children can reason. If what they see contradicts what they think, however, they believe what they see. For this reason, they have been described as "perception bound." It makes them easy to fool, so they can look smart one minute and dumb the next. We don't dismiss them as unable to reason, however. Instead, we consider that their ability to reason is still weak, or they that lack confidence in what reason tells them. Great apes also show the smart–dumb pattern and may use "seeing is believing" rules, so the same interpretation may apply.

Even considering dedicated skeptics, most experts now agree that all the great apes achieve rudimentary abilities to reason. You might say they have finally made it across Artistotle's divide. In fact they made it across 14–12 million years ago. The only thing new is that humans finally looked long enough and fairly enough to see it.

What's It All About?

The long and the short of it is, Western thinkers have attended to what they *want* great apes to be more than to finding out what they are. History offers ample evidence that Western views of animal intelligence do not apply to great apes. The scientific credentials behind these views are, in fact, extremely shaky. The wherewithal to build informed views has been available for several centuries. Bontius' and Tulp's descriptions of orangutan behavior were accurate. Rousseau correctly recognized the orangutan's natural habitat to be the tropical forest and inferred that they are primarily fruit eaters, generally peaceable, nomadic, lack defined territorial ranges, lack distinctive social systems, form no lasting sexual bonds, and, apart from infrequent copulation, have essentially inactive and hermitic lives. This information was, simply, largely ignored.

Part of the reason must have been that Europeans, on discovering equatorial Africa and Asia, faced a natural world for which their existing views did not prepare them. A world with real great apes would have upset the order of their lives, so few would have embraced the idea. Thinkers also succumbed to the tendency to explain the strange in terms of the familiar, likewise forcing great apes to fit into old concepts rather changing their ideas to incorporate such new beings: Even eminent scholars trained in science paid most heed to old written words. Finally, these particular old views are probably dying hard because they define our position relative to animals, so they largely determine our ties with other species and how we treat them.

What does it mean to change views about great ape intelligence and the human–animal divide? Some implications are far-reaching. If reason is what defines the essence of being human, for instance, then we must admit other species into the human circle and treat them accordingly. Others are less earth shattering. If reason has extraordinary significance because it revolutionized intelligence, as we have argued for 2500 years, it does not become less significant just because other species share it. All it means is that the evolutionary leap to reason occurred, at least once, in the ancestors that humans share with great apes—large, ancestral hominoids that lived in moist subtropical forests 14–12 million years ago. Most probably, what evolved uniquely with the human line were elaborations to rudimentary reasoning. Reason still contributes to what makes us human, but not to what make us unique.

Further Resources

Boesch, C. & Boesch-Achermann, H. 2000. *The Chimpanzees of the Taï Forest: Behavioural Ecology and Evolution*. Oxford: Oxford University Press.

Byrne, R. W. 1995. *The Thinking Ape*. Oxford: Oxford University Press.

Byrne, R. W. & Whiten, A. (Ed.) 1988. *Machiavellian Intelligence*. Oxford: Oxford University Press.

Corbey, R. & Theunissen, B. (Eds). 1995. *Ape, Man, Apeman: Changing Views since 1600*. Leiden: Department of Prehistory, Leiden University.

De Waal, F. B. M. 1982. *Chimpanzee Politics: Power and Sex among Apes*. New York: Harper Colophon Books.

French, R. 1994. *Ancient Natural History*. London: Routledge.

Janson, H. W. 1952. *Apes and Ape Lore in the Middle Ages and the Renaissance*. Volume 20 of Studies of the Warburg Institute. London: Warburg Institute.

Menzel, E. W. Jr. 1973. *Leadership and communication in young chimpanzees*. In: E. W. Menzel (Ed.), *Symposia of the Fourth International Congress of Primatology, vol. 1, Precultural Primate Behavior*, pp. 192–225. Basel: S. Karger.

Salisbury, J. E. 1994. *The Beast Within: Animals in the Middle Ages*. New York: Routledge.

Anne E. Russon

■ History
Niko Tinbergen and the "Four Questions" of Ethology

In the twentieth century, biologists and psychologists in increasing numbers decided to make animal behavior the focus of systematic scientific study. The most successful program of research on the biologists' side was pioneered by the Austrian naturalist Konrad Lorenz and the Dutch naturalist Niko Tinbergen. They came to call their new science "ethology."

The ethologists were keen to distinguish their approach from that of other scientists studying behavior. For example, where some psychologists were interested in studying "the animal mind," Lorenz and Tinbergen believed that the question of animal subjective experience was inaccessible to the tools of modern science. And where some investigators focused on problems of animal learning, the ethologists decided that one should first get a full knowledge of the nonlearned, "instinctive" parts of an animal's behavioral repertoire. The ethologists studied a greater number of animal species than did the psychologists. They furthermore insisted that one should study the behavior a species exhibits under natural conditions, not under the artificial and constrained conditions of the laboratory.

Niko Tinbergen provided the classic definition of ethology in 1963 in a paper he dedicated to Konrad Lorenz on the occasion of Lorenz's sixtieth birthday. Tinbergen identified ethology as "the biology of behavior." He went on to explain that to study behavior biologically is to ask four distinct kinds of questions about it:

1. What is its physiological causation?
2. What is its function or survival value?
3. How has it evolved over time?
4. How has it developed in the individual?

Tinbergen believed that the future of ethology as a scientific discipline depended on the ability of its practitioners to pursue these four questions in a well-balanced and integrated fashion. In particular, he worried that ethologists were paying too much attention to questions concerning the physiological causation of behavior patterns and not enough attention to how these functions contributed to the animal's survival. More recently, investigators have worried that the pendulum has swung too far in the opposite direction and that studies of survival value are taking a disproportionate share of studies devoted to the biology of behavior. This does not detract, however, from the importance of Tinbergen's four questions, which have served for nearly half a century now as the ethologist's primary reference when it comes to identifying what the field of ethology is all about. They are likely to continue to function as a general blueprint for how to cultivate the biological study of behavior well into the twenty-first century.

Further Resources

Tinbergen, N. 1951. *The Study of Instinct*. Oxford: Clarendon Press.
Tinbergen, N. 1963. *On aims and methods of ethology*. Zeitschrift für Tierpsychologie, 20, 410–433.

Richard W. Burkhardt, Jr.

■|Hormones and Behavior

Hormones as a Central Topic in Behavioral Biology

Hormones regulate hunger and thirst, influence love and affection and modulate attractiveness, mate choice and sex, as well as aggression and violence. These behavioral contexts make evolution tick. Therefore, behavioral endocrinology ultimately deals with the physiological mechanisms which are most directly linked with evolutionary function.

The central nervous system (CNS) integrates the input from the sensory organs with its knowledge about the environment gained through evolution and represented in the genes or collected by individual learning and represented in the brain. The decisions made by the CNS need to be communicated to the body. Motor nerves serve this function for the muscular system, hormones reach their target organ and finally, every cell in the body via the blood stream. The observable outcome of all this we call behavior. In return, feedback communication from the periphery updates the CNS on the effects of its actions and modulates its functions. Hence, just assuming that the brain is in command of an enslaved body would not be appropriate.

To achieve its control tasks in a direct way, the CNS could send a nerve fiber to every cell. However, considering that there are more cells in our bodies than stars in the universe, this would necessitate a gigantic "administrative" nervous system, way too costly to be built or maintained. The evident solution for the CNS to solve this problem was "outsourcing."

By employing the circulatory system, mainly developed for supplying cells with nutrients and oxygen and for removing waste products, chemical signals efficiently reach their targets. And there is an additional benefit to this apparently haphazard way of "humoral" (as opposed to "neuronal") communication within our bodies: By employing different types of receptor molecules at the surface of cells and within them, or by enzymatically cleaving and converting one hormone chemical into a range of other products, a single signal can be put into a diversity of actions, thereby changing the metabolism of cells, affecting behavioral dispositions, or even modulating gene expression.

Throughout development, from the fusion of egg and sperm, right up until death, all bodily and behavioral characters of individuals "develop," that is, change more or less gradually over time. These changes may be governed by genetic programs, which are themselves, modulated via a variety of environmental and bodily stimuli. Steroid hormones, for example are important and relatively direct mediators between environments and genes. Clearly, genes provide the frame, but may not have the final say in the formation of the phenotype. Hence, even a fully understood genetic code does not explain bodily or behavioral characters. Therefore, epigenesis, character formation on top of a certain genetic background, will be the big issue of the twenty-first century biology.

Hormones: What They Do, How They Do It and What They Are

Eunuchs exemplify how profoundly individuals are affected by hormones. Castrated to control their sexual desires (when employed as wardens of harems) or to preserve their soprano voices, eunuchs developed more adipose tissue and fewer muscles than intact human males; they lacked a beard and usually lived long. Sun Yaoting, for example, the last surviving eunuch of the Bejing imperial court died 1996 at the age of 94.

It was long known that removal of testicles will result in more tender and fat meat in poultry, pigs and other domestic animals. However, it was not clear that these profound differences in appearance and behavior, between intact and castrated males or between males and females of many species, are due to steroid hormones, notably testosterone and estrogen. It was a Swiss–German physiology professor at the University of Göttingen, Germany, Arnold A. Berthold, who in 1849 conducted a crucial experiment with chickens, confirming that the determining factor was produced in the testes and distributed via the blood stream (Nelson 2000). More than 150 years later we know that numerous hormones are produced by a variety of endocrine glands, which are linked with the brain and with each other via a complex network of neuronal and chemical signals and metabolic pathways.

When a relevant stimulus, for example the sight of a potential mate, reaches the brain, specific pathways relay this information also into a basal and phylogenetically old part, the hypothalamus. There, traces of specific hormones, the so-called "releasing factors" are secreted into the "portal vessel system" which connects the hypothalamus directly with a small gland at the base of the brain, the anterior pituitary. Despite its small size, this is the top command relay center of the body. It secretes the so called "trophic factors" into the blood stream, by which the pituitary commands all peripheral endocrine organs, such as testes and ovaries, adrenals, thyroid, and so on. In case of a sexual stimulus, for example, the small peptide GnRH (gonadotropin-releasing hormone) from the hypothalamus will trigger the release of the so-called "gonadotropins," the glycoproteins LH (luteinizing hormone) and FSH (follicle-stimulating hormone), which finally stimulate the synthesis of the steroid hormones testosterone in the testes, or estrogen in the ovaries. This makes the sexual system ready for action.

However, for successful mating it needs more than just an activated sexual system. Mating efforts often include contesting or even fighting competitors and going through a diversity of potentially strenuous courtship efforts. Hence, a supportive, ready-for-action body is needed, with plenty of glucose in the blood and a high blood pressure to be able to cope with the forthcoming challenges and with all unnecessary functions switched off. Therefore, in parallel to GnRH, the hypothalamus will also secrete minute amounts of CRH (corticotropin-releasing hormone), triggering the secretion of ACTH (adrenocorticotropic hormone) from cells of the anterior pituitary, which finally, stimulates the adrenal cortex to synthesize the so-called "stress hormones." Mainly the glucocorticoid steroid cortisol makes the body ready for action.

All hormone cascades are embedded into an elaborate network of feedback regulation, switching off their secretion or synthesis if no longer needed. A wealth of hormonal subsystems up- or down-regulate each other according to functional need. For example, a social stressor will up-regulate the CRH-ACTH-cortisol stress axis and will inhibit the GnRH-LH-sex steroid axis. If chronic and severe, such as in subordinates, which cannot avoid dominant individuals, this may lead to "social castration" and even to the death of the subordinate due to chronic stress Reproductive suppression by social stress may be a means of social control over the reproduction of other group members in social animals, such as wolves, or tupaias. Sexual inhibition and chronic disease by stress, for example due to high work load, or the other extreme, unemployment, may also be widespread in modern human societies. What is perceived as being stressful may be affected by evolutionary preadjustments (for example, the sight of a spider or snake), but is mainly determined by individual factors, such as experience, personality, status, and embedding in the social network.

Chemically, vertebrates have four hormone categories, which also have functional implications: 1) The peptide and protein hormones. These are chains of amino acids which can be stored in vesicles within the cell and are released upon a trigger signal. Thereby exactly timed signal peaks can be delivered. No wonder that the hypothalamic releasers and the tropins of the anterior pituitary are peptide hormones. 2) Steroid hormones are all synthesized from cholesterol, are therefore lipid soluble and may passively diffuse through membranes. Therefore, they cannot be packed and stored in vesicles and have to be produced on demand. This causes a lag of up to minutes from a triggering stimulus to a peak in the blood. Hence, steroids are less precisely timed than protein hormones, but may have longer-lasting, modulatory functions all over the body. Even though steroid hormones cannot be packed in vesicles, there is still a buffer system of inactive steroids in the blood, bound to relatively large carrier proteins. Only upon demand, the steroid is cleaved from the carrier and becomes active. Still other hormones are based on 3) monoamines, such as the "quick" stress hormone epinephrine, and 4) lipids.

Levels of Research

Neuronal and hormonal mechanisms cause behavior, just as an engine with all its subsystems makes a car run. They are, therefore, a focus of many behavioral biologists. Virtually all aspects of behavior are related to hormones, from simple housekeeping, such as the regulation of hunger and thirst, to sex differentiation and sex-specific behavior, reproductive and parental behavior, biological rhythms, social behavior, and learning and memory. And hormones affect "motivation," that is, the internal drive to perform certain behaviors. Thirst, for example is a powerful motivation to seek and ingest water. The main hormone involved is the peptide vasopressin, which is produced by brain neurons, stored in the posterior pituitary and released upon signals from hypothalmic centers, which monitor dehydration. It mainly enhances water retention in the kidneys and, together with other signals from brain osmoreceptors, stimulates drinking.

An amazing number of hormones are involved in learning and memory formation. Only at an intermediary level of arousal, mediated by epinephrine and cortisol, and concomitantly, at intermediary levels of blood glucose, learning is optimized. Also, intermediary cortisol and other steroid hormones have neurotrophic effects and further synapse formation and dendritic outgrowth. Even vasopressin, oxytocin and endogenous opiates are involved in learning. And sex-specific learning, such as spatial skills and concentration to some tasks are mediated by androgens, social orientation and verbal skills by estrogens.

However, the scientific world nowadays is no longer neatly separated in physiologically and evolutionarily orientated behavioral biologists. It is increasingly recognized that via hormonal systems, bodily functions are directly linked with evolutionary functions, that is individual reproductive success. For example, stimuli causing stress responses, the so-called "stressors" may not be the same for all individuals of a species. One of the major determinants of stress susceptibility and management is personality (i.e., the consistent style of individuals responding behaviorally and physiologically to the challenges of life) which affects the efficiency of the individual physiological machinery, because glucocorticoids mobilizes glucose from the liver and hence, channels energy into behavior. Furthermore, the personalities of individuals affect their performances within a social group and hence finally, their reproductive success.

Another area where hormones interact with evolutionary functions is individual development. For example, how genes are translated into characters is profoundly influenced by the early effects of steroid hormones. Thereby, mothers may have the potential of adjusting offspring to a predicted environment. Gulls, for example, nesting in a dense colony situation without much cover between nests, produce eggs with more testosterone in the yolk than females having a nest well covered by grass. From such high testosterone eggs, chicks hatch and defend themselves more aggressively against other colony members than chicks from low testosterone eggs. This is only one of a number of examples where it was shown that early androgen exposure modulates the personalities of vertebrates towards a more aggressive, "proactive" style of coping with the challenges of life. Via such "epigenetic" mechanisms, traits can be passed on over generations (i.e., a more aggressive female producing more aggressive and competitive offspring) without direct genomic involvement.

Early embryonic exposure to hormones will shape bodily characters, but also behavioral phenotype. Hence, early in ontogeny, hormones will mainly affect structures, whereas later, in the adult or sexually mature individuals, hormones mediate homoeostasis and modulate/activate behavior. Due to these central roles of hormones, behavioral endocrinology will be one of the hot spots of twenty-first century biological research.

Further Resources

Brown, R. E. 1994. *An Introduction to Neuroendocrinology*. Cambridge, MA.: Cambridge University Press.

Groothuis, T. & Meeuwissen, G. 1992. *The influence of testosterone on the development and fixation of the form of displays in two age classes of young black-headed gulls*. Animal Behaviour, 43, 189–208.

Koolhaas, J. M., Korte, DeBoer, S. M. S. F., VanDerVegt, B. J., VanReenen, C. G., Hopster, H., DeJong, I. C., Ruis, M. A. W. & Blokhuis, H. J. 1999. *Coping styles in animals: Current status in behaviour and stress-physiology*. Neuroscience and Biobehavioural Reviews, 23, 925–935.

Nelson, R. J. 2000. *An introduction to behavioral endocrinology.* 2nd Sunderland, MA: Sinauer Associates.

Sapolsky, R. M. 1992. *The neuroendocrinology of the stress response.* In: *Behavioral Endocrinology* (Ed. by J. B. Becker, S. M. Breedlove & D. Crews). Cambridge, MA, London: MIT Press.

von Holst, D. 1998. *The concept of stress and ist relevance for animal behaviour.* Advances in the Study of Behaviour, 27, 1–131.

Kurt Kotrschal

■|Horses

Twenty-five thousand years ago a tribal shaman drew a pair of spotted horses on a cave wall at Peche-Merle in the south of France to which were added six stencilled hand shapes. Then, in ritual trance state, the shaman reached through the membrane of rock into the spirit world of the horse beyond in search of healing and wisdom for his clan. Even at that time the lineage of *Equus* was rooted in the ancient past, far beyond the reach of ancestral memory. "Until about 1 million years ago, there were *Equus* species all over Africa, Asia, Europe, North America, and South America, in enormous migrating herds that must easily have equalled the great North American bison herds, or the huge wildebeest migrations in Africa," according to Kathleen Hunt, who has an on-line article called "Horse Evolution" (*talkorigins.org/faqs/horses/horse_evol.html#part9*).

It is easy to imagine the impact that the thundering hooves of such herds must have had on our early forefathers, but in order to gain a real understanding of the behavior of one of these species—*E. caballus*—we need to look back beyond the foundations of *homo sapiens* relationship with the horse, prise away the wrappings of myth, awe, and servitude and attempt to confront the vital nature of the animal found beneath. To make ready for the journey, we must relieve ourselves of the baggage of 6,000 years of civilization: shedding thought or image of bit or saddle, spur or stable, rider or chariot.

The animal humans might have seen approximately 6,000 years ago was a single-toed quadruped of about 12 hands in height, with a well-developed cardio-vascular capacity tapering through supple energetic loins to powerful driving hindquarters. The head is remarkably large and heavy in comparison with many other grazing species, although this doesn't necessarily mean that there is a particularly large cranial cavity because most of this is taken up by sensory organs. The weight of the head acts to pull the spine straight so that the propulsion developed in the quarters can be transmitted effeciently into explosive forward action. The large eyes are located on the sides of a face dominated by nose to such an extent that a blind spot is created. The ears are of a tidy size, extremely mobile and in constant motion. In fact the *Equus* fossil record is one of the most complete and informative of any known species. This description can very easily go on in great detail, setting out each small physical atribute and development back up the evolutionary chain, linking the rodent-like *Eohippus* with the modern horse. However, behavior leaves little physical record, and a wider range of skills are needed to report the behavior of early *E. caballus* above and beyond those of physical description.

All known present-day populations of *Equus caballus* are feral rather than wild, with the possible exception of a Tibetan pony breed, discovered during an expedition in 1995, about which very little is so far known. (Although descriptions of the pony sound like a smaller version of *Equus przewalskii* specialized for existence in the Himalayas there is, as yet, no firm evidence that this is *E. caballus*.) In order to describe naturally occuring behavior we shall have to focus on existing free-ranging and semifree-ranging groups and, with the benefit of evolutionary theory, extrapolate from there.

First and foremost *E. caballus* is and was a social animal that associates in herds. It is not found alone except when sick or injured and close to death. Herds comprise two kinds of social group—harem groups, in which a dominant stallion is accompanied by several mares plus their various progeny up to 3 years of age, and bachelor groups consisting of colts and stallions from 3 years upwards. As with all social groups there is a working pattern that controls the life of the group and in which individuals have both place and status. The pattern, which is responsible for social harmony, and sexual, parental, exogomatic and defensive strategies within the group, has two primary components, which may be characterized as genes and memes. Just as genes are the component parts that make up a genotype, memes are the components of culture.

Each animal possesses a genetic structure that predisposes it to certain basic or core behaviors that have served the previous owners of those genes well. But these genetic traits should not be equated with notions of present day "fitness." Many will have been the product of that which served their owners 10,000, 20,000 or even a 100,000 years ago—yet may have no great relevance in terms of present day conditions.

Throughout the long history of the horse's development "genes for behavior" have produced what some term "hard-wiring," as if such behaviors were set in stone; yet each of these behaviors is subject to a complex network of triggers by which they may be modified. Some have argued that the horse, as a herbivore, could not possibly sustain the dietary cost of a large brain, and therefore that behaviors must be automated rather than subject to the impact of reason or sentience, however basic. Yet the horse is so adaptable to a wide range of environments that, were this to be the case, the quantity of preprogramed information would need to be nothing short of phenomenal. It is then the interaction of genetic and environmental cues that serves as the basis for behavior—plus an "inherited" product of the environment we call culture.

We have now arrived at a didactic crossroads. Should culture be viewed as an inherently human, or at least primate, possession? Or, do many other species possess an, albeit rudimentary, cultural wealth that is passed from generation to generation? If we decide that the richness of human culture is so awesome that it is simply ridiculous to apply the same term to animals then we clearly need a new and different term to describe what other species do when behaviors are passed on within a group. Fortunately there is such a term: *memes*.

Because around 75% of the herd's time is spent grazing, it makes sense to focus on this element—and whether more or less time is spent this way is dependent on the environment, aspects of which include available plant species and pasture quantity/quality, season, and weather. Just as eating can be a social activity in the human species so it is with the horse, only more so. If we imagine an ongoing, long-distance, outdoor cocktail party, attended by the most practiced of party goers, "flowing" across the landscape while also comunally maintaining a defensive high alert and preparedness for rapid relocation, we are some way toward appreciation of *E.caballus* society. Recent research using computer modelling by Dr. Larissa Conradt and Professor Tim Roper at the University of Sussex has suggested that this type of movement is, in a sense, the product of an unconscious democracy, in which any individual may initiate movement in a new direction irrespective of age or status. Regardless of what may occur in eating behaviors in other species, the horse has grazing techniques of its own.

Before going into this in any more depth there is a need to establish some ground rules. First, the pattern is environment and group specific. What is observed in a group of horses living in flat plains territory may be very different from what would be observed were the same group to be living among rolling hills with traceries of small creeks and shallow ponds. Part of this difference will be due to the way in which the environment "triggers" innate behavior, the genetic component. Part will be due to the individuals that comprise the group, and another part of this difference will be due to the aquired knowledge the group has built up regarding the area, the "memetic" component. This compound nature of behavior introduces some restraints on what can be safely stated. It is quite reasonable to record the behavior of a set group in a particular territory and to present the observations as one variable of horse behavior. But until such time as other groups are observed to behave in very similar ways in the same type of territory it is dangerous to present the observations as "general horse behavior." If that were the aim, then a method would need to be used to remove the meme component of the group's behavior from the equation. Quite how this might be done with a reasonable level of objectivity is difficult to say.

And now, having sorted out the ground rules, the theme of grazing technique can be revisited. Take as an example a particular harem group comprising 22 individuals, grazing a gently sloping hillside over a 2-hour period from 4.00 P.M. until the onset of dusk around 6.00 P.M. on a dry evening with light winds in late autumn.

Note that the observation states a specific time, for, as with people, horse behavior is subject to change throughout the day. Weather and season also have to be taken into consideration, since both have an impact.

There are two pivotal individuals in the group, the harem stallion and head mare. Together the two form a high status nexus about which the life of the group revolves. Given

A) Low/Middle Status coalition B) Stallion and Head Mare C) Foreguard D) Highland "point" E) Lowland "point" F) Herd center G) Safe rest area H) Rearguard trio.
Courtesy of Andy Beck.

that these two are the most important members, it would be logical to expect them to be at the center, where there is the least chance of attack, and from which the safety of the group can be best coordinated—and this is exactly where they are found for the greatest portion of time. Lower status members spend decreasing amounts of time at the center and are more often to be found on the periphery of the herd. It is these lower status members that head the direction of grazing, with perhaps two or three relatively close together. Other low or middle status individuals take "point" on either side, and a minimum of two at any one time, either at the back or at the wings toward the back, form a rearguard, facing in the opposite direction to that in which the majority are moving.

As the herd flows forward the rearguard and individuals on the periphery change position periodically as one horse falls to the rear and another turns and moves closer to the center. The same kind of changes are occuring at the leading edge of the herd, although one middle status mare, which, for the sake of clarity, we might term the "pathfinder," is more often than not close to the front or actually leading.

This configuration creates an effective safety zone around the group. Information regarding threat perception is rapidly transmitted to the center, where the most trusted individual, the head mare, can then make a decision regarding flight, and the stallion, whose job it is to keep the group together, to prevent stragglers from falling behind, and, depending on the nature of the threat, to fall back and confront the source, can take up his protective role with the least possible delay.

It is virtually impossible to approach the herd without being noticed except by patient and careful use of cover, and even then the approach must be made from downwind. Nor is any one individual left to take point or peripheral sentry duty for any great period of time. Leading members such as head mare, pathfinder or stallion are protected but at the same time may cause the group to change overall direction purely by the use of subtle body language, or may actively block further movement in any particular direction.

What may previously have appeared as random movement is now revealed as the highly ordered and efficient workings of a social group that encompass both fixed and flexible enactment of role, status, discipline, centralized critical decision making, preventive crisis relocation, cooperative task sharing, risk management, and more.

Whatever the mixture of instinct and sentience may be that produces this complex result, there is no doubt that it bears great similarity to the products of human social groups. Even though it may be argued that people utilize a greater degree of sentience in social decision making, we cannot escape the conclusion that horse society, as with so many other species, is more complex than previously allowed—and that the management of domestically kept horses should acknowledge this reality.

See also Horses—*Behavior*
Horses—*Horse Training*
Horses—*Sleeping Standing Up*

Further Resources

Jackson, J. 1992. *The Natural Horse*. Flagstaff AZ: Northland Publishing.
U.S. Department of the Interior, Bureau of Land Management: Wild, Free-Roaming Horses—Status of Present Knowledge. Technical Note.

Andy Beck

■|Horses
|*Behavior*

Horses' behavior is worthy of study from a basic science perspective because they are Perissodactyla—single-hooved, large, nonruminant herbivore. They are the only domestic branch and many of their relatives such as the wild donkeys and some species of zebras are endangered as are the more distantly related tapirs and rhinoceroses, They also are charismatic megavertebrates and the most popular of the companion animals after cats and dogs. There are 5 million horses in the United States used mostly for recreational riding rather than traction or transport. In the developing world, horse and donkeys are used for plowing as well as transport of humans and goods.

Domestication

Horses were first domesticated 1500 years ago. At first, they were eaten, and only later used for traction and for riding. The first domestication probably occurred in Eurasia where horses are still an important part of the culture today. Once horses were domesticated, they were selected for different purposes. The original horse domesticated was probably relatively small, and the native ponies found today across Eurasia are of a similar type. Stallions of lines selected for a given purpose were imported to "upgrade" the native horses. The northern Europeans selected for large size, at least in part to carry a knight and his armor. These animals became the draft horse breed such as the Belgiums and Clydesdales. The North Africans chose for speed. These became the Arabian breed and later, when crossed with British horses, the Thoroughbreds. There are now hundreds of breeds of horses and ponies, but their behaviors are very similar.

Most horses are now domesticated, but there are true wild horses—Przewalski's horses or the Mongolian wild horses. These animals have two more pairs of chromosomes than domestic horses, but can interbreed and produce fertile offspring so they are closely related. The Przewalski's horse became extinct in the wild, but was preserved in zoos and recently was reintroduced into Monoglia where they live and breed successfully in a large national park.

The horses often called wild horses or mustangs are actually feral, that is, descendents of domestic horses. There are large numbers of these horses in the Great Basin area of the American West and small populations on Atlantic barrier islands from Sable Island off the Canadian coast to Cumberland Island off the coast of Georgia. The most famous are the Chincoteague ponies immortalized by Marguerite Henry in her *Misty of Chincoteague* series of children's books. These ponies still live in a national park on the eastern shore of Maryland at the Virginia border. A visit there will reveal many of the behaviors described below.

Social Structure

In the natural state, horses live in groups. There are two types of groups: adult and subadult. The adults live in family units called *bands*. Bands that share a common area are called a *herd*. The adult group is composed of one, or rarely, more than one stallion, several mares, and their offspring—colts and fillies. The colts almost always leave, but will leave sooner if there are no other colts with which to play. Most of the fillies eventually leave, too. The colts leave to join the subadult group, the bachelor band. The *bachelor band* is composed of young males, usually less than 5 years old. They spend a lot of their time playing, especially

Domestication and Behavior

The relationship between border collies and sheep illustrates how people have modified animal instincts, in this case the predatory nature of wolves, to suit their own purposes.
Courtesy of Viv Billingham-Parkes, Tweedhope Sheepdogs

Development

This young Japanese macaque learns about his environment through play and exploration.
© Fritz Polking / Visuals Unlimited.

Dominance

A male gelada baboon, with fangs bared, chases after one of his family females who strayed too close to a bachelor group. The female dodges him and shrieks for backup from other family females.
© Michael Nichols / National Geographic Image Collection.

Emotions

Both male and female cassowaries are brightly colored, and the intensity of their colors varies directly with the intensity of their moods.
© *Medford Taylor / SuperStock.*

A juvenile black-crowned night heron shows incredible patience as it waits unmoving for its prey.
© *Anthony F. Chiffolo.*

Feeding behavior

A grizzly bear forages for berries during an afternoon on the tundra.
© *Anthony F. Chiffolo.*

Hibernation

A sleeping gray squirrel.
© Kenneth M. Highfill / Photo Researchers, Inc.

Human (Anthropogenic) Effects

City raccoons (Procyon lotor) raid a garbage can in Portland, Oregon.
© Michael Durham / Visuals Unlimited.

Lemurs

Ruffed lemurs almost never walk on the ground; they prefer to remain high in the canopy.
Courtesy of Getty Images / PhotoDisc.

Parasite-Induced Behaviors

Chickens and roosters take dust baths to help control mites, insects, and other parasites that might otherwise plague them.
© Anthony F. Chiffolo.

Navigation

Monarch butterflies exhibit an amazing ability for navigating long distance journeys. They can migrate as far as 3,000 miles from eastern Canada to their overwintering sites in Mexico.
Courtesy of Corbis.

Endangered leatherback turtles, that are 6–8 feet in length and weigh as much as 1,500 pounds, have been observed 3,000 miles from their nesting grounds. The mechanisms of sea turtle navigation have been intensely investigated, but the cues or sensory systems involved are still unknown.
© Meera Anna Oomen and Kartik Shanker.

Play

It's unusual for wild adult animals to play, as these two tigers are doing.
© Mark Newman / SuperStock.

A kitten plays with a wildflower. Many young animals play with vegetation or objects, as well as with each other.
© Alan & Sandy Carey / Photo Researchers, Inc.

Opposite: Sea lions are known for their curiosity and playfulness. In this photograph, two sea lions take time to play.
Courtesy of Corbis.

Predatory Behavior

In spite of their natural predators having been extinct for more than 10,000 years, American pronghorn still maintain the skeletal and muscular development that they had when cheetahs and other large predators roamed the central prairies of North America. Extinct predators are called "ghost predators."
Courtesy of Corbis.

The reintroduction of gray wolves to Yellowstone National Park has provided an opportunity for researchers to learn about how the wolves' prey, such as this elk, react to the presence of this new threat.
© Tom J. Ulrich / Visuals Unlimited

play fighting. This prepares them for the serious fighting they will do later as adults with band stallions to obtain their own bands. Fillies usually leave their mother's band and either join another established band or encounter a young stallion and begin a new band.

Within a band there is a social organization or hierarchy. The oldest mares are usually dominant, and they will have first access to any scarce resource, a place in the shade, salt, or an especially palatable patch of grass. As new mares join the band they will be subordinate to the original mares. Within the bachelor herds, the dominant stallion is usually the first to acquire mares. If

A chincoteague stallion intimidates other ponies in his pen.
Medford Taylor / National Geographic Image Collection.

there is more than one stallion in a band, the dominant animal will do most of the breeding while the subordinate fights off any marauding bachelors.

Some fillies join their half brother's (same father–different mother) band and a few remain in their mother's band. The more closely they are related to the stallion, the fewer foals they will have as adults. Mares in their father's band have the fewest, mares in their half brother's band have an intermediate number of foals, and those in the band of an unrelated stallion have the most offspring. Inbreeding reduces the chances that the fetus will survive to be born or live to adulthood.

Communication

Horses communicate with one another using visual, auditory, and scent signals. The visual signals are the posture of the horse and his ear, lip, and tail position. An alert horse stands tall with his head up and ears pricked forward. A relaxed horse lowers his head, relaxes his ears, and his lower lip may droop. An aggressive horse flattens his ears back and lashes his tail. He wrinkles his nostrils and may thrust his head toward his victim. When even more aggressive, he will show his teeth and even bite. The frightened horse holds his ears to the side and tucks his tail close to his body. The horse who perceives something enjoyable, especially pleasant tactile sensations, will have an elongated upper lip, "the pleasure face." An exhausted horse will have an elongated lower lip as if pouting. Kicking occurs usually as a defensive aggression as a horse flees from attack, but can also be offensive when both horses turn and kick at one another. Horses may also strike with their front limbs; this is more common in stallions than mares.

When two horses meet, whether for the first time or after a separation, they greet one another by placing their nostrils together. Usually they arch their necks at the same time, and one or both may squeal and strike.

There are some special visual signals given by only one class of horses. Examples are "snapping" by immature horses, "snaking" by stallions, and the posture of a mare in heat. Snapping is also called champing or tooth clapping. The foal or young horse opens and closes his mouth, exposing his teeth, usually with his ears pricked forward and his limbs bent. He does this when he is frightened, for example, when nearby horses are fighting or

when a stallion is breeding his mother. Colts exhibit the behavior more than fillies and do so in approach–avoidance situations. A typical situation occurs when the colt, on his own initiative, interacts with the stallion. Snapping may signal that the horse is immature, and, therefore not a serious threat. It is rarely observed in stallions older than 2 years. Mares may exhibit snapping as adults, but only when they are sexually receptive and approach a stallion, another approach–avoidance situation. A more obvious sign of sexual receptivity or heat is the frequent urination with the tail held to the side and the clitoris exposed repeatedly. The latter behavior is called *winking*. The mare also holds her ears turned back, but not flattened and her lips are relaxed; this is called the *mating face*.

Stallions drive their mares away from other stallions with their heads down and ears flattened, while moving in a serpentine fashion. This particular action is, therefore, called *snaking*, but is also termed driving or herding. When displaying to a rival stallion, a horse will arch his neck (also called an arched neck threat) and trot with exaggerated stepping action.

Flehmen is an interesting facial expression of horses in which the horse raises his head, exposes his teeth, half closes his eyes, and curls up his upper lip. This is a very obvious behavior, but apparently is not a visual communication primarily, but related to perceiving chemicals in the air. Horses, in particular stallions, show flehmen when they encounter an odor, usually that of urine, which merits further investigation. What the horse is doing is drawing material into a special structure, the *vomeronasal organ*. He touches the urine with his upper lip, raises that lip so that the urine runs into his nostril and into the opening of the vomeronasal organ, a paired, hollow tubular structure that lies on either side of the nose between the hard palate and the nasal cavity. After the urine has been sucked in, presumably detected by the receptors in the organ, it is flushed out again with mucus secreted by the vomeronasal organ. The mucus can be observed dripping from a stallions' nostrils after he has exhibited flehmen. Although urine is the usual stimulus to flehmen, other unusual odors can stimulate flehmen in stallions, geldings, mares or foals. Flehmen permits the horse to investigate

"Welton Louis" displays flehmen, the curling back of the lip, at Linbury Stud Farm.
© *Kit Houghton / Corbis.*

nonvolatile substances, especially those in urine, and perception of those odors may stimulate hormone release. Stallions routinely mark urine with their urine immediately after they exhibit flehmen. The sniff–flehmen–mark sequence may be repeated several times.

Urine odors seem to be important to horses, but so are the odors of feces. In fact, horses seem to be able to distinguish sex from the smell of feces, but not of urine. Stallions will defecate on top of feces of mares, possibly to mask their sex or to signal other stallions that these mares are accompanied by a stallion. When approaching one another, two strange stallions both will defecate and smell the feces, perhaps to assess one another's strength.

There is also an odor that is produced by the skin lying between the teats of the udder. This odor, particularly if it emanates from a mare who recently foaled, seems to have a calming effect. It has been synthesized as equine appeasing pheromone.

Vocal signaling in horses is relatively simple. Horses *neigh* or *whinny* when separated from one another. It is a loud call. A much less intense call is the nicker. *Nickers* are low amplitude calls, usually given by a mare to her foal or vice versa. Some horses nicker to their human caretakers, especially for food. *Squealing*, as mentioned above, is an aggressive call given just before overt biting, striking, or kicking commences. Stallions roar, which seems to attract sexually receptive mares. Horse *snort* when they are startled.

Patterns of Behavior

Horses in a naturalistic environment spend 50% of the time grazing, 15% sleeping, 20% stand resting, and 5% in other activities such as walking, grooming, and so on. The major behavior, grazing, involves selecting a patch and a plant or two within that patch, grasping the grass with the upper lip or teeth, biting if off, and chewing it. Every few bites, the horse takes a few steps forward to reach a new patch.

Sleeping Standing Up

Katherine A. Houpt

Sleep can be classified into two types: the "sleep of the mind," slow wave sleep (SWS) or quiet sleep; and the "sleep of the body," paradoxical, active, or rapid eye movement (REM) sleep. Although overall muscle tone is very low during paradoxical sleep, the muscles of the eyes frequently contract; hence the term rapid eye movement.

The two types of sleep can best be differentiated from wakefulness and from one another by measuring brain waves using an electroencephalograph. The electroencephalogram (EEG) of the alert animal is characterized by low-voltage, fast waves that are not synchronized. Slow wave sleep is characterized by synchronous waves of high-voltage, slow activity; therefore this type of sleep is called sleep of the mind. During REM, the EEG shows low-voltage, fast activity similar to that seen in the wakeful state but with very little muscular activity; therefore, this type of sleep is called the sleep of the body. The animal is more difficult to arouse than when it is in SWS. Humans awakened from REM sleep report that they have been dreaming; the twitching of the legs (which are not completely inhibited) and whinnying during equine sleep indicate that horses may also be dreaming. We can only speculate as to the presence or content of pony dreams.

Of the domestic mammals, only horses are able to sleep standing up. This is possible because of the unique arrangement of the ligaments, tendons and muscles which stabilize the joints in their legs. This arrangement is called the *stay appartus* and allows horses to stand with minimal muscle energy. When horses rest they usually flex one hind limb, thus taking the weight off it and onto the other hind limb. The opposite hip is higher because the pelvis has tipped. When that happens the ligaments of the patella or kneecap shift so that the knee is locked in position which minimizes muscular activity even in the supporting limb. Horses can stand up during SWS sleep, but must lie down either on their side or on their sternum or chest with their muzzles resting on the ground to support their head during REM sleep. Horses very seldom lie down during the day. Instead they lie down in the middle of the night, go into REM sleep and then arise again, only to lie down a few minutes later. This probably evolved as vigilance behavior in this prey species.

In summer, the horses rest in the shade in the middle of the day to avoid the heat and insects. Grazing periods occur at the beginning and end of the day and continue at night. Horses lie down for short periods, usually during the middle of the night.

Horses groom themselves mostly by swatting at their skin with their heads, by rubbing their heads on a forelimb, by rubbing their rump on trees or fences, or by scratching their head or neck with a hind hoof. Horses also groom one another. They stand shoulder to shoulder and grasp a fold of skin between their incisors and pull. Most mutual grooming is done by preferred associates and is most common in the spring when horses are shedding. Stallions groom mares during courtship, but in that situation they usually groom around the hindquarters. Foals, especially fillies, mutually groom more than do adults. When flies irritate the horses, they may stand head to tail with another horse so that flies are displaced from both horses' faces. Not all horses do this, and what determines which pair will stand head to tail is unknown.

Sexual Behavior

Free-ranging mares will often seek out a stallion while they are in heat or estrus and display in front of him. It is unusual for free-ranging 2-year-old mares to breed and even uncommon for 3-year-olds to breed. Stallions generally do not exhibit much interest in the sexual displays of young mares. Yearling mares may show very exaggerated signs of heat and may attract males from bachelor herds. It is rare, but not unknown, for 2-year-old feral horse mares to deliver a foal, indicating that they can conceive as yearlings. It is hypothesized that the exaggerated signs of heat in young mares may be a means of attracting stallions from a distance (that is, unrelated stallions). Incest is unusual in free-ranging bands; mares either leave the band when sexually mature, are "stolen" by a stallion forming a new harem, or the herd stallion is replaced by a young stallion.

Horse are similar to humans, but unlike other domestic animals, in that they can breed a female only if they have full penile erection, and this is correlated with the degree of sexual excitement. Thus, an adequate period of sexual foreplay is essential. Stallions may tend a female for several days before she is fully sexually receptive. Nipping and nuzzling begins at the mare's head and proceeds gradually along the body of the mare to the tail area. During this testing phase, he exhibits the flehmen response. As sexual excitement increases, the male calls with neighs and roars. The female allows the male to lick her around the rear legs and back. Full erection usually develops over several minutes in the mature stallion. Several mounts are usually made before he can insert the penis and ejaculate. During copulation the stallion rests his sternum on the mare's croup and may reach forward to bite her neck.

The stallion is the sire of foals born in his band in 85% of the cases. In the other cases, the mare has been bred by bachelor stallions or the stallion of another band. In bands with more than one stallion, one horse—presumably the dominant horse—sires most of the foals.

Maternal Behavior

Afer an 11-month gestation period, mares usually withdraw somewhat from the band to foal. Labor is short and the mare lies down for expulsion of the foal. After resting on the ground, a behavior that allows blood from the umbilical cord to reach the foal before the cord breaks, she stands, breaking the cord and begins to attend to the foal. Maternal behavior begins with sniffing the foal. The mare may place her nostrils against the foal's in the

greeting ritual exhibited by adult horses. The licking behavior continues for the first hour after the foal's birth
and probably stimulates the neonate to stand as well as
drying it. This is the only time she'll do so, but it seems to
be important in identifying the foal as her own. Later, the
mare may sniff under the foal's tail when it attempts to
suckle, presumably to confirm by odor that this is her
foal. She will not let other foals nurse. During the first
hours and days after birth of the foal, she will try to keep
other horses away, including her older offspring, but later
will allow her older foals and some other mares to approach the foal.

She will nurse her foal frequently for the first year of
its life, weaning it just before the next foal is born. The
foal follows her, but when it lies down she stands over it.
During the first weeks she will simply stand over the foal.
Later she will graze as it lies, and she will graze further
from the foal as it grows older.

Foal Development

Foals are *precocial*, able to stand and walk within 1–2
hours of birth. Once they are able to stand, they seek the
udder. They will poke their muzzles into an overhanging

*A captive mother stands with her foal on a
farm in Wisconsin.*
Courtesy of Corbis.

surface which in the wild is most likely to be the mare's belly. They grasp any surface but don't
persist unless it is a hairless surface. When they do find a teat, they will suckle on it. Finding
the teat is important, not only for nourishment, but also for disease prevention. For the first
few hours after birth, the foal can absorb antibodies through its intestines. These antibodies in
the *colostrum* or first milk will protect the foal from disease until he is able to manufacture his
own. If he doesn't drink enough colostrum, he will die of an overwhelming infection.

Young foals suckle and sleep most of the time with a few bouts of exploration in between. They suckle for a minute or two every 15 minutes. This pattern will not change
much for the first 6 months, although the interval between suckling bouts gradually increases. Suckling serves two purposes: nourishment and solace. If a foal is frightened, it will
suckle immediately. Foals spend much more time lying down than adult horses do. Like
most infant mammals, they need more REM sleep.

Foals and lambs are followers, whereas calves and kids are hiders. Hiders stay in one
place, usually a well hidden spot to which the mother comes periodically to nurse the offspring. Followers follow their mother. The exception to following occurs when the foal is
sleeping when the mare stays nearby.

Foals play, usually in early morning and late evening, when the environmental temperature
is moderate. At first, their play consists of exploring their immediate vicinity and their mother
by nibbling on her mane. They gradually play further and further from the mother. Most play
consists of galloping and bucking alone or in a group. Group play differs between the sexes. Fillies are most likely to mutually groom, and colts are most likely to play fight and mount.

Foals begin to attempt to graze on their first day of life, but probably don't ingest much,
both because their teeth have not erupted and because it is difficult for them to reach the
ground; their limbs are disproportionately long. Over the next few months, they will graze

more and more of the time, reaching 50–60% of their time, the adult percentage by 6 months of age. They don't eat at random times, but instead eat when their mother eats.

Foals often eat feces, usually those of their mothers. *Coprophagia* may be a means of identifying the plants the mother has eaten or a means of introducing bacteria and protozoa into the gastrointestinal tract to allow cellulose ingestion.

The equine behaviors described are seen in free-ranging horses, but any elements of these behaviors can be seen in confined horse in corrals, pastures, race tracks and backyards. One does not have to travel far to be an equine behaviorist.

See also Horses—Horse Training
     Horses

Further Resources

Houpt, K. A. 1998. *Domestic Animal Behavior for Veterinarians and Animal Scientists*. 3rd Ed. Ames: Iowa State University Press.
McDonnell, S. 2003. *A Practical Field Guide to Horse Behavior. The Equid Ethogram*. Lexington, KY: The Blood-Horse, Inc.
Mills, D. S., Nankervis, K. J. 1999. *Equine Behaviour: Principles & Practice*. Malden, MA: Blackwell Science.
Waring, G. H. 1983. *Horse Behavior. The Behavioral Traits and Adaptations of Domestic and Wild Horses, Including Ponies*. Park Ridge, NJ: Noyes Publications.

Katherine A. Houpt

■ |Horses
|*Horse Training*

There's something inherently invasive to climbing on a horse's back, right where a predator attacks. Most riders or trainers heavily underestimate the sensitive, intelligent, and cooperative being that the horse is—all because we start out with assumptions on how a horse should behave. We rely on what we *think* all horses should know without actually knowing whether *this particular* horse knows. Much of what we do with horses relies on tradition, not on individual skill or knowledge, let alone an understanding of the individual horse. That knowledge doesn't come out of the blue, of course, but a good place to start is by studying the species.

Training is behavior modification. At one end of a continuum there is the "raw," inherent, "natural" behavior of a horse, and at the other end there is what we want to train the horse for: to behave in a way that is useful to humans.

Although talented trainers have a great intuitive understanding on how to gain a horse's cooperation, all trainers can when they understand why a horse behaves the way it does. We can learn to see the horse as an adult with responsibilities, instead of some kind of detached, withdrawn, or even social interaction-impaired child who needs to be pampered and protected because "he can't think for himself." Horses are not dumb. We want them to perform in the human world and to meet our desires, but we tend to forget how hard that is for horses—considering that they sometimes look like they come from another planet. They are, however, from this world. It's just that they are another "nation," speaking another language, but their behavior is driven by the same needs that we have: food, safety, shelter, and so on.

As Henry Beeton said in *The Outermost House*, in 1928:

> We patronise them for their incompleteness, for their tragic fate of having taken form so far below ourselves. And therein do we err. For the animals shall not be measured by man. In a world older and more complete than ours they move finished and complete, gifted with extensions of the senses we have lost or never attained, living by voices we shall never hear. They are not brethren, they are not underlings; they are other nations, caught with ourselves in the net of life and time, fellow prisoners of the splendour and the travail of the earth.

The principle to live by while training is simple: "The horse is never wrong." Horses either behave the way nature programmed them, or they behave according to what they learned from previous experiences. One way or the other, they do what they do because the behavior was reinforced in the past.

This puts all responsibility on humans: the rider or handler. If the horse doesn't do what you want—and he shows you that through his behavior—either he isn't able (yet) to do what you asked (mentally or physically), or you didn't ask the question well.

In order to know how best to react to the responses you get, humans can rephrase their next questions. For the most success in horse training, we should "know our species."

The behavior of a horse is driven largely by four fields of influence:

1. Anatomy and Senses

The horse has better hearing than humans, is mildly colorblind (according to the latest research, he lives in a mostly grey-yellowish world) with less visual acuity, but has a panoramic view; he's built for slow movement while grazing, but also for short explosions of fast fleeing; he can adapt well to changing (and low) temperatures.

Stabling a horse goes against his free roaming nature—it may be convenient for humans, but can hinder a horse's physical and mental soundness. Or take a riding example: Riding a horse on the bit requires a lot of trust in the rider since this limits the horse's view to the ground right in front of his nose. It's obvious that he can only relax in that position if he trusts his rider completely.

Horses are very emotional animals with fear as one of the top emotions. They are programmed to notice any small change in the environment, since anything unusual could very well mean that a predator is sneaking up behind them. You may not see anything, but if he explodes, the horse certainly feels that he has a reason for exploding.

2. Ethological and Instinctive Patterns

Horses are prey animals, programmed for noticing small changes and fleeing before thinking. For example, if a rider leaves the bolting horse without interference, it will stop after about 400 m (1,300 ft)—the average distance where a predator gives up the chase.

Horses hate to be on their own—safety lies in numbers, and there is a lot of social interaction as well. The horse is programmed to follow any fair leader and will show no need to question a leader's authority if this leader is quiet, dependable, and consistent. The herd hierarchy may be questioned by those ranked in the middle range, but horses also have special friends within the herd, and that choice doesn't depend on the ranking. Number one may choose number eleven as a grooming mate, but at the same time bully him away at dinner time. The message here is clear: The human has to learn to act as a quiet leader, and can be a friend at the same time. And what's more, the emphasis on being the highest rank should not be the core of training philosophy.

Horses have a lot of body language by which they communicate. Horse training is more understandable to them if it builds on ways of social interaction that they recognize.

3. Cognition

Horses have extremely good long-term memory and learn best by repetition and building routines. They may be slow starters, but once they gain knowledge they easily form concepts and remember them for a very long time. This is true of the good, but also the bad experiences!

So, take your time to learn the basics of riding and handling, and then go slowly at explaining them step by step to your horse as well. It pays off in the end, and that's not only training experience, but confirmed by scientific research.

4. Physical Condition

The only way a horse is able to tell us that he is in physical discomfort or even in pain, is through a change in his behavior. Like humans, he'll have trouble concentrating and performing. Sudden, but also gradual changes in behavior may have a physical cause.

When you train horses, remember always that you are molding them to behave in very unhorselike ways. Most things that you expect from horses go against their natural instincts. So, empathy is essential. A horse is a horse, and we must adapt to the horse at least as much as they adapt to us.

To be a good horse trainer, learn as much of the horse's natural behavior as there is to know, top that knowledge with some readings on learning theories, add a flavor of behavioral analysis, and the result is that you will be a better trainer than you thought possible.

Further Resources

Budiansky, S. 1997. *The Nature of Horses*. New York: The Free Press.
Kiley-Worthington, M. 1987. *The Behaviour of Horses in Relation to Management and Training*. London: J.A. Allen.
McGreevy, P. 1996. *Why Does my Horse. . . ?* London: Souvenir Press.
Skipper, L. 2001. *Inside your Horse's Mind: A study of Equine Intelligence and Human Prejudice*. London: J.A. Allen.
Waran, N. (Ed.). 2003. *The Welfare of Horses*. Dordrecht, The Netherlands: Kluwer Academic Publishers.
Williams, M. 1976. *Horse Psychology*. London: J.A. Allen.

Inge Teblick

■ Human (Anthropogenic) Effects
Bears: Understanding, Respecting, and Being Safe around Them

By the time you were 1 or 2 most of you had met your first bear. It was probably fuzzy and soft, with a pleasant, attractive face, and you probably cuddled it and felt warm and friendly. Entire stores are devoted to selling these critters. Each year millions of new ones are added to the North American population.

What about their wild brethren? When young, cubs look cuddly, but only their mother or siblings get this privilege. Bear mothers nurse, clean, instruct, and defend their cubs for several years from dangers such as aggressive other bears. Sometimes this defensive instinct is directed toward humans who suddenly appear nearby and threatening. Even in this context physical attack is rare. This is fortunate because bears are powerful animals with compact, well-muscled bodies and built in weapon systems in the form of jaws and teeth, and strong limbs with claws. Humans are physically weak compared to bears and in hand-to-paw encounters far more subject to injury. Even a 50 lb (22.7 kg) bear is stronger than most people.

There are three species of bears in North America: American black bears (*Ursus americanus*), brown (also called grizzly) bears (*U. arctos*), and polar bears (*U. maritimus*). Although any adult bear has the strength to seriously injure or kill a human, they seldom do this. From 1990–2004, each year an average of three people were killed in North America by a bear, far fewer than killed by dogs, bee stings, or lightning. There are about a million bears (900,000 black bears, 60,000 brown/grizzly bears, and 15,000 polar bears) in North America. Each year there are many millions of interactions between people and bears. Only a small fraction of these are aggressive interactions. The rest usually bring joy to people. Once you learn the basics of safety around bears your chances of injury can be minimized. Seeing bears can become a joy to anticipate. There are few more delightful pastimes than watching bear cubs or even adults playing as they roll, run, jump, and pounce on one another.

In 1967 I finished a Ph.D. in animal behavior. That year in Glacier National Park the first two known grizzly bear-inflicted fatal attacks occurred. There was speculation that these unusual and horrific events might have been influenced by lightning, or that the bears had been fed LSD (this was the hippy era). My research showed that a common theme in this and other similar later attacks in Yellowstone and Glacier National Parks was that the offending grizzly bears had been allowed to feed boldly on human food or garbage. Thus, aggressive, food-seeking behavior became associated with humans and their food and garbage. At some point in this learned sequence, a few bears substituted people for food and garbage. Once this was understood, parks and other areas that protected bear populations and had hiking and camping, made human food and garbage unavailable to bears. Injuries from brown/grizzly and black bears decreased dramatically. Keeping our food and garbage away from bears is fundamental both to their safety and ours.

Bear attacks can be put into two categories, offensive and defensive. Offensive attacks occur when a bear wants something, such as the right of way on a path, or people's food that it has learned can be had by acting aggressively. An offensive bear attack may also occur when the bear is after a person *as* food. Although this is extremely rare, it does occur. Black and grizzly bears eat a lot of nutritious green vegetation, roots (especially brown/grizzly bears), berries, and nuts. However, one of the best potential sources for lot of calories is animals like deer, elk, moose, or salmon. Occasionally bears are able to kill or scavenge on these animals. Very rarely does a bear treat a human like potential prey. Behaviors preceding such attacks include: following a person, quiet but fast approach, typically no vocalization, ears rotated forward, and the bear not showing signs of stress. We need to convince such a bear that we are not easy prey. This can be done by shouting, acting together, using weapons if available—including even stones and stout sticks. Cayenne pepper-based "bear spray" is marketed to repel potentially or actually attacking bears. Research has shown encouraging, but not 100% effective results. Even so, carrying and learning how to use bear spray is a good idea.

Bears don't like surprises because a nearby animal may mean an attack. Therefore, a person suddenly appearing nearby often stresses a bear, causing it to salivate, chomp its jaws, vocalize, lay its ears back, or run toward or away from the person. Rarely does an actual attack follow. Potential attacks can often be defused by standing one's ground if the bear is approaching (bears often claw or bite other bears that run away), watching the bear and backing away if it is not approaching you, using bear spray if the bear is close, and avoiding surprising grizzly bears especially. Research suggests that, if attacked by a bear acting defensively, injury is minimized by playing dead. Obviously, you must never play dead if a bear is acting as an offensive predator.

Describing such situations sounds scary, and distinguishing between a defensive and a predatory attack is difficult. Fortunately, as I have stressed, bear attacks are rare. Your behavior can help reduce the slight chance of bear attack even further. Your first action if you encounter a bear is to stop and think about the situation. You can probably just walk away watching the bear. However, if it comes in your direction you should stop. Try to determine if the bear is acting offensively or defensively and behave accordingly. Bears may also approach because they are curious. This may seem offensive and could become so if you don't stand your ground and act as if deterring an offensive bear.

Learn as much as you can about the behavior and ecology of bears. Learn where and when to expect bears and what their behavior means. This is useful in avoiding and managing unwanted encounters. During one study of grizzly bear ecology in Banff National Park, my students and I put in over 10,000 hours on the ground and in areas where grizzly bears were common. We were threatened a few times as bears acted defensively when we were nearby, but no injuries or even very close calls occurred.

Bears are a fascinating part of our natural heritage. Respecting, understanding, and supporting their needs are critical to wildland conservation. By understanding bear behavior and ecology and supporting maintenance of bear populations, we demonstrate our ability to live without destroying the ecosystems that support bears, ourselves, and all other life.

See also Humans (Anthropogenic) Effects—*Human (Anthropogenic) Effects on Animal Behavior*

Further Resources

Herrero, S. 2002. *Bear Attacks: Their Causes and Avoidance*. Guilford, CT: The Lyons Press.
Safety in Bear Country Society. 2002. *Staying Safe in Bear Country*.
   This is an excellent video, CD or DVD and is available by calling 1-800-667-1500 or e-mailing info@micworld.com

Stephen Herrero

■ Human (Anthropogenic) Effects
Edge Effects and Behavior

Habitat characteristics strongly influence animal behavior. Edge habitats, especially edges created by human habitat modifications, have generated a great deal of scientific interest because they can affect wildlife movement patterns, reproduction, habitat selection, and foraging behavior and efficiency. Edges are areas where two different habitat types meet, often referred to as *ecotones* by ecologists. Edges occur naturally, such as a pine forest bordering a grassland meadow, or may be human-caused, such as the same pine forest bordering a housing subdivision or a shopping mall, where the trees have been cut down or the meadow paved over.

The abruptness, or extent of change, along a habitat edge may influence the degree of edge effects. Edges created by natural disturbance processes (fires, floods, treefalls) are typically more gradual than human-caused edges and tend not to have the negative impacts associated with human-induced edges, probably because species have evolved in these systems over thousands of years. In contrast, human-caused edges are quite abrupt, significant in extent, and have been imposed on many landscapes fairly recently so that populations of animals have not had time to evolve to these changes.

The abrupt changes in habitat caused by human activity can have dramatic effects on how animals move through landscapes. For example, many bird species will not cross extensive gaps in vegetation, possibly because the risks of predation are too great in these open areas. Edges may function as travel corridors for some species such as raccoons, foxes, and skunks that prey upon birds' nests they encounter while moving through these habitats. Nest-parasitic brown-headed cowbirds, which lay their eggs in other birds' nests so that the hosts care for them, have increased in range and abundance with the clearing of forests. Cowbirds forage in open habitats, including irrigated lawns, and then travel to nearby forest where they tend to parasitize songbird nests at the forest edge.

Furthermore, human activities along habitat edges that provide easy access to food sources, such as garbage, pet food, and irrigated lawns, are additional factors leading to increased abundance of some human-associated species. These human-associated species include jays and crows, cowbirds, and raccoons, foxes, and skunks, among others. Numerous studies of songbirds have documented increased rates of nest predation and nest parasitism along habitat edges compared to habitat interiors, possibly because of the increased numbers of nest predators and parasites that are attracted to habitat edges.

Altering habitats and providing supplemental food sources are not the only ways that humans affect wildlife along edges. Disturbance from human activities along habitat edges such as roads, recreational trails, housing subdivisions, and parking lots may reduce habitat suitability, leading to the loss of disturbance-sensitive species from these edge habitats. However, some wildlife capable of tolerating human disturbance may benefit from a refuge effect, where they experience safer breeding conditions along certain edges when predators are displaced.

The opening of habitat along edges can change light, temperature, and moisture levels, leading to changes in plant communities. Animals often respond to these changes in habitat at edges. Edges can be areas of high species diversity and abundance because of the heterogeneity and structural diversity of plants (many different types and heights of plants) where two habitats meet. In fact, a former guideline of wildlife management, put forth by the early wildlife conservationist Aldo Leopold, was to cut gaps in forests to increase the amount of edge to attract game species. Deer and other game species are attracted to these openings because of the increased productivity of and access to ground forage, in addition to being able to forage in proximity to cover. Some ground foraging birds, which feed upon seeds and invertebrates, and flycatchers, which utilize forest gaps to catch aerial insects, may increase in abundance along edges because their foraging efficiency is enhanced. Certain snakes may utilize openings along habitat edges for thermoregulating by warming themselves in the sun.

In some cases the changes in microclimate (light, temperature, or moisture levels) can make habitats physiologically unsuitable to certain animals, and these effects can cascade through food webs. Researchers in Canada found reduced insect abundance along forest edges compared to tracts of forest farther from edges. The reduction in insect abundance may be a contributing factor in the decline and disappearance of certain insect-eating songbirds from forest edges.

Other research has shown that native wildlife, including insects, rodents, small mammals, songbirds, and raptors decline in abundance along habitat edges. These declines may be due to changes in microclimate, vegetation, increases in predation pressures, declines in prey base, or increases in disturbance levels. Some studies have suggested that house cats along suburban edges may be a major predator on native wildlife.

Most studies of songbirds indicate that edge effects of increased nest predation and parasitism extend about 50 m (164 ft) into surrounding habitats. However, evidence from other studies suggests that edges may have more far-reaching effects. Edge effects may extend miles across landscapes, particularly when top-level predators, critical pollinators, or seed dispersers are lost from these landscapes.

See also Human (Anthropogenic) Effects—*The Effect of Roads and Trails on Animal Movement*
Human (Anthropogenic) Effects—*Human (Anthropogenic) Effects on Animal Behavior*
Human (Anthropogenic) Effects—*Logging, Behavior, and the Conservation of Primates*

Further Resources

Burke, D. M. & Nol, E. 1998. *Influence of food abundance, nest-site habitat, and forest fragmentation on breeding ovenbirds.* Auk, 115, 96–104.
Knight, R. L., Smith, F. W., Buskirk, S. W., Romme, W. H., & Baker, W. L. (Eds.). 1998. *Forest Fragmentation in the Southern Rocky Mountains.* Boulder, CO: University of Colorado Press.
Lovejoy, T. E., Bierregaard, J., Rylands, A. B., Malcolm, J. R., Quintela, C. E., Harper, L. H., Brown, J., K. S., Powell, A. H., Powell, G. V. N., Schubart, H. O. R., & Hays, M. B. 1986. *Edge and other effects of isolation on Amazon forest fragments.* In: *Conservation Biology: The Science of Scarcity and Diversity* (Ed. by M. E. Soule), pp. 257–285. Sunderland, MA: Sinauer Associates, Inc.
Meffe, G. K., Nielson, L. A., Knight, R. L., & Schenborn, D. A. 2002. *Ecosystem Management: Adaptive, Community-based Conservation.* (Chapter 8) Washington, DC: Island Press.
Miller, S. G., Knight, R. L., & Miller, C. K. 1998. *Influence of recreational trails on breeding bird communities.* Ecological Applications, 8, 162–169.
Terborgh, J. 1989. *Where Have All the Birds Gone?* Princeton, NJ: Princeton University Press.

William W. Merkle

■ Human (Anthropogenic) Effects
The Effect of Roads and Trails on Animal Movement

Roads and trails. We use them daily to get to school or work, or to places to recreate and relax. As humans living in large social groupings in cities and towns, it is hard to avoid roads and trails. It turns out that many animals also cannot avoid encountering human-made roads and trails in their activities. Because they are common and often unique features of animals' habitats, roads and trails can have major effects on animals' behavior and ecology. Some studies suggest that 15–20% of the area of the United States is ecologically affected by roads. I will discuss the behavioral effects in this essay.

Before getting into examples of the effects of roads and trails it is worth clarifying terms used to describe animal movement. *Dispersal* is one-way movement, usually occurring on a regular schedule (or time of life). For example, birds often disperse from the nest where they were raised once they mature. *Migration* is a regular two-way, back and forth, movement, often for feeding or reproduction. *Irruptions* are irregular one-way movements, often as a result of overcrowding, etc. *Nomadism* is random movements of animals, for example random feeding movements within suitable habitats. Dispersal, nomadic and migratory movements are the most well studied forms in relation to the effect of roads and trails.

Roads and trails do a number of things to the environment which can affect an animal's behavior. The road itself, and fill material brought in for the road, present new substrate features. Mowing along roadways affects the plant community make-up. Disturbance from vehicles using the road brings in pollutants, including oils, exhaust fumes, deicing salts, noise, and vibration among other things. As a result, roadsides often have unique vegetative communities that include large numbers of alien and "weedy" plants. These species provide new habitats for some animals that can move in and establish themselves, while species that are adapted to native vegetation are forced to disperse or die off. As a result of the above issues, small mammals like mice and voles, some birds, beetles, butterflies, and deer are all known to have larger populations near roads.

The noise near roads is thought to drive off many types of birds who are either stressed by the disturbance, cannot communicate effectively with song, or are otherwise negatively affected. In some studies roads can affect bird populations for a distance of over one mile perpendicular to the road. Vibration from roads seems to induce earthworms in at least one area to attempt dispersal. Crows responded to the worms by flocking to the area for an easy meal, providing a clear example of how roads can affect species higher up the food web. Other predators also respond to increases in their prey near roads.

Roads and trails have the effect of splitting an animal's habitat or home range up into smaller sections, something scientists call *fragmentation*. If roads bisect an animal's habitat, typical migratory or feeding movement may be altered. For example, land snails are known to normally move upslope in mountainous areas to counter the downslope movement they sometimes experience when they lose contact with the ground and gravity sends them rolling downhill. This upslope movement has been found to be inhibited by trails as little as 2–3 feet wide. Small and large mammals, snakes, turtles, amphibians, spiders, and insects also find roads to be barriers to movement and turn away upon encountering them. Because roads and trails often skirt the boundary of lakes and streams, animals that make migratory movements between aquatic and terrestrial habitats can be especially affected. Lights along the sides of roads and deicing salts have both been shown to affect the natural migratory movement of frogs at night, and lights have been shown to lengthen the time when birds feed.

When roads prevent movement, animal numbers within an affected fragment may increase to artificially high levels as dispersal and irruptions are prevented. These large populations can alter the community inside the fragment and lead to increases in aggressive behavior. Individuals trapped within a fragment have fewer choices of potential mates. In fact, if they mate at all they may be forced to mate with close relatives since many animals live in family groups until their young are old enough to mate, at which time they disperse. In this case, or if the size of the population in an isolated fragment is small, the population may lose genetic diversity because gene flow is halted, and the local gene pool is smaller than that of the species as a whole. Such an effect of roads has been shown for both frogs and snails. For some animals, roads are only a partial barrier, and movement across them is only slowed relative to natural habitat. This slowing can still cause alterations to the behavior and ecology of species within the fragments, though gene flow is not likely to be affected.

Sometimes, animals move along parallel to a road or trail that is a barrier. In this way, roads and trails have been shown to be conduits or corridors for movement of species and individuals into new habitats. Studies have shown that predatory mammals, including feral cats and carrion feeders, use roads as corridors to ease their food-seeking goals, especially at night. While poorly studied, it is possible that this effect of funneling animal movement down limited corridors may lead to higher levels of competition within such species.

Road and trail fragmentation creates a lot of new "edge" habitats. Edges of habitat blocks have different microclimate features, such as wind, temperature, and sun exposure that can affect all food web levels. Behaviorally, animals near edges alter their territory sizes, and rates of interaction with predators, competitors, and parasites. Whether a particular feature increases or decreases depends on species biology.

There are solutions to the problems created by roads. Considering the impact of proposed roads on animal movement during the design and environmental impact review stage should be more common than it is today. Mitigating devices, such as natural or artificial corridors or tunnels, have been shown to allow animal movement across roads for species from frogs to large mammals. Unfortunately, use of these devices is much more common in Europe and Scandinavia than elsewhere. Further study is also needed of this area, especially how animals search for, and select, habitat patches to occupy in these situations.

See also Human (Anthropogenic) Effects—*Human (Anthropogenic) Effects on Animal Behavior*
Human (Anthropogenic) Effects—*The Edge Effect and Behavior*
Human (Anthropogenic) Effects—*Logging, Behavior, and the Conservation of Primates*

Further Resources

Ercelawn, A. 1999. *The adverse ecological impacts of roads and logging: a compilation of independently reviewed research.* Natural Resources Defense Council. http://www.nrdc.org/land/forests/roads/eotrinx.asp.

Fagan, W. F., Cantrell, R. S., & Cosner, C. 1999. *How habitat edges change species interactions.* The American Naturalist, 153, 165–182.

Forman, R. T. T. & Alexander, L. E. 1998. *Roads and their major ecological effects.* Annual Review of Ecology and Systematics, 29, 207–231.

Forman, R. T. T. et al. 2002. *Road Ecology: Science and Solutions.* Washington, D. C.: Island Press.

Riitters, K. H. & Wickham, J. D. 2003. *How far to the nearest road?* Frontiers in Ecology and the Environment, 1, 125–129.

Spellerberg, I. F. 2002. *The Ecological Effects of Roads.* Enfield, NH: Science Publishers, Inc.

Sutherland, W. J. 1998. *The importance of behavioral studies in conservation biology.* Animal Behaviour, 56, 801–809.

Trombulak, S. C. & Frissell, C. A. 2000. *Review of ecological effects of roads on terrestrial and aquatic communities.* Conservation Biology, 14, 18–30.

Dwayne Meadows

■ Human (Anthropogenic) Effects
Environmentally Induced Behavioral Polymorphisms

Polymorphisms occur when *phenotypes* (the physical expression of a trait) are discrete rather than continuous. Consider the two *morphs* (forms) of the peppered moth, *Biston betularia*: One is light colored and the other is dark colored, rather than a continuous distribution of phenotypes from very light to very dark. This polymorphism is completely genetic. Other polymorphisms, however, are environmentally induced (*polyphenism*). Because such polyphenisms have a genetic basis that influences how the environment triggers the production of each morph, they are excellent examples of phenotypic plasticity (see the diagram below). Phenotypic plasticity is a critical yet understudied aspect of animal behavior. By separating phenotypes into discrete packages, environmentally cued polymorphisms provide model systems for studying behavioral plasticity.

Many environmentally-induced polymorphisms are associated with a specific aspect of animal behavior, such as reproduction, foraging, or migration. For example, some of the most intriguing examples of behavioral polyphenism come from species exhibiting alternative male morphologies. In these examples, behavioral tactics tend to be tightly bound to physical differences. Horned beetles are a classic example of such variation. Often these beetles exhibit two alternative male morphologies, a "major" male with a long horn and a "minor" male with a small horn. Both males have longer horns than females. Such variation can be affected by environmental conditions, such as food quantity and quality, while a male beetle larva is developing. Those larvae experiencing more favorable conditions tend to become major males. Major males use their long horns to fight for access to females, with the largest male winning. Minor males do not fight, but rather attempt to sneak around major males to mate with females. In addition, in some species, major males provide more parental care than minor males.

Similar polyphenisms occur in male fish such as salmon and bluegill sunfish. In both of these groups one can observe the formation of alternative male morphologies, which correspond to alternative reproductive tactics. For example, in Pacific coho salmon, *Oncorhynchus kisutch*, males can become "hooknose" with an enlarged red body and hook-shaped mouth filled with teeth, or smaller "jacks" which are smaller and resemble females. Jacks are produced when male salmon grow quickly and mature early; slower-growing fry delay maturity and become hooknose. Hooknose fight with each other over females, while jacks attempt to sneak fertilization of the eggs while the hooknose and female are breeding.

An intriguing example of phenotypic plasticity based on foraging behavior is the cannibalistic polyphenism within salamanders. In some species, large larval salamanders will

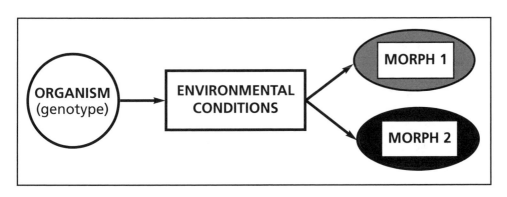

develop enlarged heads and teeth (cannibal morphs), providing them with the tools necessary to consume large prey, including smaller salamanders (typical morphs). In this case, behavior is augmented, rather than directly tied to, morphological polymorphism, because all larvae are potentially cannibalistic given the correct circumstances, yet only cannibal morphs exhibit this behavior on a consistent basis. The benefit of cannibalism is faster growth rate, increased body size, and perhaps an increased probability of survival to adulthood. Cannibal morph tiger salamanders (*Ambystoma tigrinum*) can be induced through a variety of environmental cues, including salamander density and size variation, presence of large nonsalamander prey such as tadpoles and insect larvae, and even the absence (versus presence) of siblings in experimental tanks. Other studies have shown that cannibals are selective about their prey, choosing non-kin over kin.

Foraging polyphenisms are also abundant among fish species. Pumpkinseed sunfish (*Lepomis gibbosus*), Arctic charr (*Salvelinus alpinus*), and several species of cichlids exhibit polymorphisms that are related to foraging in different environments. Research suggests that such polyphenisms may occur as a result of reduced competition. For example, pumpkinseed sunfish diversify into morphs that feed either on the bottom or in the open water, but only in lakes without the generalist bluegill sunfish (*Lepomis macrochirus*). Thus, pumpkinseed utilize polyphenism as a way of broadening their use of habitat and food in the absence of competition.

Another example of polyphenism is dispersal polymorphism in insects. In many insect species, there are two general forms: a winged, disperser morph and a wingless, less mobile morph. These morphs are induced by a variety of environmental cues, such as insect density, food quality, temperature, and photoperiod. The ability to disperse through flight is an important feature of insects that has played a key role in their evolutionary success. However, because wing production is energetically expensive, there appears to be a clear trade-off between egg and wing production in females. Recent studies suggest that there are similar trade-offs between reproduction and dispersal in males as well.

Perhaps some of the best-known behavioral polyphenisms are found among *eusocial insects* (species with separate castes, or social roles, the majority of which are nonreproductive). *Caste polyphenism* occurs when environmental conditions affect the production of different roles in the colony. Such polyphenism is common in bees, wasps, ants, and termites, where the production of workers, soldiers, or queens may occur due to changes in food level, food quality, temperature, or other environmental variables. Eusocial insects have the unique ability to manipulate the production of these different forms by actively changing the environment that larvae are exposed to in an effort to maximize the success of the colony. For example, honeybee (*Apis mellifera*) larvae that are fed "royal jelly," a nutrient-rich food, become queens, while those fed only nectar and pollen become workers.

The evolution of each of these polyphenisms may relate to the benefits of behavioral plasticity in variable environments, trade-offs associated with alternative behaviors, frequency-dependent selection, developmental constraints, or some combination of these factors. For example, alternative male polyphenism in beetles and fish, as well as wing dimorphism in insects, appear to be maintained in part through trade-offs between different fitness components (traits that provide evolutionary advantages, such as increased survival and reproduction). However, frequency-dependent benefits could also be important. For example, as the frequency of hooknose males increases, hooknose mating success should decline through competition with other hooknose, while the success of the few jacks should increase because of the multiple hooknose with which jacks could attempt to sneak fertilizations.

Few studies have adequately measured the fitness consequences of environmentally induced behavioral polymorphisms, and for many polyphenisms the environmental triggers underlying this variation are still being explored. Thus, behavioral polyphenisms are ripe for future research, and by understanding the evolution of these polymorphisms, we will gain important insight into the evolution of behavioral plasticity in general.

See also Human (Anthropogenic) Effects—*Genetically*
        Modified Fish
        Human (Anthropogenic) Effects—*Human*
        (Anthropogenic) Effects on Animal Behavior

Further Resources

Andersson, M. 1994. *Sexual Selection*. Princeton, NJ: Princeton University Press.
Alcock, J. 1998. *Animal Behavior: An Evolutionary Approach*. 6th edn. Sunderland, MA: Sinauer Associates.
Emlen, D. J. 1994. *Environmental control of horn length dimorphism in the beetle* Onthophagus acuminatus (Coleoptera: Scarabaeidae). Proceedings of the Royal Society of London, Series B, 256, 131–136.
Evans, J. D. & Wheeler, D. E. 2001. *Gene expression and the evolution of insect polyphenisms*. BioEssays, 22, 62–68.
Gross, M. R. 1991. *Salmon breeding behavior and life history evolution in changing environments*. Ecology, 72, 1180–1186.
Whiteman, H. H., Sheen, J. P., Johnson, E. B., VanDeusen, A., Cargille, R. & Sacco, T. W. 2003. *Heterospecific prey and trophic polyphenism in larval tiger salamanders*. Copeia, 2003, 56–67.
Zera, A. J. & Denno, R. F. 1997. *Physiology and ecology of dispersal polymorphism in insects*. Annual Review of Entomology, 42, 207–231.

Howard H. Whiteman

■ Human (Anthropogenic) Effects
Genetically Modified Fish

Genetic engineering, or *transgenes*, has given rise to new possibilities in the study of animal behavior. By adding or deleting genes, we have the potential to study effects of certain genes on the behavior of the animal. This will allow us to better understand how genes translate into behavior, and how and why organisms have evolved into what they are today. However, making transgenic fish is a laborious and complicated task, and in most cases behavioral effects have been neither the aim nor the focus of this practice. Consequently, little is known about the behavior of genetically modified fish to date. In addition, studying transgenic fish is important in itself because it may help us predict how these organisms will affect the lives of other species should transgenic fish end up in the natural environment.

Some 40 species of fish have been modified in such different ways as increasing cold tolerance and growth rate, or changing color, all of which may have effects on behavior. However, so far, behavior has been studied only in fish genetically modified for increased growth rate. These fish, mainly salmonids, catfish and tilapias, are provided with extra gene segments coding for growth hormone, a natural protein involved in regulating growth. The result is an increased production of growth hormone, which, if sufficient food is available, results in faster growth, and in some species a larger size.

Laboratory studies have shown that growth hormone transgenic fish generally have an increased appetite, and respond by trying to feed more, just as a hungry or starved fish will do. This also makes the transgenic fish less discriminating when deciding what to feed on, and they are more likely to try novel prey. Enhanced hunger also increases aggression, and the transgenic fish are more competitive when fighting with normal fish over food. However, feeding and fighting involves activity, which may attract larger predatory fish, mammals or birds. Indeed, experiments on transgenic salmon confirm that they are more willing than normal fish to take risks to obtain food.

It is not yet clear how the altered behavior of growth hormone transgenic fish will affect their survival and reproductive success in nature. Most likely it will depend on several factors such as the abundance of prey and predators, as well as their ability to acquire mates. The transgenic state seems to be advantageous only when the fish acquire enough food to outgrow normal fish. If food is limited in nature, increased hunger and aggression may therefore not pay off, and transgenic fish may suffer higher mortality from predation since they are more risk prone.

We currently have a very poor understanding of the complete effects of inserted genes in fish. For example, in addition to direct effects on behavior, it has been observed that growth hormone transgenes can affect many aspects of morphology and physiology (e.g., head shape, swimming ability), and these in turn can have indirect effects on the behavior of the organism. Although the behavior of genetically modified fish is still poorly known, the potential of greater knowledge is increasing with new types of modified fish becoming available. The study of behavior of transgenic fish in the future will certainly yield much information both for basic science examining the relationship between genetics and behavior as well as for science-based environmental risk assessments.

See also Human (Anthropogenic) Effects—*Human (Anthropogenic) Effects on Animal Behavior* Human (Anthropogenic) Effects—*Environmentally Induced Behavioral Polymorphisms*

Further Resources

Abrahams, M. V. & Sutterlin, A. 1999. *The foraging and antipredator behaviour of growth-enhanced transgenic Atlantic salmon.* Animal Behaviour, 58, 933–942.

Devlin, R. H., Johnsson, J. I., Smailus, D. E., Biagi, C. A., Jönsson, E. & Björnsson, B. T. 1999. *Increased ability to compete for food by growth hormone-transgenic Coho salmon* Oncorhynchus kisutch (Walbaum). Aquaculture Research, 30, 479–482.

Gong, Z. Y., Wan, H. Y., Tay, T. L., Wang, H., Chen, M. R. & Yan, T. 2003. *Development of transgenic fish for ornamental and bioreactor by strong expression of fluorescent proteins in the skeletal muscle.* Biochemical and Biophysical Research Communications, 308, 58–63.

Guillén, I., Berlanga, J., Valenzuela, C .M., Morales, A., Toledo, J., Estrada, M. P., Puentes, P., Hayes, O. & de la Fuente, J. 1999. *Safety evaluation of transgenic tilapia with accelerated growth.* Marine Biotechnology, 1, 2–14.

Lee, C. G., Devlin, R. H. & Farrell. A. P. 2003. *Swimming performance, oxygen uptake and oxygen debt in adult transgenic and ocean-ranched coho salmon (Oncorhynchus kisutch, Walbaum).* Journal of Fish Biology, 62, 753–766.

Sundström, L. F., Devlin, R. H., Johnsson, J. I. & Biagi, C. A. 2003. *Vertical position reflects increased feeding motivation in growth hormone transgenic coho salmon* (Oncorhynchus kisutch). Ethology, 109, 701–712.

L. Fredrik Sundström, Jörgen I. Johnsson & Robert H. Devlin

■ Human (Anthropogenic) Effects
Human (Anthropogenic) Effects on Animal Behavior

Humans are here, there, and everywhere. We are a curious lot, and our intrusions, intentional and inadvertent, have significant impacts on a wide variety of animals and plants, as well as water, the atmosphere, and inanimate landscapes. When humans influence the behavior of animals the effects are referred to as being *anthropogenic* in origin. Often our influence on the behavior of animals and the unbalancing of nature is very subtle and long-term. Often we become at odds with the very animals with whom we choose to live when they become nuisances, dangerous to us or to our pets, or destroy our gardens and other landscapes.

Many of the animals that we want to study, protect, and conserve experience deep emotions, and when we step into their worlds we can harm them mentally as well as physically. They are sentient beings with rich emotional lives. Just because psychological harm is not always apparent does not mean we do no harm when we interfere in animals' lives. It is important to keep in mind that, when we intrude on animals, we are influencing not only what they do but also how they feel.

In my home state of Colorado, many people enjoy the outdoors, and many people also work to protect a wide variety of animals. Many of us live in one place and travel elsewhere to experience nature. Our understanding and appreciation of wildlife result from various types of research and "just being out there."

Some examples of behavior patterns influenced by various research methods and other forms of human intrusion include nesting and other reproductive activities (abandonment of nests, increased egg loss, disruption of pair bonds), mate choice, dominance relationships, the use of space, vulnerability to predators, patterns of vigilance, foraging, resting, and feeding and caregiving behaviors. Often animals are so stressed that they are unable to acquire the energy they need to thrive and to survive. Intrusions include such activities as using various devices and instruments to study behavior, marking and handling animals, censuring animal populations, visiting nests, urbanization (urban development and sprawl, the development of bodies of water, changes in vegetation, installing power lines, the need for more electric power), and recreational activities, including the use of snowmobiles and other off-road vehicles, environmental pollution including oil spills, photography, travel, and ecotourism. In November 2003, it was reported that only 2 years after a new finch-like bird, the Carrizal blue–black seedeater, was discovered in Venezuela, its habitat was destroyed so that a hydroelectric dam could be built!

Models that are generated from these studies can be misleading because of human intrusions that appear to be neutral. It is ironic that often our intrusions preclude collecting the data we need to answer specific questions. I have picked representative studies to show how widespread human influences can be and the diversity of species that are affected. Many of these findings apply to other situations and species. The topic of human–animal interactions is relevant to studies of applied ethology. Detailed ethological studies are needed because we need to take into account just how our research influences the behavior of other animals, otherwise we risk drawing the wrong conclusions. Also, it is increasingly important to conservation efforts to understand how humans influence and change the behavior of animals.

Research Effects

Patterns of finding food can be affected by human intrusions. The foraging behavior of Little penguins (average mass of 1,100 g or about 2.4 lb) is influenced by their carrying a small device (about 60 g or 2 oz) that measures the speed and depth of their dives. The small attachments result in decreased foraging efficiency. Changes in behavior such as these are called the "instrument effect." In another example of the influence of humans on penguins, researchers discovered that tourism and nest site visitation caused behavioral and hormonal changes.

A tourist has a close encounter with a chinstrap penguin in Antarctica. Researchers discovered that tourism and nest site visitation caused behavioral and hormonal changes.
Gordon Wiltsie/National Geographic Image Collection.

Marking animals also influences their behavior. Placing a tag on the wing of ruddy ducks leads to decreased rates of courtship and more time sleeping and preening. In this case, data on mating patterns, activity rhythms, and maintenance behaviors would be misleading.

Mate choice in zebra finches is influenced by the color of the leg band used to mark individuals, and there may be all sorts of other influences that have not been documented. Females with black rings and males with red rings had higher reproductive success than birds with other colors. Blue and green rings were especially unattractive on both females and males. Leg-ring color can also influence song tutor choice in zebra finches and mate-guarding in bluethroats.

Fitting animals with radio collars can affect their behavior. For example, the weight of radio collars influences dominance relationships in adult female meadow voles. When voles wear a collar that is greater than 10% of their live body mass, there is a significant loss of dominance. Here, erroneous data concerning dominance relationships would be generated in the absence of this knowledge. Radio collars can also influence pair bonding and breeding success in snow geese. However, when female spotted hyenas wear radio collars weighing less that 2% of their body weight, there seems to be little effect on their behavior. Similar results have been found for small rodents, for which small radio collars do not increase the risk of predation by birds.

Methods of trapping can lead to spurious results. Trapping methods can bias age ratios and sex ratios in birds. For example, mist nets capture a higher proportion of juveniles, whereas traps captured more adults. Furthermore, dominant males tend to monopolize traps that are baited with food, leading to erroneous data on sex ratios. These are extremely important results because age and sex ratios are important data for many different researchers interested in behavior, behavioral ecology, and population biology.

It also is known that capturing and recapturing large grey mongooses influences their use of space. It is important to ask if the use of space really is the use of space by individuals avoiding traps or avoiding human observers. If, for example, cages are being designed to take into account animals' movement and activity patterns, then data that are used to make decisions about designing enclosures need to be based on information that reliably indicates what the animals typically use and need in the wild.

Not only do research methods influence a wide variety of behavior patterns, but they can also influence susceptibility to infection. For example, ear tagging white-footed mice led to higher infestations by larval ticks because the tags impeded grooming by these rodents. Thus, for researchers interested in grooming and maintenance behavior, the presence of ear tags could influence results.

Discerning the effects of human intrusions, even those that are meant to help animals, can be extremely complicated. A highly-disputed example of the possible effects of human interference into wild populations concerns the plight of African wild dogs. Interference into the lives of wild dogs involved vaccinating them against rabies and canine distemper. Whereas some scientists maintain that handling the dogs and inoculating them was indirectly responsible for their decline because the handling weakened the dogs' immune system making them less resistant to stress, others conclude just the opposite, namely that handling and inoculating were not the cause of their decline. Here we have a example of extremely competent scientists, all of whom care deeply about African wild dogs, not being able to discern what caused their decline. This is because the problems are so incredibly difficult. Some questions that these researchers pondered included: Should the researchers interfere and possibly cause animals to die or let nature take its course? If the rabies and distemper were introduced by domestic dogs who would not have been there in the absence of humans (an anthropogenic cause), are we more obligated to try to help the wild dogs than if the rabies and distemper were natural? There are no simple answers to these questions, and they are an example of the sorts of questions that are raised when humans intrude into the lives of other animals.

The above examples stem from research on mammals and birds, but there also are indications that human disturbance can influence the behavior and movement patterns of numerous insects (bedbugs, termites, yellow jackets, and ants), skinks (lizards), and can also delay reproduction in snakes. Undoubtedly, future studies will show that humans influence the behavior of numerous diverse species.

Just Being There

Not only does "hands on" contact have an influence, but so might "just being there." Mere human presence influences the behavior of many different animals' behavior. In the early 1900s titmice in England learned to pry off the lids of milk bottles that were delivered to peoples' homes. Elk and numerous other animals avoid skiers. Research performed by my students and myself showed that humans have a large influence on prairie dogs such that individuals who have a lot of contact with humans are less wary of their presence than individuals who do not. The same is true for various species of deer. Similarly, magpies not habituated to human presence, spend so much time avoiding humans that it takes time away from essential activities such as feeding. Researchers interested in feeding patterns must be sure that their presence does not alter species-typical behavior, the very information they want to collect.

A young bull moose strolls through an Anchorage neighborhood as he makes his way to a wooded area. As winter approaches and food becomes scarce, urban moose will be cruising the neighborhoods looking for food and handouts.
© *AP / Wide World Photos.*

People often enjoy watching animals from cars, boats, or airplanes. However, the noise and presence of vehicles can produce changes in movement patterns (elk), foraging (mountain sheep), and incubation. In swans, the noise and presence of cars result in increases in the mortality of eggs and hatchlings. Once again, these effects are not obvious when they occur, but data show they are real.

Adélie penguins exposed to aircraft and directly to humans showed profound changes in behavior including deviation from a direct course back to a nest and increased nest abandonment. Overall effects due to exposure to aircraft that prevented foraging penguins from returning to their nests included a decrease of 15% in the number of birds in a colony and an active nest mortality of 8%. There are also large increases in penguins' heart rates. Here, models concerning reproductive success and parental investment would be misleading, once again because of the methods used. Trumpeter swans do not show such adverse effects to aircraft. However, the noise and visible presence of stopped vehicles produced changes in incubation behavior by trumpeter females that could result in decreased productivity due to increases in the mortality of eggs and hatchlings. Data on the reproductive behavior of these birds would be misleading.

Many people adore young animals and try to get close to nests or dens without disturbing residents. In various research projects it is important to observe parental behavior and to count eggs and young so as to learn about patterns of survival and morality. However, when the nests of some birds (for example, white-fronted chats and ducks) are visited regularly by humans, these birds often suffer higher predation than when their nests are visited infrequently. It has been suggested that some animals may become so accustomed to intrusions by humans who do not kill them that they subsequently allow other animals, including natural predators who will kill them, to get too close.

Activities such as mountain climbing are also intrusive. Climbers can influence activity patterns of birds such that they fly more and perch less and consequently waste energy. In Boulder, Colorado, and other places, climbing is restricted during nesting season. In a study of the effect of climbers on grizzly bears in Glacier National Park, Montana, where bears forage for moths, researchers discovered that climber-disturbed bears spent about 50% less time searching for moths and about 50% more time avoiding climbers. It was recommended that climbers be routed around areas where bears live to minimize disturbance and associated caloric losses.

Recreational trails built by humans are also associated with changes in behavior and mortality. In Boulder, within forest and mixed-grass ecosystems, nest predation is greater near trails, but we do not know if it is the trail itself, trail use by humans, trail use by predators, or all three that are responsible.

Aquatic habitats also find themselves being intruded on by gawking and intruding humans. Dolphins have become sort of a "cult" animal, and people often visit dolphins in order to swim with them. Numerous studies have discovered that swimming with dolphins in the wild may be harmful to entire groups, especially when swimmers try to be close to dolphins, who, unbeknownst to the swimmers, are resting, feeding, or mating. Swim programs may also be risky to humans. Even experts agree that we really need more detailed information about the effects of swimming with dolphins. Many researchers proffer that when we do not know the negative effects we should err on the side of the animals and leave them alone.

Dolphins and other animals are often fed by humans. Feeding (and harassing) wild dolphins is illegal in the United States, and there are severe penalties for engaging in these activities, but this is not so for other countries. There are documented instances of wild dolphins being fed

firecrackers, golf balls, plastic objects, balloons, and fish baits with hooks (so that hooked dolphins can be caught). Provisioning dolphins with fish has been associated with a change in the social behavior of free-ranging bottlenose dolphins in Monkey Mia, Australia. Dolphins who have been fed also change their foraging behavior and frequent heavily trafficked harbors and marinas. Some get struck by boats. People have also been seriously injured trying to feed wild dolphins. The National Marine Fisheries Service and other organizations are mounting highly visible campaigns to stop the feeding and harassment of wild dolphins. It also has been noted that some problems associated with feeding terrestrial mammals (changes in foraging patterns and hunting skills) are relevant to concerns about the feeding of dolphins.

Provisioning dolphins with fish has been associated with a change in the social behavior of free-ranging bottlenose dolphins in Monkey Mia, Australia. Here a local woman feeds a wild dolphin.
© Paul A. Souders / Corbis.

Low frequency active sonar (LFAS) that is used to detect submarines by the United States Navy (and by other countries) can be fatal for marine life including whales, turtles, and some fish. Low frequency active sonar can be carried as many as 400 miles through water, and animals can suffer 140 decibels of sound pressure, about the level of noise produced by an earthquake. Less powerful sonar has been responsible for whale strandings in the Bahamas. Fish are known to suffer internal injuries, eye and auditory damage, and temporary stunning due to low frequency active sonar.

In his book *Sperm Whales: Social Evolution in the Ocean*, whale expert Hal Whitehead notes that sperm whale populations are vulnerable to many threats caused by human activity. These include the increasing use of harpoons to capture whales, collisions with ships, debris such as plastic that might resemble squid who are eaten by whales, entrapment in fishing gear, noise, chemical pollution including heavy metals that enter the food chain, and global warming. Whitehead suggests that because sperm whales seem to use culturally transmitted information about their environment, oceanic changes due to global warming and other human activities might mean that this information becomes outdated or irrelevant. And small changes in survival can influence sperm whale numbers. Many anthropogenic effects are unpredictable, and this causes concern among researchers because humans might unknowingly be doing damage that is irreparable.

There also are observations of humans causing seal pups to stampede and be trampled, and humans sometimes strike and injure individual pups with boats.

Habituation and Behavioral Flexibility Due to Human Presence

Habituation of wild animals to the presence of humans is another a major problem. Individuals of many species are known to habituate to the presence of humans rather than flee from them. They become less wary, less secretive and more visible. For example, black-tailed prairie dogs studied by my students and myself in and around Boulder are less wary and more tolerant in urban areas, where they have a lot of contact with humans, than in more remote habitats where there are fewer human intrusions. Urban prairie dogs show reduced flight distances to humans; they allow people to get closer to them and generally are less disturbed by human presence. Birds such as great crested grebes, ospreys, greylag geese, and great blue herons also show high levels of habituation in areas of high human activity.

Cougars (also called pumas and mountain lions) living in areas with dense human populations also exhibit a wide range of behavioral changes. In his book *The Beast in the Garden*, David Baron explores the behavior of cougars who repopulated their ancestral homeland in and around Boulder, Colorado. Historically, cougars were very elusive, secretive creatures. They avoided humans, tended to fear dogs, and were rarely seen during the daytime. Since cougars began to repopulate the open lands around Boulder in the late 1980s, however, biologists and ethologists have noted many behavioral changes. The cougars of Boulder tend to be less fearful of people, often coming into people's yards and even onto their decks. They have lost their fear of dogs, and some cougars around Boulder have even begun preying on dogs.

Many of these changes in behavior are not surprising to biologists and animal behaviorists who stress that the behavior of animals is malleable and subject to change. For example, biologists know that animals are opportunistic and will identify new sources of food, as some Boulder cougars did when they began preying on domestic dogs. It is harder to understand why these cougars lost their age-old fear of dogs, although biologist Maurice Hornocker believes it may be that in areas where wolves have been exterminated, cougars no longer learn to fear canines because their ancient enemy the wolf has disappeared. It is also easy to understand why Boulder cougars tend to be more active during the daytime than their ancestors. The German ethologist Eberhard Curio has noted that predators tend to synchronize their predatory activity with their prey. Because the deer of Boulder, Colorado, are more active during the day than deer in true wilderness areas, the Boulder cougars are also more active during daylight hours. It is also not surprising that the cougars, following the deer into the lowlands of Boulder, Colorado, began turning up in Boulder neighborhoods, at times even killing and stashing deer in people's backyards.

Lee Fitzhugh, a wildlife biologist at the University of California at Davis, has studied cougar–human interactions, including fatal attacks in California and other areas in the Rocky Mountain West. He has noted that in certain circumstances cougars can come to view humans as prey, whereas they had previously avoided them. Victims who survived attacks by cougars noted that the cougars crouched and swept their tails while eyeing them, which indicates that the attacks were predatory in nature, rather than defensive. The cougars were not fearful, territorial, or curious. They were assessing humans as potential meals. In his book, Baron opens with the story of a jogger in Idaho Springs (a rural community just west of Denver, Colorado) who had been killed and partially eaten by a cougar. It is important to stress that cougar attacks, while very rare, are not necessarily the result of the cougars being injured or sick. Habituation to people has led to changes in their behavior, and healthy cougars will on occasion attack humans.

That the behavior of cougars and other animals can change at what is called the wild/urban interface is fascinating, for it shows that behavior can be flexible and influenced by social factors such as the presence of humans. Many animals, including urban foxes, bears, coyotes, cougars, and deer show changes in space use and activity rhythms that can influence such activities as hunting and mating. Spotted hyenas also show variations in space use as a result of human activity. We still know very little about how the behavior of animals changes when they are forced to compete with intruding humans.

Predators and prey also show behavioral flexibility and often their activity patterns are linked to each another. For example, mule deer around Boulder are more active during the daytime than are deer in the areas where there are fewer humans. The cougars around Boulder, whose primary source of food is mule deer, are also more active during the daytime. Both deer and cougars have become habituated to humans and to domestic dogs. Also, changes in vegetation can alter the spacing behavior of herbivorous animals, and these environmental modifications also can have far-reaching effects on a wide variety of animals.

Many animals show changes in behavior or alterations in their ecological niche due to anthropogenic activities that are *not* associated with habituation to humans. Some of the species that are affected are unfamiliar to many people and these changes go unnoticed except by researchers. For example, Preble's Meadow jumping mice began using waterways constructed by humans over 100 years ago with the advent of agricultural irrigation. The use of these waterways changed the spatial distribution of these mice and, as a result, their patterns of interactions with other species. The giant floater mussel in Colorado was forced to change its ecological niche when humans began damming rivers. Damming increased sedimentation and the water became unsuited for the mussels. Nowadays, giant floaters live only in human-made reservoirs built since 1940. And, what is also very interesting, is that the only way that giant floaters are able to get to these reservoirs is by their larvae being carried in the gills of fish that are raised in hatcheries by the Colorado Division of Wildlife and then transported and stocked in these reservoirs. The fate of these mussels has been influenced by two separate anthropogenic activities.

One obvious conclusion that can be drawn from these and other examples is that many aspects of behavior are extremely flexible because of our presence in the lives of these animals. This is an important lesson to recognize. Just because some of the behavior patterns that are used in predation or mating appear to be hard-wired or innate, does not mean that they cannot be modified due to environmental influences. And humans are part of the environment of many animals.

While there are many problems that are encountered both in laboratory and in field research, the consequences for wild animals may be different from and greater than those experienced by captive animals, whose lives are already changed by the conditions under which they live. This is so for different types of experiments that do not involve trapping, handling, or marking individuals. Consider experimental procedures that include visiting the home ranges, territories, or dens of animals; manipulating food supply; changing the size and composition of groups by removing or adding individuals; playing back vocalizations; depositing scents (odors); distorting body features; using dummies; and manipulating the gene pool.

All of these manipulations can change the behavior of individuals, including movement patterns, how space is used, the amount of time that is devoted to various activities including hunting, antipredatory behavior, and to various types of social interactions including caregiving, social play, and dominance interactions. These changes can also influence the behavior of groups as a whole, including group hunting or foraging patterns, caregiving behavior, and dominance relationships, and also influence nontarget individuals. There also are individual differences in responses to human intrusion. All these caveats need to be considered when a specific study is being evaluated. And perhaps these changes are long-term and open to evolution via natural selection.

How Can Animals Be Studied Effectively?

Clearly, when behavior and activity patterns are used as the litmus test for what is called "normal species-typical behavior," researchers need to be sure that the behavior patterns being used truly are an indication of who the individual is in terms of such variables as age, gender, and social status. If the information used to make assessments of well-being is unreliable, then it is likely that the conclusions that are reached and the animal models that are generated are also unreliable and can mislead current and future research programs. And, of course, human errors can have devastating effects on the lives of the animals being studied. Many believe that, as students of animal behavior, our research ethic should require that we learn about the normal behavior and natural variation of various activities so that we learn just what we are doing to the animals we are trying to study.

In addition to learning about how our intrusions influence the lives of animals, it is important to share this knowledge so that we do not inadvertently change them. Sharing involves disseminating information about what is called the "human dimension" to administrators of zoos, wildlife theme parks, aquariums, and areas where animals roam freely so that visitors can be informed of how they may influence the behavior of animals they want to see. Tourism companies, nature clubs and societies, and schools can do the same.

It is important to stress that what appears to be relatively small changes at the individual level can have wide-ranging effects in both the short- and long-term. On-the-spot decisions about what study techniques should be used often need to be made, and knowledge of what these changes will mean to the lives of the animals who are involved deserve serious attention. Many researchers believe that one guiding principle might be that the lives of the animals whom humans are privileged to study should be respected, and when we are unsure about how our activities will influence them we should err on the side of the animals and not engage in these practices until we know (or have a very informed notion about) the consequences of our acts. This precautionary principle will serve us and the animals well. Indeed, this approach could well mean that exotic animals that are so attractive to such institutions as zoos and wildlife parks need to be studied for a long time before they are brought into captivity. For those who want to collect data on novel species that are to be compared to other (perhaps more common) animals, the reliability of the information may be called into question unless enough data are available that describe the normal behavior and species-typical variation in these activities.

There is a continuing need to develop and improve general guidelines for research on free-living and captive animals. These guidelines must take into account all available information. Professional societies can play a large role in the generation and enforcement of guidelines, and many journals now require that contributors provide a statement acknowledging that the research conducted was performed in agreement with approved regulations. Guidelines should be forward-looking as well as regulatory. Much progress has already been made in the development of guidelines, and the challenge is to make them more binding, effective, and specific.

Humans are a force in nature, and obviously we can change a wide variety of behavior patterns in many diverse species. Often, and paradoxically, these changes might make it difficult to answer reliably the research questions in which we are interested. Coexistence with other animals is essential. By stepping lightly into the lives of other animals, humans can enjoy their company and learn about their fascinating lives without making them pay for our interest and curiosity. Our curiosity about other animals need not harm them.

The power that we potentially wield to do anything we want to do to animals and to nature as whole is inextricably coupled with compelling responsibilities to be ethical humans beings, responsible stewards, and responsible researchers. The data that we collect not only informs us about the behavior of the animals we study, but also about the similarities and differences between their behavior and ours. It is essential to collect the most reliable data possible. An important result of being as nonintrusive as possible is that all animals, nonhuman and human, will benefit.

See also Conservation and Behavior—*Species Reintroduction*
Conservation and Behavior—*Wildlife Behavior as a Resource Management Tool in the United States National Parks*
Conservation and Behavior—*Preble's Meadow Jumping Mouse and Culverts*
Human (Anthropogenic) Effects—*Bears: Understanding, Respecting, and Being Safe around Them*
Human (Anthropogenic) Effects—*Edge Effects and Behavior*
Human (Anthropogenic) Effects—*The Effects of Roads and Trails on Animal Movement*
Human (Anthropogenic) Effects—*Environmentally Induced Behavioral Polymorphisms*
Human (Anthropogenic) Effects—*Logging, Behavior, and the Conservation of Primates*
Human (Anthropogenic) Effects—*Pollution and Behavior*

Further Resources

Baron, D. 2003. *The Beast in the Garden: A Modern Parable of Man and Nature.* New York: W. W. Norton & Company.
Bekoff, M. 2000. *Field studies and animal models: The possibility of misleading inferences.* In: *Progress in the Reduction, Refinement and Replacement of Animal Experimentation* (Ed. by M. Balls, A.-M. van Zeller and M. E. Halder), pp. 1553–1559. Amsterdam: Elsevier.
Bekoff, M. 2002. *Minding Animals: Awareness, Emotions, and Heart.* New York: Oxford University Press.
Bekoff, M. & Jamieson, D. 1996. *Ethics and the study of carnivores: Doing science while respecting animals.* In: *Carnivore Behavior, Ecology, and Evolution*, Volume 2. (Ed. by J. L. Gittleman), pp. 15–45. Ithaca, NY: Cornell University Press.
Festa-Bianchet, M. & Apollonia, M. (Eds). 2003. *Animal Behavior and Wildlife Conservation.* Washington, D.C.: Island Press.
Goodall, J. & Bekoff, M. 2002. *The Ten Trusts: What We Must Do to Care for the Animals We Love.* San Francisco: HarperCollins.
Herrero, S. & Higgins, A. 1999. *Human injuries inflicted by bears in British Columbia: 1960–97.* Ursos, 11, 209–218.

Herrero, S. & Higgins, A. 2003. *Human injuries inflicted by bears in Alberta: 1960–98*. Ursos, 14, 44–54.

Traut, A. H. & Hosteller, M. E. 2003. *Urban lakes and waterbirds: Effects of development on avian behavior*. Waterbirds, 26, 290–302.

Marc Bekoff

■ Human (Anthropogenic) Effects
Logging, Behavior, and the Conservation of Primates

The merging of evolutionary theory and classical ethology produced the field of behavioral ecology a little over a quarter century ago. But, despite obvious areas of common interest, there has been surprisingly little mixing of behavioral and conservation biology. Many conservation projects do in fact include aspects of animal behavior, but only in a trivial way. For example, many conservation studies determine home range and diet of a species of conservation concern. But rarely are these integrated into the body of theory that is now available within behavioral ecology. Yet the importance of considering behavioral and ecological data in conservation programs has been demonstrated in a variety of animals. In the future, conservation efforts on behalf of endangered species will ultimately depend on this kind of data to develop informed management plans and policies.

There is probably no mammalian order whose global fate is more closely tied to the fate of the world's rainforests than the primates. Most of the world's nonhuman primates are found only in the tropics. Primates are believed to have evolutionary ties with tropical rainforest, and approximately 90% of living species are found primarily in rainforest environments. In many rainforests, primates are the predominant component of the tree-living vertebrates. The tree-living vertebrates of tropical rainforests are known to play an important role in the ecosystem. A large majority of rainforest trees are adapted for animal dispersal of their seeds, and primates have long been recognized as important dispersal agents. The fate of the surviving primates and of tropical forests are thus inevitably linked.

The great tragedy is that tropical rainforests are currently being lost more rapidly than ever before in humankind, and this loss is due to the direct activities of humans. Humans have been exerting selective pressures on tropical forests since the dawn of prehistory. Age-old practices that have altered the forests include hunting and gathering, collecting forest products for trade, and agriculture. More recently the widespread addition of selective logging and clearcut deforestation for agriculture has caused a dramatic increase in the pace and scale of forest alteration, with the result that very little so-called primary forest has survived. Various estimates place the annual loss of tropical rainforest at 160,000 to 200,000 km^2 (about 62,000 to 77,000 mi^2), an area greater than the nation of Greece or the state of Florida.

Logging accounts for about 30% of tropical deforestation (agriculture accounts for the other 70%). The selective removal of trees from forests (as opposed to the complete cutting down of all trees in a logged region) has become the normal practice in tropical rainforests, but selective cutting practices that are typical of logging operations in the tropics can also significantly damage the forest. Logging operations often use only a small fraction of the forest, but damage much of what remains. For example, a general review of logging damage to trees in Southeast Asia's rainforests indicated that 10–75% of all trees greater than 15.0 cm

(6 in) were damaged (with the average slightly greater than 50%) in order to remove only 3% of the trees for market.

Continuing pressures on the world's tropical rainforests have significantly affected the status of primates and their habitats. Of some 300 species of primates recognized today, more than half are classified as threatened or endangered based on the small sizes and fragmented distributions of their populations. That logging can reduce the size of primate populations is well established. An important review of primate studies in logged forests in the 1980s concluded that of the 38 primate species examined throughout the tropics, 71% showed an appreciable decline in numbers with forest disturbance. However, very little is known of the behavioral mechanisms causing population declines following logging. For example, population densities can drop because of the disappearance of whole groups, or by reduction in their size. There appear to be four different patterns of response for primates in logged forest:

1. lower group densities of equivalent group size;
2. higher group densities of smaller group sizes;
3. roughly equivalent group densities of smaller group sizes; and
4. lower group densities of smaller group sizes.

Slash and burn agriculture in the rainforests of Latin America.
Courtesy of Corbis.

It is not clear why some primates reduce the size of their populations in one way, and other species of primates reduce the size of their populations in other ways. However, social strategies directly affect population growth rates by affecting the reproductive potential of females within groups, and thus an understanding of how this mechanism works may help conservation planners in predicting and monitoring the effects of habitat disturbance on the growth of threatened and endangered primate populations. The responses of Sulawesi crested black macaques to logging is illustrative of how an understanding of behavior can aid in conservation planning.

The eastern Indonesian island of Sulawesi (pronounced "Sulawaysee") holds one of the world's unique radiations of primates. In a land area less than half that of California, 7 of the 19 species (37%) of macaques, a genus of Old World primates, occur. The Sulawesi crested black macaque (*Macaca nigra*) is known only from a small geographic area on the eastern tip of Sulawesi's northern peninsula, and recently, from the small island of Bacan (pronounced "Batchan") in the North Moluccas, or Spice Islands, of Indonesia. Crested black macaques are fruit-eating primates that live in large multimale/multifemale groups, averaging 27 individuals within a group, but ranging from the low 20s to groups as big as 90 individuals.

In Sulawesi, all populations of *M. nigra* are threatened by hunting and land clearing. However, habitat degradation resulting from commercial logging is the primary threat to crested black macaques on Bacan. A study was begun in 1992 to determine the effects of logging on crested black macaque populations on Bacan. The study found population densities of crested black macaques in logged forest to be approximately 21% less than population densities in unlogged forest.

Habitat degradation is a threat to many different species, including lion-tailed macaques in India.
© *Adam Jones / Visuals Unlimited.*

The lower density of crested black macaques in logged forest on Bacan is best explained by the lower carrying capacity found in the logged forest. Logged forest contain fewer fruit (food) trees, smaller fruit trees, and fewer species of trees compared to unlogged forest. These factors suggest that the habitat quality for crested black macaques is lower in logged forest than in unlogged forest. As would be predicted from these features of the habitats, primate abundance is lower in logged forest than in unlogged forest.

However, although the observed reduction in primate abundance in logged forest is important, it is the details of the finding that are perhaps the most interesting. Population reduction in logged forest results from a decrease in the number of groups found per square kilometer (6.8 in unlogged forest compared to 4.9 in logged forest). However, the macaques in logged forest on Bacan Island form similar-sized or larger groups compared to unlogged forest (27.4 in logged forest compared to 24.9 in unlogged forest). Although it may be understandable that lower food abundance in logged forest explains lower macaque densities, it is not necessarily easy to determine the ecological factor, or combination of factors, that is responsible for the lower density being a result of a decrease in the number of groups, with an increase, rather than a decrease, in the size of groups. Why do group sizes remain large in logged forest if food resources are indeed reduced? What are the negative effects of decreased group size for crested black macaques?

Predation and food resources are often suggested to be the most important ecological determinants of primate group size. The benefits of living in a group include safety from predators and a competitive advantage against other primate groups in access to food resources. The primary cost of group living in primates is competition within the group for resources, which can lead to increased mortality and lower female reproduction. Ultimately, group size in primates is determined by a trade-off of between the costs and benefits.

The quantity and quality of food patches will determine the amount of competition for food resources within a group. The size of food patches sets a limit to the number of animals able to use a patch, since any additional members to the group beyond this limit increases competition between group members to a point where an individual will not be able to obtain enough food to survive. Additionally, the number and spread of trees influence feeding efficiency by affecting the amount of energy it takes to travel between trees. If there are few trees and they are spread far apart from each other, it will take more energy to find the same amount of food compared to trees that are close to each other and readily available. Group size then, should reflect some balance between the benefits of foraging as a group and the cost of feeding competition. As group size exceeds this optimal balance, feeding competition should cause a group split in order to become smaller. Following this line of reasoning, food patches that are small in size (rapidly depleted), or that are of low density and scattered distribution (incur high travel costs) should favor small groups. These forest characteristics are precisely what are found in logged forest, leading to an expectation of smaller group sizes compared to unlogged forest. *Macaca nigra* prefers to feed in large canopy fruit patches, such as those found in unlogged forest, in order to reduce feeding competition among individuals within the group. If food availability alone were responsible for limiting group size, it would be reasonable to expect the population of crested black macaques in logged forest to have smaller group sizes than those found in unlogged forest.

However, observations of encounters between groups of crested black macaques on Sulawesi over fruit resources suggest a clear advantage to large group size. Two macaque groups will often encounter each other at key fruit resources. This will result in a fight for control of the fruit tree and exclusive access to the food within it. The outcome of these encounters is almost always determined by group size. Big groups win. This suggests a clear advantage to large group size in Sulawesi macaques. Despite the increased competition for food among group members that results from smaller and fewer fruit trees in logged forests, cooperative defense of fruit trees against other groups may be more important in determining group size. If, because of within-group feeding competition, the size and distribution of preferred fruit patches were the primary determinants of group size for crested black macaques, then a much smaller average group size would be predicted in logged forests than is observed. This suggests that the cost of within-group competition in large groups may be balanced by the benefits of resource defense against other macaques in intergroup encounters, and may drive group size in Sulawesi crested black macaques in logged forests.

Since social strategies directly affect population growth rates by affecting the fertility of females within groups, understanding how changes in food supplies resulting from logging affects social strategies is a necessary step in developing informed conservation management plans on behalf of endangered and threatened populations. An understanding of how these mechanisms operate may help conservation planners in predicting and monitoring the effects of habitat disturbance on the growth of threatened and endangered primate populations. To date, detailed tests of the ecological factors affecting these mechanisms are limited to a relatively few studies. With many questions remaining and the fate of many of the world's primates in jeopardy, it is clear that investigations attempting to understand how habitat disturbance affects animal behavior will be an exciting area for future research.

See also Feeding Behavior—*Social Foraging*
Human (Anthropogenic) Effects—*Edge Effects and Behavior*
Human (Anthropogenic) Effects—*Effects of Roads and Trails on Animal Movement*
Human (Anthropogenic) Effects—*Human (Anthropogenic) Effects on Animal Behavior*

Further Resources

Boinski, S. & Garber, P. A. (Eds.) 2000. *On the Move: How and Why Animals Travel in Groups*. Chicago: University of Chicago Press.

Caro, T. (Ed.) 1998. *Behavioral Ecology and Conservation Biology*. New York: Oxford University Press.

Fimbel, R. A., Grajal, A. & Robinson, J. G. (Eds.). 2001. *The Cutting Edge: Conserving Wildlife in Logged Tropical Forests*.

O'Brien T. G. and Kinnaird, M. F. 1997. *Behavior, diet, and movements of the Sulawesi crested black macaque* (Macaca nigra). International Journal of Primatology, 18 (3), 321–351.

Rosenbaum B., O'Brien, T. G., Kinnaird M. F., & Supriatna, J. 1998. *Population densities of Sulawesi crested black macaques* (Macaca nigra) *on Bacan and Sulawesi, Indonesia: Effects of habitat disturbance and hunting*. American Journal of Primatology, 44 (2), 89–106.

Barry Rosenbaum

■ Human (Anthropogenic) Effects
Pollution and Behavior

Pollution refers to human-made substances in the environment, and includes a wide range of petroleum products, solvents, organochlorines, organophosphates, metals, plastics and trash. (See the following figure.) Animals are exposed to these pollutants externally, or internally through inhalation, ingestion, or direct dermal contact. Any of these pollutants can alter behavior that ultimately affects the ability to reproduce, longevity, or survival. Behavioral deficits can occur at exposure concentrations of one to several orders of magnitude lower than lethal levels. These behavioral deficits can, under some conditions, be manifest as changes in incubation and chick-rearing behavior, locomotion, thermoregulation, and individual recognition, as well as impaired social interactions or increased vulnerability to predators. All of these deficits can affect reproductive success and survival.

While the vast majority of ecotoxicology studies with animals in the wild focus on death as the endpoint, understanding sublethal alterations in behavior—that is, those that cause behavior change but not death—is critical for assessing environmental effects, and as

---

Major Forms of Pollution That Can Affect the Behavior of Animals

Petroleum products

These are complex mixtures of aliphatic and organic compounds, including oil and gasoline.

Solvents

These include in particular short chain chlorinated aliphatics such as trichloroethylene, tetrachloroethylene, and formerly carbon tetrachloride, as well as aromatic solvents such as toluene and xylene.

Organochlorines

Many of the chlorinated pesticides (e.g. DDT) or their breakdown products are highly persistent in the environment and in the body.

Organophosphates

These substances exert mainly acute nervous system toxicity by interfering with acetyl-cholinesterase. Some evidence of prolonged and even delayed neurotoxicity in survivors.

Polychlorinated di-aromatic compounds (PCBs, dioxins)

Highly persistent chemicals, which vary greatly in their toxicity. Effects on the nervous system of some of these compounds are secondary.

Metals and metalloids

Many metals have potent effects on the nervous system. For example, mercury, lead, manganese, and chromium can alter behavior.

Plastics and trash

Plastic fishing lines, soda packs, and other trash can impale young and adult birds at the nest and when foraging, can disrupt digestion in sea turtles and fish, and can snare a variety of small mammals.

an early warning of impending environmental damage. Understanding the role of contaminants in the disruption of behavior and reproductive success of animals is essential for maintaining healthy ecosystems and for forecasting future problems with viable animal populations (e.g., behavioral changes such as poor parental incubation behavior, lower ability to detect prey or to capture prey, and less ability to perceive predators).

It is important to understand potential effects of chemicals on behavior and to assess the magnitude of these changes so that effects due to chemicals are not ascribed to other natural stressors; mechanisms of effects can be used to understand the underlying mechanisms of behavior; and laws and regulations can be instituted when necessary to protect animal populations. Some behavioral abnormalities, once ascribed to natural variation, are now known to be caused by contaminants.

Behavioral effects involving sensation, perception, cognition, integration, coordination, and motor function can lead to deficits in feeding, predator avoidance, and reproductive success, resulting in shifts in population dynamics, and population declines. Behavioral effects may result directly from the action of neurotoxic chemicals on one or more targets of the nervous system, or secondarily from chemical effects on the endocrine system that in turn modifies behavior.

Susceptibility

Although all animals are exposed to pollution and suffer a variety of effects, more information is available on birds than any other group because they are abundant, conspicuous, diurnal, and of interest to the general public. Scientists, students, managers, and the general public have an "idea" about how birds behave in nature and are likely to notice unusual behavior that may be indicative of the effects of chemicals. Birds have had an important role as sentinel species for monitoring of chemical exposure. It was the dramatic mortality of passerine birds such as robins, and the decline of birds of prey such as bald eagles following extensive use of the pesticide DDT that led to public concern and the subsequent banning of DDT.

Part of the difficulty with understanding the importance of contaminants in behavioral changes is that there are differences in how individual species within a group (such as birds) respond, as well as differences in how different groups respond. Eagles, hawks, and pelicans suffered behavioral disruptions and lowered reproductive success when exposed to lower concentrations of DDT than did the chickens used to test these chemicals in the laboratory. Invertebrates such as clams and mussels experience behavioral alterations when exposed to much lower amounts of cadmium than do vertebrates. There are also individual differences in how a given chemical affects the individuals within a species. These differences are the rule rather than the exception.

Following are some examples of how specific contaminants affect behavior, but such behavioral disruptions are not limited to the groups of animals described. In most cases the effects of pollutants on different animals groups have not been examined. Some of the effects of contaminants have been described only from laboratory studies, and in other cases the effects were studied in animals living in the wild. Sometimes the effects of chemicals on behavior occur in parents, their offspring, and their offspring's offspring; the effects are truly multigenerational.

Behavior Deficits Caused by Metals

In the last 80 years, lead has entered our environment from lead paint and from lead used in gasoline. Although both sources have been outlawed in the past 20 years, lead paint still sloughs off the walls in old apartments and lead remains in the soil near highways and gas stations. Lead poisoning has been known for at least 2,500 years, although it gained prominence because of the mortality in ducks caused by lead shot from hunters. In the 1960s and 1970s Frank Bellrose, Ronald Kendall and others showed that lead causes loss of weight, greater susceptibility to hunters, smaller clutch size, increased egg-laying interval, reduced hatching success, and a variety of physiological effects in birds. Lead depresses feeding behavior in a wide range of species of birds and mammals, and shortens the migration of birds. It also disrupts the normal behavior of earthworms.

At the same time, laboratory experiments indicated that lead disrupts learning and cognitive abilities in a variety of primates. Subsequently, work by Herbert Needleman and others over the last 30 years indicated that lead exposure causes learning impairment in human children, leading to poor school performance and increased rates school dropouts and juvenile delinquency, as well as job losses later in life.

In a series of field and laboratory experiments with birds over the last 20 years, Joanna Burger and Michael Gochfeld developed a paradigm to use environmentally relevant concentrations of lead in blood to examine the effect of varying time of exposure and dose on neurobehavioral development birds. Unlike many laboratory experiments, they used behaviors that are relevant to survival in the wild. They found that lead impairs balance, locomotion, thermoregulation, depth perception, endurance, begging and feeding, learning of a running task, and parental and sibling recognition.

Similar behavioral effects were found for chromium, manganese, and mercury. Mercury has severe effects on the nervous system, particularly in vertebrates, and these effects are multigenerational. Tin causes feeding suppression, weight loss, collapse, convulsions and eventual death.

Behavioral Deficits and Organochlorines

A wide range of organochlorines produce similar behavioral and physiological effects in several classes of vertebrates in laboratory experiments. Similar behavioral effects have been observed in the field, but wild animals are not normally exposed to only one contaminant, thus making it difficult to isolate the specific cause of an observed effect. Another difficulty is that effects may be noted, but the mechanism may not be identified.

The first clear demonstration of behavioral disruption caused by organochlorines was in colonial birds. In the 1960s and 1970s the biologist George Hunt and others found that gulls exhibited skewed sex rations at breeding colonies. There were more females than males, leading to female–female pairs and trios of two females with one male, leading to nests with six eggs instead of the usual clutch size of three. These abnormalities were associated with exposure to organochlorines such as DDT, PCBs and presumably TCDD. DDT was used primarily as an insecticide in the 1950s and early 1960s to kill noxious insects, and PCBs are used primarily as an insulating oil in transformers in power stations and large buildings. Oil carrying PCBs gets into our streams and rivers, and eventually into foods, such as fish and shellfish. Like metals and DDT, PCBs persist in the environment for a very long time.

In general, PCBs are developmental neurotoxicants producing effects on activity, neurological development, courtship, predator avoidance, and cognitive function. Exposure to organochlorines has been associated with cessation of nest building and egg incubation, and destruction and cannibalism of eggs in birds, leading to lowered reproductive success and population declines in the Great Lakes region. For some species, increased levels of PCBs were associated with declines in colony site tenacity; birds simply did not return to breed in the colonies they had used in the previous years. This is a problem because the birds then have to relearn where to forage and how to avoid predators in the new colony site. Researchers Glen Fox, Michael Gilbertson and others found that decreases in reproductive success in terns, gulls and cormorants in the Great Lakes were due to deficits in parental care; parents simply failed to incubate their eggs all of the time, and eggs failed to hatch or were preyed upon. Eggs from uncontaminated colonies, transferred to these birds, also failed to hatch, indicating the problem was deficient incubation behavior.

The behavior and physiology of fish and other vertebrates are also affected by PCBs and other organochlorines. Laboratory animal exposure studies show that PCBs can cause neurobehavioral effects in exposed monkeys and rodents.

Endocrine Disruption

There is a close relationship between endocrine disturbances and behavioral alterations. DDT is the classic example of an endocrine disrupter—a chemical that mimics a hormone, thereby disrupting normal behavior. Birds exposed to high levels of DDT laid thin-shelled eggs, which broke when they were incubated, ultimately leading to low reproductive success in fish-eating birds. Some species, such as pelicans, had no reproduction in some years, leading to drastic population declines. DDT also disrupts nest defense in birds, leading to increased rates of predation. The constellation of behavioral abnormalities found in colonial nesting birds in the Great Lakes is often ascribed to endocrine disrupting chemicals.

Biologist Lou Guilette's work with American alligators of Lake Apopka, Florida, is one of the most cited examples of a wildlife population affected by environmental contaminants. Lake Apopka is polluted with DDT and its breakdown products. Alligators hatched from Lake Apopka had abnormal gonadal development, small testis, female oocytes were multinucleated (instead of having only one nuclei), and reproduction decline markedly. The abnormalities were associated with abnormal sex steroid levels in both males and females, apparently due to environmental contaminants, which affected mating behavior. Mosquitofish from Lake Apopka also showed decreased male courtship behavior.

Other compounds like dieldrin and PCBs, that are potential endocrine disrupters, can alter the concentrations of brain neurotransmitters following exposure. Other organochlorines may cause behavioral effects through mechanisms involving endocrine disruption (disruption of steroid hormone action), thyroid hormone metabolism or by affecting homeostasis of vitamin A. Regardless of the mechanisms, these chemicals have the ability to disrupt behavior, leading to both lowered reproductive success and lowered survival.

Behavioral Deficits and Oil

Increases in oil transport and off-shore oil drilling have resulted in increased oil pollution of coastal areas from tanker accidents, bilge washings, and slow seepage of oil from well sites at sea. Fish and marine invertebrates are impacted because the oil and its breakdown products filter down through the water column, disrupting the transport of gases over sensitive membranes. The behavior of invertebrates living in the tidal zone is also disrupted. Fiddler crabs exposed to oil spills decrease their feeding rates, dig fewer burrows, and dig their burrows shallower than crabs not exposed to oil. With fewer burrows, crabs have to move over greater distances to find a burrow when exposed to high tides or predators, and many perish. If burrows are not deep enough, they are inundated during high tides, or the animals freeze during the winter.

Marine mammals, seabirds and shorebirds that feed at the water–land interface are most affected because they plunge through the oil slick on the surface of the water. Oiled birds spend more time preening to remove the oil, and less time foraging or watching for predators. Oiled birds also are more aggressive than non-oiled birds. In breeding colonies, time devoted to preening reduces the time devoted to incubation and care for their young, reducing their reproductive success.

See also Human (Anthropogenic) Effects—*Human
(Anthropogenic) Effects on Animal Behaviour*
Human (Anthropogenic) Effects—*Edge Effect
and Behavior*
Human (Anthropogenic) Effects—*The Effects
of Roads and Trails on Animal Movement*
Human (Anthropogenic) Effects—*Urban Wildlife
Behavior*

Further Resources

Bellrose, F. C. 1959. *Lead poisoning as a mortality factor in waterfowl populations.* Illinois Natural History Survey Bulletin, 27, 235–288.
Burger J. & Goichfeld, M. 2004. *Effects of lead and exercise on endurance and learning in young herring gulls.* Ecotoxicology and Environmental Safety, 57, 136–144.
Burger, J., Kurunthachalam, K., Giesy, J. P., Grue, C. & Gochfeld, M. 2002. *Effects of environmental pollutants on avian behavior.* In: *Behavioral Ecotoxicology* (Ed. by G. Dell'Omo), pp. 337–375. New York: Wiley & Sons.
Colborn, T. & Clement, C. (Eds). 1992. *Chemically induced alterations in sexual and functional development: The wildlife/human connection.* Princeton, NJ: Princeton Scientific Publications.
Environment Canada. 1991. *Effects of contaminants on wildlife species.* In: *Toxic Chemicals in the Great Lakes and Associated Effects.* Vol. II, Part 2. Canada: Environment Canada.
Fox, G. A., Gilberston, M., Gilman, A. P. & Kubiak, T. J. 1991. *A rationale for the use of colonial fish-eating birds to monitor the presence of developmental toxicants in Great Lakes fish.* Journal of Great Lakes Research, 17, 151–198.
Fox, G.A., Seseloh, D. V. W., Kuviak, D. V. & Erdman, T. C. 1991. *Reproductive outcomes in colonial fish-eating birds: A biomarker for developmental toxicants in Great Lakes food chains.* Journal of Great Lakes Research, 17, 153–157.
Guillette, L. J. Jr., Gross, T. S., Masson, G. R., Matter, J. M., Percival, H. F. & Woodward, A. R. 1994. *Developmental abnormalities of the gonad and abnormal sex hormone concentrations in juvenile alligators from contaminated and control lakes in Florida.* Environmental Health Perspectives, 102, 680–688.
Hoffman, D. J., Rattner, B. A., Burton Jr., G. A. & Cairns, Jr., J. (Eds.) 1995. *Handbook of Ecotoxicology.* Boca Raton, FL: CRC Press.
Hunt, G. & Hunt, M. W. 1977. *Female–female pairing in western gulls* (Larus occidentalis) *in southern California.* Science, 196, 483–486.
Kendall, R. J. & Scanlon, P. F. 1981. *Effects of chronic lead ingestion on reproductive characteristics of ringed turtle doves* (Streptopelia risoria) *and on tissue lead concentrations of adults and their progeny.* Environmental Pollution, 26, 203–213.
National Research Council. 1999. *Hormonally Active Agents in the Environment.* Washington, D.C.: National Academy Press.
Needleman, H. L., Schell, A., Bellinger, D., Leviton, A. Allred, E. N. 1990. *The long-term effects of exposure to lead in childhood: An 11-year follow-up report.* New England Journal of Medicine, 322, 83–88.

Joanna Burger

■ Human (Anthropogenic) Effects
Urban Wildlife Behavior

For anyone interested in the mysteries of animal behavior, cities are noteworthy, if not exactly obvious, places in which to work. Many animal species have adapted to urban life—gray squirrels in Washington, DC, rooks in London, and rhesus monkeys in New Delhi are but a few examples. All share the fundamental trait of being tolerant of living in proximity to humans, a characteristic identified by the term *synanthrope*. All may be accessible to study in ways that their more rural counterparts are not. To the wild animal, the city is a real environment to which it must adapt in order to survive. Urban habitats present both expected and unexpected challenges for which both routine and novel behavioral responses may be required. Cities frequently present wild animals with resources and opportunities not available elsewhere, such as new and abundant sources of food and refuge. Among the greatest challenges facing any wild animal attempting to survive in an urban habitat must be predicting the behavior of humans, who might alternately be compassionate providers, neutral bystanders, or heartless antagonists.

Even though people and animals have been living together in urban environments since the rise of the first cities more than 5,000 years ago, the idea of an "urban" wildlife deserving of study in its own right is still little accepted. Few scholars interested in animal behavior think to turn to the city as a research site, following a bias whose roots are part historical and part cultural. The earliest studies of animals in the West were focused on the discovery and categorization of living forms, activities that largely took place in unexplored wilderness. Developing cities were naturally inhospitable to many wild animals, if for no other reason than that many were looked on as food, not subjects of study. What early colonizers of urban habitats could be found were likely those species regarded as nonedible, often "vermin" or pests such as the commensal rodents. Eradication, not study, was and mostly still is the orientation of most humans toward such species. Even today, many resources may be devoted to killing programs while few are given to elucidating the behavior of problem species. Logic dictates, however, that the most successful approaches to controlling our conflicts with wildlife will involve behavior change, as appropriate technologies are used to intervene in problem-causing activities in ways that are environmentally sound. An age-old attitude that sees humans as holding absolute dominion over nature still seems to encourage in some the certitude that force alone is sufficient to create solutions. Indeed it would be, if this were not a world in which people now choose to take moral objection to anthropocentric bias.

Black skimmers nest in the parking lot of Dow chemical plant in Freeport, Texas.
© *Dan Guravich / Corbis.*

By the nineteenth century, some naturalists had oriented their studies toward the sort of descriptive narrative characterized by the term "rambles," allowing expanded observations on such aspects of animal life as behavior and life histories. One of these naturalists, the physician John Godman, wrote about his walks in and around Philadelphia and the animals he observed there as early as the 1830s. But it was more than a century later before the first works exclusively focused on urban wildlife were published. Combining the enumerating approach of the catalogers with the more expansive style of rambles, Richard Fritter and John Kieran produced the first books written for popular audiences on the natural history of cities. Research papers on urban wildlife published in professional journals began appearing in the 1960s, and today a number of promising areas of inquiry have begun to emerge. These include the study of behavioral phenomena associated with colonization, comparative studies focusing on the differences in behavior and social organization of urban and nonurban populations of the same species, behavioral adaptations to novel or rapidly changing environments, and behavioral responses to anthropogenic stress. Undoubtedly more exist.

Urbanization has been characterized as a "massive, unplanned experiment," a statement whose acceptance would compel us to lament many already missed opportunities. Consider the small bird best known by its common name as the English or house sparrow (*Passer domesticus*). Actually a type of finch, this is one of more than a dozen Old World bird species deliberately introduced into North America in the late nineteenth century. Moreover, it was reintroduced when early efforts failed to take hold, and it was for some time propagated to create an exportable population that could be shipped to cities throughout the United States. All of this was for the purpose of insect control in cities, a dubious proposal to begin with, given that house sparrows are primarily seed-eaters and intensively forage for insects only when raising young. The record shows many such cases where both native and non-native species have been introduced, propagated or promoted without even a superficial understanding of their behavior and potential to create problems, sometimes with results that have ecosystem-wide consequences.

In the case of the house sparrow, people found, only after it was too late, that this was an extremely aggressive colonizing species. The undesirable result is that today this bird is so widespread and numerous that many consider it a pest. But, however abundant house sparrows may be now, they were likely to have been more so in the nineteenth century at a time when horses and other draft animals were the principal means of transport and conveyance. The seeds available in animal droppings, the ready availability of bedding straw that could be used for nests, and the numerous recesses in stables and barns where nests could be built undoubtedly combined to create ideal conditions for sparrows to thrive and expand. The rise of the automobile and demise of the horse should have dealt sparrow populations a telling blow. Typically, a species goes into decline when the main subsistence base for a species is radically changed. If such a blow landed, the house sparrow may have taken it and moved on. Backyard feeders might have compensated for food lost elsewhere, and nesting habitat could have actually increased with the urban building boom. Even the nemesis may have become provider, as sparrows learned to glean insects from the road, killed by passing automobiles the night before. Cities are laboratories in which long-running, natural experiments in animal adaptation abound. We need only focus on their potential to see the opportunity they can provide for the study of animal behavior.

A special form of behavior involves innovation. Cities may be especially provident places to observe this phenomenon, if only because there are so many human eyes and ears ready to observe and document change. In England just after the close of World War II, titmice

Margaret Morse Nice, Backyard Ornithologist

John Hadidian

A unique combination of drive coupled with skill, opportunity, and education marked the research career of Margaret Morse Nice, allowing her to persevere in a seminal, multi-year study of song sparrows (*Melospia melodia*) in suburban Columbus, Ohio, that has become one of the lasting foundations upon which modern ornithology is based. Working in the 1930s and 40s, Nice seems to have been almost as challenged by contemporary mores as by the daunting scientific task she set herself to. Staying home to raise her children as dictated by the times, however, freed her for many hours to conduct "backyard" observations on the little bird she had chosen to devote her life and work to. By carefully and painstakingly following individuals over the course of years, Nice found that the same birds were returning to Columbus year after year to set up territories in virtually the same locations. She observed the offspring of these individuals returning to establish territories of their own, and proved to the world of science that her sparrows were distinguishable as individuals, with unique and personal life histories. One example of her tenacity as a field researcher stands out in explaining how she was able to unlock so many of the secrets these birds held. On May 11, 1935, Nice determined to count all of the songs that would be sung by one of her resident males, whom she identified after banding as 4M. He started singing at 5:50 a.m. and only stopped at dusk. By that time he had sung a total of 2,305 times. 4M had been returning to the area where she had banded him for 7 years, making him at least 8 years old at the time.

learned to open the cardboard tops that sealed the milk bottles left, in those days, on front stoops. Once opened the birds would help themselves to the rich top layer of cream, exploiting an entirely new food resource. This adaptation, arising perhaps from a predisposition to manipulate bark-like objects, was nonetheless a novel invention, and because it spread over a wide geographic area, it also qualified as a learned tradition. The impact of this finding, at least in anthropological circles, was to challenge the concept of culture in a way that led scholars to redefine this supposedly sacrosanct human institution. Other examples of innovative behavior in urban birds include the growing use of rooftops for nesting by different species of gulls and the almost complete shift of winter roosts among crow populations in the United States from country to city, a phenomenon that has occurred only within the last 30 years. These indicate that innovative behaviors are neither rare nor restricted in scale, but they do not tell us yet whether urbanization is an especially active catalyst of these or not.

The red fox (*Vulpes vulpes*) may be the best studied of any urban animal, thanks to the many different researchers who have focused on this carnivore and the wealth of comparative information derived from their studies. Our understanding of foxes especially benefits from a unique long-term study taking place in Bristol, England, where Stephen Harris, along with students and colleagues, has been collecting information since the late 1960s. At a glance, the foxes of Bristol seem largely to be like their counterparts in more rural areas. They may be slightly more nocturnal, exploit a wider food base, and enjoy opportunities for denning under structures that are not as available in the country, but they are still basically foxes. It is when the population surveys are examined that some of the potentially significant differences come to light. For about the first 30 years (1960–1990) of study, researchers in Bristol found that the number of fox social groups was high compared to nonurban populations, but that the size of the average group was well within the species norm. Overall, in the main study area numbers averaged about 14 adult foxes for each square kilometer. Over a 4-year period from 1990–1994 the density of the foxes, and the average group size rose dramatically, to a point where the adult density was about 26 animals per square kilometer and as many as 10 individuals could be identified in a single social group. Such fox densities had never been recorded before, and it was clear (if necessarily inferred) that significant changes had to be taking place in behavioral mechanisms that dictated social relationships and the tolerance of individual foxes for one another.

In the summer of 1994, sarcoptic mange, a disease that can be deadly to foxes, decimated the population. By the winter of 1995 the fox density had fallen to 0.9 per square kilometer. Correspondingly, the size of the territories occupied by remaining foxes increased dramatically, a finding that would not be at all surprising except that the researchers were able to determine that territories of this size were not necessary to provide subsistence needs. All along, humans had been playing an apparently major role in supporting high fox numbers. The population increases of the 1990s were linked to the abundance of human-derived foods ("scavenge"), which comprised as much as 60% of the diet of some animals. Some people persisted in feeding and maintained high levels of provisionization even when densities were lowest—hence, the "contradiction" between territory size and resource need. As the researchers themselves note, more questions about the complex pattern of population and behavioral changes exist than have been resolved. From the answers is likely to come not only a better understanding of foxes but of the nature of social organization in mammals as well.

For anyone interested in directly observing animal behavior, cities can provide a unique opportunity to view animals that became not only acclimated but often indifferent to people. Observations can often literally be made by looking out the kitchen window, and they can be made in ways that do not take away from, but add to, daily routines. Ironically, the vast majority of urbanites not only fail to take advantage of such opportunity, they are unaware it even exists. Schools could certainly provide children with more connection to the animals with whom they share their immediate environment. Children live in a world where here and now is their orienting reality. Yet we send them on field trips to zoos to watch animals in cages that completely constrain natural activities, while free-ranging pigeons swarm at their feet unconstrained by any bars. Simple and easily taught techniques of behavior observation, such as enumerating patterns of behavior (ethogram research) are rarely, if ever, present in school curricula. These lost opportunities are especially unfortunate if, as many posit, there is a direct connection between immediate experience and the heightening of awareness and sensitivity to the natural world.

The lack of good understanding of the behavior of urban species affects not only public understanding, but clouds as well the approaches taken by professional wildlife managers, with almost certain implications for animal welfare. Canada geese (*Branta canadensis*) have increasingly taken up occupancy in urban habitats, following an intense period of state and federally sponsored artificial propagation and translocation programs aimed at increasing the size of their populations. Spurred, it seems clear, by an interest in increasing hunting opportunities, plans to have these populations accessible to hunters went awry when many introduced geese failed to exhibit migratory behavior and "chose" to live year-round within the relative safety of urban and suburban neighborhoods. The consequence of this has been increasing calls for "control" of urban goose populations, with the increasingly preferred solution being

Canada geese scatter in a parkland in the city of Vancouver, Canada.
© Gunter Marx Photography / Corbis.

the round up and killing of hundreds, sometimes, thousands, of birds when they are flightless during the annual molt. Preliminary plans issued by federal agencies charged with wildlife management now call for a reduction of the continental population of "resident" geese by more than a million birds, meaning that many times that number will have to be killed over the 10 years in which reductions are targeted. What little research on urban geese that has been conducted indicates some potentially very different solutions may exist. Work under way on the behavior of some "resident" geese in southeastern Michigan indicates that a high percentage of females experiencing nesting failure will undergo a "molt migration," not only moving out of the urban area, but traveling North to summer on the Canadian tundra as much as 1,500 miles away. Verging on a massive campaign to depopulate the country of "resident" geese, the wildlife managers seem to come full circle in disregarding behavior as the key to dealing with wildlife populations. As house sparrows, and even geese themselves, were once introduced without a sound understanding of the behavioral consequences, so now geese will be killed without similar knowledge.

Cities are natural laboratories for the study of animal behavior at many different scales, and they are also the places where people and wild animals are likely to come into greater and more sustained contact than anywhere else. Urban environments are and will increasingly come to be important habitats for both people and wild animals. Human life can only be enriched by greater contact and better understanding of the natural world, and with some effort and foresight, after centuries of domination and exploitation, we may be able to look forward to a new era of harmony in living with our wild neighbors. The foundation on which this understanding must be built will be animal behavior.

See also Human (Anthropogenic) Effects—*Edge Effects and Behavior*

Further Resources

Baker, P., Newman, T. & Harris, S. 2001. *Bristol's foxes—40 years of change*. British Wildlife, August, 411–17.

Fritter, R. S. R. 1945. *London's Natural History*. London: Bloomsbury Books.

Godman, J. D. 1833. *Rambles of a Naturalist*. Philadelphia: Thomas A. Ash.

Harris, S. 1994. *Urban Foxes*. London: Whittet Books.

Kieran, J. A. 1959. *Natural History of New York*. Boston, MA: Houghton Mifflin Company

McDonnell, M. J. & Pickett, S. T. A. 1990. *Ecosystem structure and function along urban–rural gradients: An unexploited opportunity for ecology*. Ecology, 71(4), 1232–1237.

Michigan Department of Natural Resources. 2000. *Controlling Canada goose conflicts in Michigan: Activities conducted under federal permit, 1999*. Lansing, MI: Department of Natural Resources.

Nice, M. M. 1964. *Studies in the Life History of the Song Sparrow, Volumes I & II*. New York: Dover Publications.

John Hadidian

■|Infanticide

Infanticide occurs when an individual kills a juvenile of the same species. Victims of infanticide have been hatched (insects, amphibians, reptiles, birds, and so forth) or born (mammals), but some biologists think that the term infanticide also should include destruction of eggs (*ovicide*) and killing of embryos and fetuses (*abortion*).

To avoid detection and retaliation by protective parents, infanticidal individuals often are sneaky and fast. Documentation of infanticide by human observers is therefore elusive. Verification of infanticide among black-tailed prairie dogs, for example, required dawn-to-dusk observations of marked individuals and excavations of long, deep burrows. Despite such difficulties, biologists have documented infanticide for hundreds of species from scores of taxonomic groups.

For animals such as African lions, hanuman langurs (also called temple monkeys), and red colobus monkeys, killers are adults, and victims of infanticide are usually genetically unrelated (or only distantly related). For black eagles and masked boobies (seabirds), by contrast, killers are juveniles, and victims are siblings. For black-tailed prairie dogs, adult females commonly kill juvenile grandoffspring and juvenile nieces and nephews.

At least five reasons explain why infanticide occurs.

1. For flour beetles, house mice, and European rabbits, infanticide sometimes results from overcrowding under laboratory conditions. Overcrowding in the wild also can lead to infanticide. In dense colonies of northern elephant seals, for example, infanticide commonly results when adult males crush juveniles who interfere with territorial defense.

2. Parents sometimes kill their own offspring to improve the survivorship of themselves or other offspring. When closely pursued by a predator, for example, a hill kangaroo mother sometimes ejects a joey from her pouch so that she can run faster. When food is limited, mothers in some human societies kill their babies to improve probability of survival of their older offspring.

3. Infanticide among white pelicans, short-eared owls, and blue-footed boobies probably results from sibling rivalry. After killing its sibling, the infanticidal juvenile experiences less competition when parents return to the nest with food.

4. Cannibalism commonly follows infanticide. For spotted hyenas, lesser black-backed gulls, Belding's ground squirrels, and Utah prairie dogs, increased nutrition via cannibalism is probably the primary benefit from infanticide. Additional nutrition enhances both survivorship and reproductive success.

5. Killing of nursing offspring after invasion by a new male is common for animals such as African lions, purple-faced langurs (monkeys in Sri Lanka), and mountain gorillas that live in social groups with several breeding females and one breeding male. The payoff for infanticidal males is that victimized mothers mate and become pregnant more quickly than do mothers that continue to suckle their offspring.

Predictably, maternal defense against infanticide is often impressive. Black-tailed prairie dog mothers, for example, vigorously defend territories around burrows that contain their unweaned offspring, and mothers also plug secondary burrow entrances (back doors) that lead to their offspring.

Because juveniles are pivotal to the continued survival of any species, the prevalence of infanticide in so many species might seem puzzling. But natural selection works at the level of individuals, not species. If killers consistently produce more offspring than nonkillers, then natural selection will favor the evolution of infanticide.

See also Siblicide

Further Resources

Hausfater, G., & Hrdy, S. B., (Eds.). 1984. *Infanticide: Comparative and Evolutionary Perspectives.* New York: Aldine.

Hoogland, J. L., 1995. *The Black-tailed Prairie Dog.* Chicago: University of Chicago Press.

Sherman, P. W., 1981. *Reproductive competition and infanticide in Belding's ground squirrels and other animals.* In: *Natural Selection and Social Behavior.* (Ed. by R. D. Alexander & D. W. Tinkle), pp. 311–331. New York: Chiron Press.

Van Schaik, C. P., & Janson, C. H., (Eds.). 2000. *Infanticide by Males and its Implications.* Cambridge: Cambridge University Press.

John L. Hoogland

▮|Laterality

Laterality (or lateralization) refers to differences between the left and right sides of the brain. It is sometimes manifested as side biases in the behavior of the animal. The left and right sides of the brain (usually the hemispheres) may be different in structure, or they may process information differently and control different functions. The latter is also known as hemispheric specialization. Handedness in humans is the most commonly recognized example of laterality expressed in this way but, as the examples to follow will show, laterality can include responding differently to a stimulus according to whether it is on the animal's left or right side or to a preference for turning in one direction.

Laterality was first discovered in humans, in relation to right-handedness and the control of speech by the left hemisphere (i.e., in the majority of people). In fact, until about two decades ago, laterality was thought to be a unique characteristic of humans, underlying our ability for language and using tools; hence, apparently, explaining our superiority over other animals. We now know that this assumption was incorrect and, in fact, that laterality might be a characteristic of all vertebrates.

In some cases, individual animals are lateralized but there is no bias for the group or population to be lateralized in the same direction (e.g., in some strains of laboratory bred mice and rats, half of the individuals prefer to use their left paw to reach into a tube to obtain food and the other half prefer to use their right paw). In other cases, laterality is expressed as a population bias (e.g., the majority of cockatoos prefer to use their left foot to hold food), and this is the form of laterality of most interest to us here.

Some forms of laterality are expressed in one species and not in another, although there are some common patterns of laterality, as we will see below. Even within a species some forms of laterality may show no population bias, whereas other forms are population biased (e.g., rats showing no population bias for paw preference do show a population bias for many other brain functions). This means that we cannot think of a species as simply being lateralized or not lateralized, but instead we must specify the behavior or function that we are considering.

Discovering Whether an Animal or a Species Is Lateralized

A number of methods have been used to determine the presence of laterality. One approach has been to damage the left or right hemisphere of the brain, either by lesioning it or by injecting a chemical that disrupts its function (e.g., the neurotransmitter glutamate), and then testing to see whether the behavior of the left- versus right-treated animals is different. The first discoveries of laterality in animals used such invasive procedures.

There were three such early discoveries:

1. Canaries with a lesion placed in the higher vocal center of the left hemisphere cannot sing, whereas those with the same lesion on the right side can sing perfectly;
2. Damage to the entire right hemisphere of the rat impairs its emotional (or affective) behavior, whereas the same treatment of the left hemisphere does not; and

3. Glutamate treatment of the chick's left hemisphere impairs its ability to learn to find grain scattered on a background of small pebbles (the pebble–grain test), whereas glutamate treatment of the right hemisphere has no such effect.

Although the studies showing this for the first time were important, for ethical reasons scientists are now less inclined to use such invasive techniques.

In fact, in many species simply covering the left or right eye temporarily can reveal lateralized visual behavior. Chicks tested monocularly on the pebble–grain test can learn well when they use their right eye (left eye covered), but not when they use their left eye (right eye covered). This result fits with the finding that the left hemisphere is involved because most of the visual information received by the chick's right eye is processed by its left hemisphere, and most received by the left eye is processed by the right hemisphere. As this eye-to-opposite hemisphere connection is common in most species with their eyes positioned laterally on the sides of their heads, monocular testing is a useful technique for showing lateralization.

Other forms of laterality are revealed by monocular tests: Chicks tested using their left eye attack more than those using their right eye, and they perform better on tasks requiring use of spatial information from their surroundings. Rats also perform better on spatial tasks when they use their left eye compared to their right eye.

Lateral Biases in Intact, Untreated Animals

Can laterality be seen in animals that have not been treated in any way, and when they are using both eyes? This is an important question because the answer will help us to decide whether laterality has any relevance to the animal's behavior in its natural habitat. Animals do show side biases of this kind. For example, toads have a bias to strike at prey on their right side (as it enters the right half of their visual field) and to attack another toad on their left side. They are also more responsive to a predator lunging at them on their left side than one on their right side (a model snake was used in the experiments showing this). Lizards and chicks, too, are more likely to attack another member of their species on their left side and to be more responsive to a predator approaching on their left side. Even gelada baboons are known to be more aggressive to another member of their species when it is on their left side. These side biases must have consequences for the animals' survival in their natural environment.

Animals also show eye and ear preferences when they attend to certain stimuli. Chicks turn their head to look at more distant stimuli with the left or right eye depending on which hemisphere they prefer to use to process the information they are receiving. Also, when a chicken hears its species-typical alarm call that signals that there is a predator overhead (e.g., a raptor), it will tilt its head to look upward with its left eye (and so use its right hemisphere). The same has been found to be true in the Australian magpie.

Kookaburras perched on power lines turn their head to the right to look below with their left eye as they scan the ground for prey; perhaps they do so because they are using spatial information. A striking example of laterality in a wild species is that of the new Caledonian crow: When it is cutting tools from pandanus leaves, the crow positions itself so that it can use its right eye to see where it must use its beak to cut the leaf.

Ear preferences for attending to stimuli are determined by the ingenious method of playing a sound from behind the animal and scoring which way the animal turns its head to listen: Such tests have shown that rhesus monkeys prefer to listen to their typical vocalizations using the left ear. The same left ear preference is also typical of the common marmoset, a small New World monkey.

Turning preferences have been seen in several species of schooling fish but not in those that do not form schools. Each fish was tested alone, and its direction of turning recorded when it reached a barrier. Individuals of the schooling species preferred to turn in the same direction. This form of laterality might be essential for keeping the school together. Fish also show a preference to swim with another member of their species on their left side (and so monitor it using the left eye). This was discovered by testing fish, one at a time, in a long tank with mirrors on the left and right sides. At one end there was a predator fish, and the test fish would approach the predator to inspect it, as they sometimes do in their natural habitat. The fish tested preferred to swim next to the mirror on its left side.

Hand and Limb Preferences

Preferred use of one limb over another is another form of laterality. There are many ways to measure hand preferences. The strict definition of handedness is the preferred use of one hand over the other in a range of different tasks and in the majority of individuals in the group or species.

Whether or not apes show handedness is currently a topic of debate. Some researchers claim that apes kept in captivity show right-handedness only because they have learned this from humans. So far there is little evidence for handedness in apes in the wild, whereas captive ones do show hand preferences at a group level. However, orangutans in their natural environment have been found to prefer to use their left hand when they touch and manipulate parts of their face, such as when they rub their eyes or clean their teeth.

The best way to determine handedness is to look at tasks requiring collaborative (or coordinated) use of both hands and to see whether one hand is dominant over the other. Very few studies of this kind have been conducted even for the most obvious task of feeding. Nevertheless, observations of coordinated hand use in wild orangutans and chimpanzees during feeding indicate that they do not exhibit handedness, although there is recent evidence for captive chimpanzees showing a population bias for right-handedness during coordinated bimanual feeding.

Lower Old World primates (prosimians) show a preference to use the left hand to reach out to grasp or pick up food, and New World primate species display the full spectrum, some species being right-handed, others left-handed or ambi-preferent. But limb preferences are not limited to primates. Many species of parrots and cockatoos have a strong left-foot preference for holding food (only one, the crimson rosella, is known to prefer the right foot), and toads prefer to use their right paw to remove objects from their head or snout.

Although hand and other limb preferences are interesting in their own right, there has been rather too much emphasis on them at the expense of looking for other forms of laterality in both wild and captive animals. A brain can be lateralized without the individual showing a hand preference. In fact, once animals had hands, they could reach to the left and right sides with a single limb/hand and so their forelimbs were to some extent freed from lateralization at the level of the brain.

General Pattern of Lateralization in Vertebrates

In a range of vertebrates including humans, the left hemisphere is specialized for processing information in a serial way, categorizing stimuli and controlling responses that are made after the animal has considered some alternatives. The latter is shown in the case using the right hand (and left hemisphere) for manipulating objects and also by the crow using its right eye to guide cutting tools from leaves. Also, if an animal approaches a stimulus that it will manipulate, it will use its right eye and left hemisphere. For example, if a chick approaches a food dish with a lid that it must remove by grasping a small piece of string in its beak, it will approach at an angle that allows it to use its right eye. If it is simply given a food dish with no lid, it approaches from an angle allowing use of its left eye: In this case, no planned response or manipulation is needed.

The right hemisphere is specialized for the expression of intense emotions (including fear and aggression), control of rapid responses, and processing spatial information using maps. The examples mentioned above, of left-eye preference for agonistic or attack responses in the chick, toad, lizard and baboon, illustrate this. Looking at this in another way, humans and monkeys express intense emotions, such as fear, more strongly on the left side of their face (controlled by the right hemisphere).

Evidence of the right hemisphere's specialization for spatial maps is shown by testing rats in a large water-filled arena requiring them to swim until they locate a platform just below the surface of the water. To find the hidden platform they need to use the spatial cues that they can see outside the tank. They can do this if they are tested with a patch placed on their right eye but not if the patch is on their left eye. Similarly, chicks tested with a patch on their right eye can locate hidden food using spatial cues but not when the patch is on their left eye. These experiments show that the right hemisphere is used for spatial tasks which depend on diffuse or global attention to features of the environment. In humans, too, the right hemisphere is specialized for spatial cognition.

In fact, the general pattern of these lateralities is remarkably similar to that of humans. It evolved early in vertebrates and has been retained.

Disadvantages and Advantages of Being Lateralized

Since laterality in the form of side biases is common in vertebrates, we may assume that it confers some kind of advantage to the individual and/or the species. However, it is difficult to see how responding to a predator more readily when it is on the left than when it is on the right could be beneficial to an animal. We can only surmise that such a deficit on one side is counterbalanced by some advantage, but what might this advantage be?

For some time, it has been assumed that the advantage of having a lateralized brain is to increase the brain's capacity. Only recently has some experimental evidence in support of this become available. One example involved giving chicks two tasks to perform at the same time, one task demanding use of the left hemisphere (finding grain) and the other demanding use of the right hemisphere (detecting a predator). Each chick was tested on its own on the pebble–grain task and, while the chick was feeding, a model predator shaped like the silhouette of a raptor was moved over the cage. Strongly-lateralized chicks were compared to weakly-lateralized ones. The strongly-lateralized chicks had superior ability on both tasks, finding the grain and detecting the predator.

An entirely different task has shown that chimpanzees with stronger hand preferences are more efficient in fishing termites from a mound. They "fish" the tasty insects from their nest by inserting a stick into the holes so that the termites grasp it. Then they pull out the stick and wipe it across their other arm while they consume the dislodged termites.

Both of these examples are evidence that having a strongly lateralized brain might be a distinct advantage for survival in the wild. The first is an example of population (or species-typical) laterality and the second an example of individual laterality. These are just the beginnings of research investigating the function of laterality in, or relevant to, the animal in its natural environment.

Inheritance and Development of Laterality

The development of lateralization depends on both genes and experience. Two different studies have shown the importance of experience in early life. The first was a study with rats showing that if the rat pups are handled during the first 3 weeks of life (meaning that they were taken away from their mother for a few minutes each day), their lateralization is stronger in later life. The part of the brain affected by handling was found to be the corpus callosum, a tract of nerve cells connecting the left and right hemispheres. Handling causes the tract to become larger and this, it seems, means that one hemisphere is better at suppressing the activity of the other and, as a consequence, laterality is expressed more strongly. The sex hormone levels circulating in the rat pup also affect the size of the corpus callosum and the strength of lateralization.

The second study showing the influence of experience on lateralization was carried out using chicks. It showed that exposure of the chick embryo to light for as little as 2 hours once during the last 3 days before hatching establishes some forms of lateralized visual behavior. For example, chicks hatched from eggs incubated in the dark are not lateralized for finding food scattered on pebbles or for attack responding. Two hours of light establishes the right eye superiority for the pebble–grain task and the left eye advantage for attack. This effect of light has been traced to asymmetry in the visual pathway from the part of the brain known as the thalamus to the Wulst region of the forebrain: In light-exposed chicks, the left side of the thalamus (which receives input from the right eye) has more projections to the forebrain than does the right side of the thalamus (which receives input from the left eye). The reason for this is that the chick embryo is turned inside the egg so that the chick's left eye is occluded but the right eye is next to the eggshell. Hence the right eye can be stimulated by the light entering through the shell. The same has been shown to be true of the pigeon, although in this species the asymmetrical development of another visual pathway is affected by the light exposure.

Also, certain hormones have a role in the development of this form of visual lateralization. If the levels of the sex hormones in the embryo, estrogen and testosterone, are elevated by injecting them into the egg, no asymmetry develops even though the embryo is exposed to light. The stress hormone, corticosterone, has the same effect. Since hens deposit different amounts of these hormones in their eggs, this result may indicate the importance of hormones, laterality, and behavior in natural conditions. We know that in some species (e.g., canaries) the first egg laid in a clutch has less testosterone than the next and so on for each egg in the clutch. This might mean that the first laid embryo has less lateralization than the next and so on, but this has not yet been confirmed. We do know that the later laid eggs give rise to more aggressive chicks. Since attack responses are lateralized, there may well be a connection here.

Conclusion

There are many examples of laterality in animals, but so far a few species have been studied in much more detail than others. Those model species have shown us that there is a consistent pattern for certain types of lateralization, that such lateralization might well have an evolutionary advantage, and that genes, hormones and experience are all influential in the development of laterality.

See also Laterality—*Laterality in the Lives of Animals*

Further Resources

Bisazza, A., Rogers, L. J. & Vallortigara, G. 1998. *The origins of cerebral asymmetry: A review of evidence of behavioral and brain lateralization in fishes, reptiles and amphibians.* Neuroscience and Biobehavioral Reviews, 22, 411–426.

Denenberg, V. H. 1981. *Hemispheric laterality in animals and the effects of early experience.* Behavioral and Brain Sciences, 4, 1–49.

Hauser, M. D., Agnetta, B. & Perez, C. 1998. *Orienting asymmetries in rhesus monkeys: The effect of time–domain changes on acoustic perception.* Animal Behaviour, 56, 41–47.

Hunt, G. R. 2000. *Human-like, population-level specialization in the manufacture of pandanus leaves by New Caledonian crows* Corvus moneduloides. Proceedings of the Royal Society London B, 267, 403–413.

MacNeilage, P. F., Studdert-Kennedy, M. J. & Lindblom, B. 1987. *Primate handedness reconsidered.* The Behavioral and Brain Sciences, 10, 247–303.

McGrew, W. C. & Marchant, L. F. 1997. *On the other hand: Current issues in and meta-analysis of the behavioral laterality of hand function in nonhuman primates.* Yearbook of Physical Anthropology, 40, 201–232.

Rogers, L. J. 2000. *Evolution of hemispheric specialisation: Advantages and disadvantages.* Brain and Language, 73, 236–253.

Rogers, L. J. 2002. *Lateralization in vertebrates: Its early evolution, general pattern and development.* In: *Advances in the Study of Behavior, Vol. 31* (Ed. by P. J. B. Slater, J. Rosenblatt, C. Snowdon & T. Roper), pp. 107–162. San Diego: Academic Press.

Rogers, L. J. & Andrew, R. J. 2002. *Comparative Vertebrate Lateralization.* N.Y.: Cambridge University Press.

Lesley J. Rogers

■ Laterality
Laterality in the Lives of Animals

Laterality refers to a tendency to use one side of the body preferentially, or differently, from the other. The most obvious example is the strong right- or left-handedness that most people show when writing, throwing a ball, etc. In fact, until recently, most scientists felt that laterality was one of the defining characteristics of human beings and not typical of other species.

The strong handedness that people show in writing appears to be related to the fact that one side of the human brain (the left) is specialized for language comprehension and production. Thus, since language is also a defining characteristic of humans, the outlook in the past provided a neat little package: namely that language and laterality were interconnected traits that humans possessed and that other animals did not.

In recent decades, however, some scientists began to question that assumption and looked for evidence of laterality in other species. Four questions have guided these efforts:

Do individual animals show evidence for consistent laterality comparable to the right- and left-handedness shown by individual humans?

For some species the answer appears to be yes. For example, when mice have been tested in conditions where they need to reach with one paw into a tube to retrieve morsels of food, some mice showed a consistent preference to use their left paw while others consistently preferred to use their right.

However, there are other species in which similar "pawedness" appears not to exist. For example, in dogs most individuals switch back and forth between using their left and right forelimbs showing little consistent preference.

And the answer can be even more complicated. This is because forelimb preference is not the only way to assess laterality in animals. For example, rats (unlike mice) do not usually cooperate by reaching with one paw for food; so they do not usually show pawedness in the classic sense. However, individual rats do show clear preferences to turn left or to turn right when exploring an open field. Thus, even though rats do not show a trait corresponding exactly to human handedness, they show another form of laterality that is consistent in its own way.

Do any animal species have population-wide biases comparable to the predominant right-handedness in humans?

A great example of species with a clear population-wide trend are the fiddler crabs of the west Pacific Ocean. In all fiddler crab species, males characteristically have one very large claw that is used for competition with other males and one small claw that is used for feeding. In most species there are equal number of right-clawed and left-clawed individuals. But in the West Pacific nearly all the males have their large claw on the right side. This is one of the most extreme forms of laterality in the animal kingdom.

In the fiddler crab, the laterality in behavior is matched by a pronounced laterality in anatomy as well. But this is not always so. For example, the bodies of gray whales are largely symmetrical. Yet, they too show a population-wide asymmetry in behavior. These animals make their living by diving to the bottom of shallow seas, rolling on one side and scooping up mouthfuls of copepods from the bottom sediment. Interestingly, individual whales have strong preferences for which side of their mouth they use in feeding in this way, and 90% of them prefer to plow the right side of their mouth into the sediment, not the left side.

On the other hand, other species show no consistent population trend in their laterality. For example, although individual California sea lions have consistent lateral preferences in the use of their flippers when they are running on land (some have a tendency to twist their hindquarters to the left while others typically twist to the right), across the population, there are approximately equal numbers of left twisters and right twisters. Similarly, although individual mice demonstrate a consistent right- or left-pawedness as mentioned above, there are equal numbers of lefties and righties in the population as a whole. Thus, for these species—and it appears at this point that this is characteristic of the majority of animal species—even when there are consistent individual lateral preferences (comparable to individual right- or left-handedness in humans), there is no clear direction in the population of this laterality (comparable to the predominant right-handedness that exists in the human population).

California sea lions consistently twist to one side when running.
Courtesy of Michael Noonan.

Does laterality in the behavior of animals reflect a coherent cerebral lateralization the way it does for humans?

For most species, we simply do not yet have enough information to answer this question. And, the little information we have acquired so far suggests that the answer is not going to be a simple one.

One potential insight comes from an assessment of consistency across different measures of laterality within the same species. In humans, most people who are right-handed also preferentially use their right foot and their right eye, and neuropsychologists interpret this as a sign of an overall left-brain dominance that exists in most people.

However, in species as widely divergent as laboratory rats and California sea lions, there is essentially no cross-behavior consistency in laterality. That is, in these species, if we know that an individual animal is a consistent left turner, we can conclude nothing about his lateral preference shown in other circumstances. Rats, for example, which have a consistent preference to step with their right foot when placed on a balance beam, are just as likely to turn left in a swimming test as they are to turn right. All in all, this picture is somewhat puzzling. There must be asymmetries in the brain mechanisms which underlie each of these individual behaviors. But there appear to be separate and independent asymmetries in each of the different brain subsystems that underlie each of the separate lateralized behaviors. In other words, there appears to be no single lateral brain dominance that ties them all together in the way that handedness, footedness, and eyedness are evidently connected in the human brain.

Why does laterality exist in the first place?

There are at least three advantages that scientists can envision for laterality. One stems from the need for rapid and decisive responses. When a human being is faced with a task that requires one hand to be used, he or she typically does not hesitate. The dominant hand (usually the right) reaches out and the brain–behavior system understands that the nondominant hand will only come into play if additional help or support is needed. In the same way, an animal faced with an opportunity to grab insect prey would be better off if it already had an established decision about which paw would be used to grab and which paw would be used to hold on for support. There would presumably be a disadvantageous delay if there was an internal debate between the two sides of the brain each time one paw were chosen.

A second advantage may pertain to the problem of left–right confusion. The predominant symmetry of most animals and the interconnectedness of the two sides of the brain presents a unique problem—distinguishing left from right. Conversely, it can be argued that the greater the lateral asymmetry in an animal, the less that problem will exist. Presumably this is the reason that adult humans can tell their left from right more easily than children (there is separate evidence that the brains of adult humans are more lateralized than those of children), and presumably this is why humans are better at distinguishing left from right than animals of most other species. The two sides of the human brain are so distinctly different that they confer a distinctly different feeling on the two sides of the body. Laterality presumably conveys that advantage to any species that evolves it.

The third advantage is suggested by the reasoning offered for the benefit of specialization of function generally. Just as it is valuable to have one portion of the brain specialized to regulate breathing, another specialized to regulate hormones, and so on, it is presumably also advantageous to have different sides of the brain specialized for different cognitive processes. In the case of the human, having the left hemisphere alone specialized for linguistic processing allows the circuitry of the right hemisphere to be programmed for another specialty—spatial perception.

In recent years, some evidence for lateral specialization of function has also been uncovered in certain animal species. To give just one example, there is evidence that one hemisphere in each rat's brain is specialized for perseveration (the tendency to hold onto learned behaviors when circumstances change), while the other hemisphere is more labile and ready to abandon old stimulus–response associations in favor of new ones. It is presumably advantageous for the rat to have both styles of learning well represented in its brain, and it appears to do so by perfecting each learning strategy in a different hemisphere.

Whether these or other possible advantages of laterality can ever be confirmed experimentally is a question that deserves considerable attention in the future. And that suggests the best way to think of laterality in animals at this point—a fascinating topic that still is only beginning to be explored, and one that can lead to new insights into the nature of animals at many levels.

See also Laterality

Further Resources

Bock, G. R. & Marsh, J. (Eds.). 1991. *Biological Asymmetry and Handedness*. New York: Wiley & Sons.

Palmer, A. R. 1996. *From symmetry to asymmetry: Phylogenetic patterns of asymmetry variation in animals and their evolutionary significance*. Proceedings of the National Academy of Sciences, 93, 14279–14286.

Michael Noonan

Learning
Evolution of Learning Mechanisms

Defining Learning

Learning involves the acquisition, storage, and retrieval of information that can potentially affect behavior. Learning is not the same as behavioral change. Behavior can change for reasons other than learning (maturation of the nervous system), and learning can occur in the absence of immediate behavioral change (early learning may affect adult behavior). Acquisition, storage, and retrieval are psychological concepts that refer to specifiable neural operations (circuitry, neurotransmitters, and cell–molecular processes) that may vary across species. To understand the evolution of these operations, comparative psychologists study the learning skills of different species and make inferences about the way in which such skills are implemented in each species.

Understanding exactly how experience can leave a trace in the brain turned out to be a rather complex task. Researchers opted for the intense study of a few species, rather than a superficial description of learning in a wide variety of animals. Accordingly, what is known today about learning comes predominantly from such animals as rats and pigeons. More recently, other species have joined this list, including monkeys, rabbits, goldfish, honeybees, marine slugs, and soil worms. These animals belong to four different phyla (chordates, arthropods, mollusks, and nematodes) that were already differentiated in the Cambrian period, about 500 million years ago. Thus, their study can yield insights into deep homologies and divergence in learning mechanisms.

Comparative Methodology

Learning is assessed by repeated exposure of an organism to a particular situation. A behavior is measured in each of the trials and any change in this measure—beyond changes observed in some control condition—is considered to reflect learning. Learning is inferred from behavior, much like Gregor Mendel (the father of modern genetics) inferred the genotype of garden peas from phenotypic characters.

In comparative studies of learning, the goal is to make analogous measurements in different species and to evaluate their similarities and differences. Despite its apparent simplicity, the interpretation of results is complicated because behavior can be influenced by factors other than learning, and because species may differ in these nonlearning factors. Species differences in sensory capacities, motor skills, response strategies, motivation, and so on, may yield differences in behavior even when the underlying learning mechanisms are the same. Consider a hypothetical example.

A comparative psychologist wants to study learning in rats and turtles. The experiment involves a straight alley, one trial per day, and food reinforcement for walking from one end of the alley to the other. Walking speed provides a behavioral measure. Species of similar body size are chosen. All animals are equally deprived of food, reinforced with five food pellets, and given equal exposure to the alley for familiarity. After 20 trials, walking speed is faster in rats than in turtles.

Are rats more efficient learners than turtles? Despite the careful matching of variables, it is impossible to answer this question. First, the difference may reflect the fact that rats generally walk faster than turtles. Second, perhaps rats walk faster because they are more motivated by five pellets than are turtles. Third, the foraging styles of these species may make the task easier for the rats. Rats actively look for distributed food, whereas the turtle's food tends to be concentrated in an area. Turtles may perform better than rats if the task were to involve a less active strategy to obtain food. Whereas it is possible to equate many variables across species, it is impossible to ascertain whether each species responds in the same manner to each variable. Instead, researchers assess the extent to which a particular learning phenomenon is affected by the same variables in different species. If such variables as the number of pellets, number of trials, length of the runway, and so on, have similar effects in both species, then one may conclude that the species share the same learning mechanisms. If, however, these variables have different effects across species, then learning mechanisms may have diverged as a result of evolution. The following sections describe some examples of divergence in learning mechanisms in vertebrates.

Learning and Attention

Learning requires that we focus on some events at the expense of others. This is called attention. Situations that involve novelty or excitement, such as a challenging teacher or an exciting topic, attract our attention. When a situation becomes boring we learn to turn down our attention. A phenomenon called *latent inhibition* (LI) appears to involve attentional decrement. In a LI experiment, a stimulus is repeatedly presented during many trials and, later, that stimulus is used as a signal for an important event. For example, a hungry rat is exposed to many presentations of a light and, subsequently, the same light is used as a signal for food. Normally, rats learn rapidly to respond to the light when it signals food, but the LI procedure retards such responding because the organism has initially learned to be inattentive to the light.

Latent inhibition has been conclusively demonstrated only in mammals. Experiments with goldfish and Atlantic salmon demonstrate that learning occurs equally fast in animals that were preexposed and those that were not preexposed to the stimulus. This was observed with visual, auditory, and chemical stimuli; with electric shock or food as reinforcers; with variation in the number of preexposure trials; and so on. In fact, sometimes preexposure actually improves subsequent learning in fish.

Do fish demonstrate other attentional effects? This answer also appears to be negative. In mammals, for example, it is possible to direct attention at a particular dimension of a complex stimulus (such as its color), and turn down attention to another dimension of the same stimulus (such as its shape). If a subsequent problem is presented to the same animals with new colors and new shapes, rats and primates learn faster when the same dimension continues to be relevant (color to color), than when the other dimension is now the relevant one (color to shape). Thus, intradimensional transfer is easier than extradimensional transfer. Interestingly, goldfish learn these complex transfer problems at the same speed, whether the transfer is intra- or extradimensional.

All together, these results suggest that the ability to control attention involves mechanisms that are not present in the fish brain, but that evolved in the mammals.

Learning and Frustration

When somebody mentions a familiar restaurant, we immediately get a mental image of the place, the food they serve, and whether or not we like it. These internal responses are called *expectancies*. Suppose we decide to go to this restaurant because we love their food, but, when we arrive, we find that it has gone out of business. Such unexpected losses induce frustration, and these conditions can be reproduced under controlled conditions in the laboratory. Suppose some animals receive a large amount of food in each of several trials and then, unexpectedly, they receive only a fraction of the original amount. Mammals exposed to such surprising reward reductions reject the small amount of food more than control animals exposed only and always to the small reward. This effect, called *successive negative contrast* (SNC), has been observed in humans, monkeys, rats, mice, and opossums in many different situations. SNC is associated with the release of stress hormones, a fact suggesting that frustration is stressful.

When similar experiments are carried out with nonmammalian vertebrates (fish, amphibians, reptiles, and birds), no evidence of SNC is observed. In these species, behavior is sensitive to the difference in reward size, but a downshift in reward magnitude is not followed by a rejection of the small reward. For example, a group of goldfish reinforced with 40 worms swam faster in a straight alley than a group reinforced with 4 worms, but exhibited no change in behavior when shifted from 40 to 4 worms.

Nonmammalian vertebrates also fail to exhibit other related effects. Interestingly, these effects are also absent in infant rats, a fact suggesting a parallel between comparative and developmental studies of learning. Adult rats can be induced to exhibit a performance analogous to that of nonmammalian vertebrates and infant rats by several treatments, including the administration of anxiolytics (drugs that relieve anxiety). These results suggest that the mechanisms responsible for the acquisition of information about the organism's own emotional reaction to reward loss are unique to mammals.

Learning and Problem Solving

Early experiments suggested that new skills are acquired gradually—a process called *trial-and-error learning*. Cats placed in a box gradually learn, across many trials, to operate a latch that opens the exit door to reach for food. In contrast, other experiments showed that new skills are sometimes acquired abruptly—a process called *insight*. In a famous experiment,

Insight

Colleen Reichmuth Kastak

People are sometimes surprised by an instant of clarity amidst confusion, a moment where they're tempted to call out "Ah-Ha!" as they solve a challenging new problem or suddenly figure out the solution to a puzzle. This phenomenon is called *insight*, a term that is often associated with concepts such as innovation, creativity, and intelligence. While most examples of problem solving reveal a slow, gradual learning process based on trial-and-error experiences, insightful behavior involves an immediate change in performance as a problem is suddenly and appropriately solved.

There are surprisingly few documented examples of insightful performance in animals. The first was a report by the German psychologist Wolfgang Köhler, entitled *The Mentality of Apes* (1925), which described the responses of a captive group of chimpanzees to a variety of problems created by Köhler, all of which involved obtaining a difficult to reach food reward. For example, after initial futile attempts to leap to fruit that was placed out of reach, the chimps sometimes paused and then used assorted objects found in their enclosure (boxes, poles, sticks, and strings) in new ways and combinations to order to acquir the desired reward.

The performance of the animals on Köhler's puzzles led him to believe that the problems were solved by the chimp's insightful perception of the relevant relationships between the food, the various objects, and the motor behaviors used to connect the two. If this was the case, the notion of insight might explain the sudden and appropriate changes in behavior. If not, how were the solutions to the problems formulated? Later investigators discovered that solutions to similar problems were not readily demonstrated in the absence of prior experience. Chimpanzees often need practice in pushing and climbing boxes, waving sticks and linking them together, jumping with the leverage of poles, and pulling strings, and they get can get this practice through spontaneous exploration and play with the objects found in their environments.

Chimps with a history of such object-oriented experiences are more likely to succeed in overcoming obstacles than those who lack these prior experiences. Direct training on elements of a problem can also facilitate performance. For example, pigeons can be taught to climb on a box to peck at an arbitrary target, such as a plastic banana, and independently taught to push a box from one spot to another. Following training of these behaviors, pigeons will spontaneously push the box to a position under the target and then hop on the box to in order to peck it. Although the elements of the problem-solving behavior are taught to the subject, it still must combine what it has learned into a new behavioral sequence when presented with an unfamiliar situation.

Currently, the study of insight is still a highly controversial topic in the field of animal behavior. Modern studies attempt to improve our understanding of insight by combining analyses of experiential learning with tests that require innovative solutions; for example, *stimulus equivalence* predicts how animals may mentally structure their environmental experiences so that novel behaviors can emerge.

See also Cognition—*Imitation*
Learning—*Social Learning*

Further Resources

Birch, H. G. 1945. *The relation of previous experience to insightful problem-solving.* Journal of Comparative Psychology, 38, 367–383.

Epstein, R., Kirshnit, C. E., Lanza, R. P., & Rubin, L. C. 1984. *'Insight' in the pigeon: Antecedents and determinants of an intelligent performance.* Nature, 308, 61–62.

Köhler, W. 1925. *The Mentality of Apes.* New York: Harcourt, Brace, and Company.

a chimpanzee named Koko was shown an inaccessible piece of fruit hanging from the cage ceiling and a box placed on the floor. Initially, Koko attempted (unsuccessfully) to reach the fruit, but paid no attention to the box; then he directed his attention to the box, throwing it, sitting on it, and directing violent attacks to it; finally, he suddenly turned toward the box, seized it, dragged it beneath the fruit, stepped on it, and took the fruit.

Problem-solving abilities similar to Koko's are observed in animals extensively exposed to various objects and subsequently tested for insight. Suppose a pigeon is separately trained to peck at a stimulus hanging from the ceiling and to push a box placed on the floor by pairing each behavior with food. Evidence shows that, like Koko, this pigeon would first try to unsuccessfully reach for the hanging stimulus and push the box, until it would suddenly exhibit the correct sequence: push the box, step on it, and reach for the stimulus. Insightful problem solving is facilitated by extensive familiarization with a particular set of objects.

Is there any evidence of species differences in the ability to transfer acquired information to new situations? In a series of experiments, several species of primates were trained to discriminate between two stimuli until a criterion of correct responses was reached and, then, additional trials were administered to overtrain the discrimination. How would these animals adjust to a reversal of the original discrimination in which the previously correct and incorrect stimuli are now incorrect and correct, respectively? Overtraining interfered with reversal learning in primitive primates (lemurs, New World monkeys), but it facilitated reversal learning in advanced primates (Old World monkeys, apes). Although all species were able to learn the reversal problem eventually, primate species differ in the way in which they learn. Primitive species learn discriminations by acquiring response tendencies (always reach for a specific stimulus). Response tendencies are difficult to reverse, thus causing negative transfer. In contrast, advanced species acquire rules (choose one stimulus and stay with it if reinforced, but choose the other if nonreinforced). Positive transfer follows from the fact that both the original discrimination and its reversal follow the same rule.

Learning and the Brain

The tendency from negative to positive transfer across primates reviewed previously correlates positively with the species' brain size. Results such as these suggest that having a relatively larger brain may allow an animal to perform more complex forms of information processing and to exhibit greater behavioral flexibility. This also applies to restricted brain areas. For example, if a species is biased toward the use of visual information (as in humans), then the visual system will be relatively larger and more complex than the visual system of a species biased toward a different sensory system.

This so-called *principle of proper mass* also applies to learning. A well-studied example involves species of birds that store food for periods varying from weeks to months. These birds eventually recover the cache, usually during the winter when food is scarce, by memorizing the location where they have deposited food. Interestingly, birds that exhibit food-storing behavior have a larger hippocampal size than birds that do not store food. The hippocampus is a part of the brain involved in spatial learning and memory in all vertebrates.

Mammals also have a brain size that is approximately 10 times larger than that of a reptile, even in species of similar body size. Since mammals evolved from reptiles, it seems possible that the learning skills described in previous sections (attentional modulation, frustration, and transfer effects), are based on additional neural tissue and brain modules that are simply unavailable to vertebrates with smaller relative brain size. Exactly what these neural processes may be remains to be determined.

See also Behavioral Plasticity
Learning—*Insight*
Learning—*Social Learning and Intelligence in Primates*
Learning—*Social Learning in Animals*

Further Resources

Bitterman, M. E. 1975. *Comparative analysis of learning.* Science, 188, 699–709.
Domjan, M. 1998. *The Principles of Learning and Behavior.* 4th edn. Pacific Grove, CA: Brooks/Cole.
Heyes, C., & Huber, L. (Eds.) 2000. *The Evolution of Cognition.* Cambridge, MA: MIT Press.
Macphail, E. M. 1982. *Brain and Intelligence in Vertebrates.* Oxford, UK: Oxford University Press.
Papini, M. R. 2002. *Comparative Psychology. Evolution and Development of Behavior.* Upper Saddle River, NJ: Prentice Hall.
Papini, M. R. 2002. *Pattern and process in the evolution of learning.* Psychological Review, 109, 186–201.
Shettleworth, S. J. 1998. *Cognition, Evolution, and Behavior.* Oxford, UK: Oxford University Press.

Mauricio R. Papini

■ Learning
Instrumental Learning

Instrumental learning (or instrumental conditioning) helps organisms adapt to their environment, shaping behavioral responses to optimize their outcome (seeking to maximize reward and minimize pain). Anyone who has used a piece of food (the outcome) to train a dog to sit (the response) has performed instrumental training. Food represents a desired outcome, and animals will often learn a new response to gain a reward. Events that the animal does not like can also be used to alter behavior. For example, you may have decreased the frequency of an unwanted behavior in your pet by applying a mild punishment each time the behavior occurred.

Instrumental learning represents a behavioral principle that applies across a wide range of species and situations. Complex social behavior in humans can be shaped by reinforcement. We reward a child for behaving appropriately and administer punishment when their behavior is disruptive. At the other extreme are simple inborn reflexes designed to promote approach to food and withdrawal from a harmful event. Neurons within your spinal cord can, in the absence of any input from the brain, produce a withdrawal response (a *spinal reflex*). The vigor and duration of a spinal reflex can be modified by altering the response–outcome relationship, and this instrumental learning can occur without any direction from the brain. Similarly, simple invertebrates are sensitive to instrumental reinforcement.

To demonstrate instrumental learning, we must show that the change in behavior is due to a particular response–outcome relationship. While this may seem simple, there are a variety of factors that can complicate our analysis. For example, the occurrence of just the response or outcome *alone* might bring about a behavioral change. Repeatedly performing a response can cause fatigue which may decrease response vigor. Likewise, continued exposure to an outcome could alter its capacity to reinforce behavior. A cookie would not be a very effective reinforcer for a child that has just finished eating a gallon of ice cream. Because these changes do not depend on the *relationship* between a response and an environmental event (the outcome), they do not represent examples of instrumental learning. Other factors can also mislead us into believing that instrumental learning has occurred when it has not. For these reasons, researchers have derived formal criteria that can be used to help judge whether a behavioral modification is due to instrumental conditioning. These criteria are listed in the following figure. The first two criteria require that we show that imposing a particular response–outcome relationship brings about a behavioral change, and that this effect is neurally mediated. The latter is needed to rule out effects that might be ascribed to factors such as muscle fatigue.

Learning implies a form of memory—that is, the consequence of the experience is retained over time. If the behavioral change produced by our response–outcome relation disappears as soon as the relationship is removed, no learning has occurred. Many of us have repeatedly experienced a headache (the outcome) while eating ice cream (the response in this example). Though the outcome brings about a temporary lull in ice cream eating, the relationship between these two events generally has no long-term effect on our behavior—you are not any less likely to eat ice cream in the future because it sometimes causes a painful outcome. In this case, it appears that the key relationship is not encoded over time

Criteria for Instrumental and Operant Learning

Minimum Criteria

1. Instituting a relationship between the response and an outcome produces a change in behavior (*performance*).

2. The effect is neurally mediated.

3. The modification outlasts (extends beyond) the environmental contingencies used to induce it.

4. The behavioral modification depends on the temporal relationship between the response and the outcome.

Advanced Criteria

5. The nature of the behavioral change is not constrained (e.g., either an increase or decrease in the response can be established).

6. The nature of the reinforcer is not constrained (a variety of outcomes can be used to produce the behavioral effect).

Adapted from Grau, Barstow & Joynes, 1998.

in a way that brings about a lasting change in behavior, therefore no instrumental learning has occurred. To demonstrate instrumental learning, we must show that the consequences of training are retained over time (criterion 3) and that the response–outcome relationship was encoded (criterion 4).

Any example of learning that meets criteria 1–4 can be considered an example of instrumental learning. In this sense, instrumental conditioning represents a very broad term that can be applied across a wide range of training situations, from the modification of reflexive behavior to complex social interactions. Though the key relation remains the same, there are factors that separate the extremes. Simple reflexes are biologically constrained and operate under a narrow range of parameters. We can show that neurons within the spinal cord can learn to exhibit a stronger leg lift response to reduce exposure to a painful shock, providing an example of instrumental learning that meets criteria 1–4. However, the behavior observed is very constrained—there is no evidence that we can arbitrarily train the spinal cord to exhibit either a leg lift or a leg extension using the same reinforcer. Likewise, the behavioral change can only be induced using a limited range of outcomes, all of which activate pain fibers. At the other extreme, a rat can learn to run through a complex maze for a variety of reinforcers. It will run for many different types of food, and it will run to avoid a painful event. Further, if one path is blocked, the rat can flexibly switch to an alternative path. The response observed is not constrained to a single path. Similarly, we could train you to go from one house to another using a variety of reinforcers (money, food, entertainment, escape from a painful situation) and you could reach your goal in a variety of ways. If one entrance was blocked, you could flexibly discover an alternative path. At least two factors appear to distinguish examples of complex instrumental behavior from the modification of a reflex arc. The first involves the range of solutions that can be flexibly brought to bear. The second concerns the range of effective reinforcers. A simple reflex may be modified by a response–outcome relation, but both the nature of the change produced and the range of effective outcomes are limited. More sophisticated examples of instrumental learning allow the organism to operate on its environment in a variety of ways (criteria 5) and can be trained using many different types of reinforcement (criteria 6). Skinner referred to this type of learning as *operant* conditioning. Within the current framework, operant learning represents a sophisticated form of instrumental conditioning.

See also Learning—*Evolution of Learning Mechanisms*

Further Resources

Grau, J. W., Barstow, D. G., & Joynes, R. L. 1998. *Instrumental learning within the spinal cord: I. Behavioral properties*. Behavioral Neuroscience, 112, 1366–1386.

Hearst, E. 1975. *The classical–instrumental distinction: Reflexes, voluntary behavior, and categories of associative learning*. In: *Handbook of Learning and Cognitive Processes: Conditioning and Behavior Theory* (Ed. by W. K. Estes), pp. 181–223. Hillsdale, NJ: Erlbaum.

Hilgard, E. R., & Marquis, D. G. 1940. *Conditioning and Learning*. New York: Appleton-Century-Crofts.

Konorski, J. A., & Miller, S. M. 1937. *On two types of conditioned reflex*. Journal of General Psychology, 16, 264–273.

Skinner, B. F. 1938. *The Behavior of Organisms*. Englewood Cliffs, N. J.: Prentice-Hall.

Thorndike, E. L. 1898. *Animal intelligence: An experimental study of associative processes in animals*. Psychological Review, Monograph, 2, 8.

James W. Grau

Learning
Social Learning

Imitative and Nonimitative Social Learning

Behavioral scientists with an interest in animal social learning have been concerned with one of two quite different issues. Many psychologists and primatologists who study social learning want to know whether animals, other than humans, can imitate, can learn to do an act simply by watching another perform that act.

On the other hand, most biologists who study social learning are more interested in discovering how interaction with others contributes to development of adaptive patterns of behavior in animals living in natural circumstances. Such researchers are usually not too concerned with whether the social learning that facilitates development of adaptive behavior is truly imitative or results from a nonimitative form of social learning.

An animal watching another behave can learn several quite different things. The observer can learn about the behavior of its model, about aspects of the environment that would otherwise be hidden from it, or that the environment can be changed in some way. For example, one chimpanzee might learn from watching another use a stick to pry open a termite mound and eat termites that there is food inside termite mounds. Alternatively, the observing chimpanzee might learn that sticks can be used to break into termite mounds. Or, the observer might learn to insert a stick into a termite mound and apply a prying motion. Only, as in the last case, when the observer learns directly about the behavior of its model, is the learning referred to as *imitative*. Learning about the environment or about possible effects of manipulating the environment is conventionally described as *nonimitative*.

Why is imitation considered special? Because often an observer cannot see its own movements when it imitates the movements of another. For example, if I see you bow and then bow myself, there is no way I can imitate your bow by directly matching what I see when I bow with what I saw when you bowed. Consequently, imitation seems often to require that an observer match signals produced by movement of its own body (*proprioceptive signals*) with representations in memory of visual images of movements made by another. Such *cross-modality matching*, that is, matching proprioceptive signals to visual signals, would seem to require considerable cognitive sophistication. It is this potential sophistication in manipulation of representations that has captured the interest of scientists for more than 100 years. There has been considerable progress. For example, structures in the brains of both humans and monkeys have been discovered recently that respond similarly both to seeing an act performed and to performing the act oneself. These *mirror-image neurons*, as they are called, may provide an important first clue as to how imitation is possible.

Nonimitative Social Learning

Nonimitative social learning, which is the focus of this article, is so common in animals and involves so many different types of interaction between a model and a social learner that an entire book devoted to the topic would be needed for a comprehensive review. We have space here to mention only a few examples out of many hundreds of instances of non-imitative social learning now known from the study of animals from insects to primates.

Nonimitative Social Learning in Rats

Following is a detailed discussion of a single example: nonimitative social learning of food preferences in rats (animals often used as subjects in laboratory studies of behavioral development) as a typical case of social nonimitative social learning.

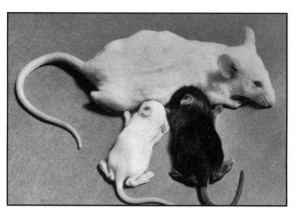

Baby mice suckling off their mother.
© Al Fenn / Time Life Pictures / Getty Images.

Food choices of rats. It is well established that young rats tend to eat the same foods that adult members of their colony have learned to eat, and considerable effort has gone into describing the ways in which interactions between adult rats and their maturing young result in their having similar food preferences. Results of these efforts have revealed a range of social interactions occurring at various stages in development that are important in transmission of food preferences from one generation to the next (*vertical transmission*).

The milk of a nursing mother rat (or human) contains flavors reflecting the taste of the diet she has been eating. These flavor cues allow her suckling young to learn what foods she has been eating. Young rats choosing their first meals of solid food prefer foods with flavors they have experienced previously in their mother's milk.

When young rats leave the safety of their burrow to seek food in the outside world for the first time, they look for adults, approach them and, if the adults are feeding, eat where the adults are eating in preference to other potential feeding sites. The young often crawl under the body of a feeding adult and emerge right under the adult's chin to eat. So the feeding behavior of young rats can be directed toward specific foods by a feeding adult. Wild rats are hesitant to eat any foods they haven't previously eaten (that is why rats are so difficult to control with poison baits), and once a young rat has been introduced to a food by interaction with adults, many days or weeks may pass before it samples other foods.

As adults eat a food, they mark both the food itself and the area around it with residual chemical cues that are attractive to their young, and young rats prefer marked foods and feeding sites to unmarked alternatives. Further, adults returning from a feeding site to their burrow deposit scent trails leading to the place where they have eaten, and their young will follow such trails from the burrow to food.

Adult rats are also amazingly tolerant of attempts by their young to steal food while the adults are eating it, and young rats subsequently prefer a food they have taken from an adult's mouth to other foods they have eaten.

Last, but not least, after a rat (an observer) interacts with another rat that has recently eaten some food (a demonstrator), the observer has a greatly enhanced preference for whatever food its demonstrator ate. Observer rats smell the food that a demonstrator has eaten on the demonstrator's breath, and experience of the scent of a food together with a chemical produced by the demonstrator causes the observer to have an increased preference for the food that its demonstrator ate. Such socially induced food preferences are very powerful. They can last for weeks, and even cause observer rats that have interacted with a demonstrator rat fed foods laced with pepper (that rats normally avoid eating) to prefer pepper-flavored food.

Social influences of a sort can also determine how rats eat. In much of the world, pine forests are inhabited by squirrels that live on pine seeds that they remove from the pinecones that grow everywhere in the forest. However, there are no squirrels in Israel, and rats living there have occupied the ecological niche that squirrels fill elsewhere.

Extracting pine seeds from pinecones is not easy for rats. To recover more energy from the pine seeds than is used to extract them from under the tough, non-nutritious scales that protect them, rats must use a rather special technique that takes advantage of the structure of pinecones. The scales surrounding the base of a cone must first be removed. Then, the spiral of scales that circle the pinecone's shaft from its base to its tip must be removed one after another in sequence.

Studies in captivity of several hundred rats taken from areas in Israel other than pine forests revealed that only a handful ever learned for themselves to use the spiral pattern of scale removal that permits a rat to maintain itself on a diet of pine seeds and water. Most rats taken from outside of pine forests either ignored pinecones or gnawed on them in ways that produced few seeds in return for much work.

On the other hand, all rats captured in pine forests knew how to extract seeds from pinecones efficiently, and so did rats that were taken as infants from mothers living outside pine forests and given to mothers who knew how to open pine cones and were living on a diet of pine cones and water. Further studies revealed that young rats allowed to finish stripping scales from pinecones started appropriately by an adult rat became efficient exploiters of pinecones, and so did young rats given pinecones a human had stripped of scales, mimicking the early stages of scale removal used by successful adult rats.

Social learning by rats of the efficient method of opening pinecones is particularly interesting because here social learning opened a new ecological niche to a species, allowing rats to thrive in a portion of the environment that was otherwise closed to them.

More Examples of Nonimitative Social Learning

As we have seen, rats can learn socially by nonimitative means what to eat, where to eat, and how to eat difficult foods. Of course, animals other than rats also use nonimitative social learning to increase the efficiency with which they learn to exploit resources. Perhaps surprisingly, honeybees are among the most sophisticated social learners on our planet. Like the rats discussed above, honeybees have several ways to communicate about foods. A successful forager that has returned to its hive with a load of nectar performs a dance on the honeycomb within the hive. The dance provides fellow foragers with information as to the direction and distance to the place where the returning forager has gotten the nectar it is carrying as well as its sugar content. Bees also mark rich food sources with a scent that is attractive to their fellows and carry food odors back to the hive on their bodies that other foragers can use to identify the place where the returning forager has been feeding.

A bee with a number attached to it "dances" to communicate a feeder location to other members of the swarm during an experiment at Michigan State University.
© James L. Amos / Corbis.

Birds can use the behavior of others not only to locate potentially profitable places to feed and to select appropriate items to ingest, but also to decide when food in an area has been exhausted, and it is time to move on to forage. Fish also often feed socially, as do insect-eating bats, though relatively little is known of the details of their social feeding.

Animals learn about many things other than food by interacting with their fellows. Naive young monkeys, blackbirds and kangaroos have been shown to learn to identify potential predators by watching the responses of knowledgeable individuals to potentially dangerous objects. Both female birds and female fish learn about the desirability of potential mates by watching other females of their species choose a partner, and male birds and fish can appraise the fighting ability of potential opponents by "eavesdropping" on them while they engage in aggressive interactions with others.

The list of behaviors that have been demonstrated to result from nonimitative social learning is long and growing. And, as anyone who regularly watches nature shows on television surely knows, the list of behaviors of animals—from chimpanzees, using twigs to fish for termites, to dolphins, holding sponges in their mouths while feeding—that might be acquired by interaction with others is even longer.

Why Learn Socially?

There are three ways that animals can develop adaptive behavioral repertoires. First, their behavioral development may be highly constrained so that behaviors that are typical of a species (i.e., instincts) develop in essentially any environment. Second, an individual may learn to behave appropriately from trial-and-error interactions with its physical environment, and third, an animal may learn socially, through interaction with others of its species.

Within an individual's life span, instincts cannot change in response to changing conditions. Trial-and-error learning can track environmental changes, but is relatively time-consuming, and as its name implies, involves making mistakes that may be life threatening. An animal whose individual learning about the environment is biased by interaction with others of its species enjoys the best of two worlds. Such an animal can respond adaptively to the environment in which it matures without incurring all the costs associated with learning independently about what works and what doesn't work. Learning by interacting with more knowledgeable others should be advantageous in many circumstances.

Learning by imitation appears to require neuronal systems that are probably expensive both to construct and to maintain. Consequently, animals that can learn socially by nonimitative means may be able to realize the benefits of social learning without incurring costs associated with building and maintaining a nervous system able to learning by imitation.

See also Behavioral Plasticity
          Communication—*Honeybee Dance Language*
          Learning—*Insight*
          Learning—*Social Learning and Intelligence*
            in Primates
          Social Organization—*Social Knowledge in Wild*
            Bonnet Macaques

Further Resources

Galef, B. G., Jr. 1996. *Social influences on food preferences and feeding behaviors of vertebrates.* In: *Why We Eat What We Eat* (Ed. by E. D. Capaldi), pp. 207–231. Washington, D.C.: American Psychological Association.

Heyes, C. M. & Galef, B. G., Jr. 1996. *Social Learning in Animals: The Roots of Culture*. San Diego, CA: Academic Press.

Seeley, T. D. 1995. *The Wisdom of the Hive*. Cambridge, MA: Harvard University Press.

Zentall, T. R. & Galef, B. G., Jr. 1988. *Social Learning: Biological and Psychological Perspectives*. Hillsdale, NJ: Lawrence Erlbaum Associates.

Bennett G. Galef, Jr.

■ Learning
Social Learning and Intelligence in Primates

Our quest to understand human origins encompasses a long-standing interest in the roots of human intelligence. Humans are set apart from other animals for their high level of cognitive functioning, particularly evident in their dealings with their social world. Notably, humans are specialized in forging and handling complex social relationships, predicting and manipulating the behavior of others, and maintaining social balance and group cohesion. In addition, humans have perfected means of communicating about themselves and others (i.e., by using language). Lastly, humans form mental representations of themselves and others, and are able to reflect on their own mental states (e.g., desires, fears) as well as those of others (i.e., through mind reading and empathy). To better understand the evolution of human cognition, anthropologists have adopted the classic comparative approach, turning their attention to the cognitive skills of other anthropoid primates, such as monkeys and apes. By relating the variation in mental abilities present across species to generalized patterns of morphology, ecology, or behavior, primatologists have formulated several hypotheses to account for the seemingly exceptional cognitive abilities of primates. This discussion presents the major candidates, focusing on the *Social Intelligence Hypothesis*, which postulates that cognition evolved in response to the demands imposed by the animal's social environment.

Despite surprisingly little evidence to support the assumption that cognitive capacities and brain size are linked, primate intelligence traditionally has been attributed to a relatively large brain. The degree of "encephalization" (the proportional size of the brain relative to some other body measure) provides a gauge of the expected brain size for any given body size, and primate quotients exceed those predicted for their body size. More specifically, primate cognitive capacities have been ascribed to structural reorganization of the brain, for instance, through greater investment by anthropoid primates in the visual system (as opposed to the olfactory system). Some researchers also have proposed that human cognition and language result from delayed maturation of the brain; however, the comparably slow developmental trajectory of newborn great apes suggests that the extent of human brain immaturity at birth and subsequent developmental delay has been exaggerated. Nevertheless, if factors such as brain size or organization account for human intelligence, it still leaves open the question of why these changes in the brain occurred.

Because increasing brain size also correlates positively with certain life history variables, such as an extended period of infant dependency, delayed sexual maturity (or increased juvenility), and longevity, other theorists have proposed that primates can achieve greater cognitive feats because of their slower progression through longer life stages. In other words, young animals have more supervised time at the bosom of their social group and under the tutelage of older family members, which provides the necessary circumstances for them to learn difficult skills.

Alternately, because primates face various ecological demands and adopt diverse feeding strategies, some authors have credited primate cognitive development to various aspects of their foraging behavior, in which difficulties associated with obtaining an adequate diet sometimes involve the specialized use of hands. For instance, to "know" when and where trees will be in fruit, a *fruigivore* (fruit-eater) presumably has an increased need, compared to a *folivore* (leaf-eater), to "map" its environment in time and space. Similarly, an *extractive forager* (one that must obtain hidden food items from crevices or tough encasings) must perfect certain complex skills in manual food processing or tool use, which require some degree of dexterity or hand coordination. Familiar examples of such behavior include excavating a grub from a tree hole, husking a leafy shoot, cracking open a nut with a stone hammer and anvil, or even forging a tool to extract termites from their mound. These hypotheses (i.e., temporal–spatial mapping, extractive foraging, and food processing) follow a similar line of argument: The need to exploit and manipulate inconvenient food resources requires greater brainpower and is therefore the main selective force driving intelligence.

Although additional explanations for primate cognitive evolution have been proffered, none have been as widely championed as the Social Intelligence Hypothesis, which suggests that cognitive benefits arose from social living: The social environment of primates supported the evolution of large brains and complex forms of social learning as an adaptation to predict, interpret, and exploit the behavior of group members. The rationale involves a feedback loop, whereby sociality led to an increase in brain size, which in turn allowed for further development and sophistication of social organization and control, until survival in such a complex society demanded greater cognitive skills and further brain enlargement.

The Social Intelligence Hypothesis rests on the hallmark complexity of primate societies, derived in part from the repetitive nature of social interactions with known individuals. Biologists describe two distinct types of societies, which they refer to as *anonymous* and *individualized*. Members of anonymous societies (illustrated, for example, by the social insects) do not recognize each other as individuals, per se, but rather as group (or colony) members. Although members may perform different roles (e.g., drone or queen), they are not characterized as having individual personalities. By contrast, members of individualized societies (illustrated, for example, by primates) recognize each other and have their own personalities. Moreover, these individuals form their own social bonds or attachments, attend to the interactions and social relationships between other group members, and develop complex and enduring social networks. These networks are usually highly structured—organized along family lines, alliances, or formal dominance hierarchies—and the members possess several efficient means of communication (e.g., vocal, visual, olfactory, and gestural communication).

Another core aspect of social intelligence involves social learning. In its simplest form, social learning involves the modification of individual behavior or performance through the mere

A mother aye-aye eating a piece of fruit as her baby reaches for the food.
Courtesy of David Haring.

presence of others (e.g., the audience effect). Further attending to and possibly copying the behavior of others (e.g., through opportunism, observational learning, or imitation) can influence the acquisition, expression, and transmission of knowledge. When animals use each other as "social tools," information can spread between members of the same age class or across generations (e.g., from mother to offspring), potentially leading to the formation of behavioral traditions. One of the best known examples of primate "protoculture" involves the sweet potato washing and wheat washing behavior of Japanese macaques (*Macaca fuscata*), whereby, after a young female monkey first displayed the innovative washing behavior, similar behavior gradually emerged in members of her family and eventually spread throughout the troop. These and other examples (such as local variation in chimpanzee tool use) are believed by many to be the precursors of human culture.

When knowledge of the social domain is combined with cunning, evidence of social learning within individualized societies becomes increasingly complex, with social strategizing and score keeping further spurring intelligence. At its extreme, the exploitation of members of your own species becomes "Machiavellian," evidenced in tactical deception as an example. Accordingly, social interactions are viewed much like moves in a chess game in which each player strives to foretell and outwit the maneuvers of its opponent. The resulting competition is a never-ending struggle that drives intelligence. According to this hypothesis, intelligent behavior in all aspects of daily life, including the nonsocial realm, are by-products of the skills developed to compete in social interactions.

Recent comparative investigations across numerous primate species provide evidence that proxies for social complexity (e.g., group size) or social learning (e.g., frequencies of social transmission, innovation, and tool use) are positively correlated with proxies for intelligence (relative neocortex size or "executive" brain size). These results are consistent with the predicted link between sociality and intelligence. To further test this link, biologists have turned their attention to members of other individualized societies, such as cetaceans, elephants, and social carnivores. Such broadly comparative studies of social learning and intelligence across different species and taxonomic groups promise to shed further light on the origins of human intelligence and culture.

See also Cognition—*Imitation*
Learning—*Evolution of Learning Mechanisms*
Learning—*Social Learning*

Further Resources

Byrne, R. W. & Whiten, A. 1988. *Machiavellian Intelligence: Social Expertise and the Evolution of Intellect in Monkeys, Apes and Humans.* Oxford, UK: Oxford University Press.
de Waal, F. B. M. & Tyack, P. L. 2003. (Eds.). *Animal Social Complexity: Intelligence, Culture, and Individualized Societies.* Cambridge, MA: Harvard University Press.
Humphrey, N. K. 1976. *The social function of intellect.* In: *Growing Points in Ethology* (Ed. by P. P. G. Bateson & R. A. Hinde), pp. 303–317. Cambridge, UK: Cambridge University Press.
Reader, S. M. & Laland, K. N. 2002. *Social intelligence, innovation, and enhanced brain size in primates.* Proceedings of the National Academy of Sciences, USA, 99 (7), 4436–4441.
Tomasello, M. & Call, J. 1997. *Primate Cognition.* Oxford, UK: Oxford University Press.

C. M. Drea

■|Learning
|*Spatial Learning*

Among recent studies exploring abstract concepts in nonhuman animal species, experimental approaches to spatial learning (SL) in animals have addressed a variety of questions, some of which have parallels in the human literature on spatial learning. The diversity of tasks and hypotheses tested in animals may be due, in part, to the variety of methodological and theoretical approaches in studies of similar spatial processing by humans. However, although there has been considerable cross-fertilization among SL studies with many species, including humans, the diverse approaches make cross-species comparisons difficult.

As early as the 1930s, Edward Tolman, an experimental psychologist, studied the ability of rats to learn the spatial organization of mazes. As the animals explored each maze, they made various turns, sometimes down blind alleys, and eventually located the goal box. Tolman proposed that during the course of their exploration, the rats came to represent the spatial configuration of the maze internally as a kind of "cognitive map," a term that is still used today to refer to some type of mental representation that may or may not be topographic to the space it represents. Whatever its form, the question remains as to whether or not animals in their natural habitat are able to form some type of mental image of their territory, the landmarks within it, sites where food has been cached, or a host of other spatially mediated information that may contribute to the animal's survival. Since Tolman first proposed the idea of a cognitive representation of the external environment as being within the capacity of rats, scientists from numerous disciplines, including comparative psychology, neuroscience, cognitive science, biology, and ethology, among others, have designed many types of experiments to test their ideas about what types of mental representations might exist among a wide range of species. These studies have included testing of many types of passerine birds, a variety of rodents, nonhuman primates, as well as comparative studies with children and adult humans from differing cultures.

There is little question that there are important features of an animal's internal and external environment that are critical to their survival, including the daily use of time, space and number. In fact, many species have specific adaptations for making maximal use of these abilities within their natural habitat. Consequently, studies of spatial navigation and learning can provide a wealth of clues toward our understanding of the basic mechanisms that support processing of spatial information. Most animals must be able to acquire flexibly both immediate and long-term information about their location and orientation within their territory relative to foraging for food, seeking mates, migrating, selecting nest or den sites, caring for offspring, storing food and avoiding potential predators, all of which may make great demands on spatial learning abilities and memory.

Among these mechanisms, foraging and food storing in animals, particularly studies of wild populations of caching birds that can be compared and contrasted with well-controlled experimental studies in captivity using artificial food sites, represent a particularly burgeoning literature in the animal cognition field over the past decade. Hoarding food and relocating it later, sometimes months after the original caching, places exceptional demands on spatial memory. Among the caching birds, which are those who store food for later retrieval, several species of jays and nutcrackers all store large numbers of nuts and seeds, and are dependent upon finding stored food in order to survive the winter months. Although there are other ground mammals that also store food (such as squirrels and some species of rats, as well as other birds such as chickadees), bird species that are members of the jay family have been the most extensively studied under both field and captive conditions. During experimental laboratory studies, efforts to control possible odor and visual clues did little to decrease the birds' accuracy in locating cached food. However, displacement of the cached food to a site a short distance away resulted in the birds' inability to locate it. In addition, although there have been some suggested alternative explanations for the birds' success, including mere chance, their ability to locate and retrieve hoarded food has repeatedly not been able to be attributed to chance encountering of the sites because the animals' performance was far better than would have been predicted by chance alone. Such findings have been replicated in numerous laboratories with several different species under comparable conditions with the same results

With respect to spatial orientation, a variety of strategies have been documented in animals with which they are able to find their way home or identify some location in space. Among these mechanisms are the use of: landmarks, global reference systems, path integration, and cognitive maps. The use of *landmarks* refers to orientation and/or navigation based on the use of stationary objects or surfaces that permit identification of a location, and provide both distance and directional information for the animal. *Global reference systems* refer to processes such as celestial or magnetic navigation strategies which provide compass-like information known as a reference bearing and allow for the individual to orient to a position based on input from the stars or geomagnetic forces. *Path integration* represents the use of feedback information that is integrated during movement in time and space. This, in turn, allows an animal to determine distance and direction from the origin of the path, permitting direct return to a starting point. Finally, as noted previously, the use of a *cognitive map* refers to some form of representation of the spatial relationships among the various locations and/or landmarks in a spatial location that is constructed cognitively or possibly inferred from a partial representation or knowledge of the spatial relations among the location's features.

Similar to the recent findings that have revealed species' differences in spatial memory in birds, concurrent studies with rodents have demonstrated both species and sex differences in spatial learning, related to the species' particular social structure and mating system. For example, males and females of a species that mate exclusively, and thus are considered monogamous, live in the same territory together, and consequently, both sexes use similar spatial abilities for getting around their range. On the other hand, species that are polygynous, that is, those whose males have access to more than one female, have different requirements for keeping track of females' locations within the territory, avoidance of rival males, and a host of other critical features that might require better spatial processing abilities than a monogamous male. For example, among polygynous species, males must range over much larger areas in order to maintain contact with each female who occupies her own separate territory. These hypotheses have been explored in studies of sex differences in neural structure and spatial cognition in different rodents, specifically prairie voles, a monogamous species, and meadow voles, which are polygynous. Thus, males and females of these two closely related species were compared on their performance on a series of mazes in order to test their spatial abilities. As predicted, sex differences in ranging behavior were associated with male–female differences in maze performance, with polygynous males outperforming monogamous males. Interestingly, in some species, there are seasonal sex differences in the use of space, which can result in unique opportunities for a variety of interesting experimental studies. It is likely the mechanisms and processes that subserve spatial learning and memory in animals will continue to be an active area in the field of animal cognition, as a host of critical and intriguing questions remain about how animals acquire and utilize spatial information.

See also Navigation—*Spatial Navigation*
Social Organization—*Dispersal*
Social Organization—*Territoriality*

Further Resources

Drickamer, L. C., Vessey, Stephen H., & Jakob, Elizabeth M. 2001. *Animal Behavior: Mechanisms, Ecology, Evolution.* 5th edn. New York: McGraw-Hill Science/Engineering/Math.
Gallistel, C. R. 1990. *The Organization of Learning.* Cambridge, MA: MIT Press.
Golledge, Reginald G. 1999. *Wayfinding behavior: Cognitive Mapping and Other Spatial Processes.* Baltimore: Johns Hopkins University Press.
Shettleworth, S. 1998. *Cognition, Evolution and Behavior.* New York: Oxford University Press.

Sarah T. Boysen

■ Lemurs
Behavioral Ecology of Lemurs

Lemur species throughout Madagascar are able to change their behavior and social groups when their habitat is not sufficient and they are not able to find enough food. One lemur species in particular, the black-and-white ruffed lemur (*Varecia variegata*), is found in areas ranging from undisturbed primary forest to disturbed secondary forests in both mountain and lowland biomes characterized by food that may be hard to find due to seasonal availability or distribution.

The black-and-white ruffed lemur
(Varecia variegata*), has surprisingly
varied social organization as well as
environment.*
Courtesy of Noel Rowe.

When compared with other lemur species, the black-and-white ruffed lemur requires a greater portion of its diet (70–90%) to include fruit. The size and richness of fruit patches may, therefore, be linked with ruffed lemur density, group size, and dispersal opportunities. The black-and-white ruffed lemur has a varied social organization including male–female pairs, small multimale–multifemale groups, and communities with up to 16 individuals.

It is not clear why these lemurs' social organization is so varied, but there is strong evidence that groups fission into temporary subgroups when fruit production is low. When patches of ripe fruit from preferred food trees become more scattered (by distance and/or by time), black-and-white ruffed lemur groups split into smaller foraging parties until desirable food patches are located. Once a ripe fruit patch is located, individuals often lead other group members to the patch, or vocalize until all individuals converge on the area. Because visual signals are not an effective means of communication when a group fissions, or splits up, in the forest, it has been hypothesized that this species uses vocalizations to coordinate movements between temporary subgroups to improve foraging efficiency.

The species has several reproductive and behavioral traits making it unique among diurnal primates. Although some nocturnal primates have two to four offspring at a time (e.g., Galagos, mouse lemurs and dwarf lemurs), the black-and-white ruffed lemur is the only member of the prosimian suborder that is diurnal (active during the day) and that has two sets of mammary glands and consistently produces litters of two or more young. The infants cannot cling to their mothers' fur, but instead are carried in the mouth and parked in trees or epiphyte tangles and attended to by group members. This type of lemur is the only known medium-sized primate that builds nests in preparation for the arrival of its young. Among other prosimians greater than 2 kg (4.4 lb) in size, only the nocturnal aye-aye (*Daubentonia madagascariensis*) builds nests. The aye-aye, however, produces only single offspring, and its nests are used throughout the year for sleeping.

When food is scarce, black-and-white lemurs produce no offspring, but as soon as ecological conditions are favorable, one female may successfully raise up to four offspring in a year. The infants grow rapidly and mature early. The species' unique pattern among lemurs in how they adjust their reproductive output in response to environmental conditions may be the key to their behavioral plasticity. Larger groups may provide benefits such as enhanced guarding of multiple young or protection of group members from predation. Among the lemur taxa, mobbing behavior toward threatening situations has only been reported for the black-and-white ruffed lemur.

See also Behavioral Plasticity
Lemurs—*Learning from Lemurs*
Social Organization—*Dispersal*
Social Organization—*Monkey Families and Human Families*
Social Organization—*Social Dynamics in the Plateau Pika*
Social Organization—*Socioecology of Marmots*
Social Organization—*Territoriality*

Further Resources

Balko, E. A. 1998. *A behaviorally plastic response to forest composition and logging disturbance by* Varecia variegata variegata *in Ranomafana National Park, Madagascar*. Ph.D. Thesis. SUNY–College of Environmental Science and Forestry, Syracuse, New York.
Morland, H. S. 1991. *Preliminary report on the social organization of ruffed lemurs* (Varecia variegata variegata) *in a northeast Madagascar rainforest*. Folia Primatologica. 56, 157–161.
Ratsimbazafy, H. J. 2002. *On the brink of extinction and the process of recovery: Responses of black-and-white ruffed lemurs* (Varecia variegata variegata) *to disturbance in Manombo forest, Madagascar*. Ph.D. Thesis. SUNY-Stony Brook. Stony Brook, NY.
Vasey, N. 1997. *Community ecology and behavior of* Varecia variegata rubra *and* Lemur fulvus albifrons *on the Masoala Peninsula, Madagascar*. Ph.D. Dissertation, Washington University, St. Louis.

Elizabeth Balko

■ Lemurs
Learning from Lemurs

Lemurs are endearing creatures (see photos), but they are also of particular interest and value to students of behavior. Along with monkeys, apes and humans, lemurs belong to the primate family, but have a more ancient ancestry. They are indigenous to Madagascar, a land mass once connected to Africa but which drifted away from that continent. The isolation

Due to the isolation Madagascar has afforded, different species of lemurs have adapted widely different behaviors.
Courtesy of Peter H. Klopfer.

from competing species this afforded allowed lemurs to occupy a wide range of habitats and to diversify into numerous species, from mouse-sized to large cat-sized species, and even some meter-high giants, now extinct. Some of the roughly two dozen still extant species are adapted to life in the rainforest on the island's northeast coast, others to the dry forest of the west, and still others the desert-like conditions of the southeast. The coming of humans to Madagascar, probably about a millennium ago, led to the disappearance of some species no longer represented today.

Not only do lemur species look different from one another, they also display considerable variation in behavior. Some, such as *Varecia variegatus*, the ruffed lemur, almost never walk on the ground, remaining high in the canopy. Brown lemurs, *Eulemur fulvus*, may spend some of each day foraging on the forest floor, whereas the ringtailed *Lemur catta* will be on the ground for as much as one-third of its waking time. Parental care patterns vary, too. The ruffed lemur parks its infants in a nest and visits them sporadically to nurse, whereas brown and ringtailed lemurs carry their infants across their fronts. The brown lemurs rarely allow others of their group to act as babysitters, but cattas do so frequently.

Parental habits vary between species: The canopy-living species don't usually carry their young with them, as their ground-dwelling relatives tend to.
Courtesy of Peter H. Klopfer.

Social organization varies, too, from the solitary, through family groups, to large organized groups of up to two dozen animals. Since the species have a common ancestry, it is likely that the causes for these differences are related to the particular habitats in which they have settled. Possibly, chance events that transpired early in the separation of one species from another played a role. For example, the ruffed lemurs, which not only live high in the canopy but also have multiple young, would be hard-pressed if they had to carry their young around rather than to deposit them in a tree cavity. For the more earth-bound brown and striped lemurs, toting babies is easier. However, the orientation of the infants' body with respect to the mother differs between brown and striped lemurs: In one case the infant is usually at a right angle to the mother's axis, in the other, parallel. It is hard to imagine why this should be so, and this may just possibly be a case of a difference that resulted from chance alone. Perhaps studies of the biomechanics of movement while carrying an infant will clarify the issue.

Whatever the explanations for the differences between lemur species, it should be apparent why lemurs are so useful: Many species living in a variety of quite different (and some in quite similar) habitats, but all with a common ancestry. Darwin would have been delighted had he had such a marvelous model for comparative studies available to him. Darwin never made it to Madagascar, however, though the lemurs allow us an opportunity to test some of his ideas on the mechanisms underlying evolution.

See also Lemurs—*Behavioral Ecology of Lemurs*

Further Resources

Doyle, G. A, & Martin, R. D. (Eds.). 1979. *The Study of Prosimian Behavior*. New York: Academic Press.

Kappeler, P. M. & Ganzhorn, J. U., 1993. *Lemur Social Systems and their Ecological Basis*. New York: Plenum Press.

Martin, R. D., Walker, A. C., & Doyle, G. A. (Eds.). 1973. *Perspectives in Prosimian Biology*. London: Duckworth Publications.

Peter H. Klopfer

■|Levels of Analysis in Animal Behavior

Why does an animal perform a certain behavior? Or, why does individual A perform behavior X? This question is deceptively simple. Answering it is challenging because behavior X can be explained from multiple complementary perspectives or *levels of analysis*. Two of these perspectives provide *proximate* (or immediate cause) explanations and two others provide *ultimate* (or long-term cause) explanations. Proximate explanations focus on physiological and cognitive *mechanisms* underlying a behavior and on how it *develops* in individuals; whereas ultimate explanations consider how a behavior affects its performer's *reproductive success* and its evolutionary *history* in the phylogeny of the performing species.

Actually, this pluralistic conceptual framework can be used to develop and test hypotheses about any organismal attribute, whether it is morphological, physiological, or behavioral. As a behavioral example, imagine that you are observing a dog who smells or sees a companion (canine or human). The dog freezes, orients toward the approaching individual, raises its tail, and swishes it from side to side. Let's consider the question, "Why does this dog wag its tail?" at each level of analysis.

First, regarding causation, we would ask, "What makes the wag happen?" This is primarily a question of behavioral physiology, one whose answer requires knowing how the dog's neuronal, hormonal, skeletal, and muscular systems interact to bring about the behavior. This includes determining how the dog senses its companion, how its hormonal system adjusts responsiveness to this stimulus, and how the dog's skeletal–muscular system generates tail movements. One can also consider immediate causes, or proximate causation, from the perspective of cognition; that is, by investigating how the dog processes information in connection with greeting a companion by wagging its tail. This includes determining how the dog senses the companion, recognizes him or her as a friend, and decides to wag its tail, and how the dog may even have intentions, beliefs, and self-awareness as parts of its mental make-up.

Second, with respect to ontogeny (or development), we would ask, "How did the wag develop during this dog's lifetime?" The answer derives from understanding how the sensory–motor mechanisms that produce the behavior were shaped as the dog matured from a puppy to an adult. The shaping of this behavioral machinery can involve both internal and external influences, so understanding the development of tail wagging behavior requires investigation of the influences of both the dog's genes and environment (i.e., its lifetime experiences, both social and nonsocial).

Third, regarding fitness effects, we would ask, "How does tail wagging contribute to genetic success?" This is a question of behavioral ecology, one whose answer depends on determining the effects of the behavior on canine survival and reproduction; that is, mating success and offspring production or how tail wagging helps our dog overcome obstacles to survival and, ultimately, perpetuate its genes.

And fourth, with respect to evolutionary history, we would ask, "How did wagging evolve from ancestral forms?" This is a question of behavioral evolution, one whose answer requires inferring antecedent behaviors and the step-by-step evolutionary pathway leading to the current behavior. In essence it involves reconstructing the sequence of tail movements and transitions between them, from the origin of canid tail wagging to the present. Since tail wagging, like most behaviors, leaves no traces in the fossil record, comparative studies of living relatives of dogs, such as wolves and coyotes, might be used to draw evolutionary inferences. Not surprisingly, this is the most difficult (and contentious) level of analysis for behaviorists to investigate.

It is worth noting that the study of animal behavior, like all branches of the biological sciences, has an extra dimension compared to the physical sciences. Both the biological and physical sciences seek explanations of natural phenomena in physico–chemical terms. In biology, such explanations are called *proximate* causes (i.e., our first two levels of analysis). The extra dimension of biology is that it also seeks explanations for natural phenomena in terms of *ultimate* causation. Thus in biology we can legitimately ask, "What is the use of a particular life process today?" and "How did it come to be this way over evolutionary time?" Provided the words *use* and *come to be* mean "promoting genetic success" and "evolving through natural selection" (i.e., our third and fourth levels of analysis), we may legitimately ask these questions in biology generally, and in animal behavior particularly. In physics and chemistry, however, they are out of bounds. For example, it is reasonable to ask, "What use are the movements of a dog's tail?" whereas it is metaphysical to ask, "What use are the movements of an ocean's tides?"

Every hypothesis in biology is subsumed within the levels of analysis framework. This means that there are multiple types of "correct" answers to any question about biological causality. Answers at one level complement, but do not supersede, those at another. No level is inherently more important: Ultimate answers are not superior to proximate answers. Which category of answer is most satisfactory or interesting is largely a matter of individual training and taste. In the case of our dog, some persons may be content to know how the muscles and nerves coordinate the tail wag, others may be satisfied knowing that the wag indicates a "happy" state of mind, whereas still others may wish to know only how wags affect mating success.

Regardless of which level(s) of analysis a person favors, the cardinal principle is that *competition between alternative causal explanations appropriately occurs only within and not among levels.* Many acrimonious debates in behavioral biology originated as supposedly alternative hypotheses which, actually, were on different analytical levels. A famous example of this problem, in which protagonists talked past each other for years, is the "nature–nurture" controversy. Here, fitness effects were inappropriately pitted against ontogenies. The reason these debates have persisted is that protagonists each were at least partially correct, but that did not jeopardize any hypotheses at different analytical levels. For example, a particular dog may have learned to wag its tail by observing its mother greet a friend and, simultaneously, wagging may be beneficial for reducing aggression or increasing mating success.

Up to now, we have considered each level of analysis separately. However, complete understanding of any behavior requires integration of answers at all four levels. To accomplish this, explanations at each level of analysis are treated as complementary, not mutually exclusive, to those at other levels, and explanations at each analytical level are used to inform, not stifle, hypothesis development and testing at other levels.

See also Behavioral Phylogeny—*The Evolutionary Origins*
        of Behavior
    Sociobiology

Further Resources

Alcock, J. & Sherman, P. W. 1994. *The utility of the proximate–ultimate dichotomy in ethology*. Ethology, 96, 58–62.

Armstrong, D. P. 1991. *Levels of cause and effect as organizing principles for research in animal behavior.* Canadian Journal of Zoology, 69, 823–829.

Bass, A. H. 1998. *Behavioral and evolutionary neurobiology: A pluralistic approach.* American Zoologist, 38, 97–107.

Holekamp, K. E. & Sherman, P. W. 1989. *Why male ground squirrels disperse.* American Scientist, 77, 232–239.

Koenig, W. D. & Mumme, R. L. 1990. *Levels of analysis and the functional significance of helping behavior.* In: *Interpretation and Explanation in the Study of Animal Behavior. Volume 2. Explanation, Evolution, and Adaptation* (Ed. By M. Bekoff & D. Jamieson), pp. 268–303. Boulder, CO: Westview Press.

Mayr, E. 1961. *Cause and effect in biology.* Science, N.Y., 134, 1501–1506.

Sherman, P. W. 1988. *The levels of analysis.* Animal Behaviour, 36, 616–619.

Tinbergen, N. 1963. *On aims and methods of ethology.* Zeitschrift für Tierpsychologie, 20, 410–433.

Paul W. Sherman

■ Lorenz, Konrad Z.
(1903–1989)

Konrad Z. Lorenz was born in Vienna in 1903. He is regarded as the primary founder of European ethology. As a boy, Lorenz loved to keep a menagerie of diverse pets—the roots of his later research. He became an animal behaviorist to the chagrin of his physician father, who wanted Konrad to follow in his footsteps. Like most scientists credited with founding disciplines, Lorenz had a wonderful gift for creative writing and for attracting adherents. His genial manner with people hid the sharp mind of both a gifted animal keeper and a philosopher. In 1973, he shared a Nobel Prize with Niko Tinbergen and Karl von Frisch. Lorenz's image has been tarnished in recent years for alleged associations with the Nazi party, and his theories on the innate character of aggressive behavior in humans have stirred some controversy. Lorenz died in 1989.

Donald A. Dewsbury

■ Lorenz, Konrad Z.
Analogy as a Source of Knowledge

Nobel Lecture, December 12, 1973

1. The Concept of Analogy

In the course of evolution it constantly happens that, independently of each other, two different forms of life take similar, parallel paths in adapting themselves to the same external circumstances. Practically all animals which move fast in a homogeneous medium have found means of giving their body a streamlined shape, thereby reducing friction to a minimum. The "invention" of concentrating light on a tissue sensitive to it by means of a diaphanous lens has been made independently at least four times by different phyla of animals; and in two of these, in the cephalopods and in the vertebrates, this kind of "eye" has evolved into the true, image-projecting camera through which we ourselves are able to see the world.

Thanks to old discoveries by Charles Darwin and very recent ones by biochemists, we have a fairly sound knowledge of the processes which, in the course of evolution, achieve these marvellous structures. The student of evolution has good reason to assume that the abundance of different bodily structures which, by their wonderful expediency, make life possible for such amazingly different creatures under such amazingly different conditions, all owe their existence to these processes which we are wont to subsume under the concept of adaptation. This assumption, whose correctness I do not propose to discuss here, forms the basis of the reasoning which the evolutionist applies to the phenomenon of analogy.

2. Deducing Comparable Survival Value from Similarity of Form

Whenever we find, in two forms of life that are unrelated to each other, a similarity of form or of behaviour patterns which relates to more than a few minor details, we assume it to be caused by parallel adaptation to the same life-preserving function. The improbability of coincidental similarity is proportional to the number of independent traits of similarity, and is, for n such characters, equal to 2^{n-1}. If we find, in a swift and in an airplane, or in a shark, or a dolphin and in a torpedo the striking resemblance illustrated in Fig.1, we can safely assume that in the organisms as well as in the manmade machines, the need to reduce friction has led to parallel adaptations. Though the independent points of similarity are, in these cases, not very many, it is still a safe guess that any organism or vehicle possessing them is adapted to fast motion.

There are conformities which concern an incomparably greater number of independent details. Fig. 2 shows cross sections through the eyes of a vertebrate and a cephalopod. In both cases there is a lens, a retina connected by nerves with the brain, a muscle moving the lens in order to focus, a contractile iris acting as a diaphragm, a diaphanous cornea in front of the camera and a layer of pigmented cells shielding it from behind—as well as many other matching details. If a zoologist who knew nothing whatever of the existence of cephalopods were examining such an eye for the very first time, he would conclude without further ado that it was indeed a light-perceiving organ. He would not even need to observe a live octopus to know this much with certainty.

3. The Allegation of "False Analogy"

Ethologists are often accused of drawing *false* analogies between animal and human behaviour. However, no such thing as a *false* analogy exists: an analogy can be more or less detailed and hence more or less informative. Assiduously searching for a really false analogy, I found a couple of technological examples within my own experience. Once I mistook a mill for a sternwheel steamer. A vessel was anchored on the banks of the Danube near Budapest. It had a little smoking funnel and at its stern an enormous slowly-turning paddle-wheel. Another time, I mistook a small electric power plant, consisting of a two-stroke engine and a dynamo, for a compressor. The only biological example that I could find concerned a luminescent organ of a pelagic gastropod, which was mistaken for an eye, because it had an

Note: The illustrations in the original paper have not been reproduced here. To view them go to
http://nobelprize.org/medicine/laureates/1973/lorenz-lecture.html

epidermal lens and, behind this, a high cylindrical epithelium connected with the brain by a nerve. Even in these examples, the analogy was false only in respect of the direction in which energy was transmitted.

4. The Concept of Homology

There is, in my opinion, only one possibility of an error that might conceivably be described as the "drawing of a false analogy" and that is mistaking an *homology* for an analogy. An homology can be defined as any resemblance between two species that can be explained by their common descent from an ancestor possessing the character in which they are similar to each other. Strictly speaking, the term homologous can only be applied to characters and not to organs. Fig. 3 shows the forelimbs of a number of tetrapod vertebrates intentionally chosen to illustrate the extreme variety of uses to which a front leg can be put and the evolutional changes it can undergo in the service of these different functions. Notwithstanding the dissimilarities of these functions and of their respective requirements, all these members are built on the same basic plan and consist of comparable elements, such as bones, muscles, nerves. The very dissimilarity of their functions makes it extremely improbable that the manifold resemblances of their forms could be due to parallel adaptation, in other words to analogy.

As a pupil of the comparative anatomist and embryologist Ferdinand Hochstetter, I had the benefit of a very thorough instruction in the methodological procedure of distinguishing similarities caused by common descent from those due to parallel adaptation. In fact, the making of this distinction forms a great part of the comparative evolutionist's daily work. Perhaps I should mention here that this procedure has led me to the discovery which I personally consider to be my own most important contribution to science. Knowing animal behaviour as I did, and being instructed in the methods of phylogenetic comparison as I was, I could not fail to discover that the very same methods of comparison, the same concepts of analogy and homology are as applicable to characters of behaviour as they are in those of morphology. This discovery is implicitly contained in the works of Charles Otis Whitman and of Oskar Heinroth; it is only its explicit formulation and the realization of its far-reaching inferences to which I can lay claim. A great part of my life's work has consisted in tracing the phylogeny of behaviour by disentangling the effects of homology and of parallel evolution. Full recognition of the fact that behaviour patterns can be hereditary and species-specific to the point of being homologizable was impeded by resistance from certain schools of thought, and my extensive paper on homologous motor patterns in *Anatidae* was necessary to make my point.

5. Cultural Homology

Much later in life I realized that, in the development of human cultures, the interaction between historically-induced similarities and resemblances caused by parallel evolution—in other words between homologies and analogies—was very much the same as in the phylogeny of species and that it posed very much the same problems. I shall have occasion to refer to these later on; here I want to illustrate the existence of cultural homology. Fig. 4 illustrates the cultural changes by which the piece of medieval armour that was originally designed to protect throat and chest was gradually turned, by a change of function, into a status symbol. Otto Koenig, in his book *Kulturethologie*, has adduced many other examples of persistent historically-induced similarity of characters to which the adjective "homologous" can legitimately be applied.

Ritualization and symbolisms play a large role in traditional clothing and particularly in military uniforms in their historical changes, so that the appearance of historically-retained similarities is, perhaps, not very surprising. It is, however, surprising that the same retention of historical features, not only independently of function, but in clear defiance of it, is observable even in that part of human culture which one would suppose to be free of symbolism, ritualization and sentimental conservativism, namely in technology. Fig. 5 illustrates the development of the railway carriage. The ancestral form of the horse-drawn coach stubbornly persists, despite the very considerable difficulties which it entails, such as the necessity of constructing a runningboard all along the train, on which the conductor had to climb along, from compartment to compartment, exposed to the inclemency of the weather and to the obvious danger of falling off. The advantages of the alternative solution of building a longitudinal corridor within the carriage are so obvious that they serve as a demonstration of the amazing power exerted by the factors tending to preserve historical features in defiance of expediency.

The existence of these cultural homologies is of high theoretical importance, as it proves that, in the passing-on of cultural information from one generation to the next, processes are at work which are entirely independent of rational considerations and which, in many respects, are functionally analogous to the factors maintaining invariance in genetical inheritance.

6. Deducing Function from Behavioural Analogies

Let me now speak of the value of analogies in the study of behaviour. Not being vitalists, we hold that any regularly observable pattern of behaviour which, with equal regularity, achieves survival value is the function of a sensory and nervous mechanism evolved by the species in the service of that particular function. Necessarily, the structures underlying such a function must be very complicated, and the more complicated they are, the less likely it is, as we already know, that two unrelated forms of life should, by sheer coincidence, have happened to evolve behaviour patterns which resemble each other in a great many independent characters.

A striking example of two complicated sets of behaviour patterns evolving independently in unrelated species, yet in such a manner as to produce a great number of indubitable analogies is furnished by the behaviour of human beings and of geese when they fall in love and when they are jealous. Time and again I have been accused of uncritical anthropomorphism when describing, in some detail, this behaviour of birds and people. Psychologists have protested that it is misleading to use terms like falling in love, marrying or being jealous when speaking of animals. I shall proceed to justify the use of these purely functional concepts. In order to assess correctly the vast improbability of two complicated behaviour patterns in two unrelated species being similar to each other in so many independent points, one must envisage the complication of the underlying physiological organization. Consider the minimum degree of complication which even a man-made electronic

Austrian scientist Konrad Lorenz swims with a trio of Graylag geese, Bavaria, Germany, 1964.
© Nina Leen / Time Life Pictures / Getty Images.

model would have to possess in order to simulate, in the simplest possible manner, the behaviour patterns here under discussion. Imagine an apparatus, A, which is in communication with another one, B, and keeps on continuously checking whether apparatus B gets into communication with a third apparatus C, and which furthermore, on finding that this is indeed the case, does its utmost to interrupt this communication. If one tries to build models simulating these activities, for example in the manner in which Grey-Walter's famous electronic tortoises are built, one soon realizes that the minimum complication of such a system far surpasses that of a mere eye.

The conclusion to be drawn from this reasoning is as simple as it is important. Since we know that the behaviour patterns of geese and men cannot possibly be homologous— the last common ancestors of birds and mammals were lowest reptiles with minute brains and certainly incapable of any complicated social behaviour—and since we know that the improbability of coincidental similarity can only be expressed in astronomical numbers, we *know for certain* that it was a more or less identical survival value which caused jealousy behaviour to evolve in birds as well as in man.

This, however, is *all* that the analogy is able to tell us. It does not tell us wherein this survival value lies—though we can hope to ascertain this by observations and experiments on geese. It does not tell us anything about the physiological mechanisms bringing about jealousy behaviour in the two species; they may well be quite different in each case. Streamlining is achieved in the shark by the shape of the musculature, in the dolphin by a thick layer of blubber, and in the torpedo by welded steel plates. By the same token, jealousy may be—and probably is—caused by an inherited and genetically fixed programme in geese, while it might be determined by cultural tradition in man—though I do not think it is, at least not entirely.

Limited though the knowledge derived from this kind of analogy may be, its importance is considerable. In the complicated interaction of human social behaviour, there is much that does not have any survival value and never had any. So it is of consequence to know that a certain recognizable pattern of behaviour does, or at least once did, possess a survival value for the species, in other words, that it is *not pathological.* Our chances of finding out wherein the survival value of the behaviour pattern lies are vastly increased by finding the pattern in an animal on which we can experiment.

When we speak of falling in love, of friendship, personal enmity or jealousy in these or other animals, we are *not* guilty of anthropomorphism. These terms refer to functionally-determined concepts, just as do the terms legs, wings, eyes and the names used for other bodily structures that have evolved independently in different phyla or animals. No one uses quotation marks when speaking or writing about the eyes or the legs of an insect or a crab, nor do we when discussing analogous behaviour patterns.

However, in using these different kinds of terms, we must be very clear as to whether the word we use at a given moment refers to a concept based on functional analogy or to one based on homology, e.g. on common phyletic origin. The word "leg" or "wing" may have the connotation of the first kind of concept in one case and of the second in another. Also, there is the third possibility of a word connoting the concept of physiological, causal identity. These three kinds of conceptualization may coincide or they may not. To make a clear distinction between them is particularly important when one is speaking of behaviour. A homologous behaviour pattern can retain its ancestral form and function in two descendants, and yet become physiologically different. The rhythmical beat of the umbrella is caused by ondogenous stimulus generation in many hydrozoa and in larva (*Ephyrae*) in other medusae.

In adult *Scyphomedusae,* however, it is caused by reflexes released through the mechanism of the so-called marginal bodies. A homologous motor pattern may retain its original physiological causation as well as its external forms, yet undergo an entire change of function. The motor pattern of "inciting" common to the females of most *Anatidae* is derived from a threatening movement and has the primary function of causing the male to attack the adversary indicated by the female's threat. It has entirely lost this function in some species, for instance in the Golden-eyes, in which it has become a pure courtship movement of the female. Two non-homologous motor patterns of two related species may, by a change of function, be pressed into the service of the same survival value. The pre-flight movement of ducks is derived from an intention movement of flying, an upward thrust of head and neck, while the corresponding signal of geese is derived from a displacement shaking of the head. When we speak of "pre-flight movements of *Anatidae*" we form a functional concept embracing both. These examples are sufficient to demonstrate the importance of keeping functional, phylogenetical and physiological conceptualizations clearly apart. Ethologists are not guilty of "reifications" or of illegitimate anticipations of physiological explanations when they form concepts that are only functionally defined—like, for instance, the concept of the IRM, the innate releasing mechanism. They are, in fact, deeply aware that this function may be performed by the sensory organ itself—as in the cricket—or by a complicated organization of the retina—as in the frog—or by the highest and most complicated processes within the central nervous system.

Deducing the Existence of Physiological Mechanisms from Known Analogous Functions

Recognizing analogies can become an important source of knowledge in quite another way. We can assume with certainty that, for instance, the functions of respiration, of food intake, of excretion, of propagation, etc., must somehow be performed by any living organism. In examining an unknown living system, we are, therefore, justified in *searching* for organs serving functions which we know to be indispensable. We are surprised if we miss some of them, for instance the respiratory tract in some small salamanders which breathe exclusively through their skin.

A human culture is a living system. Though it is one of the highest level of integration, its continuance is nevertheless dependent on all the indispensable functions mentioned above. The thought obtrudes itself that there is one of these necessary functions which is insufficient in our present culture, that of *excretion*. Human culture, after enveloping and filling the whole globe, is in danger of being killed by its own excretion, of dying from an illness closely analogous to uraemia. Humanity will be forced to invent some sort of planetary kidney—or it will die from its own waste products.

There are other functions that are equally indispensable to the survival of all living systems, ranging from bacteria to cultures. In any of these systems, adaptation has been achieved by the process, already mentioned, which hinges on the *gaining of information* by means of genetic change and natural selection, as well as on the storing of knowledge in the code of the chain molecules in the genome.

This storing, like *any* retention of information, of knowledge, is achieved by the formation of *structure*. Not only in the little double helix, but also in the programming of the human brain, in writing, or any other form of "memory bank," knowledge is laid down in structures.

The indispensable supporting and retaining function of structure always has to be paid for by a "stiffening," in other words, by the sacrifice of certain *degrees of freedom*. The structure of our skeleton provides an example; a worm can bend its body at any point, whereas we can flex our limbs only where joints are provided; but we can stand upright and the worm cannot. All the adaptedness of living systems is based on knowledge laid down in structure; structure means *static* adapted*ness*, as opposed to the dynamic process of adaptation. Hence, new adaptation unconditionally presupposes a *dismantling* of some structures. The gaining of new information inexorably demands the breaking down of some previous knowledge which, up to that moment, had appeared to be final.

The dynamics of these two antagonistic functions are universally common to all living systems. Always, a harmonious equilibrium must be sustained between, on the one hand, the factors maintaining the necessary degree of *invariance* and, on the other, the factors which tend to break up firm structures and thereby create the degree of *variability* which is the prerequisite of all further gaining of information, in other words, of all new adaptation. All this is obviously true of human culture as well as of any other living system whose life-span exceeds that of the individual, e.g. of any species of bacteria, plants or animals. It is, therefore, legitimate to search for the mechanisms which, in their harmonious antagonism of preserving and dismantling structures, achieve the task of keeping a culture adapted to its ever-changing environment. In my latest book *Die Rückseite des Spiegels*, I have tried to demonstrate these two antagonistic sets of mechanisms in human culture.

The preservation of the necessary invariance is achieved by procedures curiously reminiscent of genetic inheritance. In much the same manner as the new nucleotids are arranged along the old half of a double helix, so as to produce a *copy* of it, the invariant structures of a culture are passed on, from one generation to the next, by a process *in which the young generation makes a copy* of the cultural knowledge possessed by the old. Sheer imitation, respect for a father-figure, identification with it, force of habit, love of old ritualized customs and, last not least, the conservativism of "magical thinking" and superstition—which as we have seen influences even the construction of railway carriages—contributes to invest cultural tradition with that degree of invariance which is necessary *to make it inheritable at all.*

Opposed to these invariance-preserving mechanisms, there is the specifically human urge to curiosity and freedom of thought which with some of us, persists until senescence puts a stop to it. However, the age of puberty is typically the phase in our ontogeny during which we tend to rebel against tradition, to doubt the wisdom of traditional knowledge and to cast about for new causes to embrace, for new ideals.

In a paper which I read a few years ago—at a Nobel symposium on "The Place of Value in a World of Facts"—I tried to analyse certain malfunctions of the antagonistic mechanisms and the dangers of an enmity between the generations arising from these disturbances. I tried to convince my audience that the question whether conservativism is "good" or "bad," or whether the rebellion of youth is "good" or "bad," is just as inane as the question whether some endocrine function, for instance that of the thyroid gland, is "good" or "bad." Excesses as well as deficiency of any such function cause illness. Excess of thyroid function causes Basedow's disease, deficiency myxoedema. Excess of conservativism produces living fossils which will not go on living for long, and excess of variability results in the appearance of monsters which are not viable at all.

Between the conservative representatives of the "establishment" on the one hand and rebelling youth on the other, there has arisen a certain enmity which makes it difficult for each of the antagonists to recognize the fact that the endeavours of *both* are *equally* indispensable for the survival of our culture. If and when this enmity escalates into actual *hate*, the antagonists cease to interact in the normal way and begin to treat each other as different, hostile cultures; in fact they begin to indulge in activities closely akin to tribal warfare.

This represents a great danger to our culture, inasmuch as it may result in a complete disruption of its traditions.

See also Frisch, Karl von (1886–1982)
History—*History of Animal Behavior Studies*
Nobel Prize—*1973 Nobel Prize for Medicine
or Physiology*
Tinbergen, Nikolaas (1907–1988)

◼ Maynard Smith, John (1920–2004)

April 18, 2004, saw the death of one of the greatest evolutionary biologists of the twentieth century, Professor John Maynard Smith. A professor at the University of Sussex since 1965, Dr. Maynard Smith succumbed to mesolthelioma, and died peacefully in his sleep at age 84. Although Maynard Smith—known to friends and colleagues as JMS—officially retired in 1985, he was actively involved in research until the very end, having published many books after his so-called retirement.

Maynard Smith was educated at Eton (a prestigious British preparatory school), and then received his degree in engineering at Trinity College in 1941. After working as an engineer during World War II, Maynard Smith's ever curious mind was no longer satisfied, and he pursued a lifelong interest in evolution by joining the lab of J.B.S. Haldane. His early work focused on aging, using the fruit fly as a model system.

Dr. Maynard Smith's contributions to evolutionary biology were many, but he will most likely be remembered as the man who pioneered what is known today as evolutionary game theory. Evolutionary game theory's fundamental principle—that actions taken by an one individual have effects on the fitness of others and that all such effects must be accounted for when examining the evolution of a trait—has revolutionized the study of evolution and behavior.

Central to evolutionary game theory is the idea of an *evolutionarily stable strategy* (ESS). Maynard Smith defined an ESS as "a strategy such that, if all the members of a population adopt it, no mutant strategy can invade." Here *mutant* refers to a new strategy introduced into a population, and successful invasions center around the relative fitness of established vs. mutant strategies. If the established strategy is evolutionarily stable, the payoff from the established strategy is greater than the payoff from the mutant (new) strategy. For example, if *cooperation* is the strategy adopted by individuals in a population, we can easily envision a mutant *cheater* strategy emerging. If cooperation produces a higher fitness payoff than cheating, it is an ESS and should be maintained in the population. If cooperation is not an ESS, we expect to see a greater and greater frequency of cheaters over time.

The idea of an ESS has not only been the centerpiece of hundreds of papers on behavior and evolution, but has surpassed the boundaries of biology and can often be heard being uttered by the likes of political scientists, mathematicians, and psychologists. It is certainly rare for mathematical terminology created by evolutionary biologists to be adopted across such an array of disciplines, and this stands as a testament to the influence of Professor John Maynard Smith.

Maynard Smith's legacy is not confined to his work on the evolutionarily stable strategies. He made fundamental contributions to our understanding of speciation, extinction, parental investment, altruism, kinship, animal signalling, sexual reproduction, aging, the evolution of flight (his engineering background is evident here), the evolution of early life forms on earth, and the evolution of complexity.

Those who were fortunate enough to know John Maynard Smith, knew that he was not only a great scientist, but a gentleman, in the true sense of the word.

Lee Alan Dugatkin

■|Memes

The concept of a meme was first introduced by Dr. Richard Dawkins in his book *The Selfish Gene*. Dawkins created this term to capture what the *cultural* equivalent of a gene might look like. If we think of genes as the units of genetic transmission, then memes are the units of cultural transmission.

The simplest definition of a *meme* is "whatever is passed on by imitation." What makes memes of fundamental interest to the study of evolution and behavior is that they possess a set of attributes that for a long time we believed to be unique to genes. Memes are replicators, where a replicator has three properties: *fidelity* (good copies are made), *fecundity* (lots of copies are made), and *longevity* (copies are made for a long time).

Replicators, like memes and genes, are unique in that they are the "beans" of the evolutionary accounting system, and replicators have only one "objective," and that is to make more copies of themselves. They are not here *for anything else*, just to make copies of themselves.

A solid meme has the three characteristics replicators need—fidelity, fecundity, and longevity. To make the idea of a meme more concrete, consider birthday parties. When was the last time you went to such a party and failed to hear everyone sing "Happy Birthday to You"? In fact, take a moment and see if you can even think of a rival song that you sometimes hear at parties. The tune "Happy Birthday to You" is a meme, and a good one at that.

To see why "Happy Birthday to You" is a meme, recall the three characteristics that define a replicator—fidelity, fecundity and longevity. You don't hear many people, even little children, making mistakes when singing "Happy Birthday to You." What's more, when young children are taught "Happy Birthday to You," they are all taught virtually the exact same song. Fidelity (how accurately a replicator is copied) is not a problem for "Happy Birthday to You." Add on to this the fact that virtually every child will learn "Happy Birthday to You" early in his life, and that this has been the case for many years, and you will see that "Happy Birthday to You" meets the fecundity and longevity criteria of a replicator as well. The Introduction to Beethoven's Fifth Symphony (Da Da Da Daah, Da Da Da Daah) and "In the beginning G-d created the Heavens and the Earth . . ." are also good candidates for memes, as they have been around for about 250 and 3000 years, respectively.

While some people have argued that memes may be unique to humans, there is evidence that animal memes may exist as well. Two good examples involve antipredator behavior in birds and mate selection in fish. In blackbirds, individuals appear to learn what qualifies as a dangerous predator by watching and learning how others respond to possible predation threats. The meme here might be something akin to, "Predators look like X—so stay away from X." In terms of mate choice, female guppies appear to use the preferences of other females to determine, in part, who is a suitable mate and who isn't. In this case, the meme in question might be, "Males that possess X are attractive—choose such males as mates." Obviously, in neither the guppy nor the blackbird case do we need to imagine that the animals think like humans think—they need only act in a way that seems to satisfy our criteria for a meme.

How many animal memes truly exist is a question that is just now being examined seriously. Whether the answer turns out to be "very few" or "a lot," the idea of the meme has raised new and interesting questions for animal behaviorists to ponder.

Lee Alan Dugatkin

■ Methods
Computer Tools for Measurement and Analysis of Behavior

Paper, Pencil, and Stopwatch

The classical method of data collection in animal behavior research is systematic observation: A human expert watches the animal(s) and records the occurrence of relevant behavioral events and interactions, together with the time when these are observed. The traditional tools with which these data are recorded are paper, pencil and stopwatch. However, this has severe drawbacks: Looking away from the animal to take notes leads to events being missed, and transcription of the data into a computer for analysis is very time-consuming and error-prone.

Using a Computer as an Event Recorder

Fortunately, computers can improve the quality of such measurements and automate significant parts of the data collection and analysis process. Since the 1980s, paper-and-pencil recording methods have gradually been replaced by computer event-recording programs, such as The Observer®. This program allows the user to define a list of subjects and behaviors relevant to the study (the ethogram) and map these to keyboard codes. During the observation an event is scored by pressing the corresponding key(s), something a trained observer can do without looking away from the animal. The software thus simplifies the data-entry process and increases the accuracy of collected data by validating each entry against the pre-defined coding scheme. The program also automatically adds a timestamp to each event, using the internal clock of the computer. Coding schemes can range from a single list of mutually exclusive categories pertaining to a single focal subject, to very elaborate configurations with multiple subjects, multiple classes of behaviors, and modifiers for objects, intensities, directions or receivers of behavioral acts. After an observation session has ended, the data are immediately ready for analysis, thus saving the researcher a laborious transcription process. Applications of The Observer vary from time budget studies of individual insects or predator–prey interactions to the analysis of social relationships in large colonies of monkeys.

Mobile Data Collection with a Handheld Computer

The Observer software runs on standard desktop or notebook PCs. Special versions are available for handheld computers, which are available with waterproof, dustproof, and shockproof housing and can be used for data collection in the field. These have either a keyboard (e.g., the Psion Workabout), or a touch screen operated with a pen (e.g., the Pocket PCs such as HP iPAQ or Dell Axim), or both (e.g., Panasonic Toughbook 01). Researchers have used such devices for behavioral data collection in outdoor environments ranging from Antarctica to the tropical rainforest of central Africa.

Scoring Behavior from Video

Scoring behavior while observing, especially with a handheld computer, makes the researcher highly mobile and does not require large investments. However, the downside of live observation is that no permanent record of the behavior is made, so there is no possibility to

An entomologist observing and collecting data on the behavior of insects, using The Observer software installed on a Psion Workabout handheld device.
Courtesy of Lucas Noldus.

review it afterwards. Furthermore, when events occur in rapid succession or many events happen at once, live data entry will lead to missed events and timing errors. Therefore, for detailed measurements, scoring behavior from video is the preferred option. When doing so, time information is extracted from the video source, which can be an analog tape, digital tape, a digital media files on disk (AVI, MPEG) or a DVD with movie titles. The software synchronizes the observational data with the video recording, which allows posttest review and editing of behavioral records. With The Observer, detailed observations can be made from video recordings at any playback speed without loss of time information. This technique is indispensable for accurate measurement of precise behaviors. Another advantage is that it facilitates the use of continuous focal sampling, which, compared to interval sampling, has the advantage that it registers all behaviors, including behaviors that occur rarely or momentarily. Furthermore, every aspect of behavior can be studied this way, including vocalizations. Recent advancements in digital video technology have been extremely helpful in increasing the quality and feasibility of behavioral observations, with costs going down in the mean time. With the latest version of The Observer (5.0 at the time of writing), behavioral data can be collected from CD, DVD and any video device connected to a computer via a FireWire interface (e.g., a digital camcorder hooked up directly to a notebook PC).

The Need for Automation

Direct observation of animal behavior and manual event recording offer maximum flexibility to the researcher, because any behavior that is observed can be recorded. However, prolonged direct observation is highly time-consuming and thus costly. Furthermore, scoring by a human observer is by definition a subjective process, which means that the inter-rater reliability of observational data is always an issue. This problem can be worsened by observer fatigue or drift. Therefore, there is an ongoing desire to automate behavioral observations. Automation provides significant advantages, of which the time-saving aspect is most prominent. Furthermore, behavior is recorded more reliably because an automated system always works in the same way, so observations can continue almost infinitely.

Using Computer Vision to Track Animal Movement and Behavior

Various techniques for automated measurement of animal behavior and movement have been developed in the past two decades, ranging from activity meters based on infrared photobeams, ultrasound, or Doppler radar, to versatile video tracking and behavior recognition software. Miniaturization and price reduction of sensors and transmitters continue to create new opportunities for animal monitoring and tracking. Inexpensive passive infrared detectors (commonly used in burglar alarm systems) can be used to detect gross activity levels of animals, but these do not provide information about body posture, behavioral patterns or movement. GPS receivers have become small enough to be carried by birds for studies of ranging behavior. Of all available techniques, computer vision offers most potential for automated behavioral observation. An example of a commercially available instrument based on this technique is EthoVision®, a general-purpose video tracking, movement analysis, and behavior recognition system. EthoVision was designed to automate behavioral observation and movement tracking of single or multiple animals against a variety of backgrounds. A video camera records the area in which the animals are (i.e., an experimental enclosure in laboratory, farm or field). The video signal is fed into the computer and digitized. During data acquisition, EthoVision processes up to 30 video images per second. The software analyzes each frame in order to distinguish the object(s) to be tracked from the background, on the basis of either their grey scale (brightness) or hue and saturation (color) values. The next step is object identification (when tracking more than one animal per arena) on the basis of either size or color differences. Having detected the objects, the software extracts relevant image features, including the coordinates of the geometric center, the surface area, and the number of pixels that changed position since the previous frame. The tracking process is displayed on the screen. After a data acquisition run, calculations are carried out on the features to produce quantified measurements of the animal's behavior. For instance, if the position of an animal is known for each video frame, and the whole series of frames is analyzed, the average speed of locomotion or the proportion of the time spent moving can be calculated. If multiple animals are present in one arena, the distance and orientation of movement between several individually identified animals can be computed for each frame, to quantify social or aggressive behavior. In addition, if certain regions are identified as being of interest (the center and edges of a circular arena, for example), the proportion of time spent by the animal in those regions can be determined. EthoVision was designed as a generic tool that can be used in a wide variety of different setups and applications. It is widely used in laboratory studies with rats and mice, to evaluate the effect of pharmacological treatments or to characterize different genotypes. The system is also used in research on insects, fish, birds, primates and other mammals.

Computer-Aided Gait Analysis

Tracking the whole body movement of an animal with a video tracking system does not reveal the details of movement of individual limbs, which are often invisible when looking from above. For the study of gait patterns and walking disorders, each foot must be tracked individually. Traditionally, methods to assess gait patterns are either qualitative (e.g., the Bristol gait score for chickens) or quantitative, but manual, such as the analysis of the prints of inked paws of a rat or mouse on a piece of paper. As an alternative for these subjective and laborious methods, computer tools have been developed. For example, one can use a *pedobarograph* to measure pressure patterns for various regions of the foot, or a force plate to measure the ground reaction force (GRF) in one or more directions (vertical, craniocaudal or mediolateral), from which one can derive step length and many other descriptors of walking style. Another technique is the CatWalk™ system for gait analysis in laboratory rodents, which enables easy visualization of foot floor contact and automates the quantitative analysis of many gait parameters.

Infrared Thermography

Advances in computer vision contribute to automatic classification of behavioral patterns. However, certain behavioral states of animals cannot be distinguished visually. It is known that animals exposed to stressors show various physiological responses, including changes in body temperature. Monitoring the body temperature of an animal may thus yield additional information about the state of the animal. The body temperature of a freely moving animal can be monitored by means of an implanted transmitter (biotelemetry). Infrared thermography offers a noninvasive alternative. A thermographic camera converts thermal radiance, emitted by the body, into a video image. Thermographic cameras have evolved significantly in recent years. Earlier models required cooling with liquid nitrogen, which was very cumbersome and severely limited the positioning and handling of the device. Modern thermal imagers are noncooled, can be calibrated in degrees Celsius (accurate to 0.1°C) and produce a standard video signal, allowing computerized image processing. The latter is necessary to extract the temperature values pertaining to the objects of interest, that is, the animal's whole body or a specific body part. Combining such a camera with a video tracking system results in a unique integrated and noninvasive system that can track movement, behavior, and body temperature simultaneously. This technique has been used successfully in studies on rats.

Using Computers to Analyze Animal Sound

The study of animal vocalizations has also benefited from advances in digital technology. Modern PC-based sound analysis tools, such as SIGNAL (Engineering Design, *www.engdes.com*) or Avisoft-SASLab (Avisoft Bioacoustics, *www.avisoft.de*) help researchers to unravel the acoustical structure of an animal's vocal repertoire and to study the relationship between experimental variables and vocal behavior. Vocalizations can be quantified by means of, e.g., call *intensity* (loudness), *pitch* (sound frequency), number of calls, mean length of calls, and number of notes per call. Calls can be visualized as a *sonogram*, a graph in which sound frequency, intensity and temporal structure are integrated, and spectral analysis can be used to classify specific call patterns. For those who do not need such a level of detail, UltraVox® offers a low-end alternative. This tool serves as an "acoustical event recorder": It simply records the onset and offset of calls that occur within a certain frequency band and above a preset amplitude threshold. This program is used, for example, in lab studies to count the number ultrasonic calls produced by rat pups when separated from their mother.

Data Analysis

Once behavioral data have been collected, software tools can be used for analysis. For instance, The Observer includes functions for the creation of time–event tables and graphical plots, calculation of frequency, and duration of behaviors (how often did a behavior occur, how long did it last), lag sequential analysis (how often is behavior A followed by behavior B) and reliability analysis (how does the data set of one observer compare to that of another). EthoVision turns movement tracks into a wide range of time-or distance-related parameters (e.g., distance traveled, velocity, number of entries into a certain area) or path shape descriptors (e.g., turning rate, heading, meander). Furthermore, it is able to detect specific individual behaviors or social interactions. Programs like The Observer and Etho-Vision can also work together: For instance, EthoVision can serve as a preprocessor for The Observer by automatically scoring very long video recordings (up to 60 hours or more) and generating a shortlist of episodes in which certain animal activity occurs, so that the researcher needs to inspect only those with The Observer. One can also export the data to other programs, that is, general-purpose spreadsheets (e.g., Excel) or statistics packages (e.g., SAS, SPSS) or specific behavioral analysis programs such as MatMan™ (to calculate behavioral profiles of individuals or dominance hierarchies in groups) or Theme™. The latter program searches sequential data sets for hidden patterns, which cannot be fully detected by visual inspection or standard analytical methods. It considers not only the order and relative timing of behavioral events, but also their hierarchical organization. It is particularly good at detecting small "signals" buried deep within large amounts of "noise."

Computers continue to create new opportunities for animal behavior research. They take away the burden of routine observations and constantly offer new measurement and analysis methods that allow researchers to widen their scope. In the coming years we will see the arrival of more advanced computer vision systems capable of recognizing more and more body postures and behaviors, and systems that integrate the measurement of behavior and physiology. From a technical perspective, the future of animal behavior research is very bright!

Up to date information about The Observer, EthoVision, UltraVox, MatMan and Theme can be found on the website of Noldus Information Technology, *www.noldus.com*.

Further Resources

Lehner, P. H. 1996. *Handbook of Ethological Methods*. 2nd edn. Cambridge, UK: Cambridge University Press.

Magnusson, M. S. 2000. *Discovering hidden time patterns in behavior: T-patterns and their detection*. Behavior Research Methods, Instruments & Computers, 32, 93–110.

Martin, P. & Bateson, P. 1993. Measuring Behaviour: An Introductory Guide. 2nd edn. Cambridge, UK: Cambridge University Press.

Noldus, L. P. J. J., Trienes, R. J. H., Hendriksen, A. H., Jansen, H. & Jansen, R. G. 2000. *The Observer Video-Pro: New software for the collection, management, and presentation of time-structured data from videotapes and digital media files*. Behavior Research Methods, Instruments & Computers, 32, 197–206.

Noldus, L. P. J. J., Spink, A. J. & Tegelenbosch, R. A. J. 2001. *EthoVision: A versatile video tracking system for automation of behavioral experiments*. Behavior Research Methods, Instruments & Computers, 33, 392–414.

Lucas P. J. J. Noldus

■ Methods
Deprivation Experiments

In its general form, a *deprivation experiment* (DE) involves housing research subjects in situations that lack one or more sets of environmental elements that would generally be found in a "normal" environment. The purpose of this type of experiment is to assess the impact of environmental deprivation on the behavior and/or physiology of the organisms involved. The type of deprivation differs based upon the experimental purpose and may be primarily sensory (e.g., reducing the length of a light cycle), social (e.g., rearing alone or with fewer than normal number of conspecifics), or both. Clearly, when social deprivation takes place, sensory stimulation is also necessarily reduced. The DE typically involves several separate groups of subjects that differ with respect to the degree or nature of the deprivation under study. After some predetermined period of time, the behavior of the deprived subjects is compared to that of controls housed in a "normal" environment. When the environment altered is the one in which the animal is reared, the research is concerned with studying the effects of early experience on adult behavior. Deprivation experiments have been carried out on animals ranging from invertebrates to primates including humans.

The DE gained experimental importance during the initial phase of what is called the "Nature–Nurture" controversy. Beginning in the 1930s, many students of behavior, like the prominent ethologist Konrad Lorenz, became very interested in the question of whether a certain behavior was "innate" or "learned." In this theoretical scheme, behaviors classified as innate were assumed to have been genetically "wired in" by the process of evolution and required only the maturation process to unfold to become part of an organism's repertoire. On the other hand, learned behaviors, like returning to a specific location for food, were understood as being acquired during the course of an animal's life and were "shaped" by contemporary experiences such as reward or punishment. Lorenz and his followers were initially convinced that all behaviors could be placed into one category or the other. Further, he believed that the study of innate behavior was the proper province of biologists and learned behavior was the purview of comparative psychologists.

Therefore, in this historical context, if a researcher questioned whether the form of a particular bird song was innate or learned, they would likely employ a DE. In this case the researcher might incubate fertilized eggs and raise young hatchling birds in an environment that systematically eliminated the sound of adult species-typical song as it would be experienced by normally raised birds. When the hatchlings reached the age when the song is normally produced they would be released from the deprived environment and their songs would be compared to that of controls. If the bird song of the deprived birds was equivalent to that of normally reared birds, the conclusion would be that the behavior was innate. If the song was nonexistent or significantly different from controls, the conclusion would be that the behavior belonged to the category of behavior that required learning and experience in order to develop. To illustrate how important these questions were seen to be at the time, consider the design of an experiment concerned with the question of innate behavior in humans. In 1935 Wayne Dennis, a prominent behavior analyst, attempted to determine whether reaching, standing, and sitting behavior in human infants was innate or required learning to become established. To conduct this experiment Dennis recruited a pair of female fraternal twins from a local family. The children were removed from their parents at 36 days of age and housed in an experimentally controlled bedroom until the fourteenth month of life. During that time the children lived in an environment that was devoid of graspable objects and toys, and they were dressed in such a way to prevent exploration of their garments. They were also kept on their backs to prevent practicing standing or sitting. Further, the caretakers did not handle (except for required bathing and feeding), speak to, or play with the children. The children were tested for the presence of the targeted behaviors at points in time when the behaviors would be present in normally reared children. Dennis found that the appearances of the behaviors were all significantly delayed but developed rapidly once the restrictions were removed. This evidence was interpreted as supporting the contention that the behaviors were not instinctive, but required learning to develop. Over time, the use of the DE changed for three reasons: issues of experimental interpretation, a change in the concept of the nature–nurture controversy, and ethical concerns about the impact of deprivation on the experimental subjects.

First, researchers began to appreciate that there were problems inherent in the interpretation of the results of the DE that interfered with the clear black and white conclusions originally assumed possible. For example, if a particular behavior were present after deprivation rearing, one could not so easily assume that the behavior was innate. Rather, the specific deprivation that was used may not have excluded the type of environmental input or learning experience that was necessary for the behavior to develop. On the other hand, if a specific behavior were absent following deprivation, it was possible that instead of it being properly classified as requiring learning and experience, it might be the case that the deprivation had the effect of disrupting the physiology necessary for the development of the behavior. In other words, the rearing may have resulted in atrophy of the underlying mechanisms necessary to produce the behavior in question. For example, it has been demonstrated that rearing primates in low levels of light results in the partial destruction of the neurological systems that support sight. An additional problem concerning the interpretation of DEs is that the behavior changes seen after DE might not implicate the origin of the behaviors but be a function of the discrepancies in novelty, complexity, or stimulus intensity between rearing and postrearing stimulation. When these discrepancies are large, as they are in the DE, the behavioral changes may be due to what has been referred to as "emergence trauma."

Second, researchers began to emphasize the point that the question of whether a specific behavior was learned or innate was perhaps the wrong, or at least an incomplete, question. Surely the behavior doesn't exist in the zygote. As it became clearer that genetic, learning, and experience factors interacted with one another to produce behavior, the focus shifted from innate versus learned to how a behavior develops. In this context, if rearing an organism in a DE altered its behavior significantly, this was seen as potentially shedding light on the developmental process rather than on just the innate or learned quality of the behavior. For example, Harry F. Harlow (Harlow & Harlow 1965) reared rhesus monkey infants in environments which deprived the animals of contact with mothers and peers for periods of time up to one year. Control monkeys, while raised in the laboratory, were given extensive experience with mothers and age-mates. Harlow found that upon release from the deprived environment, compared to the controls, the deprived animals exhibited higher levels of aggression and self-directed behavior, lower levels of curiosity and affiliative behavior, and in most cases were unable to execute appropriate mating, social, or maternal caregiving behaviors. Again, at this point in time, these data were not seen to provide definitive information about the nature and nurture issue, but about the process of development in general. In addition, because experiments of this sort frequently produced behavioral pathologies that were both striking and

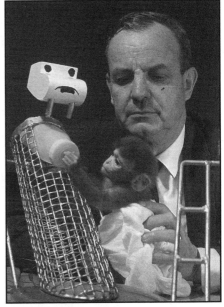

Harry F. Harlow with a baby rhesus monkey and its wireframe mother.
© *Nina Leen / Time Life Pictures / Getty Images.*

similar in appearance to several behavioral pathologies (e.g., depression and anxiety) that affected humans, the DE experiment came to be used as a tool to create animal models of human abnormal behaviors.

Finally, beginning in the 1970s and continuing into the present, some researchers, animal protectionists, and philosophers have questioned whether the balance between the scientific contributions of the DE and the harm to the animals exposed to severely deprived environments justifies their continued use (see Singer 1975). These concerns were further emphasized by findings demonstrating that many animals were clearly capable of experiencing pain, distress, and suffering. While the DE continues to be used in the tradition of the Harlow experiments discussed above, the length of the deprived exposure is at least briefer.

See also Welfare, Well-Being, and Pain—*Psychological Well-Being*
Nature and Nurture—*Baldwin Effect*
Nature and Nurture—*How to Solve the Nature/ Nurture Dichotomy*
Nature and Nurture—*Nature Explains Nurture: Animals, Genes and the Environment*

Further Resources

Harlow, H. F. & Harlow, K. K. 1965. *The affectional systems.* In: *Behavior of Nonhuman Primates, Vol.II* (Ed. by A. M. Schrier, H. F. Harlow, & Stollnetz), pp. 287–334. New York: Academic Press.
Lorenz, K. Z. 1965. *Evolution and the Modification of Behavior.* Chicago: University of Chicago Press.
Singer, P. 1975. *Animal Liberation.* New York: Avon Books.

John P. Gluck, Tony DiPasquale, & Charlene McIver

■ Methods
DNA Fingerprinting

Observations are an important method used to determine animal behavior, but animals do not always cooperate in making their behaviors known, especially in their natural habitats. Mammals are often nocturnal, living under cover of darkness. Birds take flight when disturbed, leaving the observer behind. Amphibians and fish hide underwater, while reptiles slink from view into protective burrows. Insects and other invertebrates are often small and live in habitats that are difficult to navigate such as soil or water.

Since observations can be difficult to obtain, behaviorists need to be part scientist and part detective. Like a good forensic expert, they must use all the tools at their disposal. One of the more useful techniques developed in the last few decades is DNA fingerprinting and its associated genetic techniques. DNA can provide behaviorists with information about parentage and mating choices, family and kin relationships, and population dynamics and movement patterns.

Even though blood and semen are often mentioned as sources of DNA, nuclear material can be obtained from any body cell. Behaviorists take advantage of this fact and have become quite creative at obtaining cell samples, often without disturbing the animals they are studying. Skin samples, including skin biopsies, follicle skin on hairs or feathers, fish fins or shed snake skins are good sources of DNA. Saliva contains cells from inside the mouth and primatologists have been able to recover DNA from food wadges, the spit out remains of fruit. The intestinal tract is continuously sloughing old cells; therefore, feces are also a good, if a not so appealing, source of DNA.

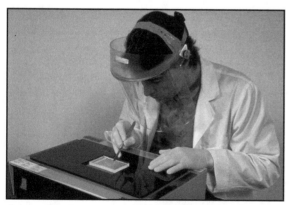

A lab scientist prepares DNA samples. After the DNA is cut, the pieces are separated by size in a gel through a process known as electrophoresis.

Once tissue has been collected, the DNA is extracted from the cells. To do this, the cells are ruptured with a soap-like solution. Then, using a double layer of liquids, the DNA is separated from the other cellular parts. If the sample is small, the DNA can be amplified or copied using a technique known as PCR (Polymerase Chain Reaction).

After the extraction process, the DNA looks like long stringy fibers. At this stage the DNA does not convey any information. To get a picture of the structure of the nucleotide bases making up the DNA, the long DNA strands are cut with a molecule called a restriction enzyme. There are many different restriction enzymes which cut the DNA at specific locations. After the DNA is cut, the pieces are separated by size in a gel through a process known as electrophoresis. At this point the DNA is spread out in an organized column, but it is invisible to the naked eye.

To make the DNA visible, it is exposed to either a radioactive or phosphorescent probe. There are many different probes that attach to regions of DNA known as minisatellites or microsatellites. These regions are made up of repeated sequences of nucleotide base pairs which are not actually involved in the production of proteins, but are more variable than the gene regions of the DNA, making it easier to tell individuals apart. After the probes attach to specific regions of the DNA, x-ray film is used to make the radioactive probe visible (phosphorescent probes glow). The pattern of visible marks made by the probe are the fingerprints. Since these regions are inherited, they can be used to determine parentage and genetic relatedness.

Probably the most surprising discovery behaviorists have made with DNA fingerprinting involves birds. Most birds are socially monogamous, having only one mate during the breeding season. Before the availability of DNA fingerprinting, behaviorists had to assume that the mated pair were the parents of the offspring. Unless extrapair copulations were observed, the male helping at the nest was considered the father. Even if extrapair copulations were observed, paternity was unknown. By the late 1990s, 72 species of birds had been studied using DNA fingerprinting. The number of nestlings from extrapair copulations varied greatly among species: black vultures = 0%, red-winged blackbirds = 28%, and fairy wrens = 76%. Seventy-four percent of the species studied showed some degree of extrapair paternity.

Dr. Alec Jeffreys, creator of the DNA fingerprinting process, examining a DNA fingerprint in his office at Leicester University, England.
© Terry Smith / Time Life Pictures / Getty Images.

Since most animals are not monogamous, DNA fingerprinting is useful in determining female choice and paternity. Behaviorists have identified fathers of calves born to promiscuous humpback whale females. They have discovered that black bear cubs born in the same litter can have different fathers. In vampire bat colonies, behaviorists once assumed dominant males fathered most of the offspring, but found they only fathered 45%. Fingerprinting analysis was used to determine paternity in the northern water snake, a species that mates in large groups, and also revealed that green sea turtle eggs were fertilized by as many as four males.

Behaviorists using DNA analysis to sort out relationships of animals living in groups, found that some animals give up their reproduction to help others raise kin. Social wasps are more related to the queen than they are to each other, so they repress each other's reproduction to help raise offspring of the colony. They also found that some individuals become helpers so they have a chance to mate secretly. Dunnocks are a small sparrow-sized birds in which two males may be found taking care of the nestlings, the primary mate and a helper male. Behaviorists discovered with DNA fingerprint that males helping at the nest fathered some of the nestlings.

DNA is also used to assess the genetic diversity of populations and determine movement patterns between populations, a particular concern these days with shrinking habitats and stressed populations. Behaviorists have found that pilot whale males mate outside their pod (a good way to insure genetic diversity in the population). They also found that horseshoe crabs move at a rate of five individuals per generation between Florida Atlantic and Gulf populations, and that chimpanzee females in isolated groups in Guinea mate with neighboring males, which prevents inbreeding.

Although observations are essential in the study of animal behavior, new tools such as DNA fingerprinting are allowing behaviorists to discern behaviors they would have difficulty observing in the wild. As genetic analysis becomes more available and easier to conduct, it can only lead to additional exciting discoveries in the future.

See also Methods—*Molecular Techniques*

Further Resources

Burland, T. M. & Wilmer, J. W. 2001. *Seeing in the dark: Molecular approaches to the study of bat populations.* Biological Reviews, 76, 389–409.

Cockburn, A. 1998. *Evolution of helping behavior in cooperatively breeding birds.* Annual Review of Ecology and Systematics, 29, 141–177.

Fridell, R. 2001. *DNA Fingerprinting: The Ultimate Identity.* New York: Franklin Watts.

Kempenaers, P. M. 1998. *Extra-pair paternity in birds: Explaining variation between species and populations.* Trends in Ecology and Evolution, 13, 52–58.

Kirby, L. T. 1990. *DNA Fingerprinting: An Introduction.* New York: Stockton Press.

Mitton, J. B. 1994. *Molecular approaches to population biology.* Annual Review of Ecology and Systematics, 25, 45–69.

Pamelo, P., Gertsch, P., Thoren, P. & Seppa, P. 1997. *Molecular population genetics of social insects.* Annual Review of Ecology and Systematics, 28, 1–25.

Westneat, D. F. & Sherman, P. W. 1997. *Density and extra-pair fertilizations in birds: A comparative analysis.* Behavioral Ecology and Sociobiology, 41, 205–215.

Susan U. Linville

Methods
Ethograms

There is a dazzling array of animal behavior free for the taking. Watch awhile and it is easy to become entranced with the impassioned dramas that mark life in the animal kingdom. When a person watches animal behavior long enough to get to know it well, they have an informal vocabulary of the species' body language in mind. Formally, this is known as an ethogram. The *ethogram* is a dictionary that names and defines all the individual behaviors that make up a species' body language. Like learning a language, building an ethogram is an ongoing joy.

People use ethograms all the time. Anyone who knows sports has an unwritten ethogram of all the basic moves in mind, be it football or gymnastics. Professional gamblers and casino security have thorough mental ethograms of the various games of chance. In behavioral research, the ethogram is assembled at the start of a study as a collection of behaviors to collect as data, and is an essential and basic analytical tool.

The Ethogram as Dictionary of Body Language

As a dictionary, an ethogram identifies and provides standard terms for species-typical behaviors. It lists behaviors by name along with descriptions or pictures that allow someone who has never seen the species to get a clear idea of each behavior. An ethogram creates a common language that enables researchers to talk about the same behaviors the same way every time, and equips them with a shorthand way of taking data.

Ethograms are usually developed to serve a particular research project so they vary in size and detail (*http://www.lpzoo.org/ethograms*). A complete ethogram is a thorough inventory of *all* the typical behaviors every species member can display, regardless of gender or age and exclusive of idiosyncrasies. Most are less-detailed, partial ethograms restricted to behaviors pertinent to a particular study. The ethogram is built from as much observation of a species' behavior as necessary to complete an ethogram with the level of detail appropriate to the study.

How to Recognize Behavior for the Ethogram—Behavior Units

An animal never stops behaving. Different activities occur and blend into one another: A dog flops down, yawns, sleeps, wakes up, stretches, and so on. Ethologists call this the *constant stream of behavior*. The task of research is to subdivide the constant stream of behavior into meaningful units to collect meaningful data.

Animals behave with a certain consistency. This consistency stems in part from the repeated occurrence of small, brief behaviors called behavior units, events, elements, or action patterns. Behavior units are to the stream of behavior what words are to conversation. Just as words remain the same but can communicate an infinite number of ideas depending on which are combined, different combinations of behavior units create species-typical behavioral states. Behavioral states are broad, ongoing activities like feeding and traveling. Behavior units are strings of activity that make up behavioral states or become behavioral states when they are exhibited for a period of time.

A behavior unit is recognized every time because it has the same basic physical characteristics regardless of who is performing it. For example, a human behavior unit is walking. Walking can be described by how it looks, "upright bipedal locomotion performed by one foot alternately placed in front of the other to move forward in space," or by what it does, moves someone in space. No matter who walks, we recognize walking. Walking is typical of humans and would go on a human ethogram as a type of locomotion.

As you observe your study species closely and note a recurring behavior, record the physical characteristics that let you to recognize it as unique from other behaviors. For example, human walking is always characterized by moving in space. If someone is not moving in space, they can't be walking. To be walking instead of skipping, walking involves specific leg movements with a particular flexion of the knees. Universal characteristics form the basic description of a behavior unit and are the components that would comprise the creation of virtual behaviors on computers.

Basic descriptions are supplemented with obvious variations in the way the behavior is performed. For example, a person can power walk at speed in a straight line or meander along a twisting path. Basic descriptions are refined with details like characteristic velocity, duration, amplitude, orientation, direction, physical shape or topography, location, or impact on the environment until the level of accuracy the study requires is achieved.

As familiarity with animal behavior improves, it becomes increasingly easy to see universal patterns, whether in a dog or a wild dolphin. It's fascinating! But, breaking the stream of behavior into separate pieces can also be a challenge. For instance, some dolphin behavior occurs in obvious units, like breaching. Others are vague. A "steep dive" differs from a "forward progress dive" because the tailstock is visible at the water surface. However, whether the dolphin uses the dives differently, and therefore the dives should be separate on an ethogram, must be determined by research.

Another challenge is called "lumping versus splitting." For every two people, there are two opinions about what constitutes a species-typical behavior. One person will say, "That's a run and *that's* a run and so is that third one. Okay, add 'run' to the ethogram." The other person will say, "No, the first run was traveling. That second run was a leg–kick run, a threat. That third run was a play–invitation run. Add three different runs to the ethogram." The first person is a *lumper*, consolidating all similar behaviors into one big behavior. The second person is a *splitter*, separating similar behaviors based on more subtle differences. Ideally, one should lump or split animal behavior the way the animals do. Because lost data cannot be retrieved, one should split initially and lump later in analysis.

An ethogram should be exhaustive and based on mutually exclusive behaviors. An exhaustive ethogram is not exhausting to construct. It is one that accounts for every behavior during data collection by having a behavior or category for everything. If the animal can disappear from view, the exhaustive ethogram has a "not visible" category. If social behavior is of interest, it includes a "nonsocial behavior" category to account for solitude. Mutually exclusive behaviors are units that are discrete and distinct from other behaviors, just as words are individual units of language with distinct meanings. Mutually exclusive units do not overlap (although they can be simultaneous). There should be no confusion about which to use, such as the mutually exclusive behaviors of sneezing and coughing.

How a species' stream of behavior is broken into its constituent elements has a profound influence on the outcome of the research,* and the many ways to accurately subdivide the stream depends on the goals of the research, the study species, and the researcher. Familiarity with recurring patterns is a direct result of the amount of time spent observing. Observe extensively. Behavior units are increasingly easy to see and verify when the study species is repeatedly observed in many contexts. The validity of any particular subdivision should be judged by how much it increases our knowledge of the causes, functions, evolution, and ecological role of behavior.

How to Collect Information for the Ethogram

Go out, observe behavior closely, and describe your observations clearly. Ethogram construction is exploratory, so observe as often and whenever possible throughout the day or night to get the broadest view of the species' behavior; otherwise, there is the risk of disconnected results when an observer who is unfamiliar with a species narrows his or her

---

* The question of whether behavior units are natural or figments of observer imagination is far from trivial but beyond the scope of this essay.

view to a single question. Schedule at least one full day or night of observation to see how animals do different things at different times. Because focused observation can be demanding (especially for inexperienced observers), initially schedule observations for short periods to remain mentally fresh. Stay longer and watch when the animals are doing interesting things, or leave if they are inactive. Let the behavior of greatest interest determine when, how often, and how to observe: Schedule observations to the period when the species is most active for the goals of the research.

Observe animals from a position that does not disturb them. Remain hidden or get the animals used to your presence (i.e., habituate them). The study species and budget will determine the observational platform (e.g., land, boat, air, underwater, treetop, cage, or poolside). Develop the ethogram from this vantage.

Begin by watching the animals in a free-form way, an unstructured study called "ad libitum" or ad lib observations. Describe and record everything they do. Reject the idea that only active animals are behaving. Immobile or static behavior often contains important information, such as the sick animal that does not move normally.

There are three ways to collect descriptive data from ad lib observations for the ethogram. One, write or type the animal's behavior as completely and accurately as possible during observation. Choose this method depending on the complexity of the behavior being described. It works well for simple behavior like hopping in toads, but not for complicated behavior like play in young primates.

Two, describe behavior verbally into a tape recorder. Dictation is good for fast-moving or complicated behavior because the animal can be observed continuously. With practice, an observer can describe behavior as fast as it occurs. Dictation is accurate, inexpensive, and maximizes observation time, but is labor-intensive due to transcription time. Transcribe observations as soon as possible while they are fresh in your mind.

Three, videotape behavior. This method is ideal when the behavior needs to be reproduced for further analysis or shown to other people, but may be too complicated for ethogram building. Animals use a limited number of behavior patterns. Once the same behaviors are seen repeatedly, videotaping becomes a time-expensive way to see new behaviors. Remember that cameras intimidate many animals. Give them time to habituate.

Which method to use depends on the amount of time allotted for this phase of investigation. Use dictation unless time is not a factor. Some behaviors occur frequently, and a good description

Augment the ethogram with illustrations, photographs, or film images whenever possible. This pair of polar bears illustrates several play behaviors, including, left to right from top; (1) head shake, erect ears, and look away; (2) approach and playface; (3) sit and mouth wrestle; (4) play pouncing; (5) arm wrestle; (6) bipedal stance, playface, and cuffing. Courtesy of Ann Weaver.

is quickly obtained. Other behaviors are rare and take many sessions to recognize and record. Note any idiosyncrasies; ethograms typically exclude behaviors peculiar to one animal. Most of the behaviors have been observed when there are few new behaviors per unit of observation time, an asymptote that can take tens to thousands of observation hours to reach. Stay aware of your thoughts during observation. Separate observation from interpretation. Record fleeting impressions during observation as potential sources of valuable insights and hypotheses, but keep them separate from direct observations in some way (perhaps by bracketing them). Resist the temptation to continue refining a complete ethogram (one that can be used to answer research questions) instead of getting on with data collection.

How to Describe Behavior Units— Description versus Interpretation

Descriptions have to be accurate and detailed enough so that someone else can "see" the behavior, as in a capuchin tail shiver: "rapid, small-amplitude shudder of the tail." Often, simple descriptions suffice, as in bonobo grooming: "close visual inspection and manual manipulation of hair or body parts." To avoid redefining standard terms in every description, name typical postures and body parts at the front of the ethogram and use these terms in descriptions. For example, name the part of the dolphin's body between its forehead (melon) and dorsal fin, the "foresection." Use the term "dorsal position" to refer to the typical dolphin posture of back toward the sky and stomach toward the seabed. Then, use them to streamline descriptions, as in describing a darting surface: "a sudden upward and outward thrust of the foresection over the water surface by a small animal abruptly surfacing in the dorsal position."

To construct descriptions of complicated behaviors, dictate events like a sportscaster objectively dictating the events of a game to a radio audience. Or describe the behavior as it looks frame-by-frame in film. Describe each step in a behavior sequentially, such as "this happens first, this happens second," and so on. Move a model of the animal through each behavior unit, describe this series, and the description is complete.

If the description is lengthy, summarize the behavior in a single statement at the beginning and add details after that. Be concise and objective. Use language carefully. Consider the inferences based on changing one word: "The parrot raised her feathers *to* draw attention to herself" versus "The parrot raised her feathers *and* drew attention to herself." The first statement involves an interpretation that is hard to verify. The second involves objective description. Exclude opinion, for example, "That looks like fun!" Otherwise, the real purpose of the behavior might never be understood. Description requires accuracy, objectivity, and precision.

Behavior can be described physically, functionally, or both. A physical description of behavior says what the behavior looks like: "The dog is on its side with its eyes closed." This is describing behavior by its operation or structure and using an empirical approach. A functional description of behavior says what the behavior is for or how it serves the animal: "The dog is resting." This is describing behavior by its consequences and using a functional approach. Ethograms often end up combining types of descriptions because it can be hard to "see" a purely empirical description.

The example of human walking was a physical or empirical description of the motor pattern of walking that excluded any explanation of *why* the human was walking; for example, going for coffee or leaving in a huff. The operational description of a behavior only lets us visualize what the behavior looks like, not what it is for. When unfamiliar with a species, describe behavior physically without interpretation. As familiarity increases and understanding is gained about *why* a behavior is performed, interpret it by providing a functional description.

Say a researcher were examining wolf pups at a den, and a growling adult approached, staring eyes in slits, ears forward, muzzle wrinkled, lips withdrawn, teeth bared and clenched. (A picture would clarify this description immensely.) The adult wolf is clearly upset, but an operational description only portrays what the wolf looks like, not what she is doing. The researcher could scan her ethogram and say calmly, "Ah, yes, wrinkled face behavior," without interpreting it as a threat. If instead she gulped "threatening wolf" as she carefully backed away, she'd be describing how the behavior served the wolf, that is, functionally: "Back off or I attack." If this behavior were only described functionally on the ethogram as a "threat face," it carries the assumption that all readers know what a threatening wolf looks like. Rather than assume, provide an accurate and precise description of what the behavior looks like, that is, its motor pattern.

Name behavior units as they are recognized during observation, using shortened descriptions when possible to give an immediate but general idea of the behavior without interpretation, for example, body scratch or tailslap. Write down the first name that occurs when the behavior is recognized as a unit. It can be refined later, but the most accurate recording is done on the scene. Names that are instant descriptions improve the accuracy of data collection because they are the easiest to recall. Since data collection is likely to involve a shorthand way to record behaviors (like two- or three-letter abbreviations or codes), try to name behaviors with abbreviations in mind. Never use the name of the behavior in the description to define it. Use other terms.

How to Organize the Ethogram— Functional Categories

Organizing behavior units on an ethogram is like starting dinner by getting all the meat items together and all the salad items together, and so on. Organize behaviors into an actual ethogram by putting them into functional categories. Functional categories reflect the basic things animals do—eat, sleep, mate, groom, fight, and socialize. See the following table for examples. Put all the behavior units that relate to finding, capturing, preparing, and eating under *Feeding*. Group all behaviors that relate to staying clean under *Self-Maintenance*. Use the functional categories that most accurately reflect your study species. You may not need them all. Behavior whose functions are unknown can be categorized by similar motor patterns. For example, dolphin behaviors associated with forward movement can be categorized as *Types of Swims*.

Animal behavior is a dynamic science where new discoveries are made regularly, so describe and interpret with care. Each researcher must scrutinize his work for subjectivity, misleading language, and the inappropriate imposition of human characteristics.

Conclusion

Ethograms bring great benefit. This is because it *is* confusing when an observer first watches animal behavior. It is *not* always obvious what animals are doing, or why. If new observers begin with a focus on ethogram construction and let the rest of the mind go, something wonderful happens: They develop a tremendous amount of understanding of the species. The founders of ethology called this one's *gestalt* for the species, a deep understanding of its behavior as vastly more than the sum of its parts. In the restricted regimen of ethogram observations, natural curiosity takes over. For the lucky ones, it is embraced by endless wonder.

Behavioral Categories for Grouping Behavior Units on an Ethogram

Functional Category	Description
Agonistic Behavior*	Negative social interaction connected to contest or conflict. Includes noncontact conflict, avoidance,aggression, fighting, dominance and submission, defensiveness, and freezing.
Allelomimetic Behavior*	Mutual mimicking or the tendency to behave as a group. Includes schooling among fish, flocking among birds and mammals.
Caregiving or Epimeletic Behavior*	Care given to dependent young (parental care, maternal care, paternal care, nurturant, attentive, or protective behavior) or to injured or distressed animal (standing by or succorant behavior).
Care-Soliciting or Et-Epimeletic Behavior*	Activities designed to obtain care (young calling for food or throwing a tantrum) and/or attention (courting pairs). The overall term is et-epimeletic since this behavior occurs among adults as well.
Communicative Behavior	Signals sent or received between individuals. Includes acoustic (vocalizations), visual (body postures), and chemosensory (pheromones) senses.
Developmental Behavior	Changes in behavior over time. Usually connotes youth.
Eliminative Behavior*	Defecation and urination.
Ingestive Behavior*	Eating or drinking. Includes drinking mother's milk (nursing), crop milk, or regurgitants; looking for food (foraging, hunting); finding, capturing or gathering, preparing, and feeding; chewing the cud.
Investigative Behavior*	Exploration or sensory examination.
Locomotory Behavior	Activities resulting in movement through space.
Maintenance or Comfort Behaviors	Grooming, preening, stretching, yawning, shaking, scratching, sneezing, sandbathing, dusting, and changing position.
Marking Behavior	Leaving a physical or chemical sign.
Play Behavior	Immaturely- or loosely-performed adult behaviors.
Reproductive Behavior*	Courtship, sex, and pregnancy.
Stationary and Resting Behavior	In-place postures. Includes supine posture, reclining,lying down, sitting, crouching, squatting, standing still.
Shelter-seeking Behavior*	Finding or constructing optimum environmental conditions.
Social Behavior	Interaction between two or more individuals.

*From Scott (1958).

Begin a behavioral study by just learning to tell the animals apart and assembling an ethogram. The animals themselves will reveal what to study. No matter how well an observer gets to know a species' behavior, the animals will *always* reveal a new secret. Nothing is more satisfying than the study of animal behavior.

See also Communication—Vocal—*Social Communication in Dogs: The Subtleties of Silent Language*
Methods—*Zen in the Art of Monkey Watching*
Zoos and Aquariums—*Studying Animal Behavior in Zoos and Aquariums*

Further Resources

Drummond, H. 1981. *The nature and description of behavior patterns.* In: *Perspectives in Ethology*, Volume 4. (Ed. by P. P. G. Bateson & K. Klopfer), pp. 1–33. Cambridge, MA: Plenum Press.

Lehner, P. N. 1979. *Handbook of Ethological Methods.* New York: Garland STPM Press.

Müeller, M., Boutiere, H., Weaver, A., & Heut, N. 1998. *Ethogram of the bottlenose dolphin, with special reference to solitary and sociable dolphins.* Vie et Milieu (Life and Environment), 48, 89–104.

Scott, J. P. 1958. *Animal behavior.* Chicago: University of Chicago Press.

Weaver, A. 1997. *Infant development.* In: *The Care and Management of Bonobos* (Pan paniscus) *in Captive Environments.* (Ed. by J. Mills, G. Reinartz, H. De Bois, L. van Elsacker, & B. van Puijenbroeck). Milwaukee, WI: Zoological Society of Milwaukee County.

Weaver, A. 2003. *Conflict and reconciliation in captive bottlenose dolphins,* Tursiops truncatus. Marine Mammal Science, 19, 836–846.

Weaver, A. & de Waal, F. 2002. *An index for measuring relationship quality based on attachment theory.* Journal of Comparative Psychology, 116, 93–106.

Ann Weaver

■ Methods
Molecular Techniques

Laboratory-based molecular genetic technology increases the ability of researchers to infer behavior in many species. Information held in the chromosomes allows a researcher to access hidden moments in an organism's life such as a secret copulation between a male and female superb fairy wren or the destruction by a worker honeybee of another worker's young. These tools rely on the fact that genetic material is inherited, and therefore the similarity of the genetic code between two individuals indicates their relatedness, a measure of how closely related they are to one another. Because much of animal behavior theoretically involves decisions based upon kin relationships (e.g., parent to offspring) these tools have proven useful in untangling how these choices are made. The inherited nature of genetic material also allows for relationships among populations and species to be discerned and information to be gathered regarding how individuals move among populations and how behavior is subdivided. Finally, molecular genetic technology can be used in the search for the genes responsible for behavior. Ultimately, most successful studies use a combination of these techniques with observation and experiment.

The main genetic material used in molecular technology is deoxyribonucleic acid or DNA. *DNA* is the cell's template for creating the proteins that build and run the machinery of the organism. It is primarily composed of four nucleic acids: adenine (A), guanine (G), thymine (T), and cytosine (C). These nucleic acids are arrayed along the length of a DNA strand, forming a code or sequence. Nucleic acids on one strand pair with complementary nucleic acids (A with T, G with C) on another strand, forming a double helix. Particular sequence locations in the genome are called *loci* and variations in particular loci between individuals are called *alleles*. Since the molecules that make up DNA are inherited (usually half from the mother and half from the father), relatives share a larger proportion of the code, and more similar alleles, with one another than do nonrelatives. Similarly, members of the same population share more alleles with one another than other populations, and members of a species share more alleles with other individuals of the same species than with individuals of another species.

For molecular genetic analysis, a sample of DNA or protein is extracted from material such as blood, tissue, hair, or feces. This sample is then subjected to one of several forms of available analysis. Early on, researchers made comparisons between proteins, or allozymes, extracted from the tissue. Though allozyme analysis is still used, the limited amount of variability inherent in these markers generally makes them less useful than DNA analyses. The most detailed analytical approach available is the comparison of sequences of particular DNA loci. The loci used depend upon the question being asked since the mutation rates and the mechanism of evolution differs among loci. Researchers can use sequences from the chromosomes, thus accessing genomic DNA, and/or sequences from DNA found in the mitochondria. Mitochondrial DNA is generally inherited only from the mother and therefore possesses different information than genomic DNA. All the DNA found in the mitochondria is considered a single locus as opposed to genomic DNA, which has many loci. Since it is best to use several loci in most analyses, researchers tend to use mitochondrial sequences in studies where they also have access to genomic sequences. The expense in time and money of sequencing makes it less often used in the studies of animal behavior than other analysis techniques. It is possible, however, that in the foreseeable future full genome sequencing of individuals for screening variation in the genetic material will make sequencing a more often used tool.

A variety of alternative approaches to analyzing DNA, apart from sequencing, are presently available. These include the comparison among individuals of variable nucleotide tandem repeats (VNTRs), restriction fragment length polymorphisms (RFLPs), amplified fragment length polymorphisms (AFLPs), randomly amplified polymorphic DNA (RAPDs) and single nucleotide polymorphisms (SNPs). The most commonly used approaches in animal behavior are VNTRs (also called mini- and micro-satellites). These markers are repetitive sequences found throughout the genome that mutate rapidly and therefore allow for discrimination among individuals. VNTR alleles are codominant so, theoretically, both alleles for each individual show up in the analysis. Minisatellites represent several loci and, while the proportion shared indicates the level of relatedness, their patterns of inheritance are not well understood. Each microsatellite marker is a single locus piece of repetitive DNA, typically defined as four or more repeats (e.g. AGAGAGAGAGAG). The inheritance pattern of microsatellites is clearer than that of minisatellites.

The type of analysis used depends upon the level of comparisons (e.g., comparisons between species versus comparisons between individuals), the questions being asked, and the resources available. In pair-wise comparisons such as parentage and other kinship surveys, the technique of choice must exploit portions of the genome with high mutation rates so that differences can be resolved among individuals. Loci used in sequencing have a variety of mutation rates, and determining the appropriate loci to sequence may be expensive and time-consuming. Markers such as mini- and microsatellites tend to have high mutation rates because of their structure and therefore these markers provide the information needed for these studies. Conversely, these high mutation rates make these techniques less useful for analyses at larger scales, and therefore sequencing tends to be used for species comparisons. Further-more, the evolutionary processes involved in marker analysis are less clear than are the processes involved in sequencing, and thus, in studies where evolutionary process is being considered, such as examinations of species relationships, sequence data is much more useful. Since sample sizes tend to be smaller in species and population comparisons than in kinship analysis, the higher cost per sample for sequencing is less of a concern for these studies.

For all of these techniques, the analysis is not quite as straightforward as simply nu-merically counting up similarities and declaring individuals "sister" or "cousin," "related" or "not related" or even "same species" or "different species." This is because unrelated indi-viduals may share many of the same pieces of DNA by chance. In parentage studies, for ex-ample, low marker resolution or number limits researchers to excluding an individual as a possible parent (if there is a mismatch at some loci), whereas higher resolution allows in-cluding an individual as a possible parent. In determining these relationships, researchers must use probability theory and the distribution of similarities and differences in a popula-tion or species to determine what is the probable relationship between the two individuals. Statisticians and genetic researchers continue to develop ways of examining the probability that an individual will share the same genetic patterns, and there are several computer pro-grams already that have been developed to analyze the data.

One type of kin relationship that has been subject to particular scrutiny is parent and off-spring, especially in bird species. Since copulations are often unobserved, molecular markers allow researchers to infer what happens during some of these moments by indicating the adults involved in producing a particular offspring. Mini- and microsatellite markers have been the technique of choice for studying parentage, while the potential for SNPs, AFLPs, and sequencing in the elucidation of parentage has yet to be explored. Studies using molecular techniques in birds demonstrate that in many seemingly monogamous species, such as east-ern bluebirds, bobolinks and indigo buntings, individuals copulate regularly outside of their pair bonds, engaging in extrapair copulations or EPCs. These data have led researchers to view mating systems in bird species as individual decision making in light of cost and benefits rather than as characteristics of a species as a whole. This way of thinking combined with molecular approaches has been used, for example, to demonstrate benefits to multiple mating by female red-winged black birds and Mexican jays, such as access to more resources and extra provisioning and protection (mating areas) from extrapair males. In species that do not form pair bonds, such as those that establish leks (mating arenas), molecular data help test hypothesis about female choice. For example, a high proportion of female ruffs produce young sired by multiple males, perhaps to facilitate genetic diversification of their clutches.

Studies examining mating behavior using molecular techniques also continue to expand our understanding of reproductive decision in a variety of nonavian taxa. For example, these molecular approaches have helped researchers determine that hermaphroditic planarian worms trading sperm do not trade fertilizations equally, and that female grain beetles benefit from mating with several males, but not from mating several times with a single male. In species with multiple mating, the importance of mating order can also be evaluated using molecular markers, such as in spotted salamanders where the first males have the advantage.

Another area of study in animal behavior that has been greatly affected by molecular tech-niques is the study of the effect of kin relationships on social behavior. Theory suggests that kin should be more likely than nonkin to perform seemingly altruistic acts for one another. Since the actual pedigrees of populations of many species are unknown, molecular techniques allow researchers to estimate relatedness indirectly by calculating the amount of DNA that is shared between two individuals. As with parentage studies, mini- and microsatellite markers tend to be the techniques of choice for these studies. Much of the work using relatedness values has focused on the behavior of social insects. Studies have generally supported the potential for kin selection to act in these groups, including wasp and bee species. However, in some cases, such as the wasp *Polistes dominulus*, the lower than expected relatedness among helper and reproductives indicate that forces acting on decision making other than kin selection must be invoked. The results of tests of the potential for kin selection to act in other species using these tools range from some support in species such as African lions to a lack of support in species such as Seychelles warblers. Once the potential, or lack thereof, for kin selection to drive behavior has been established in a species, researchers can combine molecular data with experiments and observations to test its actual influence and to examine other potential forces driving decision making.

Molecular techniques such as microsatellites, AFLPs, and sequencing are also used to examine behavior through the lens of population and species comparisons. Dispersal behavior among populations can be revealed through a summary of relatedness over entire populations, or gene flow between populations. For example, these tools have shown that males but not females tend to disperse in Lake Malawi cichlids, and that urban stray cats rarely disperse among colonies. At a higher taxonomic level, these tools allow researchers to determine the rate of hybridization between species such as white and golden collared manakins where sexual selection seems to be driving this pattern. Finally, at a still higher taxanomic level, molecular tools can be used to build models of relationships among groups of species, independent of behavioral traits. These traits can then be mapped onto these phylogenies. In some species groups, for example, this approach has generated the hypothesis that preferences for sexually selected traits, such as tail attributes in swordtail fish, may exist in a species prior to the trait itself.

A final area where researchers use molecular techniques to study animal behavior, is the study of the genetics of behavior. These studies require large and very precise data sets and utilize a variety of data-intensive techniques including quantitative trait mapping (QTLs) and the study of candidate genes. In QTL mapping techniques the segregation of markers relative to a behavior is studied. QTL mapping has yielded information about the genetic architecture involved in a variety of behaviors from foraging in honeybees to parental care in mice. However, few studies have pinpointed the actual gene location for these behaviors. In studies of candidate genes, the genes identified as probable sources of behavioral variation such as dopamine receptors and attention-deficit and novelty-seeking behavior in humans, are examined. The second technique requires that candidate genes are known through knowledge of the species or of its relatives, and at present not enough is known about the genetics of most species or their relatives to provide these candidate genes.

Ultimately, until full genome sequencing becomes more widespread, the most common use of molecular genetics in the study of animal behavior will remain in the ability of these tools to clarify relationships among individuals. With these tools researchers can peer into unseen moments in an organism's life—the female California quail copulating with a male that is not her mate, or the male white footed mouse leaving his natal group. These techniques give researchers the ability to test hypotheses that were never before amenable to testing. They also lead to a reevaluation of basic assumptions and the generation of new theory and hypothesis that are, hopefully, closer to the truth. Recent genome projects and the increased accessibility of molecular techniques promise to increase their use in the study of animal behavior. In the end, molecular genetics, when combined with observation of the whole organism, promise to continue to lead to great advances in our understanding of human and nonhuman animal behavior.

See also Methods—DNA Fingerprinting

Further Resources

Avise, J. C. 2002. *Genetics in the Wild*. Washington, D.C.: Smithsonian Institution Press.

Parker, P. G., Snow, A. A., Schug, M. D., Booton, G. C. & Fuerst, P. A. 1998. *What molecules can tell us about populations: Choosing and using a molecular marker*. Ecology, 79, 361–382.

Queller, D. C., Strassmann, J. E. & Hughes, C. R. 1993. *Microsatellites and Kinship*. Trends in Ecology and Evolution, 8, 285–288.

Sunnucks, P. 2000. *Efficient genetic markers for population biology*. Trends in Ecology and Evolution, 15, 199–203.

Zhang, D. & Hewitt, G. M. 2003. *Nuclear DNA analyses in genetic studies of populations: Practice, problems and prospects*. Molecular Ecology, 12, 563–584.

Jennifer Calkins

■ Methods
Research Methodology

Science is structured knowledge and the processes used to generate it. This knowledge facilitates explanation, prediction and control of the real world. Systematic procedures for obtaining this knowledge are known as scientific methods, and a tremendous variety of them exist, even within a single discipline such as animal behavior. Here only a few key aspects of research methodology are discussed.

Questions and Hypotheses

While walking through a forest, one might suddenly hear loud bird vocalizations and discover a group of small birds repeatedly flying toward and then away from a tree. On closer inspection, an owl is seen resting on a branch. After learning that the observed behavior is termed *mobbing* and reading about it, one may decide to go further and carry out his own study.

Initially one might want to make more precise observations and answer questions such as: How many bird species were involved? Were same-species birds of same sex? Same age? Did all come equally close to the owl, running the same risk of being attacked? Were they all calling at same rate? Then broader questions would arise: How had the birds recognized an owl? Why did different categories of birds behave differently? What might have been the adaptive value of the behavior in this instance? Tentative answers to these questions, formulated within the context of existing knowledge about mobbing, would be *research hypotheses*.

It also is possible to devise research hypotheses without making one's own observations of particular phenomena. There are no rigid rules for the finding of good research hypotheses.

Three Types of Studies

There are three main complementary methodologies for conducting scientific research: theoretical, descriptive, and experimental. Investigations that combine them can be especially informative.

In following a *theoretical approach*, one might think about a subject and develop new ideas about it, and then test the ideas by referring to existing knowledge, by using logic and, often, by using mathematical models. One could, for example, develop a mathematical model that described how mobbing behavior affected choice of nest sites by owls. The mechanisms that induce a given behavior are represented by mathematical equations. These generate predictions which can then be compared with observed behavior. Models can be very useful in suggesting ideas for further observational or experimental studies.

Observational or descriptive approaches are a second major type. They are designed to gather more information on a phenomenon, often for the purpose of testing hypotheses, and can be simple or complex. One might use time-lapse photography as an aid to quantify how close birds come to the owl and how frequently, or use recorders to measure the frequency and nature of birds' calls, and repeat these measurements at different times of day, at different seasons, or in different locations. One could also use a dummy owl placed on a branch to see how mobbing behavior varied as a function of time of day or other variables. Observational studies require careful consideration of what the sampling design and basic sampling unit should be.

Preference trials are common in animal behavior. Examples would include presenting female subjects with different types of males and then documenting which types are selected as mates; or presenting different types of food items to animals and then recording the frequency with which each item is selected. Such preference trials are a type of observational study similar to sample surveys of human populations.

Manipulative experiments are the third major methodological approach. These entail manipulation of some experimental variable (or treatment factor) by the experimenter for the purpose of measuring its effect on one or more response variables. Their particular advantage is that they allow direct determination of causal relations. For example, field observations or theory might suggest that owl size may affect the occurrence or intensity of mobbing by other birds. To test this, one could use stuffed owls of different sizes and place these in trees on alternating occasions or at different sites and measure the responses of other birds. A disadvantage of manipulative experiments is that they usually cannot be carried out on large spatial or temporal scales that are often of great interest and importance.

Specific Methodologies

In any given observational or experimental behavioral study, there are numerous other aspects of research methodology. These are the specific field and laboratory methods, equipment, techniques, and protocols involved in selecting field sites, finding and maintaining animals, applying experimental variables, measuring responses, and recording and analyzing data. These naturally will be very different for each study, so useful generalization about them is not possible.

Experimental Design

Experimental design is sometimes used to refer to all the methods, procedures and operations involved in the conduct of a *manipulative experiment*. As mentioned above, these vary so much from one field to another and from one study to another, useful generalizations about this sense of design are difficult. The more precise and useful meaning of experimental design is the logical structure of a manipulative experiment. The purpose of such an experiment is to assess the effects of one or more experimental variables or treatment factors on one or more properties of the experimental unit. This unit can be an individual organism in a cage or tank, a group of organisms, a plot of ground, an entire lake, a bird nest, or any of a variety of other systems.

An experimental design has four aspects: *treatment structure, treatment replication, design structure*, and *response structure*. These can be defined and illustrated with an experiment to study how a fish species changes its territorial behavior according to food availability. Different levels of food supply would constitute the experimental treatments. To set up a manipulative experiment one would supply tanks with fish, add different quantities of food to different tanks, and then record fish behavior. A minimum of two groups of tanks would be needed, one set of tanks receiving a larger food ration and the other receiving a smaller one. Let's assume that we have four tanks for each treatment, and that each tank will have three fish. Ideally we assign the food levels or treatments to the tanks at random, to avoid the possibility of experimenter bias.

Treatment structure is the set of experimental treatments or treatment combinations used and how they relate to each other. In the territoriality experiment there are two treatments or levels, low and high food availability, of one treatment factor. This is the simplest treatment structure possible, since a manipulative experiment always has at least two treatments. An example of a more complex treatment structure would be if one used light intensity as a second treatment factor, using three different intensities, with a separate set of tanks set up for each of the six food availability–light intensity combinations (3 light levels x 2 food levels). When two or more treatment factors are used, the design structure is said to be (multi)factorial.

Treatment replication refers to the number of experimental units that will be subjected to a treatment. Often, but not always, this number is the same for all treatments, as in the experiment on territoriality where four tanks were established for each food level.

Design structure refers to the manner in which treatments or treatment combinations are assigned to experimental units. There are three basic design structures. The simplest would be a *completely randomized design*, where, for example, the six light–food level combinations would be assigned at random to the, say, 18 tanks available in a single array in an aquarium room.

If we only had a total of six tanks available, one might use a *randomized block design*. This would entail setting up six tanks with fish, assigning one tank to each of the six treatment combinations, and recording their observations on fish behavior. Then one would discard the fish and water from these tanks, wash them out, set the six tanks up again with new fish, re-randomize the assignment of treatments to tanks, and repeat the imposition of treatments and recording of observations. This could be repeated any number of times; each run or set would constitute a *block*.

The third basic type of design structure would be a *split-unit design*. In such there are always two or more treatment factors and the experimental unit would actually be defined differently for different factors. For example, one might have available six chambers with light controls, two at each of the three light levels. In each chamber one could place two fish tanks and assign one to the high food and one to the low food treatment.

The *response structure* consists of the list of response variables to be measured and the sampling plan that specifies when, where, and on what components of the experimental unit one will make and record observations and measurements. Each of these individual components is an *evaluation unit*. In the territoriality experiment, our principal observations would be on individual fish that were monitored for specific periods of time on specific occasions over, say, one week. One might define specific types of behaviors and record the frequency of each, estimate the size of each fish's territory, and measure variables such as quantity of food left unconsumed, concentration of dissolved oxygen, and so on. Repeated measurement of a given response variable on each experimental unit represents a *repeated measures* response structure. The sampling plan often is quite different from one response variable to another.

Experimental Design

Stuart H. Hurlbert & Celia M. Lombardi

The term *method* is commonly used within a philosophical framework, and the term *design* refers to the actual arrangement of variables used in experiments. A *variable* is anything that can change its value, and experiments have two main sorts of variables: the independent (also called experimental variable, or treatment factor) and the dependent variable. Since the latter variable in animal behavior is invariably behavior, the independent variable is whatever a researcher does to produce an effect on behavior. Although the terms treatment and dependent variable are often used with reference to both descriptive and manipulative designs, they are only properly applied to the latter, which meet the conditions required by the most widely used types of statistical analysis.

A *manipulative experiment* aims to determine, within a certain degree of probability, the effect that one or more treatments exert on one or more properties (behaviors) of some particular system (experimental unit). The experimenter must have full control over the assignment of treatments to experimental units. A thorough description of this type of design requires specification of three aspects: the design structure, the treatment structure, and the response structure.

The procedure followed for the allocation of treatments to experimental units specifies the *design structure*. When the selection procedure is at random, the design is termed completely randomized design. In the mobbing example (see the Methods—Research Methodology essay), the only recognizable difference between experimental units (small birds in this particular case) is provided by the treatments applied. However, if the researcher has some idea about inherent variation of the selected units, s/he could control it by means of what is termed blocking. In the above example, we may suppose that males and females may make up different classes whose units may behave similarly within each class. Assigning treatments at random to units of different sex results in a randomized complete block design.

Regarding *treatment structure*, there are many different forms of goals for an experiment. The investigator studying mobbing may be interested in comparing, for example, the birds' responses to aerial predators as opposed to terrestrial predators. S/he may also desire to discern whether responses change over time, or whether different degrees of the predator's dangerousness have any effect. The many different ways in which treatments may relate to one another, would make up an experiment's treatment structure.

Lastly, a design's *response structure* is specified by how evaluation units are related to experimental units. It must be decided which behaviors (dependent variables) will be measured, and how this will be done. For instance, the duration of a specified behavior or its frequency per time unit may be measured. In the mobbing example, birds may be individualized, enabling various measurements to be recorded on the same individuals (several evaluation units per experimental unit). Measuring several times the number of tail flicks each bird makes represents a repeated measures response structure.

Preference experiments are seldom manipulative but rather a type of observational study, very common and useful in animal behavior, akin to sample surveys of human populations. To illustrate, they arise when one confronts a female subject with different types of males to disclose which type she selects as a mate, or when animals are exposed to different types of seeds to study a species' diet.

Further Resources

Cook, T. D. & Campbell, D. T. 1979. *Quasi-Experimentation: Design & Analysis Issues for Field Settings*. Boston: Houghton Mifflin Company.
Cox, D. R. 1958. *Planning of Experiments*. New York: John Wiley & Sons.
McFarland, D. 1981. *Laboratory studies*. In: *The Oxford Companion to Animal Behaviour* (Ed. by D. McFarland), pp. 327–332. Oxford: Oxford University Press.
Mead, R. 1988. *The Design of Experiments*. Cambridge: Cambridge University Press.

The details of the above aspects of an experimental design are important to understand and specify clearly, because they determine the specific types of statistical procedures that would be appropriate for analyzing the data.

Sampling Design

For observational or descriptive studies, regardless of their complexity, the parallel area is that of sampling design. This may be simply defined as the logical structure of an observational study; that is, the way in which sampling units on which measurements or observations are to be made are selected from or distributed over the sampling universe of interest.

Sampling designs come in a very wide variety. This is due in part to the fact that, unlike manipulative experiments where the focus usually is on a single scale—that of the experimental unit—many observational studies have an interest in several scales simultaneously. A nesting behavior study, for example, could be aimed at studying variation among individuals within a local population, at variation among populations in a given region, and at variation among regions.

For each scale, different types of formal sampling designs are available. The three principal ones are simple random sampling, stratified random sampling, and cluster sampling. In the *simple random sampling*, one would simply locate all nests in a local population, give each a number, and then pick at random however many nests were thought to be needed for the study. In *stratified random sampling*, if one third of the nests were in tree species A and two-thirds in tree species B, then one might select the nests to be observed so that they came from the two tree species in corresponding proportions. In *cluster sampling*, one might select, perhaps for reasons of convenience, three different points at random in the forest used by the local population, and then select for observation all the nests that were present within a 100-meter radius of each point.

Of course, applying such formal sampling designs to real animal populations can be very difficult. Often one must simply make do with whatever nests, flocks, or individuals one can find. Nevertheless it is important to understand the principles of formal sampling design because they relate to how data should be analyzed and interpreted.

Statistical Analysis

Statistical methods are logical and mathematical procedures designed to help us separate the "signal" from the "noise" in data. They help us distinguish real patterns and trends from ones that are only apparent and the result of measurement or sampling error or the inherent variability of our subjects. Well carried out, statistical analyses increase the clarity, conciseness, and objectivity with which results are presented and interpreted.

So many statistical methods are used in animal behavior studies that no brief summary of them can be useful. Readers may want to browse through some elementary statistics textbooks, but should keep in mind that, unfortunately, errors abound in many of them.

Pseudoreplication

Pseudoreplication is a serious type of statistical error that is unfortunately common in all the sciences. It was originally defined in the context of manipulative experiments, but can also occur in observational studies.

With experimental data, *pseudoreplication* occurs when measurements made on multiple evaluation units, or multiple times on a single evaluation unit, in each experimental unit are treated statistically as if each represented an independent experimental unit.

How pseudoreplication might be committed in a simple behavioral study can be demonstrated by reference to the experiment on fish territorial behavior described earlier where there were four tanks under each treatment and three fish in each tank. As the measure of territorial behavior or response variable, one might record the number of aggressive acts by each individual fish over some period of time. These data would allow calculation of the mean number of aggressive acts for each tank, and the mean number of aggressive acts for each treatment or food level (i.e., for each set of four tanks).

To determine whether fish behaved differently when they were supplied with additional food, a statistical test would be applied to compare the means of the two treatments. A valid test would entail assessing whether the difference between the two treatment means was large relative to the variation among the means for individual experimental units (tanks) within treatments. If, however, one used in such a test the variation among *evaluation units* (the 12 fish in each treatment), then they would be committing *sacrificial* pseudoreplication, the commonest form of pseudoreplication: Information on variation *among* experimental units is mixed up with that on variation *within* experimental units.

If only a single tank of three fish were set up under each food level, and if one carried out a similar statistical test for a treatment effect, then he or she would be committing *simple* pseudoreplication.

The usual consequence of pseudoreplication is exaggeration of both the strength of the evidence for a real difference between treatments and of the precision with which any difference that does exist has been estimated.

Another example shows the form that pseudoreplication might take in an observational study. A researcher wishes to estimate for a 2 km² (.77 mi²) lake the mean density of nests of a fish that creates conspicuous nests as depressions on the lake bottom in shallow water. She selects one 100 m (328 ft) section of shoreline, randomly select six points along it, and establishes six band transects each 1 m (3.3 ft) in width and extending to deep water. She then swims along each of these with scuba gear, counts nests, estimates nest density for each of the six transects, and then calculates mean nest density and its standard error. This would constitute pseudoreplication if she claimed or implied that the standard error so calculated estimated the precision of her estimate of *lakewide* nest density when in fact it only reflects the precision of mean nest density estimated for the one 100 m (328 ft) section of shoreline used. One might say she had treated replicate subsamples (transects) as if they could serve as substitutes for replicate sampling units (shoreline sections) of the sort appropriate to the stated objective. To calculate the standard error appropriate to an estimate of lakewide nest density she would need to swim transects established at two or more portions of shoreline randomly selected from the lake's entire shoreline.

Pseudoreplication

Stuart H. Hurlbert & Celia M. Lombardi

Pseudoreplication is a serious type of statistical error. It was originally defined in the context of manipulative experiments, but can also occur in observational studies. With experimental data it occurs when measurements made on multiple evaluation units, or multiple times on a single evaluation unit, in each experimental unit are treated statistically as if each represented an independent experimental unit. An experimental unit is the smallest system or entity to which a single treatment is assigned and applied by the experimenter independently of other such systems. An evaluation unit is the specific component of an experimental unit on which an individual measurement is made.

Let us consider how pseudoreplication might be committed in a simple behavioral study. We wish to study how a fish species changes its territorial behavior according to food availability. Different levels of food supply would constitute the experimental treatments. To set up a manipulative experiment, we would supply tanks with fishes, add different quantities of food to different tanks, and then record fish behavior. We would need a minimum of two groups of tanks, one set of tanks receiving a larger food ration and the other receiving a smaller one. Let us assume that we have four tanks for each treatment and that each tank will have three fish. Ideally we assign the food levels or treatments to the tanks at random, to avoid the possibility of experimenter bias.

As our measure of territorial behavior (response variable), we might record the number of aggressive acts by each individual fish over some period of time. These data would allow to calculate the mean number of aggressive acts for each tank and the mean number of aggressive acts for each treatment or food level, for each set of four tanks.

To determine whether the fish behaved differently when they were supplied with additional food, we would apply a statistical test to compare the means of the two treatments. A valid test would entail assessing whether the difference between the two treatment means was large relative to the variation among the means for individual experimental units (tanks) within treatments. If, however, we used in such a test the variation among *evaluation units* (the 12 fish in each treatment), then we would be committing *sacrificial* pseudoreplication, the commonest form of this error. Information on variation among experimental units is mixed up with that on variation within experimental units. If we set up only a single tank of three fish under each food level and carried out a similar statistical test for a treatment effect, then we would be committing *simple* pseudoreplication.

The usual consequence of pseudoreplication is exaggeration of both the strength of the evidence for a true difference between treatments and of the precision with which any difference that does exist has been estimated.

Further Resources

Hurlbert, S. H. 1984. *Pseudoreplication and the design of ecological field experiments*. Ecological Monographs, 54, 187–211.

Jenkins, S. H. 2002. *Data pooling and type I errors: A comment on Leger & Didrichsons*. Animal Behaviour, 63, F9–F11: http://www.academicpress.com/anbehav

Kroodsma, D. E. 1989. *Suggested experimental designs for song playbacks*. Animal Behaviour, 37, 600–609.

Mead, R. 1988. *The Design of Experiments*. Cambridge: Cambridge University Press.

Further Resources

Hurlbert, S. H. 1984. *Pseudoreplication and the design of ecological field experiments.* Ecological Monographs, 54, 187–211.

Hurlbert, S. H. & White, M. D. 1993. *Experiments with freshwater invertebrate zooplanktivores: Quality of statistical analyses.* Bulletin of Marine Science, 53, 128–153.

Kroodsma, D. E. 1989. *Suggested experimental designs for song playbacks.* Animal Behaviour, 37, 600–609.

Lehner, P. N. 1979. *Handbook of Ethological Methods.* New York: Garland STPM Press.

Martin, P. & Bateson, P. 1986. *Measuring Behaviour.* Cambridge: Cambridge University Press.

McFarland, D. 1981. *Classification of behaviour.* In: *The Oxford Companion to Animal Behaviour* (Ed. by D. McFarland), pp. 63–64. Oxford: Oxford University Press.

Mead, R. 1988. *The Design of Experiments.* Cambridge: Cambridge University Press.

Stuart H. Hurlbert & Celia M. Lombardi

Sonic Tracking of Endangered Atlantic Salmon

Fred Whoriskey

The Atlantic salmon (*Salmo salar*), once numerous in East coast North America, has now declined to the point that the species is at risk of extinction in the southern third of its range. In a desperate conservation effort, biologists are turning to hatcheries (a.k.a. live gene banks) to try and bolster the endangered populations. In hatcheries, food is plentiful and there are no predators, so large numbers of fish can be reared rapidly for population supplementation. However, fish in hatcheries adapt both physically and especially behaviorally to the captive environment, and may have a difficult time of survival when released to the wild.

In theory, a promising salmon live gene bank strategy is to allow fish to develop in captivity up to maturity, then to release them to rivers to spawn. In this approach, you gain the survival benefits of a hatchery program, and although the adults are "hatchery" fish to some degree, their progeny will grow up fully "wild." For the method to work, the adults must have enough "wild" behavior in them to survive, find a spawning ground, and mate.

Jim Martin releasing adult salmon.

Courtesy of Fred Whoriskey.

My colleagues and I have been running an experiment (Magaguadavic River, New Brunswick, Canada) to see if hatchery-reared adult salmon will show normal reproductive behavior once they are released to the wild. The study fish were released after being surgically fitted with acoustic transmitters. Receivers at fixed locations in the river let us track individuals. We were asking:

1. Do hatchery fish, released at normal river entry times for Atlantic salmon (July and August), move to spawning areas in time for the October spawning period?

2. Will upriver movements be different if the fish are released singly and have to face predators and other threats alone, or liberated in groups (3–5 fish)?

3. Will the fish consort with kin, or avoid them and the possibility of inbreeding?

4. Are wild juvenile salmon numbers improved by the introduction of these adults?

Initial tracking results showed that being in a group seemed unimportant for upriver movements. Fish released with companions dispersed and made their way alone upstream at the same rates that single fish moved. Most disappointing was the fact that instead of heading to spawning grounds, virtually all the gene bank fish peeled off from the main river into a tributary where they congregated through the spawning season, near the water discharge pipe from a local hatchery. Clearly their hatchery experience somehow overrode what should have been natural spawning behavior.

Fortunately, we had kept some fish in reserve back in the hatchery. These we trucked upriver in October and released them directly onto spawning areas. Preliminary tracking results showed that they generally stayed where released, or if they moved it was to other spawning sites in the headwaters. We received reports of spent fish being captured by anglers, so at least some mated. We will be monitoring the river for the presence of juvenile salmon from the spawning of these fish in upcoming years, and use DNA profiling to determine what, if any, inbreeding occurred.

Further Resources

Anderson, J. M., F. G. Whoriskey, F. G. & Goode, A. 2000. *Atlantic Salmon on the brink.* Endangered Species Update, 17(1), 15–21.

Carr, J. W., Lacroix, G. L., Anderson, J. M., & Dilworth, T. 1997. *Movements of non-maturing cultured Atlantic salmon* (Salmo salar) *in a Canadian river.* ICES Journal of Marine Science, 54, 1082–1085.

Lacroix, G. L. & McCurdy, P. 1996. *Migratory behaviour of post-smolt Atlantic salmon during initial stages of seaward migration.* Journal of Fish Biology, 49, 1086–1101.

Mills, D. (Ed.) 2000. *The Ocean Life of Atlantic salmon.* Oxford, UK: Fishing News Books.

Mills, D. 1989. *Ecology and Management of Atlantic Salmon.* London: Chapman and Hall.

Moore, A., Russell, I. C. & E. C. E. Potter, E. C. E. 1990. *The effects of intraperitoneally implanted dummy acoustic transmitters on the behaviour and physiology of juvenile Atlantic salmon,* Salmo salar L. Journal of Fish Biology, 37, 713–721.

■ Methods
Zen in the Art of Monkey Watching

> *"I use the term focal animal sampling to refer to any sampling method in which all occurrences of specified (inter)actions of an individual . . . are recorded during each sample period. . . Once chosen, a focal individual is followed to whatever extent possible during each of his sample periods."*

FROM JEANNE ALTMANN, 1974.

"The mind is like a monkey swinging from branch to branch through the forest, says the Sutra. In order not to lose sight of the monkey by some sudden movement, we must watch the monkey constantly and even be one with it."

FROM *THE MIRACLE OF MINDFULNESS*,
THICH NHAT HANH

"Who shall say what prospect life offers to another? Could a greater miracle take place than for us to look through each other's eyes for an instant?"

FROM *WALDEN*,
HENRY DAVID THOREAU

For nearly 30 years I have taught beginners how to watch monkeys. Or, as we say in the scientific parlance, how to conduct ethological observations. For almost the same number of years I have noticed that some people are truly interested in watching animals, whereas others are fascinated by the scientific issues that animal behavior allows us to explore, and still others concentrate on the mechanics of the task. The last group tends to become preoccupied with the intricacies of the gear—hand-held computers, recorders, cameras, GPS units, rangefinders, etc.—rather than the contemplation of social animals in action. Whenever I have a problem with a piece of our equipment, I look for one of our techno-primatology students. When it is a statistical issue, I look for one of our good quantitative analysts. But when I need a sense of what is going on with our monkeys—who is doing what to whom (and why) in the latest episode of our study group's daily soap opera—then I send for one of our gifted monkey watchers.

In the early days of ethology, scientists often took notes "ad libitum," simply recording whatever grabbed their attention. But it soon became obvious that this is not a good way to obtain a representative sample of behaviors for all members of the group. In many primate species, adult males are bigger and more swashbuckling than females, infants are more attractive to the eye than gawky juveniles, high-ranking individuals are more likely to be in your face than are subordinates who fade into the vegetation, and noisy fights attract more attention than do peaceful interactions. So, for the sake of better science, we developed methods to structure our attention and make our data collection more comprehensive and representative. One common and effective tool for watching animals in social groups is called the focal animal sampling technique. One individual at a time is randomly chosen from the list of all subjects and followed for a preset period during which every big and little thing that animal does is faithfully recorded by the observer in predetermined behavioral categories. The trick is that the animal must be followed in such a way as to cause as little disturbance to its natural behavior as possible while its behavior is recorded with maximum accuracy.

To stare is a threat in monkey society, and rude in most human societies, and therefore the observer must learn to watch with discretion. Successful monkey watching is an art as well as a science, an art that requires patience, skill, and empathy. So subtly are these social animals cued to the direction of attention that an observer can stand in a group of 30 monkeys, choose one as a subject, start her stopwatch and before she logs her first data entry, the chosen monkey turns its back or moves away. No wonder many monkeys go crazy in zoos under the relentless stares and camera lenses of visitors, not to mention the other bizarre behaviors some humans feel compelled to enact in front of primate cages. Good observers of animal behavior become experts at the sideways glance, the silent step, and the diagonal trajectory.

When you have committed to watch an animal persistently, not only in action but also while it sleeps, scratches, stares into space and literally does nothing, you really get to know that individual. You start to notice details—the way she holds her hands at rest, the swagger in his walk, the crook in her tail, the tremor in his calls, the unique way he assembles a bundle of grass stems like a bouquet between his molars before stripping seeds with a rapid pull on the vegetation. You notice many things you don't record because they are not officially part of the research. And yet your mind stores them away and they augment your understanding of that animal. Collecting focal data can be challenging—you have to stay completely mentally alert while discreetly watching a monkey on its daily rounds. Sometimes your focal subject leads you on a merry chase, up hill, down cliff, across stream and through the undergrowth while you try to get it all down on paper or palmtop computer. Other times, collecting focal data can be boring and frustrating. Just how many times can you stand to enter "sleep" on your data form while your subject snoozes and you can hear the intriguing sounds of a sexual encounter taking place on the other side of the tree?

Chimpanzee sitting near a researcher in the Gombe Stream National Park, Tanzania, Africa.
© *Fritz Polking / Visuals Unlimited.*

Focal data collection is well named for two reasons—we focus on one animal at a time and in so doing we focus our minds. If only we had the patience and willpower to constantly observe our world with such persistent attention to detail. Buddhists have understood for centuries the rewards of the disciplined mind, the power of paying absolute attention to what is in front of us. In this respect, ethological data collection can become a form of meditation. But like meditation, it only sets the stage for insight. Skillful recording of animal behavior is like learning to draw the bow in the Japanese art of archery. The point is to draw the bow with such experience, skill, and lack of forced effort that the arrow looses itself and hits the bulls-eye with no conscious attempt on the part of the archer. The bow and arrow are only the way to the goal, not the goal itself. Or in the case of data collection, you become so familiar with your recording system (the bow) and your subjects (the bulls-eye) that your mind (the arrow) is able to fully concentrate on the monkey in motion. At times, you find yourself knowing and predicting how your focal animal will behave, without even understanding exactly how you know this. It is because you have become so engrossed in that individual through close observation and become so familiar with its behavioral patterns that your mind automatically moves beyond data collection mechanics toward comprehension (the goal).

The good monkey watcher develops empathy with those they observe. By empathy I don't mean a desire to hug monkeys and cry over their misfortunes. Monkeys in nature do not respond well to attempted hugs from humans and don't seem to care one way or another if we weep for them. By empathy I mean the ability to see the world through the monkey's eyes. To say this requires a load of hubris on my part since we will never know exactly what a monkey sees through its eyes, what it thinks in its mind, what it feels in its body. But there is an empirically verifiable test of good monkey watching and that is the ability to predict what the animal will do next. Now this is not a magic act, and sometimes we get it wrong—that is part of what motivates us to keep watching. But there is also no way to fake this skill, it only comes from hours and hours of attentive practice. And sometimes it only comes when you put down your data sheets or computer at the end of a focal session and just move or rest with the animal. Like a sudden transformation of vision, you may see what it is like to be in some other creature's world. Surely empathy is the greatest secret of successful monkey watching and such insight its greatest reward.

The old macaque sat dozing beside me in the late afternoon shade of a cedar tree, his head hanging such that grizzled chin touched furry chest. His drowsiness allowed me to look closely at his weathered face and gnarled fingers resting on outstretched knees while he slept in a position only a monkey with sitting pads can manage to find comfortable. It was siesta time when I decided to stop data collection and relax with the adults while the kids played nearby. I tried to imagine what it must be like to have lived the life of this old male—to have seen the years go by, gradually losing fights to younger stronger males, watching juvenile females grow up to be mothers, new babies being born, sick animals growing frail and disappearing. Suddenly a noisy squabble broke out in the gang of yearlings when one pushed the other too hard in the rough and tumble game. Dai raised his eyebrows, but I thought he wouldn't budge for this daycare drama. Sure enough, he sighed, briefly ground his teeth and settled back into somnolence. Then an estrous female named Kujiro trailed by very close in front of us, brushing him with her body in what seemed a not-accidental way and leaving a lingering scent. Dai half opened his eyes long enough to register her presence then dropped his head back down into his nap. I had figured he would not be interested in Kujiro at this moment either. But when Shiro began to gecker with fear, the old male beside me suddenly sprang into action. He drove off the higher ranking female, Matsu, who had threatened Shiro, growling, lunging and slapping at the interloper, who fled screaming. After the altercation, the old male returned to the shady spot beside me with the young female Shiro, and she began to groom him, deftly parting his fur with her right hand and combing through it with the fingers of her left hand. A sense of ease slid over our napping circle again. Dai had adopted Shiro when she was orphaned at 3 months and in the decade since then they had been inseparable. Even I, a mere observer of the group, knew that to mess with Shiro meant having to deal with Dai, and I wondered how Matsu had made such a miscalculation. Perhaps she hadn't seen the old male, or judged him too sleepy to intervene. Maybe she thought her relatives would back her up, or possibly she was just too irritated with Shiro to mind the consequences. According to my roster, I was due to collect data on Matsu the next day, and I resolved to understand her better.

See also Methods—Ethograms
Zoos and Aquariums—*Studying Animal Behavior in Zoos and Aquariums*

Further Resources

Altmann, J. 1974. *Observational study of behavior: Sampling methods*. Behaviour, 49 (3–4): 227–265.
Lehner, P. N. 1996. *Handbook of Ethological Methods*. Cambridge: Cambridge University Press.
Martin, P. & Bateson, P. 1987. *Measuring Behaviour. An Introductory Guide*. Cambridge: Cambridge University Press.
Nhat Hanh, T. 1999. *The Miracle of Mindfulness*. Boston, MA: Beacon Press.
Paterson, J. D. 2001. *Primate Behavior. An Exercise Workbook*. Prospect Heights, IL: Waveland Press.

Linda Marie Fedigan

■ |Mimicry

To mimic, in general terms, means to copy or imitate. All of us have participated in this process at one time or other, perhaps in late October. With a patch over my eye, a plastic sword tucked under my belt, and a bold swagger in my step, I myself ventured out one Halloween to levy payment from the terrified citizens of my neighborhood. No one, of course, was fooled by my impersonation of Blackbeard the Pirate, but I nonetheless came home with a bag full of candy.

It is exactly because no one was deceived that most biologists would not accept my childhood anecdote as an example of "true" mimicry. According to French herpetologist Georges Pasteur, *biological mimicry* comprises at least two different individuals performing three different roles: a *model*, a *mimic*, and a target organism, or *dupe*, who is deceived by the resemblance and thus confuses the mimic for the model. Although the mimic benefits from the resemblance, the model often, and the dupe typically, suffer a fitness cost from the deception. Thus, a harmless kingsnake (the mimic) that deters the attack of a hungry red-tailed hawk (the dupe) because the kingsnake's black/red/yellow banding pattern closely resembles the aposematic (warning) banding pattern of a venomous coral snake (the model) would qualify as an example of biological mimicry; the kingsnake benefits because it was not eaten, while the hawk, mistaking the kingsnake for a deadly coral snake, loses a meal. And what about the coral snake? The kingsnake's imitation might actually put coral snakes at greater risk of predation if naive hawks learn to associate black/red/yellow banding patterns with food as a result of interacting first or more frequently with kingsnakes than with coral snakes.

The kingsnake's black/red/yellow banding pattern resembles the venemous coral snake, thus deterring predators.
© Gerold and Cynthia Merker / Visuals Unlimited.

Even though Pasteur's definition is widely accepted among scientists, the diversity of phenomena encompassed by the term *mimicry* is broad, and considerable effort has been spent trying to categorize different types of mimicry. Many readers, for example, may be familiar with the distinctions between Batesian and Müllerian mimicry. *Batesian mimicry*, named after British naturalist Henry Bates who first described this phenomenon (in 1862) among Brazilian butterflies, is the term applied to mimicry systems where an edible mimic avoids predation because it resembles a noxious or dangerous model, as in the kingsnake–coral snake system. *Müllerian mimicry*, named after German zoologist Fritz Müller, who published his ideas in 1879, is the term applied to situations where several different species of noxious or dangerous models evolutionarily converge on a common warning pattern. Müller based his ideas on a suite of commonly colored, distasteful species of butterflies he was studying in South America; another example may include the familiar black and yellow banding patterns of many bees and wasps. The benefits of such convergence is that potential predators need experience only one individual from one of the protected species to learn to avoid the entire suite of species.

Henry Bates and Fritz Müller were two of the first biologists to study mimicry in animals systematically and they certainly deserve credit for their discoveries. Some scientists, however, do not consider Müllerian mimicry to be mimicry at all, given that no deception is involved (there are no harmless mimics, only nasty models), and predators benefit by learning to associate all the models with a common aposematic signal. More important, the terms bearing Bates' and Müller's names do not capture the full diversity of mimetic systems. A katydid that looks remarkably like a dead leaf, or an Arctic hare whose white fur allows it to blend into the snow, may reduce their risk of being eaten, but not because predators confuse either the insect or the hare with some dangerous model. Indeed, the predator presumably fails to perceive the mimics at all, as they simply disappear into the background. According to British entomologist Richard Vane-Wright, such examples should be referred to as *crypsis* or *camouflage*, but not mimicry because the dupe would not even be aware of the mimic. Other biologists, myself included, find Vane-Wright's requirement that the dupe be cognizant of the mimic (or, more correctly, of the model it is imitating) too restrictive and difficult to apply; how would we know, for example, whether a an azure-hooded jay in the mountains of Costa Rica failed to perceive a leaf-mimicking katydid, or simply ignored what it assumed was an inedible dead leaf?

The term *protective mimicry* is applied to situations where a mimic avoids predation because of its resemblance to a model, regardless of whether the model is an aposematic coral snake or cryptic leaf. Thus, all of the examples presented so far illustrate the category of mimetic resemblances referred to as protective mimicry. Additionally, each is also an example of *organismal mimicry*, meaning that the roles of mimic, model, and dupe are "played" by individuals. Medical scientists have recently added yet another term to our growing lexicon of mimetic similarities, *molecular mimicry*, which occurs among cells instead of individuals. Rather than being a unique type of mimicry, molecular mimicry is simply a diabolical form of protective mimicry played between an invading pathogen and a host's immune system, with the host's B cells and T cells acting as dupes. Normally when a virus, bacterium, or parasite invades a vertebrate host, proteins on the surface of the invader (antigens) are recognized by the B and T cells as "foreign" and trigger these lymphocytes to kill the invader. Certain pathogens, however, including the parasite *Plasmodium falciparum*, which causes human malaria, have evolved antigens that resemble proteins imbedded in the membranes of the host's own cells, tricking the host's lymphocytes into treating the parasite like a normal host cell. In a

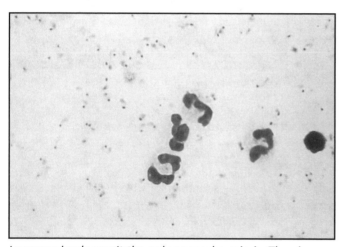

*Larger animals aren't the only ones who mimic. The virus that causes malaria (*Plasmodium falciparum*) tricks the body into thinking it's a normal cell with special proteins.*
Courtesy of the Centers for Disease Control.

sinister twist to this already insidious example, it is now believed that the antigenic mimicry of certain invading pathogens may trigger the host to attack it own cells (a medical condition called autoimmune disease), which could further speed the pathogen's invasion.

Mimics often benefit, not because they avoid being attacked by hawks or by lymphocytes, but because their imitation provides them access to scarce resources, a type of resemblance referred to as *aggressive mimicry*. The juveniles of many species of snakes, for example, have brightly colored tails that contrast with the cryptic coloration of the rest of their body. Lying otherwise motionless and camouflaged, these young snakes slowly wiggle their tails back and forth in a manner resembling the movements of a worm or insect grub, a behavior termed *caudal luring*. Terry Farrell and Ali Rabatsky have shown that young pigmy rattlesnakes in central Florida use caudal luring to attract insect-eating leopard frogs to within striking distance. Moreover, the young pigmy rattlesnakes maximize their chances of successfully deploying their lure by selecting foraging sites imbued with leopard frog odor.

Aggressive mimics may gain access to more than just food; in many cases, their deceptive resemblance provides them with access to mates. Canadian biologist Mart Gross has shown that male bluegill sunfish come in two different types, or polymorphisms: large, slow growing "parental" males who, once mature, fight with other males to establish a nesting territory in which female bluegills will hopefully lay their eggs; and small, fast-maturing "cuckolder" males who never defend territories, but instead steal fertilizations with females attracted into parental male territories. When the cuckolders are young and small, they attempt to sneak into a territory and speedily fertilize the eggs of a female just as she spawns. When a cuckolder is a bit older and approximately the size of a bluegill female, he adopts a different tactic. Rather than zipping in just as a female releases her eggs, he casually follows a female into a parental male's territory while mimicking her behavior, essentially tricking the territorial male into courting two females. When the real female deposits her eggs, the cuckolder, interposed between model and dupe, deposits his sperm.

The mimicry of females by males of the same species is a somewhat common theme, having been reported in creatures as dissimilar as diminutive marine isopods to brutish elephant seals. The benefits for these she-males range from sneaking copulations to avoiding aggression from larger, more dominant males. In red-sided garter snakes, however, the benefit appears to be the reptilian equivalent of an electric blanket on a cold winter day. Robert Mason, Richard Shine, and several of their colleagues have been studying the mating aggregations that occur just after these snakes emerge from communal hibernation in Manitoba, Canada. When a female leaves the den in spring, she is swarmed by dozens of amorous males attempting to copulate. Many newly emerged males are similarly swarmed, triggered in part by female-like pheromones exuded from the skin of these males. Garter snakes that are just emerging from their den are cold, with body temperatures around 10°C (50°F). Courting males, however, have had the opportunity to bask and they are hot, with body temperatures of 25°C (77°F) or warmer. Thus, when she-male garter snakes dupe other males into a mating swarm, the she-males warm up faster than they could have on their own, and warm snakes are better than cold snakes at avoiding their predators.

The flip side of she-males, of course, is he-females, and nature provides a few examples. Male mimicry has been reported in several species of butterflies, dragonflies, and damselflies. Female Rambur's forktail damselflies, for instance, come in two morphs; a cryptic brown-orange morph and brighter blue-green morph, the latter of which not only looks like a male but occasionally behaves like one, too. Animal behaviorist Jane Brockman and her students believe they know why. A damselfly female need mate only once to receive enough sperm to fertilize her eggs. Male damselflies, in contrast, are much more enthusiastic about mating. Female forktail damselflies that look like males can, once they have mated, rebuff continued harassment from such ardent males (saving her both time and increased exposure to predators) by adopting the aggressive behaviors males use to drive off rivals.

The terms protective mimicry and aggressive mimicry categorize different forms of mimetic resemblances according to the type of benefit accrued by the mimic; that is, protection from predators vs. access to resources. The terms female mimicry and male mimicry classify mimetic resemblances based on the model; terms such as *leaf mimic*, for the katydid mentioned above, and *flower mimic*, for a preying mantis that looks like a plant's inflorescence, are also part of this scheme. The sensory modality employed by the mimic has also been used as a tool for classification. Because humans are so visually oriented, most of our attention has focused on examples of *visual mimicry*; for example, the worm-like tails of pigmy rattlesnakes, the female-like appearance of cuckolding male bluegill sunfish, and the leaf-like camouflage of a katydid. Indeed, we rely so heavily on vision that our sensory bias permeates our everyday language; note, for example, the use of the term "focus" in the preceding sentence, or the admonition we might use if we catch someone being dishonest—"I *see* through your lies." Many organisms, however, use modalities other than vision to dupe their targets. Certain orchids, for example, produce volatile compounds that mimic the sex pheromones of female thynnine wasps; male wasps are duped by this *chemical* or *olfactory mimicry* into pollinating the orchid's nectarless flowers. Rove beetles employ *tactile mimicry*, using their antennae to tap out the signal "please regurgitate food" to unsuspecting workers in the ant colonies they parasitize. And Don Owings, Dick Coss, and I have shown that burrowing owls use *acoustic mimicry* to deceive their predators. Burrowing owls nest underground in the abandoned burrows of prairie dogs and ground squirrels; when cornered inside their burrow, the owls produce a hiss that not only

A pair of adult burrowing owls, female on the left standing just inside the burrow opening, male on the right.
Courtesy of Matthew Rowe.

sounds like the buzz of an angry rattlesnake, but also is treated as such by enemies of the owl.

Yet another classification scheme is based on the number of different species involved in playing the roles of mimic, model, and dupe. A mimetic system is said to be *disjunct* if the three parts are played by three different species, as in the kingsnake–coral snake–hawk example. A system is *partially conjunct* if two species play the three parts, as when a male thynnine wasp is tricked into pollinating an orchid that smells like a female of his own species. And a system is said to be *totally conjunct* if only a single species is involved in all three roles, as illustrated by she-male red-sided garter snakes who seek the warmth of their male conspecifics. The fact that we have so many classification schemes for mimetic resemblances attests to the diversity of phenomena that fit under this broad heading. Thankfully, the different schemes can often be applied jointly, even if the resulting phraseology is cumbersome; thus, a burrowing owl that uses a rattlesnake-like hiss to deter a hungry badger would be characterized as an example of *disjunct acoustic Batesian mimicry*.

One idiom that does not fit as nicely is *vocal mimicry*. Vocal mimicry is a term that typically has been applied to certain birds, primarily songbirds (also called passerines), who incorporate sounds they have learned from their environment into their own songs and calls. Male mockingbirds and starlings are premier vocal mimics. The sounds these mimics copy are often the calls and songs of other species of birds, but include such oddities as the croaks of tree frogs, the creaking of an old rocking chair, and even the obnoxious whine of car alarms. Many biologists, including ornithologist David Dobkin, do not consider such vocal learning to be true mimicry because no one is deceived by the imitations, and prefer the term *vocal appropriation* to describe them. Mockingbirds, for example, usually embed the note or phrase they have copied from a model (e.g., the weep call of scrub jays) into a complex, fast-paced song incorporating the phrases of many different model species. Moreover, male mockingbirds often sing their rich and distinctive jumbles from an exposed perch, a location unlikely to contribute to a dupe's confusion. Indeed, a recent study has demonstrated that scrub jays are not deceived by a mockingbird's imitation of their weep call at all. The main value of vocal appropriation, to male mockingbirds at least, is that female mockingbirds are attracted to males with larger song repertoires; older males generally have larger repertoires, and surviving to a ripe old may be a good indicator of male fitness.

One reason that many biologists like Pasteur's characterization of mimicry is that it is consistent with an emerging view of animal communication as a dynamic arms race between signalers and receivers or, as Don Owings and Gene Morton argue are better descriptors, between managers and assessors. For example, once female mockingbirds use a male's repertoire size to judge his fitness (assessment), then selection would favor those males who increase their attractiveness to females (management) through their adept incorporation of scrub jay weep calls, the squeaking noises of old rockers, and car alarms to their song list. Biological mimicry as defined by Pasteur simply adds a third player to the mix, the mimic, who capitalizes on the assessment/management interaction already established between the model and the dupe. A red-tailed hawk that avoids attacking a perfectly delicious kingsnake does so because it has previously made the connection (either ontogenetically or evolutionarily) between red/black/yellow bands and the debilitating bite of a coral snake.

In short, a mimic manages a dupe by exploiting the dupe's assessment of the model. This perspective helps explain what seem to be, for humans at least, some very poor examples of biological mimicry. Many insects, for instance, perceive ultraviolet (UV) light. In response, many plants have evolved flowers with special *nectar guides* that are highly reflective of UV radiation; insects use these guides to gather pollen and nectar and, in so doing, pollinate the plant. Fungi in the genus *Monilinia* take advantage of such insects' assessment systems in a devilishly clever way. *Monilinia* has a complex life cycle; the fungus grows on the leaves and shoots of blueberries and huckleberries, but the fungi's spores must eventually be transferred to these plant's flowers. Leaves infected by *Monilinia* appear shriveled and dead to the human eye, not at all flower-like, but are highly reflective of UV light and act as duplicitous nectar guides for many species of bees and butterflies. A butterfly that is occasionally duped into visiting a wilted, brown blueberry leaf while making its rounds among blueberry flowers is all that the fungus needs to complete its life cycle. It is the cognitive system of the dupe that helps us make sense of the resemblances between model and mimic; for species with sensory worlds quite different from the dupe's, such mimicry can appear as unconvincing as my childhood imitation of Blackbeard.

And finally, it is feedback from the dupe that helps us understand the dynamic nature of mimetic systems. A male garter snake that attempts to copulate with another male, just because the she-male smells like a female, loses time, energy, and potentially his life (if, for example, his romantic behavior draws the attention of a hungry crow). Because dupes suffer a fitness cost, we would expect the target of such deception to get better

Two juvenile burrowing owls (close to fledging) sitting just inside their burrow mouth.
Courtesy of Matthew Rowe.

(again, either ontogenetically or evolutionarily) at "seeing" through the duplicity, at distinguishing model from mimic. This refinement in the dupe's assessment capabilities would in turn select for mimics that better matched their model. In a few cases, we can find evidence of such arms races between the cognitive capabilities of dupes and the signals used by the mimic. For example, we believe that the precursor to the burrowing owl's mimetic hiss is a food begging call used not only by burrowing owls, but also by all of its close relatives, including the little owl of Eurasia. The food begging call is a pretty good match to a rattlesnake's rattle, in that the both are sibilant sounds with similar dominant frequencies. The food begging calls of burrowing owls and little owls, however, are uttered as a series of very soft, short (<0.5 sec), and repetitive hisses that sound a bit different than the loud, continuous buzz of an angry rattlesnake. Although ancestral badgers may have been fooled by the initial similarities, selection would have favored badgers who could distinguish a tasty burrowing owl from a venomous rattlesnake. Selection for more discriminating and less easily confused dupes would in turn favor burrowing owls whose hisses, when used in defensive contexts, sounded more like rattles. Burrowing owl defensive hisses are indeed both loud and continuous like a rattlesnake's rattle, similarities that have certainly been shaped by the assessment systems of their dupes.

See also Mimicry—Magpies

Further Resources

Gross, M. R. 1982. *Sneakers, satellites, and parentals: Polymorphic mating strategies in North American sunfishes.* Zeitschrift für Tierpsychologie, 60, 1–26.

Owings, D. H. & Morton, E. S. 1998. *Animal Vocal Communication: A New Approach.* Cambridge: Cambridge University Press.

Pasteur, G. 1982. *A classificatory review of mimicry systems.* Annual Review of Ecology and Systematics, 13, 169–199.

Rowe, M. P., Coss, R. G. & Owings. D. H. 1986. *Rattlesnake rattles and burrowing owl hisses: A case of acoustic Batesian mimicry.* Ethology, 72, 53–71.

Shine, R., Phillips, B., Waye, H., LeMaster, M. & Mason, R. T. 2001. *Benefits of female mimicry in snakes.* Nature, 414, 267.

Vane-Wright, R. I. 1976. *A unified classification of mimetic resemblances.* Biological Journal of the Linnean Society, 8, 25–56.

Matthew P. Rowe

■ Mimicry
Magpies

Australian birds in particular have a propensity to mimic other sounds. In his book *Bird Wonders of Australia* (1948) Alec Chisholm noted that more than 50 Australian bird species can mimic, and at least half of his claims have since been confirmed, including the well-known case of lyrebirds (*Menura* sp.) but also the Australian magpie (*Gymnorhina tibicen*).

During the breeding season, the male superb lyrebird (*Menura novaehollandiae*) uses many mimicked features in his song. He will typically incorporate birdcalls of a variety of species, but may also include sounds of other animals or even of inanimate objects. Once a sound has been adopted, that sound will take a firm (and unchanged) place in his song, and every sound will make the chain of sounds longer, but the elements of it stay in the same

Details of Species

Name: Australian Magpie (*Gymnorhina tibicen*)

Classification: Family Artamidae, subfamily Cracticinae, broadly belonging to corvidae but not related to the Eurasian black-billed magpie (*Pica pica*).

Identification: Large (40–44 cm, 330–440 g; 16–17 in, 11.5–15.5 oz), black and white bird with robust bill (about 52 mm long). Sexes similar, except for female's greyish (as opposed to white) nape (area at back of neck).

Distribution and Habitat: Throughout Australia and also southern New Guinea. Introduced to New Zealand. Open eucalypt woodlands, farmlands, urban parks and gardens.

Biology: Sedentary, territorial. Largely insectivorous, but also feeds on seed, eggs and meat. Breeds mainly late winter to mid-summer (August-January). Usually lays 2–4 eggs in stick nest. Incubation 24 days by female, 30–36 days to fledging. Parents continue to feed fledglings for 3 months. Lifespan 20–25 years.

position. Both in structure and function, the song is meant to win the favours of a female. Hence, mimicry has a specific function in this case.

The same function of mimicry cannot be attributed to that of the Australian magpie. Although an extremely accomplished songbird with unbelievable variety, rich overtones and range, the magpie's use of mimicry is at first puzzling because males and females sing alike, and they do so throughout the year. Hence, song is not associated with the breeding season, nor can one therefore ascribe that role to the mimicry embedded in their song. Through my own research I have identified 15 types of mimicry used by magpies throughout Australia. These sounds include those of familiar bird species, such

A juvenile magpie perches on Gisela Kaplan's hand.
Courtesy of Gisela Kaplan.

as the kookaburra (*Dacelo novaeguineae*), also called "jack ass," and of mammals, such as horse neighing, cat mewing, and dog barking.

Magpies use the new sounds freely and in any part of their song, and they do so in the wild and not as a consequence of being human-trained. However, magpies are also capable of mimicking human speech if they are regularly in the vicinity of humans. For an answer of possible function, we may have to look at the magpie's lifestyle. Magpies are territorial, and vocalizations go with the territory. Parts of a song may be borrowed from a neighbor, but usually no more than a quarter of its song.

So far, my research has shown that magpies mimic sounds consistently only of permanent inhabitants of their territory, excluding (seemingly consistently) any visitors (human or otherwise), transient occupants, and any occasional sounds. It has been postulated that species with large vocal repertoires may need to learn their song (as distinct from being genetically fixed) and may, at times, not learn it accurately enough. In other words, mimicry might simply be an error. However, this is unlikely because the magpies investigated showed deliberate and selective practice of mimicry. Furthermore, the accuracy of the mimicked sounds confirms that magpies have highly developed auditory perception and great musical abilities and may therefore be less likely to make "errors." In short, mimicry may not be an accident or a caprice, even if not all magpies mimic.

On the contrary, magpies may mimic deliberately as part of knowing their territory. It is conceivable that magpies, a relatively long-lived species, may form a geographic as well as a vocal map of their territory in order to defend and hold it. For all the work that has been done on higher cognition in animals, mimicry has not been featured because it is merely regarded as an exercise of copying. But what if some magpies have developed the concept of a soundscape as well as that of a landscape as part of a specific memory of territory? It may be too early to confirm this claim, but also too early to dismiss this possibility.

See also Mimicry

Gisela Kaplan

◼️|Naked Mole-Rats

"Mole-rat" is the vernacular name for 37 species of rodents (in 10 genera) that inhabit subterranean burrows like moles, but whose body shape, slender tail, and front teeth resemble rats. African mole-rats (family Bathyergidae) comprise approximately 15 species (in 5 genera). This entry focuses on one of them, *Heterocephalus glaber*—literally "other-headed smooth"—perhaps the most bizarre-looking and social mammal on the planet. The behavior of naked mole-rats is important as a "missing link" between social insects (e.g., ants, termites) and cooperatively breeding vertebrates (wild dogs, acorn woodpeckers).

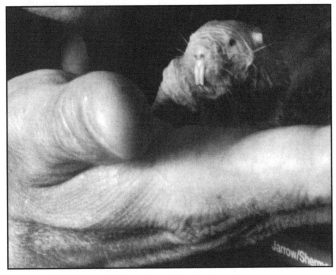

A man holds a naked mole-rat in his hand.
Courtesy of Paul W. Sherman.

Naked mole-rats are 12–16 cm (4.75–6.3 in) long and weigh 30–50 g (1–1.75 oz). They resemble overcooked sausages with buck teeth. They are not completely "naked," since they have scattered body hairs, long tactile whiskers, and hair-fringed lips and hind feet. However, their loose, wrinkled skin is nowhere concealed by *pelage* (the hairy covering of a mammal). They have tiny, nonfunctional eyes and minute ear pinnae. Naked mole-rats dig with chisel-like, procumbent incisors powered by muscular jaws. They are poikilothermic and ectothermic: Body temperature is regulated by conductive heat exchange across their thin skin.

Naked mole-rats inhabit underground burrows in Ethiopia, Somalia and eastern Kenya. Volcano-shaped mounds of soil ejected onto the ground surface during tunneling are the only evidence of their presence. Naked mole-rats feed on bulbs and tubers. Their main predators are snakes. Yet, safe in their subterranean fortresses, some individuals have survived more than 10 years; in captivity they sometimes live more than 20 years.

Naked mole-rats are renowned for their complex social behavior. They exhibit the three characteristics that define *eusociality* ("true sociality") in insects: overlapping generations, reproductive division of labor, and cooperative care of young. Naked mole-rats live in extended family groups containing 75–80 individuals on average, and 300 at the maximum. Within a colony only one female and one or two males reproduce. Nonbreeders are not physiologically sterile, but reproductive development is suppressed by physical harassment from the dominant female.

When a breeding female dies she is replaced by a colony mate, often after violent intragroup conflicts. The new breeder's body elongates, due to lengthening of individual vertebrae. She will accept a close relative as a mate and, eventually, bear four to five litters of 11–12 pups (and up to 28 pups) per year. She can rear large litters successfully because she

Naked mole rat using chisel tooth to dig.
© *Jum Jarvis / Visuals Unlimited.*

is protected and cared for by nonbreeders of both sexes. Small nonbreeders are the primary maintenance workers (carrying food, building nests, clearing tunnels), whereas larger (usually older) nonbreeders defend the colony against snakes and incursions by foreign colonies. Body size and associated behaviors vary continuously— there are no discrete worker "castes." Work is stimulated by the breeding female's aggression, and colony coordination is achieved via vocalizations (18 categories).

Naked mole-rats inhabit deserts where rainfall is sparse and unpredictable and food plants are dispersed and patchy. In brief periods after torrential rains, colonies dig cooperatively to locate sufficient food to enable them to survive through the next dry season (when tunneling is impossible). Dispersal is difficult and dangerous, so most individuals remain at home and wait for mating opportunities created by a breeder's death or colony fissioning. While waiting, individuals promote their genetic success by helping raise closely related pups. Thus both ecological factors and kin selection are implicated in the evolution of the complex social life of naked mole-rats.

Further Resources

Bennett, N. C. & Faulkes, C. G. 2000. *African Mole-Rats: Ecology and Eusociality*. Cambridge: Cambridge University Press.

Jarvis, J. U. M. & Sherman, P. W. 2002. *Heterocephalus glaber*. Mammalian Species, 706, 1–9.

Judd, T. M. & Sherman, P. W. 1996. *Naked mole-rats direct colony mates to food sources*. Animal Behaviour, 51, 957–969.

Sherman, P. W., Braude, S. H. & Jarvis, J. U. M. 1999. *Litter sizes and mammary numbers of naked mole-rats: Breaking the one-half rule*. Journal of Mammalogy, 80, 720–733.

Sherman, P. W. & Jarvis, J. U. M. 2002. *Extraordinary life spans of naked mole-rats* (Heterocephalus glaber). Journal of Zoology, London, 258, 307–311.

Sherman, P. W., Jarvis, J. U. M. &. Alexander, R. D. (Eds.) 1991. *The Biology of the Naked Mole-Rat*. Princeton, NJ: Princeton University Press.

Sherman, P. W., Lacey, E. A., Reeve, H. K. & Keller, L. 1995. *The eusociality continuum*. Behavioral Ecology, 6, 102–108.

Solomon, N. G. & French, J. A. (Eds.) 1997. *Cooperative Breeding in Mammals*. Cambridge: Cambridge University Press.

Paul W. Sherman

■ Nature and Nurture
Baldwin Effect

What is the relationship between unlearned or instinctive behavior and behavior that is intelligent, learned, or in some other way acquired by experience? This has been a perplexing question for at least two centuries, as scientists have grappled with trying to understand the diversity of behavior and the origins of psychological abilities of animals, human and nonhuman. The so-called Baldwin Effect was one of several late-nineteenth-century attempts to answer this question that has recently received renewed interest.

Some Background

As species evolved, how did new behavioral traits come into existence? One possibility was provided around 1800 by the French Biologist, Jean-Baptiste de Lamarck. Lamarck believed that animals and plants altered their habits and abilities as they met new demands in changing environments. These modified habits would be passed on directly to their offspring. This was the intuitively attractive idea of the inheritance of acquired characteristics. Thus, animals that had learned to run faster to evade predators would have offspring that would run faster as well. Lamarck thought that if you reared generations of birds in small cages in captivity, you would find that they inherited poorer flying abilities, as now found in domesticated birds. He pointed out that the many different dog breeds with their marked behavioral differences (bulldogs, water spaniels, greyhounds, and "lap-dogs") could only have evolved from wolves. The inheritance of individually acquired abilities passed on to offspring seemed to be the most plausible explanation.

A Japanese or red-crowned crane (Grus japonensis) displays itself. It holds out its wings as it struts, displaying for its mate.
© Tim Laman / National Geographic Image Collection.

Unfortunately for Lamarck, at that time there was no known mechanism of heredity and certainly none by which traits acquired by an organism in its lifetime could be passed on to the next generation. And even if possible, how could such changes spread and persist in future generations when the specific environmental change was no longer present? Charles Darwin, an English biologist, was the chief proponent of a different mode of explaining changes over generations in animal behavior. Darwin argued that natural selection would favor the survival and reproduction of animals that showed inherited variation in traits favored in certain environments, and that over generations such traits would spread and more animals would show them. Darwin thus provided a process that allowed for evolutionary change in behavior without the necessity for the direct evolution of acquired traits.

Nevertheless, Lamarckian inheritance still seemed too attractive a hypothesis to throw out. Many Darwinians and evolutionists in the late nineteenth century thought that natural selection by a "blind and accidental" process could not deal with the evolution of complex instincts and especially the change from instinctive to intelligent behavior in higher animals. Prominent among these was George John Romanes, Darwin's protégé on comparative animal behavior and comparative psychology. Romanes, Spalding, Herbert Spencer, Edwin Cope, and other evolutionists adhered to variants of the "lapsed intelligence doctrine," which held that through Lamarckian inheritance, learned behavior patterns could become instinctive. Natural selection was just not up to this task, being too slow and cumbersome to deal with the complexities of mind and behavior. However, with the acceptance of the findings in the 1880s by August Weismann, a German biologist, that reproductive (egg and sperm) cells were separate from the rest of the body's cells (soma) and not affected by changes in them, most evolutionists felt that Lamarckian inheritance was completely discredited.

Resolving the Conflict

Students of comparative psychology, very interested in the origins and evolution of behavior and mental life, began to tackle the problem and, remarkably, within a few months in 1896 three similar resolutions appeared in print. These were formulations by the comparative psychologist C. Lloyd Morgan, a friend of Romanes but much more of an experimentalist, the paleontologist Harry Fairfield Osborn, an advocate of an unyielding progressivism in evolution called orthogenesis, and James Mark Baldwin, a child and developmental psychologist. Controversy ensued over priority; the flamboyant and ambitious Baldwin "won" to the extent that the proposed resolution is now generally called the Baldwin Effect (Weber & Depew 2003). Baldwin himself termed the process "organic selection."

The effect can be described as follows.

1. Animals may react to environmental stimuli with innate, instinctive, or reflexive behavior patterns.

2. However, animals also can respond to the environment with modifications in structure, physiology, or behavior that adapt or accommodate them to enhance their survival and reproductive fitness.

3. This capacity to adapt is itself inherited, a product of natural selection, and represents an organism's innate plasticity. The specific phenotypic changes are not themselves inherited, however.

4. The animals with the acquired modifications not only are more likely to survive and breed, they will also accumulate randomly-generated genetic changes (mutations) that support the modifications (are in the same direction) and have similar adaptive consequences.

5. These genetic changes are favored by natural selection so that the original non-inheritable changes become inheritable.

6. The operation of this process is more likely in fluctuating or changing environments than in stable environments.

7. Advanced cognitive abilities, sociality, imitation, and even consciousness could have evolved in this way. As Baldwin wrote concerning a child, "the main function of consciousness is to enable him to learn things which natural selection fails to transmit."

Part of the controversy over the Baldwin effect is due to the juxtaposition of two claims: 1) that acquired traits can become heritable through an indirect process involving natural selection (as contrasted with Lamarckism), and 2) that the process also underlies the progressive evolution of intelligent behavior and consciousness. For scientists such as Baldwin and Morgan, who were interested in the evolution of mind, this concern with higher cognitive abilities was natural. However, to biologists interested in more pedestrian aspects of evolution, this approach was soon viewed as too fuzzy and speculative. On the other hand, traditional psychologists, whether experimental, social, developmental, or educational, found evolutionary issues irrelevant, unduly reductionistic, or heretical. Consequently, after an initial spate of interest, the Baldwin effect soon became marginalized in both evolutionary biology and psychology, and gene-centered evolutionary genetics became the dominant view.

Genetic Assimilation and the Baldwin Effect

In the mid-twentieth century, C. H. Waddington, a British embryologist, became a prominent advocate of the idea that phenomena like the Baldwin effect could operate in nature. Waddington argued, as did his predecessors, that random mutations and natural selection cannot alone account for complex traits and evolutionary change, especially if only the traits of reproductive adults are studied. He emphasized the importance of looking at developmental processes (*epigenesis*) and how they interact with environmental input. In fact, some traits are only seen when induced by environmental stimuli that expose the action of genes that were otherwise hidden. Today, we have many examples of genes being "turned on" by even fleeting exposure to stimuli.

The classic experiment that Waddington performed was to show that exposure to high heat 40°C (104°F) in fruit fly (*Drosophila melanogaster*) pupae at a specific time leads to the lack of crossveins in the adults. He then mated crossveinless flies with one another and also mated normal flies that did not show the effect. After a number of generations he found that some flies were developing the crossveinless trait even if they were not exposed to the heat shock. It seemed that a response to the environment could be incorporated into the genotype even in the absence of the original environmental trigger. However, unlike Baldwin and Morgan, Waddington did not think that random mutations were responsible for supporting crossveinlessness, but instead that hidden genetic variation underlying a trait became expressed and when enough of these accumulated in the direction of the induced change the environmental trigger became unnecessary. This accumulation of expressed genetic changes was termed *genetic assimilation* by Waddington.

A fruit fly (Drosophila melanogaster) *magnified 160 times.*
© Dr. Stanley Flegler / Visuals Unlimited.

These two ideas, the original Baldwin effect and genetic assimilation, lead to the same outcome: An initial phenotypic change once triggered could lead to widespread genetic change, at least when selected for in the laboratory. Recent examples of these effects and their molecular basis and possible operation in nature are discussed in Weber and Depew (2003), especially the chapters by Hall and Gilbert. More historical material on the Baldwin effect and genetic assimilation is found in Belew and Mitchell (1996). The implications of both on behavior and psychological traits are discussed at length in these volumes, by both advocates and skeptics.

The Importance of the Baldwin Effect

Today many evolutionary biologists, ethologists, and psychologists either do not know much about these two processes, the Baldwin effect and genetic assimilation, or do not see them as either plausible or important. In this respect the reception accorded these ideas is comparable to that accorded group selection. In both cases, the prominent position taken is that while the two phenomena are theoretically possible, there is little evidence of their operation in nature and, for theoretical reasons, they are at best marginal processes. However, there are examples where Baldwin-like processes may help explain some phenomena, especially in incipient species and in closely related populations. Consider the example (Burghardt 2002) of garter snakes, *Thamnophis sirtalis*, who respond at birth to chemical cues of natural prey species such as fish, worms, and frogs. These initial responses are not only innate but generally heritable. However, on one island a population was found in which it was not the initial responses of newborn snakes that was heritable, but their ability to switch to a population atypical fish diet that was heritable. This supports the notion that plasticity itself is an inherited trait, a basic postulate in Baldwin-like phenomena. It could be that the snakes in this population are evolving a plastic response that over enough generations would lead to a strong heritable preference for the new prey type.

The Baldwin effect has implications for conserving biodiversity, especially in vertebrates in which learning and domestication are major problems. If valid, the Baldwin effect points to limitations in the emphasis on genetic diversity in captive breeding populations and the ignoring of behavioral variation and other forms of plasticity. The Baldwin effect also offers a mechanism whereby complex displays arise that, such as courtship displays, incorporate behavior patterns from other contexts such as foraging, drinking, nest building, and fighting; a process called ritualization. It can also be deployed to help us understand the origins and role of extravagant sexually selected ornaments used in courtship, play, consciousness, language, and intelligence as well as more mundane traits, such as crossveinlessness in fruit flies. The growth of interest in molecular mechanisms of genetics and development may eventually reveal the role of these processes in the evolution of behavior.

See also Behavioral Plasticity
Darwin, Charles (1809–1882)
Domestication and Behavior
History—*History of Animal Behavior Studies*
Methods—*Deprivation Experiments*
Nature and Nurture—*How to Solve the Nature/
Nurture Dichotomy*
Nature and Nurture—*Nature Explains Nurture:
Animals, Genes, and the Environment*

Further Resources

Baldwin, J. M. 1896. *A new factor in evolution.* American Naturalist, 30, 441–451, 536–553.
Belew, R. K. & Mitchell, M. (Eds.). 1996. *Adaptive Individuals in Evolving Populations: Models and Algorithms.* Reading, MA: Addison-Wesley.
Burghardt, G. M. 2002. *Genetics, plasticity, and the evolution of cognitive processes.* In: *The Cognitive Animal: Empirical and Theoretical Perspectives on Animal Cognition.* (Ed. by M. Bekoff, C. Allen, & G. M. Burghardt), pp. 115–122. Cambridge, MA: MIT Press.
Weber, B. H. & D. J. Depew (Eds.). 2003. *Evolution and Learning: The Baldwin Effect Reconsidered.* Cambridge, MA: MIT Press.

Gordon M. Burghardt

■ Nature and Nurture
Nature Explains Nurture: Animals, Genes, and the Environment

Historical Background

The famous nature/nurture controversy turns around the causal contributions of genes versus environment to the behavior of both humans and animals. As such, it regularly pervades the scientific literature at least since the very beginning of evolutionary thinking and still today produces the hottest debates both in public and the academic world. For a better understanding of its ideological background it is therefore recommended to take a short look back to the recent history of modern biology. The conceptual foundations of the theory of evolution were laid down by important people such as Jean-Baptiste de Lamarck, Charles Darwin, and August Weismann during the nineteenth century. It is not by chance that these three persons stand out against the rest of the many other scientists who, at the time, were engaged in tackling the tricky problem of explaining the impressive variability of living forms. The French noble Jean-Baptiste de Lamarck, the actual founder of the idea of an "e-volution" in the living world, became famous for his "law of the inheritance of acquired characters," that is, the conviction that "everything that the individual gains or loses through the influence of the predominant use or continuous nonuse of an organ, is passed on to the offspring by reproduction." Inspired by the successful breeding methods of cattle farmers and Thomas Malthus' famous "Principle of Population" from 1798, which says that human populations grow faster than the food supply, the atheistic English theologian Charles Darwin complemented Lamarck's law by the now commonly accepted mechanism of natural selection, where only the fittest individuals of a given population survive and reproduce. Whereas Darwin was certainly the first to properly appreciate the natural origin of species through selective forces exerted both by the physical (natural selection in a narrow sense) and by the social environment (intraspecific selection, e.g., sexual selection), he was still ignorant regarding the true source of the required morphological and behavioral variability as the substrate of selection. He thus had no difficulty in agreeing with Lamarck's quite speculative hypothesis that many domestic animals had lost their capability to raise up their ears because domestication had made it unnecessary to reliably detect the treacherous noises of approaching predators. The German biologist August Weismann conducted the first extensive experiments on this controversial subject, which clearly demonstrated that there exists no inheritance of acquired characters. Not surprisingly, his severing of hundreds of mouse tails produced no visible effect (shorter tails) in the subsequent generation of young mice.

The Modern Controversy

How does the history of evolutionary theory relate to the contemporary nature/nurture controversy? If nature equals inheritance and nurture equals the influence of the environment, then ideas of Lamarck, Darwin, and Weismann are still applicable in the debates about what is today called "heritability." Lamarck noted a predominant influence of the environment, Darwin argued for a kind of dualistic influence, and Weismann was convinced of the importance of a given trait's full heritability for evolution. These three different positions still delimit the frontiers within the ongoing controversy.

1. The modern milieu theorists still stress the importance of the environment.
2. The interactionists still see the solution in a sort of dialectic compromise (half nature, half nurture); and
3. The so-called nativists still postulate that most traits must somehow be preformed from the very beginning of ontogeny.

The modern concept of heritability (h^2), developed and successfully applied by experimental psychologists such as, Hans Eysenck to promote a first quantitative assessment of the respective proportions in question, reflects this historical connection. The basic procedure is rather simple: One determines the variability of a predefined phenotypic trait in a particular population of individuals under certain environmental conditions. This trait can be complex things such as intelligence and personality in humans or any concrete behavior pattern or strategy in animals. If, during development, the environment can be assumed to be constant, or is artificially held constant, the remaining amount of variability reflects the influence of the genes. If, however, the genetics of the selected subjects can be assumed to be identical or at least similar, then the remaining amount of variability must necessarily be attributed to environmental factors. Values between these two theoretical extremes are thought to add up linearly following a simple standard model of additive effects. In other words, a heritability of 100% means that genes alone determine the manifestation of the trait under consideration, a heritability of 50% means that genes and environment have an equally important influence, and a heritability of 0% means that no genetic influence at all is involved, but instead 100% determination by the milieu (usually, there is a small additional variability of up to 10% associated with undefined "noise" which correspondingly reduces the total sum of these values).

The Twin Method

So far the logic of the method used, together with the associated statistical routines for the analysis of the data obtained (ANOVA = analysis of variance, multiple regression), are undisputed. Among other things this straightforward quantitative approach allowed modern heritability research to contribute to an important reinterpretation of human behavior in the light of evolutionary theory. Whereas during the first half of the twentieth century the predominant view was that a few simple laws of associative learning can explain practically all aspects of both human and animal behavior, it now became clear that the phylogeny of the species under consideration plays a much more important role than hitherto assumed. In the meantime it has become commonplace to know and accept that even human behavior—for the more "primitive" animal world this was only rarely contested—is, to a not-negligible degree, heritable. The hardest evidence stems from numerous studies on monozygotic human twins, who happened to be raised in completely distinct social milieus due to early adoption or other unforeseeable events (e.g., separation during war). Between 70% and 80% of their intelligence, measured as intelligence quotient IQ (individual score related to the population mean of 100), and between 40% and 60% of their personality traits seem to be in some (though in its detail still largely unknown) way influenced by the genes. This so-called "twin method" is certainly the most powerful approach to questions of heredity in humans and was first applied in a systematic manner by the psychometrist C. Merriman in 1924. Already at that time it was found that the behavior of identical twins was markedly more similar than that of fraternal twins, a result suggesting the presence of a genetic influence. The study was the first quantitative confirmation of Francis Galton's even earlier observation, in 1876, that identical and fraternal twins often develop in strikingly different ways. In the meantime, a continuous series of often quite extensive longitudinal studies has confirmed what a more detailed look at the mere external appearance by any interested layperson may reveal: The physiognomic and general morphological similarity is not only high, but often so strong that most unfamiliar persons have significant difficulty of perceiving a difference between monozygotic twins.

The situation is different in the case of animal behavior since, in contrast to human behavioral genetics, the generally preferred method here consists not of the quite reliable twin method (i.e., the comparison of fraternal and monozygotic twins), but instead more difficult and lengthy selection experiments and studies of inbred strains (which, in the long run, approaches the investigation of clones). Meanwhile, examples span the whole field of behavioral research, from genetic analyses of basic animal locomotor behavior (e.g., spatial orientation, migration) to the search for heritabilities underlying learning and problem solving capacities (e.g., conditioning, maze learning). In general, however, the measured heritabilities seem to be relatively low, ranging in most cases significantly below 30%. This is in striking contrast to the results found in human behavioral genetics where values of h^2 sometimes exceed 70%. And it is directly counterintuitive to the commonly held view that it should be primarily the animal behavior which is genetically determined, rather than the clearly more flexible and experience-dependent human behavior. At least one reason is obvious: The many heated public debates around the ever controversial human case certainly have pushed human psychologists ahead of their zoological colleagues.

The Genetic "Fate"

For a better understanding of the technical term *heritability* and how it conceptually relates to the nature/nurture debate, it is helpful to employ the idea of something like the probability of a "fate." For example, when a study provides a heritability of 40% for a concrete human behavior pattern, we then can assume that, on the condition that our parents have the trait, we as descendants would have a 40% chance of showing exactly the same trait. However, probability here precludes the idea that it is possible to divide the behavior of the individual, as it is often erroneously done, into a separate 40% genetic and a 60% environmental causation. This becomes clearer if we bear in mind that the basis of any heritability estimation always has to start with an actually existing behavioral variability within a given population. If there exists no variability, h^2 will be zero, since it is only the amount of variability which is attributed in a probabilistic manner to either the influence of genes or the milieu. Consequently, a heritability of 0% would simply mean that the chances of getting a particular trait as a member of the population are equal to zero, 50% that the chances are 50–50 and 100% that it will be impossible to escape the trait (or a particular variant of it). In this way heritability can be made understandable as a well-defined population character, which merely describes the probability of the realization of a particular trait independent of environmental influences. The latter, that is, complete independence of any milieu influence, was the original goal for developing this concept first in quantitative genetics. In particular, as long as it was completely unknown how the molecular basis of heredity was to be imagined and no concrete ideas of something like genes as the basic genetic units of organisms were present, heritability calculations represented the only feasible method of distilling out of the huge multiplicity of possible characters those traits which can be assumed to be transmittable from generation to generation, and hence be of relevance for evolution. Consequently, with increasing sophistication, quantitative genetics has been integrated into the more comprehensive field of evolutionary genetics where the simple additive model of phenotypic variance $V_p = V_G + V_E$ still forms the conceptual starting basis for all subsequent considerations (V_p: phenotypic variation; V_G: genetic variation; V_E: environmental variation).

The Variability of Heritability

If, however, h^2 is by definition a statistical population measure, its value must necessarily depend on a variety of factors. The first obvious factor is both the size and the type of the selected sample since it must be assumed that, in most cases, the resulting composition of individuals will be different. Indeed, how can we know which concrete genotypes we are confronted with when we conduct an IQ determination in a human population or a behavioral analysis in a natural animal population? In particular, size can have a considerable effect if a population is composed of individuals where the phenotypic variability of a trait itself differs from individual to individual. For example, there are cases where the variability of a given trait is even greater in an inbred (i.e., genetically homogeneous) line than in the outbred population from which this line was derived. In such a case, natural (or artifical) selection has favored, for whatever reason, a particular variant of the trait. This possibility was demonstrated in so-called *isogenic lines* of organisms (e.g., in fruitflies), where through a specific technique developed by H. J. Muller, all chromosomes in a population are made nearly *homozygous* (i.e., genetically very similar). In this case there remained a certain degree of phenotypic variability (e.g., concerning the number of bristles), which was assumed to arise from purely internal chance events. In the meantime a similar residual variability could also be proven for the orientation of inbred mice in mazes.

The second factor that influences the statistical measure of h^2 concerns the kind of species selected. Since species diverge through permanent genetic changes, it is probable that additional genes influence the *heritability* (the phenotypic expression of those genes), that produces a given common trait. In this respect, genes that are associated with behavior should be particularly concerned if they are—what is normally the rule—involved the manifestation of more than one single phenotype. Consequently, heritabilities measured for a trait in one species, of course, can, but must not necessarily, be applicable to any other related species. In general, the calculated values only very rarely coincide.

A third rather technical, but not less important, factor is the selection and definition of a phenotypic trait. As one may foresee, this is particularly difficult for complex behavioral phenotypes, as for example learning behavior. In addition to the existence of a variety of distinct learning mechanisms, from simple kinds of associative learning to complex patterns of emulation and imitation, there is often a whole sequence of distinct motor patterns coupled together in a very flexible way, not to mention the ontogenetical development of such behavior in time, which represents a problem in and of itself. As a result of these methodological difficulties, most researchers chose to simplify the problem and confined their analysis to a quantitative comparison of the starting and end points of more complicated behavior sequences. The definition of human intelligence is the best example. It is completely irrelevant how the subject proceeds in detail in his/her attempts to solve the problems posed by, if in school, the examiner or, if in a scientific investigation, the experimenter. The only relevant aspect is the attained quantitative score that alone enters the subsequent calculations. In animal learning research, the situation is almost the same. Usually, before deciding which species to investigate, the experimenter puzzles out a very general question—for example, are animals able to associate two conflicting stimuli (e.g., negative smell with visually appealing food), or are they able to develop and use a cognitive map for orienting in a given spatial area—and only then he is on the lookout for a species which potentially could fit the already predetermined setting. This procedure, in combination with a few additional purely technological constraints (e.g., breeding technique, availability, costs of raising) led to a relatively small number of test animals (mainly fruitflies, zebra fish, mice, zebra finches, and rhesus monkeys), which up to now are used for what is generally understood as the *comparative approach*. It must be clear that, by using an approach based on pure analogies rather than true behavioral homologies, it is highly unlikely that the genes underlying a specific behavior in those different species have something in common. As a result, heritabilities calculated for a variety of only superficially similar behaviors are hardly comparable across species, or even have any value with regard to a possible causal explanation of their phylogenetic origin.

How to Solve the Nature/Nurture Dichotomy
Adolf Heschl

The often heated controversy about the conflicting contributions of genes versus environment to the development and manifestation of both animal and human behavior is best understood as caused by a persistent confusion of two basically distinct processes, namely "heritability" of a biological trait (morphological structure, behavior pattern) and real genetic inheritance. Whereas the first simply describes the observable, that is, phenotypic stability of a trait during successive generations, the latter turns around the transmission of the underlying genetic information. As a rule, any kind of calculated heritability of a behavior always presupposes the existence of heritable genetic information, because truly novel phenotypic traits cannot arise without a change in the genotype (central dogma of molecular biology). This is even valid for learning and cognition, that is, behavioral strategies, which are commonly taken as going beyond the purely genetic constitution of an organism.

Phenotypic "Heritability"

Phenotypic heritability is a statistical measure for a population; it describes the probability that a given behavior is transmitted from one generation to the next and, as such, is an indicator of the strength of natural selection acting on a behavioral trait.

Typically, the results of heritability measures for most behavioral traits are not constant, but instead vary in dependence of a series of factors: definition of trait, selected population, sample size, and, most importantly, the environment. The influence of the latter is shown here with the help of a somewhat simplified example:

Environmental Stimulation	E1 Zero	E2 Low	E3 Mean	E4 High
Population A	10	20	30	40
Population B	10	30	50	70

The genetic variance between population A and B steadily increases from 0, 10, and 20, to finally 30 and this even though the environment becomes more and more favorable with regard to the behavior to be shown. In other words, a positive common milieu does not lower, but instead raises, the heritability of the investigated behavior; that is, it reveals the true amount of existing genetic variability. Such a picture is typical for rather recently evolved traits, which can be assumed to be still subject to a significant influence of directional selection, as for example many cognitive functions in both animals and humans (e.g., learning by trial and error, object permanence, concept formation, secondary representations, symbolic communication, self-awareness, Theory-of-Mind abilities, imitation of skills, tool use, abstract intelligence).

However, the vast majority of known behaviors displays a quite different pattern of phenotypic heritability:

Environmental Stimulation	E1 Zero	E2 Low	E3 Mean	E4 High
Population A	30	40	55	55
Population B	10	30	50	50

(continued)

***

How to Solve the Nature/Nurture Dichotomy (continued)

Here, the amount of genetic variance between A and B increasingly goes down as the environment becomes more and more favorable (20, 10, 5, 5). The behavior still shows some dependence on the environmental conditions—usually, this has to do with maturation and calibration processes during development—but beyond a certain level there is no more change possible. In such cases, evolution has already shaped the behavior in a way that at least a local (i.e., species-specific) optimum has been reached which is now held constant over time by a strong stabilizing selection. A significant reduction in genetic variation and thus phenotypic "heritability" is the consequence. In fact, contrary to all expectations it is primarily the ecologically important behavior patterns, such as most basic instincts and reflexes, which are characterized by rather low heritability values, that is, values ranging far below 30% (the genetic variance above is 12.5% for environment E4, heritability over all four environments is 11.1%).

Genetic Inheritance

Genetic inheritance is a material process; it describes the various modes (vegetative, sexual) of transmitting complete DNA sequences from generation to generation.

The transmission of genetic information always proceeds by discrete units; that means a gene either *is* transferred or *is not* transferred, but never at 50% as can happen without problems for a phenotypic trait after the estimation of its heritability, that is, the probability that the trait reappears in some relative (progeny, kin) in a like manner. Basically, any positive heritability value allows the conclusion that the formation of the trait is influenced by one or several (mostly many) genes. However, the opposite is not always true: The lack of heritability does not automatically mean that there is no genetic influence. There are at least two possibilities which can explain a heritability of close to zero.

1. The genetic variance of the trait has been reduced by a continuous and strong stabilizing selection (concerns over 99% of all genes of most higher animals).

2. The trait is solely determined by the environment.

Only the second point seems to confirm the equation "no heritability = no genetic influence," but even that conclusion is increasingly becoming doubtful. Learning, until now held to be the direct opposite of genetically fixed behaviors such as instincts and reflexes, is the best example. The question is: How does an organism know how to learn something? Unprecedented advances in the new field of neurogenetics unveil a steadily increasing number of genes which play a crucial role in learning mechanisms of all kinds. In fact, every single step in a learning process requiring important decisions to be made by the behaving organism seems to be subject to genetic influences. This has been recently shown in an exemplary manner for rats in which, during passing a standard learning task, the expression of no less than 30 genes is altered (possible master gene: FGF-18). On a more general level, transcription factors such as the highly variable gene families CREB, CREM, and ATF1, regulate the transition from short-term to long-term memory in a variety of animal species, from honeybees up to higher vertebrates. Finally, the discovery of the first potential candidates for genes influencing the manifestation of specifically human intelligence (a. IQ MajorGeneLocus; b. chimp/human comparison: differential expression of genes responsible for brain functions) suggests that evolution still continues in our own biological lineage.

See also Nature and Nurture—*Baldwin Effect*

The last, but maybe most important factor, which influences every heritability estimation is the environment itself. First, it is always very difficult to speak of a truly identical and hence "common" environment. As we have seen, even isogenic (i.e., genetically highly homogeneous) lines, tend to show at least some measurable morphological variability in spite of a milieu that is made artificially by the experimenter as uniform as possible. As already mentioned, such a residual variability may be caused by inevitable internal chance events, but it could equally be produced by some unknown and uncontrollable differences in the external environment. Heritability in this case must be assumed to be zero, but already a comparison with another isogenic population from the same species could possibly deliver a positive value, due to the selection of a different sample, where, for example, the fine distribution of bristles or hairs is again different.

More important for behavioral analyses is the fact that a changing environment very often changes the heritability of a given trait. This seems to be a rather surprising result since, intuitively, one would tend to expect heritability to be at least a constant measure over time, which, in addition, should be independent of the respective milieu. Or, asked differently: For what reason should heritability change if its value is commonly thought to reflect the proper genetic (i.e., immutable) part of a behavior? If, for example, one made education, in the sense of greater equality, available to everybody, the effect of heritability on intelligence or, to use a technical term, the so-called "general mental ability" g (first defined by Charles Spearman in 1904) would increase in the same proportion as the environmental influence is reduced. If one were to limit education to fewer beneficiaries (as still practiced in some societies), the opposite effect would be achieved. It was Hans J. Eysenck, one of the great pioneers of human intelligence research, who first drew attention to this fact. Hence it is not surprising that almost every new measurement of g in a new population regularly delivers new (i.e., divergent) results for h^2 (modern textbooks cite values ranging from 30–80%). In addition, the average heritability of cognitive abilities in humans—and in other primate species, it should not be much different—has turned out to change considerably with age, namely by increasing quite steadily from a relatively low value in childhood (about 30%) over a mean percentage during adulthood (about 50%) to the impressive score of more than 60% in people older than 80 years. This stands in contrast to what the convinced nativist would expect, since nativism basically says that it is mainly the earlier, more rigid patterns of behavior, which should be characterized by a high degree of genetic determination. "Innate," in its original meaning, just stresses the close connection to the situation at or at least soon after birth. The situation in animal behavior research is no different. Here too, it is only seldomly possible to transfer directly the result from one heritability measurement to another one, due to often considerable, and at the same time largely inexplicable, quantitative differences.

Heritability Is Not Inheritance

The important point now consists of understanding that, contrary to its misleading naming, *heritability* is only indirectly linked to real genetic inheritance and as such merely describes the supposed quantitative extent to which, under varying environmental conditions, already preexisting genetic information is transformed by the organism into a specific phenotypic trait. It, of course, remains an important function in measuring both the degree and the specific type of phenotypic exposure of a particular trait or behavior to natural selection, since Darwinian selection can only act on actually existing morphological and behavioral variation. This was already noticed by Francis Galton, who compared the physical height of human parents to the height of their descendants. He discovered that the mean size of a son or a daughter was often below or above the mean of the parents, if the parents were either exceptionally tall or exceptionally short (after which the technical term for this phenomenon is "regression to the mean"). In other words, the translation of genetic information into a phenotypic trait must not necessarily happen in a ratio of 1:1, which would be perfect heritability, but instead can, and in most cases does, follow much more complicated rules (e.g., dominance effects, pleiotropy, epistasis, nonadditive effects, gene–environment interaction). Concerning the effect on human body size, this comes down to a buffering of the genotype against natural selection, because even a small part of a given population is able to restore a comparatively large proportion of the original phenotypic variance due to sexual recombination processes.

Keeping in mind that phenotypic heritability is not the same as true genetic inheritance, we already foresee cases where a heritability rate of 0% for a trait does not automatically imply the conclusion that this trait is not inherited. Quite the contrary, since the overwhelming majority of all genes of most higher animals and humans (approximately 99.999%) shows no more variance in their associated phenotypic variability, potential heritability has also been reduced to close to zero. This is the logical result of the influence of natural selection which, in the course of millions of years, has eliminated the deleterious variants of most genes. Consequently, there is no more genetic polymorphism in all those supposedly perfectly adapted genes, and with it no more heritable variation. This finding results from the reversal of R. A. Fisher's "fundamental theorem of natural selection," where the rate of increase of fitness is postulated to be directly linked to the amount of genetic variance. Hence, if there is no more detectable variance, the result must be stagnation (i.e., evolutionary stasis). However, nobody would deny that these numerous functionally fixed genes are fully (i.e., 100%) transmitted to the next generation (e.g., number of extremities and fingers, types and internal arrangement of organs, basic structure of skeleton and musculature). Unlike heritability, real genetic inheritance is not a probabilistic process that could be described in percentages (apart from the distribution of genes during sexual recombination processes), but instead reflects the transmission of concrete molecular sequences that either are transmitted as a whole or not at all.

In some other cases we have again a phenotypic heritability of 0% and, at the same time, genes that simply migrate silently from generation to generation with either never (e.g., selfish DNA, noncoding "junk" DNA) or only from time to time generating phenotypic effects. With regard to the latter point, genetic imprinting in mammals is a particularly interesting case. Depending on the origin of some alleles (i.e., either from the mother or the father), they are either expressed (i.e., translated) into other molecules (RNAs, proteins) or silenced during ontogeny. Obviously, this gives the "silent" genes the big advantage to circumvent natural selection for at least one generation, but only at the price of a complete dependence on the still active part of the genome with regard to the resistance against natural selection.

Molecular Approaches

Today, the sophisticated calculations of pure quantitative genetics are increasingly complemented by a number of new molecular approaches which attempt to decipher in a more direct way the intricate causal relationships between the full range of the genetic information laid down in DNA sequences—so to speak the "deep structure" of the genome (codes for regulatory circuits)—and its milieu-dependent realization during ontogenetic development. Developmental geneticists, such as Sean Carroll, have already applied this approach with great success in the explanation of the evolution of the existing morphological diversity in two large metazoan groups of organisms (insects, vertebrates). Unfortunately, the situation for the analysis of behavioral development is somewhat different, because complexity in this case increases dramatically. Not only must the organs required for behavior (such as senses, nerves, brains, and muscles) be aptly developed, but also their permanent functioning must continue despite a continuous series of structural changes. However, there exists a strong trend within the newly founded discipline of molecular behavioral genetics showing that there seems to be no type of behavior, no level of complexity, and no degree of specificity that would not be influenced by genes. This becomes particularly evident if one looks at the investigation of learning behavior. Until quite recently, this was held to be the direct antithesis to so-called "innate" behavior (such as reflexes and other instinctive reactions), but now it turns out that practically every step of a learning process, complex as it may be, appears to be susceptible to the influence of a whole arsenal of regulatory genes. In fact, it can be justly claimed that learning itself is not only a heritable behavior, in the sense that there often exists an underlying genetic variation, but also that the detailed structure of its diverse mechanisms is fully inherited from generation to generation. Categories of associable stimuli, required spatiotemporal relationships, motor patterns to be used, in short, nearly every important parameter of a learning process, can be systematically altered by appropriate genetic manipulations. In this way, learning and cognition are a good example of how the traditional nature/nurture dichotomy is to be overcome. Whereas the influence of the environment is primarily a purely physical one (e.g., a simple shift in the local conditions or a real cognitive problem to be solved), it is up to the concerned organism itself to recruit the necessary internal information to cope in an adaptive manner with that same influence. This relationship becomes even more striking the more difficult the cognitive problem to be solved becomes. Take for example self-consciousness: Most apes (except gorillas) have no difficulty in recognizing themselves in a mirror because they are able to deduce from the visual correspondence between intended own and reflected movements—a cognitive procedure called "contingency testing" (probably derived from visually guided "branch testing," which is vital for large primates during climbing in the tropical canopy)—that the unknown conspecific in the mirror must be a picture of themselves, whereas, even after years of continuous mirror experience, monkeys never reach a correct understanding of this situation.

Conclusion

Both learning and cognition bring us back finally to the two basic mechanisms of evolutionary theory: random mutation and natural selection. If a phenotypic trait, such as learning, would allow an organism to assimilate in a directed way, truly novel information about the environment, this would, for theoretical reasons alone directly contradict the principle of random mutation. As such, this would support Lamarck's position which, in addition, assumes that the seemingly "new" information can also be transmitted to the next generation, and this not only with respect to morphological characters (e.g., calluses, strength of muscles), but also including behavior patterns (hence the misleading description of human culture as "Lamarckian"). The fact that even complex learning and cognition in animals seems to owe their proper functioning to already existing, though probably extremely intricate, genetic networks, suggests that it was Weismann rather than Lamarck and Darwin who, with regard to the question "nature or nurture?" was closer to the correct solution. There exists no new information per se coming from the environment which would additionally have to undergo a mysterious inheritance of acquired characters. The necessary information for successfully coping with the often dangerous influence of the environment must already be there in the genome. Future research in both developmental and behavioral genetics will show how far this prediction will reach.

See also Methods—*Deprivation Experiments*
　　　　　Nature and Nurture—*Baldwin Effect*

Further Resources

Books

Boake, C. (Ed.) 1994. *Quantitative Genetic Studies of Behavioral Evolution.* Chicago: University of Chicago Press.
Eysenck, H. J. 1973. *The Inequality of Man.* London: Temple Smith.
Falconer, D. S. & MacKay, T. F. C. 1996. *Introduction to Quantitative Genetics.* Reading, MA: Addison-Wesley Publication Company.
Maynard Smith, J. 1989. *Evolutionary Genetics.* Oxford: Oxford University Press.
Plomin, R. et al. 1997. *Behavioral Genetics.* New York: Freeman.

Journals

Heredity; Genes, Brain and Behavior; Behavior Genetics; Nature Genetics

Adolf Heschl

■ Navigation
Natal Homing and Mass Nesting in Marine Turtles

Eleven of the twelve families of turtles are essentially aquatic; two of these are marine, while the others are adapted to a freshwater environment. There are seven extant species of marine turtles. Sea turtles are fascinating for a number of reasons. Some, like the giant leatherbacks, foray into subarctic waters and dive down to several thousand feet in search of jellyfish, despite supposedly being cold-blooded reptiles. Others, like the loggerhead, are believed to hibernate in the mud flats off the Florida coast. The journeys some follow as they grow (*developmental migrations*) are even more impressive, such as those by juvenile loggerheads which search for feeding grounds off Baja California (Mexico) across the entire Pacific basin from their natal beaches in Japan. Most species of sea turtles undertake enormous long-distance breeding migrations, travelling thousands of miles between feeding and breeding grounds. And when ridleys undertake these migrations, they culminate in grand style, with thousands of turtles coming ashore to nest simultaneously in the space of a few days, a phenomenon known as *arribada*.

Whereas the leatherback turtle (*Dermochelys coriacea*) is the largest, measuring over 1.8 m (6 ft) in length and weighing over 500 kg (1,100 lb), the two ridleys are the smallest turtles. The Kemps ridley (*Lepidochelys kempi*), found only on the Atlantic coast of Mexico,

A nest of newly hatched turtles.
© Kartik Shanker.

is considered the most endangered of the sea turtles because a single nesting population was reduced to a few hundreds in the 1970s. There is some dispute about how different the Kemps is from the Pacific or olive ridley (*Lepidochelys olivacea*) which is widely distributed in the Pacific and Indian oceans and also at a few sites in the Atlantic.

Why do sea turtles undertake these breeding migrations? Do they return to the beaches where they were born? Where do they go once they leave the beaches as hatchlings? How many years do they take to mature, and what habitats do they occupy during their development? These are questions that have intrigued sea turtle biologists for decades. A brief glimpse into the complex life cycle of these marine reptiles may help clarify the answers.

A Brief History

Males and females begin the reproductive cycle by migrating from their feeding grounds to breeding grounds, which may be separated by several thousand miles. Courtship and mating occur primarily in the offshore waters of the breeding ground; the male mounts the female, holding her with claws in his fore flipper, and proceeds to mate. Both males and females may mate with several different individuals. Several weeks after mating, the females come ashore to nest, mostly at night. They crawl above the high water mark, find a suitable nesting site, clear away the surface sand (making a body pit), and dig out a flask-shaped nest with their hind flippers. This may be 2–3 feet deep depending on the size of the turtle. On the whole, sea turtles are remarkably stereotypic in their nesting behavior—so much so that Archie Carr, the pioneer sea turtle biologist, once remarked that "a sea turtle is a kind of turtle that never puts the same flipper back into its nest hole twice in succession."

Sea turtles are sensitive to disturbance and will return to sea without nesting if disturbed while coming ashore or digging the nest. On occasion, they may have to dig several pits to find a suitable site, and hence may have to come ashore many times before they are able to nest successfully. Once turtles start laying eggs, they fall into a *nesting trance* and are less easily disturbed. They lay about 100–150 eggs in the nest and fill it with sand; some, like the ridleys, thump the nest with their body to compact their nest. They then throw sand around the nest for camouflage and return to the sea. Most turtles nest more than once during a season, with roughly 2 weeks separating each nesting event. Whereas ridleys nest two to three times during a season, other turtles may nest more than five times during a single season. After they have completed nesting, they return to their feeding grounds until the next breeding migration. Again, ridleys are known to nest annually, though not all individuals may nest each year. Other turtles have longer remigration intervals, and some turtles may return only after several years. There is some evidence that this is determined by the amount of food reserve that the turtle is able to build up during feeding. Australian green turtles seem to nest in greater abundance after El Niño years, when nutrient upwellings are stronger in the Pacific where they feed.

The hatchlings develop in their nest over a period of 7–10 weeks. They hatch simultaneously over a period of a few days and then emerge from the nest together (in order to swamp predators), usually at night. Lower temperatures produce males, higher temperatures produce females. The pivotal temperature (i.e., the temperature that produces equal numbers of males and females) varies among species and populations, although it is usually around 28–32°C (82–90°F). The sex of the hatchling is determined during the second trimester of development. Sex ratio is likely to vary over the course of a nesting season and also between nesting beaches. Often, males are produced during one part of the season and females during the other, balancing the population sex ratio. Similarly, the sex ratio can also be maintained by spatial variation, with some beaches (dark sand beaches that have higher temperatures) producing females and other beaches (with light sand) producing males.

Hatchling emergence is nocturnal to avoid predators and sunlight. Sea finding is visual; the hatchlings seek a "brighter horizon" which is usually the moon or starlight reflecting off the surface of the sea. They also use silhouettes of sand dune and trees to orient themselves away from land and toward the sea. As soon as they enter the sea, they orient themselves to wave direction, swimming against the direction of the waves to reach the open sea. During this time, they also get imprinted on the earth's geomagnetic field. Hatchlings and adults are sensitive to both magnetic field intensity and magnetic inclination angle, and therefore have a compass sense. It is this imprinting and compass sense that enables them to migrate to their *natal beaches* (beaches where they were born) as adults.

Once in the sea, the hatchlings spend the first couple of days of their lives in a "swimming frenzy" when they use stored energy reserves to get into the open sea. Beyond this, they spend many years in a variety of juvenile habitats until they join other adults at feeding areas. Many turtle species were encountered after this pelagic phase at a stage known as "dinner plate." For decades, turtle biologists had no idea where these turtles were, and this was a period of their development known as the "lost year." This, however, is a phase that can last several years until the turtles move to developmental habitats in nearshore waters. During these years, when the turtles are lost to biologists, they spend their time floating on transoceanic currents, sheltered in seaweed rafts and fish aggregating devices. Some species (greens, loggerheads, and hawksbills) move to nearshore waters—in sheltered areas such as lagoons and coral reefs—to complete their development, after which they migrate to their natal beaches for breeding. For example, juvenile loggerheads feed in large numbers off the coast of Baja California on blue crabs, which occur in high densities. When one of these turtles was eventually tagged with a satellite transmitter, she swam straight across the Pacific Ocean to a known loggerhead nesting beach in Japan, covering a distance of about 15,000 km (9,320 mi).

The hatchlings face very high mortalities, especially in the early stages, when they emerge from the nest and swim past the nearshore waters. Predators include crabs, birds, jackals, feral dogs, and many fish once they are in the sea. Adults have few predators apart from sharks, killer whales, and humans, but occasionally nesting females may be killed by saltwater crocodile, tigers, and jaguars. Less than one in a thousand hatchlings is believed to survive to adulthood.

Natal Homing

Sea turtles have long been believed to nest on their natal beaches. For many years, this remained mere speculation. First, the hatchlings grow from a few centimeters in size to adults that are many times larger, ranging from the 80 cm (32 in) ridleys to the 180 cm (71 in) leatherbacks. Tags that would successfully last until adulthood (which could take 10 or more years) are not available. Second, considering that only about one in a thousand hatchlings survives until adulthood, the number of tags that would need to be applied to get significant results would be astronomical. Finally, research would have to be carried out for decades to demonstrate natal homing through tagging. However, in the early 1990s, a technique became available that could successfully address the question of natal homing in turtles—molecular genetic analysis.

Scientists demonstrated this technique in a landmark study on green turtles in the Atlantic Ocean. The Ascension Island is a speck in the middle of the Atlantic Ocean. The World War II pilots who refueled at this island had a saying that, "If you miss Ascension, your wife gets a pension." And yet, year after year, green turtles were known to find this island with amazing precision. Did young turtles follow older turtles to the breeding ground (*social facilitation*) or did they return to their natal beaches where they were born (*natal homing*)?

Mitochondrial DNA is maternally inherited and a particular sequence or *haplotype* is passed on from mother to offspring. Therefore, if hatchlings did not return to the same beaches where they were born, one would expect that haplotypes would be mixed between populations, especially when turtles from two nesting sites occupied the same feeding area. In this case, it was found that within the group that occupied the same feeding area off the coast of Brazil, one population migrated north to Surinam and another east to the Ascension Island in the middle of the Atlantic. When the genetic analysis was carried out, it was found that the haplotypes of the two nesting populations were completely different. This meant that green turtle females did in fact return to their natal beaches to nest as adults. Initially, this was taken as confirmation of the natal homing hypothesis for sea turtles in general, but more studies now suggest that the degree of natal homing is not as precise for all species and populations. Studies on the nuclear DNA of marine turtles shows a much lower level of population structure, indicating that there is gene flow between populations brought about by males. A recent DNA study on male green turtles in Australia shows that they also return to their natal beaches for breeding, suggesting that cross population gene flow might be a consequence of mating during breeding migrations.

Natal homing makes eminent sense for sea turtles. Feeding and breeding grounds are often separated because of necessity. The kinds of beaches that turtles need for nesting are often quite different from the areas where they find food. Since hatchlings spend many years in the open ocean, there is no certainty that they will find suitable nesting beaches when they reach adulthood. However, there is one beach that they know is almost certainly suitable for nesting, that is, the beach where they were born. These beaches may have been much closer in the past when the turtles evolved more than 100 million years ago. However, with the shifting of the continents due to plate tectonics, these beaches may have gradually moved farther and farther apart.

Mass Nesting or "Arribadas"

Some turtles return to their natal beaches not in ones and twos but in thousands. There are two species of ridley sea turtles. The Kemps ridley nests only on the east coast of Mexico. The olive ridley is globally distributed, but has mass nesting sites in Pacific Mexico, Pacific Costa Rica, and in Orissa on the east coast of India. At some of these sites, olive ridleys nest throughout the year, whereas at others like Orissa, there is a distinct nesting season which extends from November to April. Once the males and females reach the offshore waters of the breeding ground, they start to mate. In Orissa, the ridleys arrive in the offshore waters in October and November, and mating occurs largely in the nearshore waters of nesting beaches such as Gahirmatha. A few turtles nest in November and December, and nesting increases in January and February. Some time between December and March, triggered by mechanisms that are not understood, possibly environmental cues such as wind or temperature, an arribada begins. The beginning of an arribada is marked by the nesting of a large number of turtles—maybe 100 or more. The arribada peaks on the second or third day when there are turtles nesting every few feet. During this period, 50,000–100,000 turtles may nest in a single night on a beach 2–3 km (1.25–1.8 mi) long. In Costa Rica and Mexico, multiple arribadas take place during the year, and more than 10 arribadas have been recorded in some years at Playa Ostional, Costa Rica, and La Escobilla, Mexico, with annual estimates of 500,000 and 1 million nests, respectively.

Normally, sea turtles are fairly sensitive to disturbance and will return to the sea if disturbed during the nesting process. However, during an arribada, they come ashore in swarms and nest as long as they can find a suitable site. Many turtles that nest during the latter half of the arribada may dig up the nests of turtles that have nested earlier. Satellite telemetry studies show that females may not be associated with each other in the offshore waters during the internesting period, but as they approach the arribada, an aggregation forms in the nearshore waters. It has been noted both in Costa Rica and in Orissa that ridleys are often found in small congregations in the nearshore waters of the arribada beach, called reproductive patches, where courtship and mating occur. In Orissa, although turtles nest at arribada beaches separated by 300 km (186 mi) (covering several thousand square miles of nearshore waters), at least one reproductive patch off the nesting beach at Gahirmatha measured about 50 km^2 (19 mi^2).

Most sea turtles renest at intervals of 10–20 days within a nesting season. Arribadas on the other hand occur at intervals of 30–40 days. Ridleys hence have to extend the period between nesting events, and reproductive biology studies have shown that older animals may be better equipped to do so and thus nest in successive arribadas. Younger animals may be forced to nest solitarily after nesting in an arribada because their reproductive systems do not allow them to retain the eggs long enough till the next arribada occurs.

Male and female turtles that were satellite tagged off the coast of Costa Rica migrated into different areas of the Pacific Ocean, some going due west, others south and some north. In Orissa, anecdotal accounts, including reports by ship captains, have indicated that large numbers of turtles migrate along the east coast of India northward to Orissa in September and October. Recently, out of thousands of ridleys that were metal tagged in Orissa, about 20 were recovered in Sri Lanka. Of four turtles that were fitted with satellite transmitters, one migrated to Sri Lanka over a 3-month period.

Why do turtles nest in such extravagant fashion? This becomes clear 7–8 weeks after an arribada when the eggs hatch and the hatchlings emerge from the nest. Under normal circumstances, most hatchlings will be preyed upon by natural predators including, crabs, birds, and jackals. Of course, today, most are preyed upon by human-related predators such as feral dogs and crows. When each nest hatches singly, most of the hatchlings are eaten before they reach the sea. However, when hundreds of thousands of turtles emerge simultaneously, they are far more than the predators can handle, and a large proportion escape into the sea. Studies have shown that more than 50% of solitary nests may be preyed upon, compared to less than 10% of arribada nests. This "predator satiation" is a principal explanation for this extraordinary phenomenon.

Sea turtles are declining globally due to intense exploitation over many decades (in some cases, centuries) for meat, shell, and eggs. Today, many die as incidental catch in fisheries, and others are exposed to other consequences of development such as habitat loss. Artificial illumination on many nesting beaches disorients hatchlings when they emerge. Most populations are dependent on conservation programs around the world.

The Kemps ridley declined to a few hundred turtles in the 1970s and was the target of a massive rescue operation in the last 2 decades which has resulted in the recovery of the population. Similarly, the olive ridleys on the west coast of Mexico declined due to severe exploitation, but strict protection has led to an increase in the number of turtles, and about a million turtles have been recorded nesting in recent years. On the other hand, some nesting beaches in Mexico, Costa Rica, and Surinam have declined completely, and no arribadas occur any longer at these sites. The Orissa population has also been subjected to high mortality in recent years, with over 10,000 turtles counted dead on the coast each year due to fishery related incidental mortality. Not only do these species need to be conserved for posterity, but also the phenomenon of their mass nesting, which surely must rank as one of the great spectacles of nature must be preserved.

See also Behavioral Physiology—*Turtle Behavior and Physiology*
Navigation—*Spacial Navigation*
Navigation—*Wayfinding*

Further Resources

Bowen, B. W. & J. C. Avise. 1994. *Tracking turtles through time*. Natural History, 103 (12), 36–42.
Bruemmer, F. 1995. *La Arribada*. Natural History, August 1995, 36–43.
Carr, A. F. 1967. *So Excellent a Fishe: A Natural History of Sea Turtles*. Garden City, NY: Natural History Press.
Carr, A. F. 1986. *Rips, FADS, and little loggerheads*. Bioscience, 36(2), 92–100.
Lohmann, K. J. 1992. *How sea turtles navigate*. Scientific American, 266(1), 100–106.
Shanker, K. 1999. *The odyssey of the olive ridley*. Resonance, July: 68–78.

Kartik Shanker

Navigation
Spatial Navigation

Spatial navigation refers to the ability of animals to maintain whole body movement relative to a spatial reference. However, the real question is which path to follow? (For example, if you decide to leave the classroom and go to the library, you would follow a certain path, but how do you keep track of where you are going?)

Many may think that moving in space is of a simple nature, since children, or even ants, can do it. This may be because most of the complex details of spatial processing is automatic, that is, hidden from conscious experience and control. Many people will be surprised to know that spatial navigation is of interest to many scientists working in various fields ranging from cognitive psychology to molecular neurobiology. One of the main themes common to all these fields is to understand behavior and how it is controlled by the nervous system. Spatial navigation, in turn, provides one of the clearest examples of behavior that can be studied in natural and lab settings. In addition, many people, dating back to Aristotle and Charles Darwin, have reflected on some fascinating spatial behaviors, like that of migrating monarch butterflies or homing in pigeons.

Spatial Navigation Is Adaptive

The importance of spatial navigation is evident when one travels to a new place, or a nestling bird leaves the nest for the first time. In such cases, learning about the home range (known as exploration), the location of resources, what places to avoid, and the best routes to take is important for survival. One knows how a wrong turn in the desert can lead to very serious consequences. Therefore, minimizing the amount of travel, by taking the shortest routes, is essential. Thus, novel environments motivate exploration, which, in turn, reduces the novelty, as well as fear, of the traversed environment.

Short-Range and Long-Range Spatial Navigation

Short-range navigation refers to travelling through the immediate home range, from a few inches to a few miles depending on the species. Long-range navigation, like migration and homing, is travelling across unfamiliar terrain, often hundreds or thousands of miles, to reach a goal.

Migration is a very costly behavior in terms of energy and cognitive (neural circuitry) demands. This is perhaps what motivates many people and scientists to study and learn about spatial navigation. All migratory behaviors are probably adaptive (an adaptive character is one whose presence and appearance during evolution resulted in better health and more offspring). Not many would argue that bird migration from the arctic areas to overwinter in tropical and subtropical areas is adaptive.

However, the fact that animals move from a less optimal area to a more resource-rich areas is not the outstanding feature of migration. It is other aspects of migration that make one marvel at such behavior. The monarch butterfly (*Danaus plexippus* L.) will serve as a good example here. The main aspect that makes the monarch's migration legendary is that the parent generation (generation 1) that starts the journey in early summer dies out, after breeding somewhere in the central United States, leaving a new generation behind. That first offspring generation (generation 2) continues in a southwesterly direction until they, or their offspring (generation 3), reaches the specific overwintering area: the Transvolcanic Mountains just north of Mexico City. After a few months in early spring, the "frozen" monarchs become active, and a breeding episode begins. Afterwards, they start their journey back to their spring and summer homes. Again, a new generation makes it to the final destination (generation 4–5). So, unlike most migrants, where the same individual makes it back and forth, the monarchs that started their journey toward Mexico never actually see it. Also, the migrant monarchs that make it back to Canada have never actually been there before. Yet, every year, monarchs make their southward journey to end up in the same roosting location: the Transvolcanic Mountains. This is referred to as *site fidelity*. Although we now know a lot about the migratory behavior of the monarch butterflies, there is still much that we do not know. Some of the questions that scientists are trying to answer are: How did the new generations know where to proceed? Why did they not move northward or eastward, instead of southward? Why site fidelity (i.e., going back to the same location over and over)? How can they maintain their direction when sometimes it is sunny and at other times it is cloudy, windy, and so on?

Mechanisms for Spatial Navigation

As you will notice, both short-range navigation and long-range navigation are based on essentially similar mechanisms that operate on different geographic scales.

To reach a remembered goal location, an animal can rely on one of several mechanisms. The animal could simply follow a guiding cue that is closely associated with the goal. This is the same strategy that Dorothy used to reach the Wizard of Oz palace; she simply followed the yellow brick road. There are also equally fascinating examples of animals following a guiding cue towards a goal. Have you ever watched loggerhead sea turtle hatchlings (*Caretta caretta* L.) leaving their beach nest? If you did, you will notice that they move impulsively to the seashore. How do they know where to go? Well, they have an *innate* (inborn or unlearned) tendency to move toward the lower, brighter seaward horizon and away from the dark, elevated silhouettes of vegetation and dunes where they can be lost or eaten. Presumably, the same thing happens with the monarchs, where the new generations proceed southward in a predetermined path that was not learned from parents.

There are plenty of examples of using a guidance strategy in short-range navigation. One of the first and most familiar examples is following an odor trail by ants. As a matter of a fact, one rarely observes ants scampering randomly, and not marching as a procession along an edge or a boundary. The same thing can be said about mice, which are so adept at walking along walls that elders tell the common fable of mice magnets behind the walls.

The second important mechanism that animals employ for navigation is called *path integration*, or dead reckoning. When homing by path integration, a departing animal continuously monitors its displacements and uses this information to keep computing the locomotion vector that would take it back to the anchoring (start) point where the process of path integration originally started (see the first figure on p. 798). Recently, a lot of work has been done investigating the process of path integration after decades of neglect. It now appears that path integration plays a prominent role in animal navigation, as will be seen in the following examples.

The most familiar example is that of a *beeline*. Forager bees set out each morning looking for food. After a long hard search finding food, which may stretch over several miles, a bee scout heads back to the hive to recruit more foragers. The bee's path back to home is usually a straight line; hence the name beeline. This remarkable feat by this "simple," tiny creature goes almost unrivalled. How can they do it? A German scientist found that bees rely on the sun, a faraway landmark that does not change position with one's movement, to calculate how many turns it made during its search. This allows the bee to compute (a process akin to mathematical integration) a vector whose angle points toward the hive. This is called *sun-compass orientation*. However, calculating the angle home does the bee little good if she does not know how much distance was covered. A vector (solid arrow in the first figure) requires computations of both direction and distance. A bee estimates how much distance it covered by relying on how much energy (food) it consumed during the trip and possibly on *optic flow* (the speed of movement of visual images over the retina).

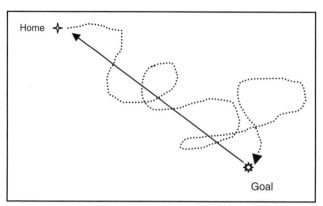

Homing by path integration requires an animal to integrate past displacements, i.e., turns and distances, and calculate a vector (solid arrow) that will take it back to home.

Courtesy of Sofyan Alyan.

Another example of setting out in a certain direction and proceeding on a certain path is that of the monarchs. Monarchs rely on winds to carry them in a southwesterly direction. However, on many occasions they drift with strong winds. They steer back and correct for deviations by relying on the sun, or, on cloudy days, on polarized light or magnetic compass cues. Homing pigeons also rely on a sun compass to proceed along a certain path. The use of a sun compass actually is widespread in all taxa within Kingdom Animalia.

One may ask, however, how can the sun be used as a compass cue when it is on the move 24 hours a day? A compass cue needs to be stationary, so that an animal can calculate accurately the angles that were made during the journey. Well, scientists are now pretty sure that animals that rely on a sun compass have an internal clock that keeps track of how much time has passed and then compensates for the sun's movement. The second figure shows that process in simple terms.

Do we now know how to decide whether a navigational mechanism is path integration or not? Yes. If orientation involves the use of a compass, counting steps and turns, or addition or subtraction, it is path integration.

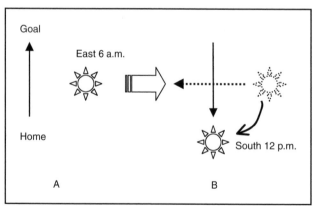

A simple example of time compensation for the sun's movement. In A, the ant moves toward the goal at 6 a.m.; i.e., the sun is in the East, to the right of the ant. After 6 hours searching and feeding, the ant heads home. The ant that did not compensate for the sun movement will keep the sun to its left and move away from home (dashed arrow). On the other hand, the ant that compensated for the sun movement (which is now in the South) will head toward the sun and home.

Courtesy of Sofyan Alyan.

The third and last orientation mechanism that animals may use to find their way is by relying on constellations of landmarks, the so-called cognitive map. A *cognitive map* simply means a stable, memory-based representation of the home range. This includes representation of distances and directions connecting recognizable places of interest, as defined by unique landmarks (i.e., unique places and their spatial relationships). The nature of navigation by a cognitive map has been always controversial. One seminal book about cognitive maps was that by John O'Keefe and Lymn Nadel (1978). In their book, O'Keefe and Nadel argued that rats can form cognitive maps that allow them to make novel (untravelled paths) shortcuts between different locations, and that flexibility in orientation is a defining characteristic of such maps. However, such flexibility has yet to be shown in any animal species. This is not like saying animals cannot form cognitive maps, but rather, there has not been one single experimental procedure from which one can reliably say a cognitive map has been formed. One of the problems that has been recognized recently is that all three mechanisms (that is, guidance, path integration, and cognitive mapping) run simultaneously, which is good. Imagine an ant following an odor trail leading to its nest, and that part of the trail was removed by wind. The ant could still go back to its nest by referring to the sun compass and surrounding landmarks. Distinguishing among the mechanisms is difficult to do in experiments. Also, the same cue, such as a light, may be used to indicate a place as part of an arrangement with adjacent lights, or it could be used as a compass cue for path integration. Another serious problem is that the defining criterion for path integration is flexibility, just like that for cognitive maps. For example, ants, unaware that they were carried to another location when at the goal, would make a return trip toward home and stop and start searching for their "nonexistent" nest at the correct distance and direction from the "unreal" goal. An ant forced into making a detour in this uncharted terrain would make a novel shortcut to reach the original path and proceed on it.

Nonetheless, there have been many studies showing landmark learning in animals. One of the first and simplest demonstrations was landmark learning in the beewolves, or digger wasps (*Philanthus traingulum*). Niko Tinbergen, one of the founders of ethology, showed that when a beewolf leaves her ground nest into which she just deposited her eggs, she hovers in the area for a while, and then takes off looking for prey. The beewolf later comes back to the nest where she leaves the paralyzed insect prey for her offspring to feed on. Tinbergen thought that the beewolf learns about the landmarks surrounding the nest, and this is why beewolf hovers before departing. To prove that, Tinbergen arranged rings of pine cones around some nests. After the beewolf wasps departed, he moved the pine rings to a nearby location, which had no nests. The wasps came back and started searching in the middle of the now displaced pine cone rings, searching, unsuccessfully, for the nest entrance. This proves that the beewolf wasps formed an image of the pine cone rings to locate the nests.

Perhaps one of the most impressive examples of using landmarks to locate a goal is that provided by the seed-caching (storing) birds. One species of seed-storing bird is the Clark's nutcrackers (*Nucifraga columbiana*). These nonmigratory birds tend to store pine and acorn seeds in the ground using their beaks. Thus, the birds can survive the harsh winter by recovering their caches often months after they stored the seeds. Some of the questions that scientists tried to answer were: Can the birds remember where they cached their seeds? If so, what is the success rate of recovering the caches, knowing that some birds may have up to 9,000 caches covering many square miles? Scientists found that the birds rely on nearby prominent landmarks, such as trees, shrubs, rocks, and others, to locate the caches. Since the ground would be covered with snow during the winter, the appearance of the cache sites themselves is unimportant. Lab experiments showed that the success rate exceeds 50–70%, which indicates quite impressive spatial memory ability in these birds.

Another interesting aspect of spatial memory deals with sex differences. Many spatial tests with mammals, including humans, found that males are generally better than females in different tests of navigational ability, especially those testing rodents in a Morris water maze. In a Morris water maze, a rodent is dropped into a tank full of cold cloudy water that has nontoxic chalk added to it. All that the rodent is required to do is to find a submerged platform within the tank, climb on it, and wait for the experimenter to take it back to its warm cage. After several repetitions, the rat becomes skillful at finding the location of the submerged platform. The speed of learning is called the *learning curve* in psychology jargon. For example, it usually takes a rat 120 seconds on the first trial to find the submerged platform, and only 5 seconds on the 10th trial.

The reason for such spatial differences between males and females could be conditions such as breeding status. Females in the breeding season do not tend to move a lot, because they have to nurse their babies and protect their nests, whereas males are on the lookout for new females to mate with.

There is one main problem associated with such conclusions from studies using the Morris water maze. The procedure tests the speed of solving the maze, and not specifically spatial ability. It usually takes females longer to solve the maze, that is find the submerged platform, as in our example. But they do eventually solve the problem. In addition, even though females in their natural habitat tend not to move or wander a lot during the breeding season, they surely wander in an area larger than the laboratory room in which they are tested, and they need to find food there! So, it seems that such sexual differences may arise due to motivational factors and may not reflect true spatial ability differences.

True Navigation

True navigation is used to refer to homing phenomena exhibited by numerous species that orient toward an imperceptible goal that is hundreds or even thousands of miles away. The criteria for true navigation are, first, that the goal or the starting point, as well as most of the traversed terrain, is totally unfamiliar to the navigating animal; and second, that the homing animal successfully returns to the same exact site when displaced (i.e., site fidelity that was mentioned above).

The monarchs, salmon, homing pigeons and green turtles are some of the true navigators. The best studied species, and for which the name true navigation was coined, is the homing pigeon (*Columba livia*). Homing pigeons have been known for centuries for their legendary site fidelity and navigational abilities. Homing pigeons were first employed as mail agents in the middle ages. However, it is only in the mid-1950s when scientists started systematic investigations into their homing abilities. Pigeons have been known to home to their natal loft (birthplace) after being displaced hundreds of miles, and there are some reports of legendary homing feats from thousands of miles. How can they do that?

Many experimental manipulations have failed to stop the heroic pigeons from homing. These manipulations have included anesthetizing the pigeons and constantly rotating them while transporting them to a distant unfamiliar location, placing dark blinders or frosted lenses on their eyes, attaching magnets to their heads, and other drastic attempts at deprivation. What this means is that homing pigeons do not need to make use of any information collected during their journey far from home, like distance, terrain landmarks, and so on. In addition, this means they can home from any release point. There is one main conclusion from all these experiments: Pigeons must have a "global" map that enables them to pinpoint home from anywhere on Earth, so to speak! This map would contain information of where the loft (nests) is relative to where they are at the moment. But what cues are used to build this map? Surely they are not visual landmarks or acoustic cues. The only manipulations that affected the homing success of pigeons were those eliminating their sense of smell (*anosmia*), like anesthetizing their "noses," filtering the air during the outward journey, or cutting the olfactory nerves. It seems that pigeons build an "olfactory map" from some unknown atmospheric gases while they are growing up at their loft. They associate the odor cue(s) with wind direction. If released in a different place, they compare the current concentration of the atmospheric olfactory gases that make up the map with the concentration at the loft, and decide in which direction to go. The main problem with this hypothesis, however, is that scientists have yet to identify the atmospheric gas(es) that could serve as a global cue for this map.

Now, once the direction to home is determined by this olfactory map, the pigeon maintains its course by relying on a sun, magnetic, or other compass cue. When the pigeon arrives at the immediate vicinity of the loft, she might then orient by relying on landmarks (cognitive map) to arrive at the loft.

Thus, one can deduce that the difference between long-range and short-range navigation mechanisms is the formation of a large-scale, geographic map, like the olfactory map of homing pigeons.

The Hippocampus and Spatial Navigation

One further evidence for the similarity between long-range and short-range navigational mechanisms are the findings of similar brain regions responsible for processing spatial behavior. One brain region, the hippocampus (Latin for seahorse), has been strongly implicated in spatial navigation and spatial memory. However, the exact involvement of the hippocampus is not well understood. Many claim that the hippocampus is the brain region where cognitive maps are formed, while others believe that the hippocampus is involved in forming flexible kinds of different memories, not just spatial ones.

See also Navigation—*Natal Homing and Mass Nesting
    in Marine Turtles*
    Navigation—*Wayfinding*

Further Resources

Dingle, H. 1996. *Migration: The Biology of Life on the Move*. New York: Oxford University Press.
Drickamer, L., Vessey, S., Jakob, J. 2002. *Animal Behavior: Mechanisms, Ecology, Evolution*. 5th edn. New York: McGraw Hill Co.
Jander, R.1975. *Ecological aspects of animal orientation*. Annual Review of Ecology and Systematics, Vol. 6, 171–188.
Papi, F. 1992. *Animal Homing*. London: Chapman & Hall.
Spatial orientation, special issue. 1996. The Journal of Experimental Biology, Vol. 199.

Sofyan Alyan

■ Navigation
Wayfinding

Home ranges of human and nonhuman animals are two- or three-dimensional spaces of various shapes wherein individuals or groups of individuals restrict their daily activities. Most bilateral animals—from "worms" (ribbon worms, earthworms, etc.) to humans—occupy temporary or permanent home ranges; only those constantly on the move do not. Sometimes home ranges are left for dispersal, or for seasonal or reproductive migrations. For home ranges to be useful, animals not only have to find their ways around therein, but they also have to know where to search for food, where to find shelters and other resources, and finally they have to know how to by-pass hazardous or impassible locations. Learning and controlling efficient and safe locomotion through a home range is the demanding task of the topographic orientation system. Such locomotion may be by swimming, crawling, walking, jumping, or flying.

Up to about a century ago, it was widely thought that animals that move away from their shelter and then easily find their way back, are not guided by a learned natural skill, but by some unexplainable, mysterious "instinct." Humans alone were to supposed command spatial intelligence. How wrong a dichotomy. Based on comparative studies, we big-brained humans have to admit that there is no evidence that our unaided skill of wayfinding in natural environments is superior to that of other large mammals or of birds. Surprisingly, some animals easily beat humans in special spatial memory tasks. As an example, take Clark's nutcracker (*Nucifraga columbiana*) in the Rocky Mountains, whose brain weighs a mere 10 grams. Each fall, this relative of jays and crows hides pine seeds in the ground at several thousand different locations. This cache will be the main food supply for months to come. Efficient retrieval of this vast supply requires that the birds not only remember the precise locations of these numerous hiding places, but that they also keep track of all previously emptied caches. Turning from birdbrains to microscopic insect brains, even the navigation skills of honeybees are hard to beat by humans. Imagine a worker honeybee in its natural environment flying several miles away from its tree nest over some roadless wilderness, then randomly meandering to visit several hundred flowers for nectar, and finally managing to fly back successfully to the very one tree among hundreds, in which the colony resides.

Given these impressive spatial cognitive faculties of animals—even when measured by standards of human spatial intelligence—it is no wonder that countless researchers have tried to find out how they manage to do it. Even though there are myriads of pertinent research publications, our knowledge is still woefully incomplete. Much of the past research on animal home-range orientation has focused on only three animal groups: the related ants and bees; rodents, with special emphasis on rats, house mice, gerbils, and hamsters; and finally, we researched ourselves. In recent years, engineers complemented this animal research by creating and building *animats*, that is, robots that emulate and test the navigational mechanisms that biologists have uncovered in animals.

The sizes of home ranges differ substantially among members of the same species and vary tremendously from species to species. Whereas a foraging fiddler crab (*Uca rapax*) on a mudflat moves up to 2m (6.5 ft) away from its burrow, the comparatively tiny wood ants (*Formica rufa* and relatives) march up to 200m (650 ft) for food, and worker honeybees (*Apis mellifera*) fly distances up to 10 km (6 mi) on a single nectar-collecting trip. Home ranges of feral house mice (*Mus musculus*) and of those of deermice (*Peromyscus leucopus*) tend to cover up to a few thousand square meters; those of wolf packs or brown bears a few hundred square kilometers; and desert elephants in North Namibia have to roam over close to 3,000 km² (1,158 mi²) in order to satisfy their enormous needs for food and water. Finally the record of all, the wandering albatross (*Diomedea exulans*), the largest flying bird, has been shown to cover several thousand kilometers on a single foraging flight away from its subantarctic island residence.

The striking similarities and minor differences in the advanced mechanisms of home-range orientation, as implemented in insects, vertebrates and animats, together with straightforward conceptual links between advanced and simple methods of home-range orientation, allow a unifying theoretical framework, as sketched in the following.

The Topomotor System, the Backbone of All Spatial Control of Locomotion

All organisms capable of active locomotion, from bacteria to people, must be able to steer their locomotion in space. Otherwise, locomotion would be useless. The universal control system required for this steering task is the *topomotor system*. Topomotor control comprises at least the quantification of forward movement and of turning. To this may be added, depending on the species and the medium, the control of backward, sideways, upward and downward locomotion.

All bilateral animals, whether or not they have home ranges, share three primitive topomotor patterns. These are:

1. straight locomotion used to move toward a perceived or expected resource or away from perils;
2. U-turns used to redirect straight locomotion in response to changing conditions; and
3. search behavior performed with meandering left and right turns in order to cover thoroughly a more or less restricted area, thereby increasing the likelihood of finding a resource.

Starting from such primitive beginnings, topomotor control increasingly developed its complexity and sophistication in conjunction with the increasing complexity of spatial orientation demands, especially those for the control of home-range orientation. The following sketch will describe this, progressing in three major steps from the simplest to the most advanced instantiations.

Guided Excursions from Home

Guidelines are common aids for guiding movements in home ranges. Trails of various types, especially self-generated odor trails, are commonly used. Other guidelines that animals sometimes follow are various channels and tunnels; edges along forests, along hedgerows or along streams of water, crestlines on top of hills, on mountains, or on fallen vegetation.

Given the three prehoming topomotor patterns mentioned above, the addition of odor-trail production and odor-trail following is all it takes to create the first and simplest home-range orientation mechanism in a primitive bilateral animal. Exactly such a primitive homing mechanism is implemented in the intertidal Seattle ribbonworm (*Paranermetes peregrina*). Moving straight out of its burrow the worm deposits a slimy odor trail. Away from home a U-turn may initiate a return; alternatively, the worm meander-searches for its annelid-worm prey and then for its trail to find the way back home.

In general, substrate-bound animals with poor vision or animals in total darkness tend to follow guidelines to find their goals in their home ranges. Most worker termites are blind and rely mainly on odor trails and tunnels when foraging away from the nest, and ant species with limited vision do so also. The virtually blind arboreal tent caterpillars (*Malacosoma americanum*) use scented silk threads to guide themselves from their communal silk tents to foraging sites and back home. Similarly, a house mouse in total darkness finds its way in flat open space by following the odor trail it left behind. For convenience, people often travel to distal goals simply by following trails, roads, or streets.

Topomotor-Based Excursions

Exclusive topomotor-based traveling and homing requires substantial, task-specific structural expansions of the primitive topomotor mechanism. A simple self test demonstrates the new requirements:

Select an open space, close your eyes, walk 3 m (10 ft) north, then 3 m (10 ft) east and then return straight to the starting point. This is an example of pure topomotor traveling

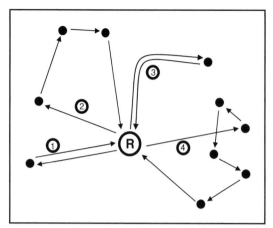

Topomotor components (arrows) of excursions of increasing complexity (1–4) away from the residence (R) to single or multiple target locations (filled-in circles).

Courtesy of Rudolf Jander.

and homing. Required for such topomotor performance is the sequential execution of distance and direction commands, the committing to memory of the unfolding route in two-dimensional space, the computation of the short-cut distance and direction based on the remembered route, and finally the execution of the computed shortcut. *Path integration* is the term commonly used to refer to such topomotor short-cut behavior.

Pure topomotor traveling and homing starts from an identified place, the anchor point, usually the home. No other place recognition or guideline use is employed. The traveled routes can be simple to moderately complex (see roundtrip 1 and 2 in the figure). The reliability of this form of locomotor control rapidly deteriorates with greater distances and with route complexity. In such cases, additional aids like guidelines and multiple place recognition have to be added. A common method for increasing the precision of topomotor-based traveling is the incorporation of a compass, commonly a light or sun compass, or a magnetic compass. The topomotor orientation of ants and bees, in particular, requires a compass (usually a sun compass), to function.

Evidence for short-range, simple roundtrips under pure topomotor control are found in fiddler crabs, cockroaches, crickets, ants some spiders and rodents. For more complex roundtrips without the benefit of guidelines (see roundtrip 3 and 4), or roundtrips covering great distances, topomotor traveling needs additional support by mapping.

Map-Based Home-range Navigation

Cartographic map-based navigation in the home range is a complex process for which only we humans have the tools and capacity. Typically, we take a cartographic map on a piece of paper, identify the scale, ego-center the map, consciously picture the map in our mind, and remember it. *Ego-centering* means, first, aligning the coordinates of the map with the axes of our body so that up on the map is the direction in front of us. Second, we take a *fix*, which is the identification of our current location in space with a location on the map. Then we look up on the map the location of our intended goal and infer the route to get there. Complex route information includes lengths, directions, and turns of paths, and the locations of place identifiers, which frequently are landmarks. Finally, we transform route information into travel (topomotor) commands, and confirm or correct performance by means of place recognition.

Is cartographic map-based navigation uniquely human? Yes and no. The cartographic map and its interpretation is uniquely human; however, the transformation of route knowledge into controlled travel commands is a skill shared and homologous among all vertebrates and has developed independently (convergently) in the evolution of a number of invertebrates, especially in bees, wasps, and ants. Such similarities in travel skills between human and nonhuman animals gave rise to the controversial notion of a shared home-range orientation tool, the so-called *cognitive map*, a map-like mental representation used in unaided traveling.

What is the grain of truth in the notion of the cognitive map? First, a simple experiment demonstrates what many, more rigorous studies confirm, that the mental-map tool for traveling without a cartographic map has properties different from a picture-like map representation in the mind. Take a floor plan of the room you are in and picture it in your mind. Take the window on your right side and the door on your left. Close your eyes and make half a turn. The mental representations of window and door are automatically updated or ego-centered by mental counter-rotation, with the window now being expected on your left side and the door on the right. Concomitantly, nothing happens to the picture of the floor plan in your mind. Only strong deliberate effort rotates the picture of the floor plan in your mind. Similarly, only by deliberate effort can you imagine the window to stay on your right side despite your rotation. Furthermore, our unaided mental travel tool—as in all animals that have it—is not only automatically ego-centered by locomotion, it is also is the tool for planning and controlling locomotion. Keeping the map metaphor, we can aptly refer to an animal's "topomotor map" as the cognitive metric map that links recognizable places by instructions of forward locomotion and turns. Concomitantly, while the subject executes such topomotor instructions, the topomotor map is continuously updated (ego-centered). How, then, is such a map acquired?

Acquisition of the Topomotor Map

Topomotor exploration is the basic learning process by which individuals acquire metric information for cognitive mapping. Starting from a home base, animals typically travel away and back over increasing distances and in different directions. Thereby they record and memorize topomotor distances and directions and link this information to recognizable places.

In the brain of vertebrates, such linkage between topomotor knowledge and place knowledge takes place in the hippocampus, a part of the forebrain whose integrity is crucial for large-scale topomotor map learning. Recording the activity of individual hippocampal neurons of rodents and humans led to the discovery of the remarkable place cells, which turned out to link topomotor and place knowledge. Specifically, neurosensory place detectors activate matching hippocampal place cells due to the recognition of particular places in which the subjects find themselves. Besides the activation by place recognition through place detectors, mere topomotor actions, the locomotion taking subjects to particular mapped places, activates such place cells as well.

With well-developed three-dimensional vision, some distances and directions to objects and places can be directly inferred and mapped, rather than be measured and mapped by locomotion. Similar spatial information may be perceptually gained by localizing objects through hearing and sonar. Such topoperceptual mapping is capable of supplementing topomotor mapping in a few animal species—us, for instance.

Social animals may acquire topographic information from other individuals. The simplest way to learn routes to useful places is to follow a knowledgeable individual. The matriarch in a herd of elephants may lead the way to a water hole, and a scout carpenter ant may lead some other workers to a rich source of food. Odor trails laid down by successful scouts in ants and termites serve a similar spatial knowledge transfer. The most amazing social spatial communication yet is the dance communication in hive honeybees.

Scout honeybees that have discovered a rich resource tell other bees about it by performing two types of communication dances in total darkness inside the hive on the vertical surfaces of combs. For the round dance the scout bee alternately circles clockwise and counterclockwise, thereby telling that a resource, such as flowers providing nectar or pollen,

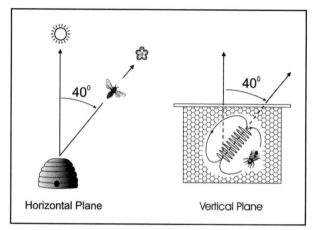

Waggle dance of the honeybee on a vertical comb inside the hive, representing the sun compass direction of the flight to the flower patch by the waggle-run orientation relative to the vertical.
Courtesy of Rudolf Jander.

is close to the hive. If the distance is greater, more than 50–100 m (165–330 ft), a waggle dance is performed. Such a waggle dance has two frequently repeated alternating components. One component is the straight wagging run that is marked by rapid lateral shaking of the rear end. The other component is the curved return to the start of the next waggle run. By alternating between clockwise and counterclockwise returns the dancing bee, performs roughly figure eights. It is information expressed in the wagging run of the dance that tells other bees that follow the dancer in what direction and at what distance to find the bounty. How, then, is this topomotor knowledge encoded in the wagging dance? The wagging-run direction relative to gravity tells the compass direction relative to the sun, to be taken when flying out. The duration of wagging, ranging from a second to several seconds, tells other bees how far to fly to reach the goal point.

See also Communication—*Honeybee Dance Language*
　　　　Navigation—*Natal Homing and Mass Nesting
　　　　　in Marine Turtles*
　　　　Navigation—*Spatial Navigation*

Further Resources

Arkin, R. C. 1998. *Behavior-Based Robotics*. Cambridge, MA: MIT Press.
Frisch, K. von. 1967. *The Dance Language and Orientation of Bees*. Cambridge, MA: Harvard University Press.
Kitchin, R. & Freundschuh, S. 2000. *Cognitive Mapping. Past, Present and Future*. New York: Routledge.
Lehrer, M. (Ed.) 1997. *Orientation and Communication in Arthropods*. Basel, Switzerland: Birkhäuser.
Redish, A. D. 1999. *Beyond the Cognitive Map. From Place Cells to Episodic Memory*. Cambridge, MA: MIT Press.

Rudolf Jander

■|Neuroethology

Animal behavior requires a functioning nervous system. The animal's nervous system uses sensory receptors to perceive, filter, and interpret environmental cues. By manipulating muscles and glands, the nervous system then produces appropriate responses. Behind all species-typical behaviors, there are neural mechanisms that produce them. Neuroethology focuses on understanding these neural mechanisms.

Neuroethology is a blending of two very different sciences, ethology and neuroscience. The blend is a difficult yet natural one for reasons discussed below.

In the early to mid-twentieth century, a group of Western European scientists gave birth to the field of ethology. These ethologists stressed the importance of studying animal behavior in the field or at least in naturalistic settings. They recognized that animal behaviors were highly adaptable, meaning that an animal's behavior would reflect not only its internal states, but also the surrounding environment in which it lived. Therefore, if one wanted to understand a given species, one should study the animal's behaviors in the context in which they evolved and developed.

The ethological philosophy contrasted with the American behaviorist school, which stressed experimental manipulations of behavior. To the ethologists, the behaviorists were studying artificial behaviors, which provided limited insight into the hows and whys of natural behavior. By providing a suitable counterpoint to the behaviorists, ethologists gave birth to, or provided insights into, such concepts as imprinting, innate behavior, instinct, motivation and drive, appetitive and consummatory behavior, and hierarchical behavioral organization, among other things. Ethologists also stressed that, because behaviors have a form, can evolve, and are species-typical, they could be compared across species, just as anatomical structures like bones could be compared across species.

In contrast, virtually all neuroscientists conduct experiments in highly controlled laboratories. Often their animal subjects are trained to perform staged or conditioned behaviors, and the animals may be restrained in various ways. Many of the behaviors that are studied do not occur naturally and would be considered pathological if they occurred in the wild. Arguably then, neuroscience has more similarities to the American behaviorist school than it does to ethology.

Hence, there are vast differences in philosophy and methods between neuroscience and ethology, and neuroethology challenges its practitioners to blend these philosophical, intellectual, and methodological differences successfully. There are other challenges for neuroethologists as well. Science is, by nature, an intellectually conservative field. Consequently, neuroethology is bounded by the historical development of both ethology and neuroscience. Because these two sciences have had different histories, the neuroethologist must work doubly hard to develop behavior studies of interest to both neuroscientists and ethologists. Neuroethologists have a dual commitment to ethological principles and practices on the one hand, and neuroscientific principles and practices on the other.

Why are neuroethologists committed to both neuroscience and ethology? Because neuroscientific methods provide a means of studying certain aspects of natural, species-typical behaviors that are not available through other means. Niko Tinbergen, one of the founders of ethology, presented the case that a scientist could ask four distinct questions about a behavior:

1. How does a behavior develop?
2. How did the behavior evolve?
3. What function does the behavior serve? and
4. What mechanisms produce the behavior?

Although neuroethology can address aspects of all four questions, its strength is in addressing the question of what mechanisms produce the behavior. Specifically, neuroethologists ask what are the neural mechanisms involved in the production of naturalistic behavior patterns?

Neuroethology provides insights that cannot be gained through other means. For instance, some neuroethologists are studying what brain regions, brain chemistry, and genes are involved in learning and memory in species as diverse as finches and sea slugs. This work has shown that behavior can activate genes in nerve cells and induce changes in neural chemistry and circuitry. This work has shown that genes do not simply determine behavior; rather, behavior can change brain circuitry and gene activity. How far behavior can induce brain changes has not been determined, but neuroethological work has opened the issue for further studies.

Most animals generate some forms of rhythmic movement, for instance, walking, flying, swimming, chewing, lapping, and breathing. Neuroethologists have studied these rhythmic movements in vertebrates and invertebrates. This work has uncovered some commonalities across species and across motor acts. A fundamental concept to come out of this work is the central pattern generator. Central pattern generators are believed to be collections of neurons that, when activated, produce the coordinated rhythmic behavior. Because stimulating these circuits produces such behaviors reliably and repeatedly, they have become a good model system for fundamental neurobiological research. One of the most surprising recent findings is that extremely simple circuits can modify their behavior output in response to environmental changes, which indicates the even very simple animals are not mere robots. Interestingly, engineers often rely on neuroethological findings to improve and design robot behavior and moveable prosthetic limbs for humans.

The neuroethological view recognizes the equally important roles played by the internal and external environments in behavior production. For example, scientists have studied the neural mechanisms involved in hunger and thirst. Angiotensin is a small protein found in the blood that is a powerful thirst promoter. When a laboratory rat is injected with trace amounts of angiotensin, it stops whatever it is doing and heads straight for water. Under natural circumstances, when an animal becomes dehydrated, angiotensin levels rise in the blood. Brain receptors sensitive to angiotensin then activate thirst centers in the brain, and the animal goes in search of water in order to rehydrate itself. Angiotensin is found in virtually all vertebrates and invertebrates studied so far. In each animal, angiotensin performs a similar water-balance function; however, the behaviors induced by angiotensin to restore water balance may be vastly different between species. It is interesting to consider the role angiotensin plays in the wildebeest, zebras, and buffalo that risk their lives to drink from crocodile-infested waters. In a very real sense, these animals are caught between angiotensin in their blood and the crocodile in the waterhole. Neuroethology offers a way to link studies of behavior strategies that an animal uses in the external environment with studies of neurobiological processes occurring in the internal environment.

It is safe to assume that neuroethology will continue to provide insights that cannot be appreciated from field observations of an animal alone. For instance, neuroethology can help identify pleiotropic effects, where a gene influences several unrelated behaviors and processes. From the previous example, angiotensin plays a role in water balance in many animals. It also appears to play a direct role in producing species-typical drinking behavior. Brain studies have also found that angiotensin is involved with memory, cognition, reproductive behaviors, muscle tone, and the regulation of immune responses. This suggests that the evolution of drinking behaviors may be tied to learning, cognition, mating behaviors, strength (and therefore social status, the ability to outrun predators or to catch prey), and an animal's ability to fight infections and heal wounds. Neuroethologists can provide important clues regarding what behaviors are linked together at the molecular level so that other biologists can study clusters of behaviors that are genetically as well as functionally linked.

Neuroethology is not a well-circumscribed field. Related fields include physiological psychology, biopsychology, and behavioral neuroscience, to name a few. However, the distinctions between these fields are also blurred, and this list by no means exhausts all the disciplines involved in studying the neural mechanisms underlying animal behavior.

See also Development—*Embryo Behavior*
Laterality
Laterality—*Laterality in the Lives of Animals*

Further Resources

Camhi, J. H. 1984. *Neuroethology: Nerve Cells and the Natural Behavior of Animals*. Sunderland, MA: Sinauer Associates.

Elliott, C. J. & Susswein, A. J. 2002. *Comparative neuroethology of feeding control in molluscs*. Journal of Experimental Biology, 205, 877–896.

Margoliash, D. 2002. *Evaluating theories of bird song learning: Implications for future directions*. Journal of Comparative Physiology, A–Sensory Neural and Behavioral Physiology, 188, 851–866.

Marder, E. & Bucher, D. 2001. *Central pattern generators and the control of rhythmic movements*. Current Biology, 11, R986–996.

Nishimura, H. 2001. *Angiotensin receptors—evolutionary overview and perspectives*. Comparative Biochemistry and Physiology, Part A, 128, 11–30.

Salzet, M., Deloffre, L., Breton, C., Vieau, D. & Schoofs, L. 2001. *The angiotensin system elements in invertebrates*. Brain Research Reviews, 36, 35–45.

Geoffrey Gerstner

■|Nobel Prize
|*The 1973 Nobel Prize for Physiology or Medicine*

The 1973 Nobel Prize for Physiology or Medicine has been awarded jointly to three zoologists: Karl von Frisch, 86 years old, of the University of Munich; Konrad Lorenz, 69 years old, of the Max Planck Institute for Behavioral Physiology at Seewiesen, near Munich; and Nikolaas Tinbergen, 66 years old, of the Department of Zoology at Oxford University for their discoveries concerning organization and elicitation of individual and social behavior patterns. The award is a new departure for the Nobel Committee of the Karolinska Institute, acknowledging for the first time major advances in the understanding of sociobiology, especially in the area of behavioral science known as ethology. At a time when studies of learning in animals were generally conducted in the laboratory, thereby posing problems largely irrelevant to their natural biology, these three men discovered in the natural behavior of animals both learned and innate patterns, exquisitely adapted to their particular phylogenetically determined ways of life. At one stroke they explained some of the most remarkable examples of the fine control of the elaborate patterns of behavior by external

Editor's Note: This entry is reprinted with permission of the American Association for the Advancement of Science from P. Marler & D. R. Griffin, "The 1973 Nobel Prize for Physiology or Medicine," *Science*, 2 November 1973, 182, pp. 464–466. Copyright © 1973 AAAS.

stimuli known to science, sometimes learned, sometimes not, while leaving no doubt in the crucial importance of genetic differences in understanding the development of behavior.

Karl von Frisch, inspired pioneer of comparative physiology, has opened our eyes to several unsuspected "sensory windows" through which animals view the world, and to complex and versatile communication behavior controlled by insect nervous systems formerly thought capable only of rigid mechanical responses. Stimulated by a distinguished family background in Vienna, including the physiologist Sigmund Exner, his boyhood enthusiasm for biology matured through studies with Richard von Hertwig, whom he later succeeded as professor of zoology at Munich. Shortly before World War I von Frisch demonstrated that, contrary to prevailing scientific opinion, fish and honeybees could discriminate colors. After the war he turned to experiments on olfaction and showed that bees could discriminate among dozens of odors, including the scents of closely related flowers. His thorough experiments in the 1920's settled in the affirmative the long-standing question whether fish could hear. Unsophisticated in the best sense, these experiments have been amply confirmed in later years with appropriate monochromators and hydrophones. An ardent Darwinian who successfully defended his views at his oral examination in philosophy against a professor who did not believe in evolution, von Frisch was motivated by a naturalist's faith that phenomena such as the colors and scents of flowers, or the Weberian ossicles of catfish, must have adaptive biological significance.

In 1923 he described as a simple language the round and waggle dances of honeybees. In that heyday of behaviorism he observed simply that round dances occurred when foraging bees brought sugar solutions into the hive from artificial feeders, whereas waggle dances accompanied the gathering of pollen. But in 1944 he found the real "Rosetta Stone" to decipher the language of bees: Round dances mean a food source nearby, waggle dances one at some distance. More important, the direction of the straight portion of the waggle dance points the way to the food, and its duration signals the distance. On a horizontal surface the dancing bee points directly toward the food, but ordinarily the dances take place inside a dark hive on a vertical surface. Here *straight up* corresponds to the direction of the sun, which serves as a directional reference point. But if the sun is obscured by broken clouds, the bees use instead the plane of polarization of light from patches of blue sky. Thus behavioral experiments that had stemmed from earlier studies of sensory physiology disclosed a new sensory channel.

Von Frisch also demonstrated that odors are very important to identify the exact food sources, and we now know that sounds or vibrations are also involved in the communication process. Bees dance only when the colony is in severe need of something, but dances are used not only for food but also for water when it is needed in hot weather to cool the hive. The most remarkable use of the dances was discovered by Martin Lindauer, one of von Frisch's leading students. When a colony of bees is swarming, scouts fly out from the teeming cluster of bees that have left their former hive and search for a cavity where thousands of bees can fly to establish a new colony. When a scout has located a suitable cavity, she signals its location by the same dance pattern used for food. Individual bees exchange information about the suitability and location of various cavities, sometimes the same bee acting alternately as transmitter and receiver of information.

Questions have been raised about the accuracy with which information is actually transmitted, and about the relative importance of the dances, odors, and sounds or vibrations. Philosophers and linguists may debate whether the term *language* is appropriate. But, for behavioral scientists, the revolutionary discovery was that an insect sometimes communicates with fellow members of a closely integrated society by flexible, iconic, graded gestures about distant objects that are urgently needed by the social group as a whole. Behavioral continuity between animals and men extends even to fruitful comparisons between animal communication and human language.

Konrad Lorenz, acknowledged founder of the science of ethology, derived his insights into the causation and organization of behavior from studying fish and birds. At Altenburg in Austria, the house of his father, a Viennese orthopedist, was always full of animals and birds. A precocious naturalist, Lorenz developed early what became a lifelong passion for raising both wild and domestic animals by hand, and for living with them in the closest quarters, and so gaining insights into the relation between genome and experience in ontogeny. Medical training at the University of Vienna was followed by excursions, inspired by Ferdinand Hochstetter, Karl Bühler, and others, into comparative anatomy, psychology, and philosophy. One senses early tension between the attractions of a career in medicine and academia, and fascination with the beauty and diversity of animals. During a two-semester stint in the Columbia Medical School in New York in 1922, he is said to have spent more time studying the inhabitants of the New York Aquarium than at lectures. Comparative ethology was deemed an inappropriate pursuit in the department of anatomy, so to his M.D. degree he added, in 1936 a Ph.D. in zoology at the University of Munich, and remained in that department until 1941. The major features of his theory of behavior were laid during that period. After World War II, under the aegis of the Institute for Marine Biology, a Max Planck Institute was established in Büldern in Westphalia for Lorenz' group, and, in 1958, it became the Max Planck Institute for Behavioral Physiology at Seewiesen in Bavaria.

Ethological findings derive much of their force and generality from insightful use of comparative techniques and subjects selected appropriate to the problem. If Lorenz has a totem animal, it is surely the greylag goose in which, with his revered teacher Oskar Heinroth, he discovered imprinting, an especially rapid and relatively irreversible learning process with an optimal critical period early in the gosling's life. Imprinting has repercussions not only on what constitutes an acceptable parental object, or companion, as Lorenz called it, but also on what becomes an appropriate sexual companion when the gosling grows up, one of many findings that have proved heuristically valuable in psychoanalysis and psychiatry.

This and other discoveries were incorporated in the panorama of ethological theory presented in 1935, and translated into English soon afterward by Margaret Nice, that was at once a treatise on the social behavior of animals and how the structure of a society relates to its component parts, and a manifesto for the objective analysis of the natural behavior of animals. A central conception complementary to that of imprinting, is the innate releasing mechanism. These were visualized as genetically determined sensory mechanisms that predispose an organism to be especially responsive to stimuli, from the environment or from companions, that have assumed special valence in the course of evolution of behavioral adaptations for survival and reproduction. They match behaviors evolved for social communication that generate key "releasing schemata" or "sign stimuli," in turn evoking or guiding particular patterns or behavior in the respondent.

A series of germinal papers over the next 15 years defined more sharply inadequacies of purely reflexive and behavioristic theories of behavior, demonstrating that endogenous changes in motivation to perform certain activities and endogenous changes in responsiveness to different kinds of stimuli cannot be omitted from a behavioral theory if it is to have any general validity.

Some of his viewpoints as expressed in the popular book *On Aggression*, which suggests an endogenous motivation to seek out opportunities for fighting in fish, and perhaps man as well, proved highly controversial. However one senses deeper roots to the outrage with which some react to analogies between animal and human behavior. In the introduction to the 1970 translation of his work, Lorenz reflects wryly:

> The fact that the behaviour not only of animals, but of human beings as well, is to a large extent determined by nervous mechanisms evolved in the phylogeny of the species, in other words, by "instinct," was no surprise to any biologically-thinking scientist. It was treated as a matter of course, which, in fact, it is. On the other hand, by emphasizing it and by drawing the sociological inferences, I seem to have incurred the fanatical hostility of all those doctrinaires whose ideology has tabooed the recognition of this fact. The idealistic and vitalistic philosophers to whom the belief in the absolute freedom of the human will makes the assumption of human instincts intolerable, as well as the behaviouristic psychologists who assert that all human behaviour is learned, all seem to be blaming me for holding opinions which in fact have been public property of biological science since *The Origin of Species* was written.

The young Niko Tinbergen, an avid naturalist from his boyhood in the sand dunes and pine forests of Hulshorst in Holland, saw the intricacies of insect behavior, especially that of digger wasps hunting other insects and provisioning nest burrows with the corpses, as a testing ground for hypotheses about the sensory control of behavior. An opportunity while a graduate student in zoology at the University of Leiden to participate in 1931–32 in an Arctic expedition added snow buntings, phalaropes, and Eskimo sled dogs to a growing list of animals into whose behavior Tinbergen was to cast profound evolutionary insights. Returning to join the zoology faculty at Leiden, a seminal meeting with Lorenz in 1936, followed by a 6-month visit to Altenberg, gave rise to their only joint paper, in 1938, on the egg-rolling behavior of the greylag goose, and to more than 30 years of mutual cooperation, criticism, and stimulation that brought the new science of ethology into full flavor. When he went to Oxford University in 1951, where he became a professor of animal behavior, he left the seeds of ethology firmly planted in Holland, and Groningen and Leiden continue as fertile research centers where aspiring students become versed in ethological discipline. One senses environmental imprinting in Tinbergen's choice of research sites while he was at Oxford. The sands of the Farne Islands, Scolt Head, and most recently Ravenglass are the Hulshorsts of England, permitting expansion of a research theme, already broached in Holland, on the social behavior of gulls, which led to fundamental insights into relationships between the behavior and ecology of animals.

From a theoretical framework established in his 1942 paper on "an objectivistic study of the innate behavior of animals," Tinbergen and his colleagues concentrated on the stimulus control of behavior. In both laboratory and field conditions, with butterflies, fish, and birds as subjects, he demonstrated that, by using inanimate models whose properties are systematically varied, experimental demonstration can replace intuitive judgment in deciding which elements of a stimulus complex control a response. New insights into how signaling behavior originates in the course of evolution were summarized, together with a general development of ethological theory in his 1951 book, *The Study of Instinct*, which introduced many English-speaking readers to the subject. Many patterns of social behavior, often with a signaling function, were understood as the outcome of social conflicts, a point of view that Tinbergen, with his wife Elizabeth, has since applied to the genesis of autistic behavior in children.

Perhaps most distinctive in the breadth of Tinbergen's research is his frontal attack in the 1950's and '60's on the problem of adaptiveness, which was for so long the subject of judgments from zoologists' armchairs. However, Tinbergen and his associates demonstrated that one can actually measure in animals preyed upon by others the cost or benefit of such traits as the color of a moth's wings or a bird's eggs, the spines of a three-spined stickleback, habits such as a gull's removal of egg shells from the nest after young have hatched, or living on the edge of a gull colony rather than in the center. The studies of gull behavior illustrate beautifully how an ecological decision made in phylogeny can reverberate through many aspects of the biology of a species. With von Frisch and Lorenz, Tinbergen has expressed the view that ethological demonstrations of the extraordinary intricate interdependence of the structure and behavior of organisms are relevant to understanding the psychology of our own species. Indeed, this award might be taken not only as fitting recognition of the outstanding research accomplishments of these three zoologists, but also as an appreciation of the need to review the picture that we often seem to have of human behavior as something quite outside nature, hardly subject to the principles that mold the biology, adapability, and survival of other organisms.

See also Frisch, Karl von (*1886–1982*)
    History—*History of Animal Behavior Studies*
    Lorenz, Konrad (*1903–1989*)
    Tinbergen, Nikolaas (*1907–1988*)

P. Marler & Donald R. Griffin

■|Parasite-Induced Behaviors

All living things need energy, and when we think of how energy is transferred from one organism to another, we may envision a classic textbook illustration: Green vegetation transforms sunlight into food, and thus becomes an energy source for herbivores, those animals—be they grasshoppers or moose—that eat green plants. Waiting in the bushes, predators lurk, anticipating *their* turn to capture energy from herbivores by eating them. Finally, scavengers gain energy by cleaning up the remains, breaking leftovers down so that they can re-enter this cycle as plant nutrients. Often missing from this picture is yet another method of acquiring energy—the way of the *parasite*.

Parasites are ticks, intestinal worms, viruses, mosquitoes—the armies of small animals that make a living on much larger organisms, their hosts. Parasites usually do not pounce, shred, munch or chase, at least not in the customary sense of the terms. In some cases, parasites kill their hosts; in others, they are almost undetectable. Like every other organism, a parasite is subject to natural selection, and the end result of the *host–parasite interaction* reflects this influence. It can be likened to an arms race that both the parasite and the host are trying to win, with the most successful players leaving the most offspring.

One challenge that the parasite faces is that of getting to a host, and one challenge that the host faces is that of avoiding (or overcoming) the parasite. The behavior of the host figures prominently in both of these challenges. For instance, most of us know that our immune systems fight off invaders (parasites) that enter our bodies, but long before our immune systems are activated, there are things that we can do, *behaviors* we can perform, to keep healthy. Our first line of defense is to simply *not get* infected by the parasite! That's why hand-washing is so important to human health. Our hands are a major connection between the environment and our bodies; they can carry all sorts of unseen visitors from doorknobs and handrails into our mouths, noses, and eyes.

Likewise, we are just beginning to realize that some behaviors of animals appear to be performed because they minimize exposure to parasites. We know, for instance, that stickleback fish choose habitats—bottom vegetation versus open areas—based on the presence or absence of a dangerous ectoparasites. (*Ectoparasites* are external parasites—in this case, a fish louse.) In the absence of these parasites, young sticklebacks prefer the bottom vegetation where they are relatively safe from predators. However, the parasites also tend to occur in this habitat, and when they are present, many sticklebacks risk the open water instead. Altered habitat choice as a result of parasite risk has been observed in many animals, ranging from ants to horses.

Behaviors that enable an animal to avoid parasites do not have to be so dramatic—sometimes a change in posture is all that is needed. Lying down seems to protect red deer from fly harassment, and many Hawaiian birds sleep with their heads and feet tucked into their feathers, thus avoiding exposure to biting flies. And of course, animals swat and slap such pests. This last behavior may not seem like much—who hasn't done that at a bug-infested picnic?—but scientists have estimated that such movements can be energetically costly. Howler monkeys in Panama were seen to spend up to 24% of their metabolic budgets in such activities.

Oddly, joining in a big group diminishes one's chances of being bitten by a foraging insect (or a foraging predator, for that matter). This is called the "selfish herd" effect, and the name reflects the fact that in a larger group, the chances of any one individual being attacked are reduced. Thus, herd sizes may change across time, and herd members may join together or separate, sometimes reflecting changes in ectoparasite intensity.

Endoparasites—internal parasites such as worms and bacteria—may also inspire avoidance behaviors. Apparent dislike of strangers commonly seen in many primate social groups may reduce exposure to parasites from outside the group. Indeed, although joining a group is an excellent way to reduce your chances of encountering an ectoparasite or predator, it can increase your exposure to endoparasites; in fact, increased exposure to disease is one cost of social behavior in general.

Many animals defecate in certain areas and are reasonably fastidious about this activity; we take advantage of this tendency when we housetrain puppies and provide cats with litter boxes. Such "hygienic" behavior may have evolved because it enhances parasite avoidance. Even mate selection has been hypothesized to play a role in parasite avoidance; careful mate selection can not only protect against sexually transmitted diseases, but may convey good, parasite-resistant genes to offspring.

What happens, however, if a parasite successfully foils these avoidance attempts and invades a host? Behavior still plays an important role as the host attempts to defeat the parasite. For instance, we are all familiar with the fever that accompanies some parasitic infections and is thought to combat them. Have you ever stopped to consider the behavioral changes that accompany fever? If you have a fever, you usually do not feel like doing much in the way of activity; you may not even feel like eating. Although that may just "feel sick" to you, in fact, your behavior has changed, and it has changed in ways that support your body, conserving energy as body temperature increases, which is energetically expensive. In fact, even so-called "cold-blooded" animals can express fever and increase their body temperatures. They do not do it with metabolism, as humans do, but they *behave* in ways that increase body temperature, often by choosing warmer environments than uninfected animals might. Thus, when Mediterranean crickets infected with a lethal microscopic parasite were allowed to stay at a temperature of their own choosing, they had a much higher survival rate than crickets that were held at other temperatures. The crickets had expressed *behavioral fever*, and it had saved their lives.

Animals can also cope with endoparasites by *self-medication*, that is, eating medicinal substances. Many plants contain chemicals that have antiparasitic properties. For instance, ailing chimpanzees chew the pith of certain plants and swallow the juice; after such episodes, worm infections (as measured by worm egg output) seem to decrease.

Animals can rid themselves of ectoparasites by grooming. Again, when we see a cat licking its fur, we do not stop to think that it may be performing a serious task, but it is. Grooming can cost a rat up to one-third of its evaporative water budget, and can take a substantial amount of time. The results, however, are impressive. When rock doves were prevented from preening, their ectoparasites loads skyrocketed. Grooming is so important that some animals even have special teeth or other structures to help with the task.

Given these and other host defenses, do parasites stand a chance of survival? There is probably not an organism on earth—including you—that is without parasites. They are remarkably successful, and again, behavior plays an important role. Intestinal parasites are of special interest here. Many have complex life cycles—they spend adulthood in the intestine of a final host, often a vertebrate, and dispense eggs through the feces. These eggs are then eaten by another animal—an intermediate host—where they develop into an infective immature form in the intermediate host. When the intermediate host is eaten by the final host, the parasite enters the intestine of the final host, where it can reproduce. Clearly, if transmission between hosts can be enhanced, the parasite benefits. This happens when parasitized intermediate hosts are easier to catch than unparasitized ones. For instance, a worm that lives as an adult in the intestine of starlings uses pillbugs (terrestrial isopods) as intermediate hosts. Infected pillbugs spend more time on light-colored surfaces, where they are conspicuous, than uninfected pillbugs do. They also prefer areas of relatively low humidity, and seem to have no preference for overhanging shelter. These altered behaviors mean that they are out in the open more often than uninfected pillbugs, and probably account for the fact that they are eaten by starlings in higher proportions.

The transmission of parasite eggs can also involve behavioral changes in hosts. For instance, what about roundworms of flies that have aquatic larvae? How can the roundworm get back to the aquatic environment? In many cases, the infected fly is attracted to stretches of water and other places where uninfected flies lay eggs; however, instead of laying fly eggs in these places, the infected fly distributes roundworm eggs, ready to infect the next generation of flies.

Some of the most devastating parasitic diseases in the world are not transmitted with food—at least not human food. They are the diseases acquired when animals, including humans, are bitten by ectoparasites that themselves are carrying endoparasites, such as the protistans that cause malaria. In this case, unsuccessful defense against such ectoparasites can expose the vertebrate host to endoparasites that are injected when the mosquito bites. As you might expect, these endoparasites have a few behavioral tricks. For instance, many endoparasites that are transmitted by biting insects interfere with the insect's feeding mechanism, causing it to feed longer and/or on more hosts—all the better to transmit the parasite. Moreover, in some cases, once in the vertebrate host, the illness that the endoparasite induces makes that host so sick that it is not very energetic in its defense against biting insects; this favors the *next* round of transmission, back to the insect, for transmission to yet another vertebrate.

This brings us back to the beginning of the story—some parasites make hosts sick, some are lethal, some are almost unnoticeable. Why the difference? As with all things in biology, we look to evolution by natural selection for the answer. Remember the worm that lives in the starling and the pillbug? The task for the worm in the bird is to live, mate, and produce eggs that are distributed, along with bird feces, over the environment, where the eggs await pillbugs. Thus, natural selection favors worms that do not necessarily harm the birds that are both home and food source to them. On the other hand, the immature form of this parasite in the pillbug has a different task, and natural selection has favored those that succeed in reaching a stage that can infect a bird, and that can manipulate the pillbug's behavior in ways that favor encounters with foraging birds.

This is only a brief introduction to the hundreds of examples of how parasites affect behavior. The next time you have a cold and sneeze, ask yourself if you are clearing out your respiratory passages so that you can breathe more easily, or if your microscopic fellow travelers, the "germs" that "gave" you the cold, are causing you to spew them onto your friends, relatives, and passing strangers?! (And for goodness sake, use a handkerchief!)

Further Resources

Desowitz, Robert, books such as *New Guinea Tapeworms and Jewish Grandmothers*.
Moore, Janice. 2002. *Parasites and the Behavior of Animals*. Oxford Series in Ecology and Evolution. New York: Oxford University Press.
Spielman, A. & D'Antonio, M. 2001. *Mosquito*. New York: Hyperion Press.

Janice Moore

■ Personality and Temperament
A Comparative Perspective

Before crouching to pet an unknown dog, a person will often ask the owner, "Is she friendly?" A zookeeper may manage a crowd's unrealistic expectations by warning visitors that a particular octopus is "shy." Primatologists chatting about the day's field observations could use terms such as aggressive, curious, and calm to describe the baboons they have been studying. Terms like friendly, shy, aggressive, and curious are often used to describe both human and nonhuman animals. The terms are used to refer to patterns of behavior that are stable across time and situations. That is, if we describe an individual as curious, we probably mean not only that she is behaving curiously now, but also that she has behaved curiously in the past and will probably continue to do so. We also mean that she will be relatively curious both in novel and familiar situations. When we use the terms in this way we are referring to an individual's personality or temperament. Researchers have been examining how personality and temperament affect human behavior for many decades, but they have only recently begun to examine how personality and temperament affect the behavior or nonhuman animals.

What Is Personality?

Researchers who study humans often make a distinction between temperament and personality; *temperament* is usually seen as the early appearing tendencies that interact with environmental influences to serve as the biological foundation for personality. Researchers who study animals have not maintained such a clear distinction between personality and temperament and often use the two terms interchangeably. Although many animal researchers prefer one term over the other, there is no clear agreement among animal researchers about how temperament differs from personality. Therefore in this chapter we use the word personality to refer to both personality and temperament.

There are many different ways to think about an individual's personality. In humans, personality can be described using a wide variety of constructs, such as motives (e.g., our goals and desires), life stories (the story we tell ourselves to make sense of our past experiences), and traits (adjectives like "friendly" and "curious" that describe broad patterns of behavior). What kinds of constructs are studied in animals? A review by Gosling in 2001 of more than 150 studies of personality in nonhuman species showed that the personality phenomena studied in animals comprise only a subset of the personality phenomena studied in humans, with the overwhelming majority of animal studies focusing on traits. Traits provide an economical way to summarize how one individual differs from another; for example, attributing the trait "dominance" to an individual summarizes a history of many different acts of dominance. It is also likely that the dearth of studies on nontrait personality constructs, such as motives and life stories, reflects the fact that animals cannot provide self reports and may not have the physical or mental faculties (e.g., autobiographical memory, conceptions of self) required by the personality tests developed for humans.

Of course an individual's behavior is also affected by factors other than its personality, such as its sex, age, and surroundings (e.g., whether it is in captivity or in the wild). An animal's species can also influence its behavior and its personality. Thus, personality can be examined both across species and within species. For example, firemouth fish can be said to be more fearful than goldfish (a cross-species level of analysis), but some firemouth fish are more fearful than other firemouth fish (a within-species level of analysis). Both types of comparisons are important. For example, before adopting a snake you probably want to know both whether the snake belongs to a particularly aggressive species (a cross-species analysis) and whether that individual snake is particularly aggressive for its own species (a within-species analysis).

Do Animals Have Personality?

What evidence is required in order to say that animals have personality? First, we need to show that personality can be measured *reliably* in animals. For example, two zookeepers should agree about the personality of the animals they both know well. Second, we need to show that personality descriptions are *valid*, and are not, as some critics claim, merely projections of human characteristics onto other animals (a process known as *anthropomorphism*). For example, personality descriptions should predict actual behavior or real-world outcomes (such as an animal's rank or health).

Research on animal personality has met the requirements of both reliability and validity. Evidence shows that observers do agree with one another about the personalities of the animals they know well, showing that personality descriptions are reliable. For example, observers agree about which rhinoceros is more fearful than the others, or which chimpanzee is the most sociable. In addition, there is abundant evidence that personality descriptions are valid. That is, they reflect real attributes of the animals, and not just anthropomorphic projections. This validity evidence comes from two major findings. First, observers' ratings predict real-world behaviors. For example, piglets rated as more sociable were also the ones who made more vocalizations and more nose contacts. Second, observers have identified different traits in different species. If observers were simply projecting human traits onto animals, they would project the same traits onto all species.

Having established that personality can be measured in animals, we can now ask which traits have been identified in which animals. Researchers have examined personality traits in a great number of species including everything from pumpkinseed sunfish and octopuses to cheetahs and chimpanzees. Almost all of the species studied have been found to possess some trait-like qualities. For example, Stephen Glickman and Richard Sroges identified various manifestations of personality in over 100 species of zoo animals. The studies are too numerous to list in full, but a comprehensive review can be found in Gosling's 2001 article. Here we present just a few examples to illustrate the breadth of species and traits that have been studied.

The trait of extraversion has been found to have considerable generality across species. *Extraversion* refers to an individual's level of sociability, energy, and assertiveness. For example, cats differ in their amount of energy, pigs in their level of sociability, and octopuses in their boldness. Similarly, the traits of agreeableness and neuroticism exist in some form in a broad range of species. *Agreeableness* refers to an individual's level of cooperativeness, trust, and lack of aggression, and *neuroticism* refers to an individual's level of anxiety, depression, and vulnerability to stress.

The scope of these traits suggests that they may be a result of general biological mechanisms that are shared across a wide range of taxa. The way these personality dimensions are manifested, however, depends on the species. For example, whereas the human scoring low on extraversion stays at home on Saturday night or tries to blend into a corner at a large party, the octopus scoring low on boldness stays in its protective den during feedings and attempts to hide itself by changing color or releasing ink into the water.

In contrast, there are some traits that are found only in humans and their close relatives. For example, conscientiousness, which refers to an individual's level of self-discipline, has been identified almost exclusively in studies of humans and other great apes. This suggests that conscientiousness may have appeared relatively recently in the evolution of Homininae, the subfamily comprised of humans, chimpanzees, and gorillas.

Although it seems that human personality varies on more dimensions than other animals' personality, there is compelling evidence that individuals in most species differ from one another on at least one personality dimension. For example, some rhinoceros are more fearful than others, some hyenas more dominant than others, and some chimpanzees are more sociable than others. In addition, animal personality research is based on a human-oriented system, so it is possible that the personalities of some species differ on dimensions for which we do not have a word or are unable to even detect.

How Has Personality Been Measured?

Animal researchers have usually measured personality in one of three ways: by coding individuals' specific, naturally-occurring behaviors, by administering behavioral tests to animals, or by rating individuals on broader, more subjective traits. Which is the better method?

Ethological coding consists of observing an animal and recording specific, naturally-occurring behavioral acts such as grooming, eating, or vocalizing. For example, Capitanio and his colleagues have used codings to measure the sociability of rhesus macaques by recording the frequency of social behaviors. Although coding has the apparent advantage of being more objective and precise than rating, there is evidence from research on humans that single behaviors are unreliable; they are difficult to assess and predict, and are not very consistent across time or situations. Nevertheless, studies using codings have found that observers' codings agree to a large extent. Thus, it may be useful to collect codings of behavior when measuring personality, particularly for animals that are in highly controlled environments where their behavior is easy to detect and unambiguous, such as in zoos or laboratories.

Behavioral tests consist of placing animals in specific scenarios designed to test and reveal a specific aspect of their personality. For example, Mather and her colleagues have used behavioral tests to measure the aggression of octopuses by prodding the octopuses with a brush and recording the octopuses' reactions. Behavioral tests are useful for animals in a captive setting such as a zoo or a laboratory, but they are often impractical or impossible with noncaptive animals. Behavioral tests are also useful when researchers are interested in measuring one specific trait, but are less useful for forming global impressions of an individual's personality.

Rating consists of having experts who know the animals well make global judgments about the animals' personalities. For example, Carlstead and her colleagues have used ratings to measure the personality of rhinoceros by having zookeepers rate the personalities of the rhinoceros they care for. Observers' ratings have historically been derided as subjective and inappropriate for scientific measurement. However, researchers have convincingly argued that observers' ratings, when averaged together, are reliable and are not influenced by the subjective biases of any one observer. Studies using ratings have also shown that independent observers agree substantially in their ratings of animals' personalities. Indeed, observers' ratings generally provide a richer and more accurate picture of an animal's personality than do ethological codings or behavioral tests.

Most researchers agree that the ideal way to measure personality is to use a combination of these three methods. Combining methods provides more information, and also enables researchers to ensure their personality measures are accurate by examining whether the different measures agree with one another. For example, Carlstead and her colleagues found that rhinoceros that were rated as more fearful by zookeepers were also the ones who reacted most fearfully in a behavioral test.

Why Is Personality Important?

Personality research is important because it helps us understand and predict behavior. For example, knowing that a chimpanzee has an aggressive personality can help researchers and caretakers predict whether it is likely to attack other individuals.

Indeed, personality ratings do predict many important real-world behaviors and outcomes. For example, assertive hyenas achieve a higher rank in their dominance hierarchy than unassertive hyenas, anxious cheetahs breed less than calm cheetahs, and sociable rhesus macaques are more likely to recover from illness than unsociable rhesus macaques. In addition, there is evidence from many species, including humans, mink, and octopuses, that personality is partly heritable. That is, personality is passed on from generation to generation, and is apparent very early in the lives of individuals. This means that personality is a strong and stable influence on an animal's life and behavior.

In summary, personality differences have been documented in a wide range of animal species. They are present early in life, are stable across the life-span, and predict important behaviors and life outcomes. For these reasons, personality is important to the study of animal behavior.

See also Personality and Temperament—*Personality*
      in Chimpanzees
      Personality and Temperament—*Personality,*
      Temperament, and Behavior in Animals
      Personality and Temperament—*Stress, Social Rank,*
      and Personality

Further Resources

Capitanio, J. P. 1999. *Personality dimensions in adult male rhesus macaques: Prediction of behaviors across time and situation.* American Journal of Primatology, 47, 299–320.

Carlstead, K., Mellen, J. & Kleiman, D. G. 1999. *Black rhinoceros* (Diceros bicornis) *in U.S. zoos: I. Individual behavior profiles and their relationship to breeding success.* Zoo Biology, 18, 17–34.

Glickman, S. E. & Sroges, R. W. 1966. *Curiosity in Zoo animals.* Behaviour, 26, 151–188.

Gosling, S. D. 2001. *From mice to men: What can we learn about personality from animal research?* Psychological Bulletin, 127, 45–86.

Gosling, S. D. & John, O. P. 1999. *Personality Dimensions in Non-Human Animals: A Cross-Species Review.* Current Directions in Psychological Science, 8, 69–75.

Gosling, S. D. & Vazire, S. 2002. *Are we barking up the right tree? Evaluating a comparative approach to personality.* Journal of Research in Personality, 36, 607–614.

King, J. E. & Figueredo, A. J. 1997. *The five-factor model plus dominance in chimpanzee personality.* Journal of Research in Personality, 31, 257–271.

Lilienfeld, S. O., Gershon, J., Duke, M., Marino, L. & de Waal, F. B. M. 1999. *A preliminary investigation of the construct of psychopathic personality (psychopathy) in chimpanzees* (Pan troglodytes). Journal of Comparative Psychology, 113, 365–375.

Mather, J. A. & Anderson, R. C. 1993. *Personalities of Octopuses* (Octopus rubescens). Journal of Comparative Psychology, 107, 336–340.

Wielebnowski, N. C. 1999. *Behavioral differences as predictors of breeding status in captive cheetahs*. Zoo Biology, 18, 335–349.

Simine Vazire & Samuel D. Gosling

■ Personality and Temperament
Personality in Chimpanzees

The field of animal personality centers upon the challenging issue of how much humans and nonhumans share a similar emotional and behavioral world. If we assume, as did Darwin, that related species may share subjective feelings and emotional expression, then perhaps human personality characteristics can, and should, be applied to nonhuman animals, particularly to other primate species. In the important early studies of wild chimpanzees by Jane Goodall, behavioral interactions are portrayed as rich accounts of the social dramas played out by distinct personalities. Later work on captive chimpanzees has documented similar levels of complexity in social relationships. Frans de Waal's account of the colony at Arnhem zoo in the Netherlands is a good example; the book's central theme concerns the interplay of different chimpanzee personalities as they struggle for dominance status and negotiate constantly changing social relationships.

It is only more recently, however, that researchers have focused more closely on the ways in which observers apply personality terms to animals. Traditional ethological and sociobiological approaches to animal behavior have tended to focus on species-specific behaviors; emphasizing the ways in which individual animals within the same age or sex class are similar, rather than the ways in which they might differ. In these accounts, animals are rarely described as active personalities with goals and intentions of their own (see Crist 1999). Yet primate infants develop within a dynamic social society, in which they need to negotiate and remember past behavioral interactions. James King and his colleagues have suggested that a basic concept of personality may allow social primates to predict the likely behavior, whether cooperative or aggressive, of their fellow group members. For this reason, it may be likely that basic human personality dimensions, such as extraversion, agreeableness or anxiety may be shared by nonhuman primate species.

Whether or not chimpanzees themselves possess a broad understanding of personality differences, many writers have, consciously or not, assumed that behind the behaviors they observe lie complex, stable personalities. In human personality psychology, much emphasis has been placed on searching for clear relationships between ratings of personality and behavior, although such trait–behavior correspondences are not always clear. Recent investigations into chimpanzee personality have compared holistic observer ratings of the personality of animals with specific measurements of social behavior, finding clear relationships between personality ratings and predicted behavior. For instance, Lilienfeld and his coworkers asked experienced observers to rate captive chimpanzees on a measure of psychopathy (a tendency to be aggressive, fearless and show a lack of empathy). Those chimpanzees rated highly on psychopathy also engaged in high levels of dominance displays, low-level aggression and teasing, suggesting that observers' subjective impressions of the chimps' personality reflected actual behavioral tendencies.

To some, personality ratings of chimpanzees have been seen as simply anthropomorphic projections of human characteristics onto nonhumans. To others, they are a useful means of summarizing complex patterns of behavior, and perhaps even explaining why animals within the same social group often develop and behave so differently. Studies of chimpanzee personality increasingly illustrate the importance of social relationships to an understanding of behavior, and are a valuable window through which researchers can glimpse the nature and quality of an individual's social life.

See also Personality and Temperament—*A Comparative Perspective*

Personality and Temperament—*Stress, Social Rank, and Personality*

Further Resources

Buirski, P., Plutchik, R. & Kellerman, H. 1978. *Sex differences, dominance, and personality in the chimpanzee*. Animal Behaviour, 26, 123–129.
Crist, E. 1999. *Images of Animals: Anthropomorphism and Animal Mind*. Philadelphia: Temple University Press.
De Waal, F. 1998. *Chimpanzee Politics: Power and Sex among Apes*. Revised edn. Baltimore and London: The Johns Hopkins University Press.
Goodall, J. 1971. *In the Shadow of Man*. London: Weidenfeld and Nicolson.
Gosling, S. D. & John, O. P. 1999. *Personality dimensions in nonhuman animals: A cross-species review*. Current Directions in Psychological Science, 8, 69–75.
Hinde, R. A. 1976. *Interactions, relationships and social structure*. Man, 11, 1–17.
King, J. E. 1999. *Personality and the happiness of the chimpanzee*. In: *Attitudes to Animals: Views in Animal Welfare* (Ed. by F. L. Dolins), pp. 101–113. Cambridge: Cambridge University Press.
Lilienfeld, S. O., Gershon, J., Duke, M., Marino, L. & de Waal, F. B. M. 1999. *A preliminary investigation of the construct of psychopathic personality (psychopathy) in chimpanzees* (Pan troglodytes). Journal of Comparative Psychology, 113 (4), 365–374.

Diane Dutton

■ Personality and Temperament
Personality, Temperament, and Behavioral Assessment in Animals

Many people believe that animal societies are organized according to *dominance hierarchies* where usually the biggest, sometimes also the oldest, individual is the boss and dictates what the others can, and cannot, do. It is also thought that dominance hierarchies depend largely on aggression. Physical attributes of an individual such as size and other characteristics (such as the presence of antlers in elk, an impressive mane in lions, a well-crested neck in stallions, brightly colored bellies in sticklebacks) are considered to be very important in determining the individual's position in the hierarchy.

The *dominance hierarchy* concept is very simple: Individuals at the top of the dominance hierarchy will have freer access to resources (eg., food, shelter, mates) than individuals at the bottom of the hierarchy. Usually an individual's position within the hierarchy remains unchanged until others die or are removed from the group. Once the hierarchy is

In primates, the young often have a similar position in the hierarchy as their mother, casting into doubt the concept of dominance hierarchies.
© *Kennan Ward / Corbis.*

stable, individuals' positions within it are maintained by threatening behaviors rather than out and out aggressive behaviors. It is undeniable that the overly simple concept of dominance hierarchies has a small place in the explanation of how groups of social animals are organized but there is a lot more to it than that. In primates, the young often have a similar position in the hierarchy as their mother, despite being younger, smaller, and not having fought successfully to earn their place. This phenomenon is known as *matrilineal inheritance of dominance*.

As more and more research into social behavior has been conducted on a lot of different species, it is becoming very clear that the concept of dominance is rather inadequate. Indeed, many instances where dominance hierarchies completely fail to explain the social relations observed within a group of animals have now been reported. Many *ethologists* (scientists who study animal behavior) believe that the personalities of the individuals present in the group are very important, whereas others also highlight the importance of temperament. This essay will look at the following areas: what *personality is*, and how it relates to *temperament*; how personality and temperament is measured in animals; and the results of personality and temperament studies in a wide range of species.

What Is Personality, and How Does It Relate to Temperament?

Wherever there is a group of animals, whether it is a herd of cattle, a band of primates, or a school of fish, even the most superficial look at their behavior, especially their social behavior, will show that all individuals within that group are not identical. They differ from each other in various ways—however large or small. Even staunch supporters of the dominance hierarchy concept (including E. O. Wilson—an eminent biologist, and Schjelderup-Ebbe—the inventor of the Dominance Hierarchy concept) admitted that differences in the behaviors of individuals should be taken into account because they help to explain how the animals behave collectively as a group. It is interesting to note that in early studies of animal social behavior, scientists removed the data of any members of the group who were behaving differently than the rest of the group. This removed information, which did not fit into an easily explained pattern, was actually extremely important because it reflected the individuals' distinct personalities. So, as mentioned above, the importance of the individuals within a group being different from each other—their personalities and their temperament—has been stressed by a lot of scientists for a wide range of animal species. Before going any further, we should define the terms *personality* and *temperament*—and also the related term, *individuality* (and *individual differences*), before setting about establishing the relationship between them.

As usual in science, and especially in the study of animal behavior, it is important to clearly define the terms being used in order to *prevent* confusion and later misuse. The three terms being used in this essay are personality, temperament and individuality. In this essay the term *personality* is used to mean the same as *individuality* where differences in personalities can be equated with *individual differences*. In other words, differences seen in the behavior between individuals are assumed to reflect their different personalities.

When terms are difficult to define scientifically, it is helpful to turn to a dictionary in order to obtain a simple, working definition that everyone can understand. The term *personality* originates from the Latin word *persona* meaning "a person's characteristics." A typical dictionary definition of personality is *the sum total of all the behavioral and mental characteristics by means of which an individual is recognized as being unique*. In other words, the personality of an individual animal depends on the way it behaves, thinks, and feels. (Don't forget though, that the way an individual's personality is expressed may also depend on the its social and physical environment.) *Temperament* is closely related to *personality* but should not be confused with it. A typical dictionary definition of temperament is *an individual's character, dispositions and tendencies; excitability, moodiness and anger; the characteristic way an individual behaves, especially toward other individuals*.

A careful look at the definitions given for personality and temperament shows that temperament involves emotion and arousal (often reflected by reactive or anxiety behavior) and also motivation (keenness to behave in a certain way), whereas personality does not. Perhaps the best species to demonstrate the difference between these definitions of personality and temperament is the domestic horse, *Equus caballus*, which interacts closely with humans. Difficult horses are commonly described as being "nervous" "reactive" (eg., are very likely to be upset by strange objects and exhibit the reactive "spooking" behavior), but tend to be very motivated (eg., to perform the tasks humans require of them, such as learning a new movement or jumping an obstacle). These horses are considered to have a "difficult" personality and a "reactive" temperament. Many people would rather misleadingly label these individuals "temperamental." On the other hand, placid horses are often described as "staid," "bomb proof" (eg., tend not to be easily aroused or upset by strange objects—no visible change in behavior is exhibited), but also tend to be less motivated to perform the tasks humans require of them. These horses are considered to have a "placid" personality and an "unreactive" temperament.

To summarize, despite people often using the words personality and temperament interchangeably, as students of animal behavior we know that personality and temperament are two separate phenomena. The personality of an individual animal is characterized by the typical behavior that it exhibits, whereas its temperament reflects its mental activity (mainly emotion, arousal and motivation). The temperament of an individual has a distinct bearing on the personality of an individual.

How Can We Measure Personality and Temperament in Animals?

We already know that within a group of animals there are likely to be clear behavioral differences between individuals reflecting their different personalities. Several people who have practically lived with species such as elephants and lions refer to this in their writings. It has also been noted that any decent animal behavior scientist will not simply discard unusual data from individuals who appear to be behaving differently than the rest of the group. Indeed, as ethologists interested in personality, we should grab this "unusual, atypical" information and examine it fully. The question is "how"? It is very difficult to measure *anything* in a species other than our own, simply because other species are nonverbal—they cannot tell us the answer/s. However, all social species communicate by some means or another, including vocalizations,

It's easy to see distinct personalities and temperament in a variety of animals, such as horses.
© Photowood Inc. / Corbis.

ear and tail movements and changes in posture. It is our job as respectable scientists to uncover ways of being able to get our questions answered in order to be able to describe the ways in which individuals differ from one another.

Scientists who study personality and temperament in humans agree that the multivariate approach is the best. This means that they look at as many different aspects of the behavior of the individuals as possible, all at the same time. It is therefore sensible to apply this approach to the study of animal personality.

Three main approaches have been taken with animals: studying their behavior when they are in their natural environment; studying their behavior when they are placed in an artificial, experimental environment; and asking their "keepers" to complete rating questionnaires about their behavioral tendencies.

Whichever approach is used it is very important to make sure that measurements are reliable and repeatable. In other words, it is important that scientists agree on what they are measuring. All of the approaches mentioned above depend upon the human observing the behavior of the animal. In an ideal world all measurements should be *objective*, in other words, each behavior has its own definition and cannot be confused with other behaviors. (Most studies of animal behavior will refer to the species *ethogram*—which is a list of all of the possible behaviors that the species can perform, complete with detailed definitions.) However, it has already been stated that personality involves temperament, which in turn reflects "mental activity." It is not possible for human observers to know directly what is going on in the animal's mind, so we become *subjective*. We cannot help it. In other words, we interpret the behavior we see based on what we know about our own emotions. It is very difficult to avoid doing this—but we should do our best to ensure that the observations of animal behavior do not become totally biased by our own thoughts. One way of doing this is to replicate observations. There are two main types of replication commonly used. When *intraobserver replication* is used, a single observer records information about the individual animals more than once. The results are then compared and, if they are similar, the data are said to be reliable, and conclusions can be drawn. The other type of replication is *interobserver replication*—where the same animal is observed by two independent observers at the same time. Again, if their results agree, it is assumed that a reliable assessment has been made of the animal's behavior.

A commonly used approach is to place the animals in experimental conditions and to monitor their behavioral responses. These are usually artificial and often mean the animal being "tested" is temporarily isolated from its mates. A very good example of this is the *backtest* used to assess the emotional response (therefore the temperament, and also the personality) of piglets. Piglets are removed from their home pens, gently laid on their sides and restrained in that position (by a human) for a short amount of time. The behaviors performed by the piglet, such as escape attempts, biting, vocalizations, are recorded.

Rating studies have also been used to assess the behavior of animals, where the keepers are asked to score the animals on personality and temperament ratings. For example, a horse owner may be asked to score his/her horse according to how "reactive------unreactive" and how "extraverted------intraverted" it is. Don't forget that ratings are unavoidably subjective but can be shown to be reliable by using either of the observer replication methods described above. An advantage of rating studies is that a lot of aspects of behavior (therefore personality) can be examined at the same time.

Examples of Personality and Temperament

We know that the temperament of an individual is an integral part of its personality, and that information about the animal's personality can be obtained from the behavior that the animal exhibits. Just as in humans, the behavior exhibited (reflecting the animal's temperament and personality) is influenced by its genetic make up (i.e., its breeding, or *nature*), its physical and social environment (*nurture*) and, crucially, its past experience. Just as in humans, an individual's personality develops as it matures and becomes increasingly socialized in its social group.

We have seen that it is possible to measure the personality and temperament of individuals within a group by means of scientific studies. Findings of such studies allow predictions to be made not only about animal behavior in general, but also the behavior of individuals, which can be very important for management purposes and welfare. Seeing how an animal reacts in one situation can allow us to predict how it will behave in other situations. For example, in a study of a group of cattle, individuals who were scared of a novel object (a brightly colored cardboard box) were also very unlikely to approach an unfamiliar human (a human they had never seen before).

The remainder of this essay summarizes some of the results of studies of individual differences (reflecting personality) in animals. Social behavior is particularly well studied. It has been found that there are larger differences between individual pigtail macaques affiliative (being nice) behavior than in their aggressive (being horrible, nasty) behavior. It has also been found that individual Japanese macaques have very different play activity profiles. When studying social behavior it is also important to examine behavior which is *received* as well as behavior which is *performed*. It has been found that cattle differ in the amount of affiliative behavior both performed and received. Similar results have been found for aggressive and withdrawing activity. Similarly, a study of a cat population revealed that some cats initiate more interactions than others. Also, some cats received a lot of behavior from many different cats, whereas others did not. These kinds of results led to the description of animals in terms of their "popularity," "sociability" and "likeability"—which is actually very similar to studies of personality and behavior in preverbal human infants. Individual differences in the distance maintained to others (individual distance) have also been noted in many species, including cattle, elephants, and horses. Closer examination of the nature of behavior being performed has also led to individuals being classified as "groomers" in cattle and "babysitters" in deer.

Piglets subjected to the "backtest" (in order to assess their emotional response/s, therefore their temperament and personality) were broadly classified according to the behavioral responses they made in order to cope with the backtest. "Proactive" piglets exhibit aggressive behavior and are very active, whereas "reactive" piglets become immobile, avoid escape, and become withdrawn.

Rating studies have also provided a substantial amount of information about personality in animals. For example, a subjective rating study of a group of rhesus monkeys carried out 4 years in a row indicated that the personality of individual monkeys could be described in terms of how confident, excitable and sociable they were. Similar studies have shown that the most timid and behaviorally responsive goats are more difficult to milk, "flighty" sows had litters that weighed less than "placid" sows and more "interested" cows made better adoptive mothers than "wary" cows. The results of a rating study on temperament and personality was used by a donkey sanctuary to match donkeys to potential owners. More "cooperative" donkeys were better suited to homes with children than "timid" donkeys.

To conclude, there is evidence that individual differences exist in a wide range of behaviors and species, over and above those that can be accounted for by variables such as age, sex, and breed, and accounted for by dominance hierarchies. This essay has described the relationship between personality, temperament and behavior and examined the three main approaches to the measurement of personality and temperament in nonverbal animals.

See also Personality and Temperament—*A Comparative*
      Perspective
      Personality and Temperament—*Personality*
        in Chimpanzees
      Personality and Temperament—*Stress, Social Rank,*
        and Personality

Further Resources

Kiley-Worthington, M. 1987. *The Behaviour of Horses in Relation of Management and Training.* London: J. A. Allen.

Syme, G. J. & Syme, L. A. 1979. *Social Structure in Farm Animals: Developments in Animal and Veterinary Sciences*, Vol 4. Amsterdam: Elsevier.

Wilson, E. O. 1975. *Sociobiology.* Cambridge, MA: Belknap Press of Harvard University Press.

Hayley Randle

■ Personality and Temperament
Stress, Social Rank and Personality

No matter how rich of an ecosystem it is, resources are not infinite, and this can produce competition among animals. Such competition could take the form of every bite of food being contested. But social species with stable groups often instead develop dominance hierarchies. In this circumstance, animals have ranks, and the high-ranking get a disproportionate share of resources without every encounter requiring bloody tooth and nail. Individuals know their place in the social system.

Dominance hierarchies, first described as "pecking orders" in chickens, take various forms. In some cases (such as female rhesus monkeys), rank is inherited, while in others, rank shifts over time (as in male rhesus monkeys). Hierarchies may be situational: Animal A might dominate Animal B only in certain settings. Hierarchies might depend on coalitions where A dominates B only when backed up by Animal C (as can occur among male chimpanzees).

In some hierarchies, every stepwise change in rank has consequences for the quality of life. In other cases, the only relevant issue is whether you're number one. This is the case with marmoset monkeys, in which only the highest ranking mate. And hierarchies can be linear, with A > B > C > D, or they can contain circularities, with A > B; B > C; C > A.

Dominance hierarchies mean that the quality of an individual's life varies with its rank. And in many species, being low-ranking is disproportionately stressful. Such individuals are subject to more physical stressors. They may work harder for their calories, and get less overall. They may be pushed to the most vulnerable positions when predators lurk. And when someone dominant is in a foul mood, they may be attacked without warning, because such "displacement aggression" can make those in foul moods feel better.

Moreover, subordinate individuals are more psychologically stressed. They lack control and predictability in their lives. Other animals can steal their food, force them from a desirable resting spot.

And they often lack outlets that reduce stress. For example, low-ranking primates are less likely to be socially groomed, an activity that appears to be stress-reducing. And, of course, a subordinate individual may lack the option to decrease stress by displacing aggression on someone weaker.

These factors may have disease consequences for subordinate animals. When an animal is acutely stressed (e.g., running for its life), its body activates the "stress response," often also called the fight or flight syndrome. Adrenaline (also known as epinephrine) and other stress hormones are released, blood pressure increases, energy is diverted to exercising muscle, long-term projects (such as growth, tissue repair, immunity and reproduction) are inhibited until more favorable times. The stress-response is critical for sustaining a life-saving sprint. But activating the same stress-response chronically because of psychological stress can cause disease. Increasing blood pressure, inhibiting growth, tissue repair, or fertility chronically, increases the risk of various degenerative diseases.

Research suggests that subordinate animals in many species are more at risk for stress-related disorders. This includes elevated blood pressure and levels of stress hormones, even when there is nothing stressful happening; low levels of "good" cholesterol and of hormones involved in tissue repair, impairments of reproduction and immunity. Seemingly, if you have a choice in the matter, you don't want to be a subordinate animal.

Naturally, there are complications to this picture.

1. *Rank tends to precede physiology.* While being subordinate and chronically stressed could increase the risk of stress-related disease, being sickly with stress-related diseases could lower an animal's rank. Which comes first? Studies generally show that it is rank.

2. *The social meaning of rank.* The picture of subordinance associated with stress and stress-related disease is the case when ranks are stable, and being subordinate entails being disproportionately stressed (e.g., in lab rats). But in some species or circumstances where hierarchies are not stable, and instead there is tense shifting and constant competition at the top of the hierarchy, it is dominant individuals who have the more stressed physiological profile (e.g., dwarf mongooses). As another exception, in some species, dominance systems are relaxed affairs, with minimal stress for subordinates, and the low-ranking are simply biding their time until maturing into dominance. In such cases (e.g., marmosets), subordinance is not associated with more stress-related disease. Thus, while rank is important, what rank *means* is more important.

3. *The personal experience of rank.* Consider a species in which subordinates are disproportionately stressed. Nevertheless, the extent to which stress hormone levels are elevated in subordinate individuals can vary depending on what it is like to be subordinate in a particular group. How often do subordinates suffer displaced aggression? How much social support do they have? This variability occurs among female rhesus monkeys. Thus, the abstract state of being dominant or subordinate is not as important as what an animal's personal experience of that rank is like in its group.

4. *The importance of personality.* Many zoologists now study personality (or *temperament*) in animals (even among fish, who differ in their willingness to explore something novel).

Among primates, to use human terms that are not anthropomorphic, individuals differ in stable ways as to how introverted or extroverted they are, whether they are calm or hot-heads when socially challenged, whether they react to the world as a glass half full or empty. And these differences predict differences in stress-related physiology. As examples, a baboon sees his hated rival threaten him from a foot away, or sees that rival take a nap a hundred yards away. Can he recognize that the former is threatening, the latter not? A baboon loses a fight. Does he mope alone or find someone to groom him? Independent of rank, animals who see rivals napping as threatening, who mope alone, tend to have higher circulating stress hormone levels. Broadly, these studies show that, independent of rank, animals who are more reactive to novelty, less capable of differentiating threat from neutral circumstances, less capable of taking advantage of social support, are more prone towards stress-related disease.

In conclusion, there is no simple relationship between an animal's rank and its risk of stress-related disease. For zoologists who revel in the complexities of their animals, this is not surprising.

See also Personality and Temperament—*A Comparative Perspective*
Personality and Temperament—*Personality in Chimpanzees*
Personality and Temperament—*Personality, Temperament, and Behavior in Animals*

Further Resources

Abbott, D., Keverne, E., Bercovith, F., Shively, C., Mendoza, S., Saltzman, W., Snowdon, C., Ziegler, T., Banjevic, M., Garland, T., Sapolsky, R. 2003. *Are subordinates always stressed? A comparative analysis of rank differences in cortisol levels among primates.* Hormones and Behavior, 43, 67.
Clarke, A., Boinski, S. 1995. *Temperament in nonhman primates.* American Journal of Primatology, 37, 103.
Creel, S. 2001 *Social dominance and stress hormones.* Trends in Ecology and Evolution, 16, 491.
Sapolsky, R. 2001. *The physiological and pathophysiological implications of social stress in mammals.* In: *Handbook of Physiology, Section 7, Volume IV: Coping with the environment: Neural and endocrine mechanisms* (Ed. by B. McEwen), p. 517. Cambridge: Oxford University Press.
Sapolsky, R. 1990. *Stress in the wild.* Scientific American, January, 116–123.

Robert M. Sapolsky

■ Play
Birds at Play

Play has been described in many mammals and birds and, as is true with humans, it is the young that spend the most time playing. The young have the most to learn—about themselves, about others and about the environment—and learning benefits the individual in cognitive as well as in social and physical ways. Individuals benefit from the exercise they get through play: It is good for their health and it strengthens their muscles. In social play, they learn about their own physical limitations vis-à-vis the abilities of others, how that places them in the group, who else will play on their level and who does not, what is the predictable response from one individual and how another's response is likely to differ. Not every playmate will react in the same way; this encourages a flexible approach to others. The young are drawn to explore and play with novel objects, thus learning the objects' uses. They repeat their actions, practicing various activities and movements, improving their motor skills, learning how to reach fruits and locate insects, or how to extract seeds or crack nuts. Through play, they may also begin to problem solve (pulling down the branch also

Cockatiels playing tug of war.
Courtesy of Jacqueline Garamella.

brings the fruit in closer), or they learn that the sibling they are chasing doesn't just disappear though she might be out of sight at the moment. Through their object play, they learn to discriminate between different appearances (the soft, yellow fruit is ready to eat, but not the hard, green one).

Play in birds has been described in about half the avian orders; the most diverse play comes from parrot species and perching birds. Many young birds play in the same ways that young mammals, including human children, play (see picture of cockatiel tug-of-war). Young song sparrows and some other perching birds play chase and they "frolic," darting about and suddenly turning, flapping their wings and jumping, much like young lambs "frisk"; keas play tug-of-war with objects they have found and roll rocks down river banks; keas, blackbirds, and ravens play king-of-the-castle; ravens and crows slide down river banks and play toss with stones and sticks, they soar and dive and glide on air currents (see picture of young crows hopping on air currents on beach), and they "dance,"

Crows play in the air currents on the shore of Lake Michigan.
Courtesy of Ken Oberlander.

leaping together with wings out, sometimes with flowers or pebbles in their bills.

Are These Activities Really "Play" for the Birds?

Why is the activity being done? Perhaps the birds are simply performing common activities without the normal stimulus that usually causes the activity (called a *vacuum activity*), or making immature efforts to be aggressive or to affiliate with others, not really playing. Play often resembles other activities, but M. Ficken suggests that there are several characteristics of play that differ from other ordinary activities: Play may be more repetitious or exaggerated; the movements may look incomplete as compared to the normal species action patterns; it may be elicited by a wide variety of different stimuli that are present. In determining if an activity is really play, it helps to have studied the individuals over time, becoming familiar with their behavior patterns, observing how the activities begin, continue, and then what the results are. Ornithologist Alexander Skutch defines play as "spontaneous, intrinsically rewarding activity, a pastime in which healthy animals who have satisfied all vital needs expend excess energy for enjoyment alone, with no ulterior motive. Whatever extraneous benefits play may yield—as in strengthening muscles, sharpening skills, improving social relations—are not its ends" (1996, p. 43). When young keas in New Zealand tussle and wrestle and jump on each other and drag each other around by the throat, they are actually gaining all three of the benefits mentioned (building strength, honing skills, and being social). Are the activities meant to be aggressive? Is anyone hurt in the rough housing? Are there any signals that tip off the playmate that it is all in fun? When keas roll over on their back, wave their feet and squeal, this encourages a partner to run over and pounce on the kea on the ground and engage in tussle play. Those who have studied the keas say these acts signal that this is play, not a fight, just as dogs signal when they want to play.

Types of Play Seen in Yellow-Crowned Parakeets

I saw a variety of play activities in a group of young yellow-crowned parakeets (kakarikis) on which I was assessing cognitive developmental levels. Many were the same play behaviors that other investigators have seen in young primates. Before the fledgling kakarikis could use their claws to grasp, they used their beaks to grasp objects as they fluttered and walked around the cage, and they explored these items in their beaks like young humans and other young primates "mouth" things to explore them. The parakeets would also repetitively hit with their beaks colored rings or hanging strips of rawhide dangling from the top of their flight cages. (Compare with babies in their cribs using their feet to keep dangling mobiles swinging over them.) One who fledged at 6 weeks kept the rawhide strip swinging back and forth continuously for 30 seconds at a time when he was just three days out of the nest box. Two weeks later his attention span was greater, and he pursued this activity for 5 minutes at a time. The kakarikis need a rather lengthy attention span and detailed, systematic searching ability in the wild because part of their diet is a type of scale insect that calls for careful and prolonged searching. I believe that play increases attention span as well as deepening the capability to "attend" to what is going on.

Aside from this object play, the kakariki fledglings also displayed locomotor and vocal play. They hopped around the cage, flying from side to side, stretched out their wings and held on to the perch exercising their wings, hung upside down on their perch by their feet and climbed around upside down on the top of the cage. They practiced their species vocalizations in a type of vocal play called "subsong" that has been compared to a human infant's babbling. For over a month, they made many strange sounds, adding high and low notes, experimenting with their vocalizations until they sounded more like their species.

As juveniles, the kakarikis also swung in small swings in their cages, standing on them and propelling themselves forward by flapping their wings or moving their bodies. They spun small tops, rolled objects on the floor, pushed a roly poly doll on a round bottom back and forth, pulled on strings activating the legs and arms of a small plastic rabbit, worked at untying knots holding elastic strings to the cage bars, played with the elastic strings, winding or pulling them up and down, did somersaults on the bottom of the cage, climbed up and down the hanging rawhide strips, and explored any tunnels they found (these birds are cavity nesters). They dropped and monitored objects from the top of the cage to the counter top and then hopped down to the counter and dropped the items again to the floor. They also spent time pushing newspapers off a high ledge to the floor.

Others have described dropping behaviors in birds such as sea gulls dropping shells for play.

The birds also had much social play. They would play bite each other's tail feathers, crowd or crawl over each other on a perch, do tug-of-war with grape pieces (but didn't necessarily eat the grape if they won it), spar over a piece of corn-on-the-cob and then share it, nibble at the claws of a sibling who was hanging upside down on a perch or else was walking on top of the cage. Play with some object by any one of the birds, like rolling or lifting and listening to a rattle toy, would attract another bird to come over and start playing with the toy. The first juvenile would not contest this loss usually, but would hop off and play with something else.

Repeating movements, exploring objects, playing with others, are activities that are important and necessary for the young in many species. From both a developmental as well as an ecological perspective, we can see why birds and mammals might have similar types of play: They have similar pressures on them to learn about their own skills and those of others, about objects, and about their surroundings. Because it is performed with more energy than purpose, play has been underestimated as a source of cognitive growth. Play is a great teacher. It seems to serve as a pleasurable rehearsal for the important events of life.

See also Play—*Dog Minds and Dog Play*
Play—*Social Play Behavior and Social Morality*

Further Resources

Bekoff, M. & Byers, J. A. 1989. *Animal Play: Evolutionary, Comparative, and Ecological Perspectives.* Cambridge: Cambridge University Press.
Diamond, J. & Bond, A. B. 1999. *Kea: Bird of Paradox.* Berkeley: University of California.
Fagen, R. M. 1981. *Animal Play Behavior.* Oxford: Oxford University Press.
Ficken, M. S. 1977. *Avian play.* The Auk, 94, 573–582.
Funk, M. S. 1996. *Problem-solving skills in young yellow-crowned parakeets.* Animal Cognition, 5, 167–176.
Gamble, J. R. & Cristol, D. A. 2002. *Drop-catch behavior is play in herring gulls*, Larus argentatus. Animal Behaviour, 63, 339–345.
Kilham, L. 1989. *The American Crow and the Common Raven.* College Station: Texas A & M.
Ortega, J. C. & Bekoff, M. 1987. *Avian play: Comparative, evolutionary, and developmental trends,* The Auk, 104, 338–341.
Skutch, A. F. 1996. *The Minds of Birds.* College Station, Texas A & M.

Mildred Sears Funk

■ Play
Social Play Behavior and Social Morality

The study of "social play," or "play," continually challenges students of behavior. I've been studying play for more than 30 years, and I am still discovering new facts about this fascinating activity—why animals play and how studies of animal play may be related to questions about the evolution of fairness and social morality. Almost all mammals who have been studied engage in play. And so do some birds, and perhaps even fish, reptiles, and amphibians. Although play is difficult to define and to study, it is recognized as a distinctive category of behavior.

What is play? The deceptively simple question has troubled researchers for many years. The following definition of social play resulted from research on play that I did with the behavioral ecologist John Byers. John studied wild pigs, peccaries, and I studied various canids (members of the dog family, including domestic dogs, wolves, coyotes, jackals, and foxes), and we (and other researchers) discovered there were many features common to the play of these (and other) mammals. *Social play* is an activity directed toward another individual in which actions from other contexts are used in modified forms and in altered sequences. Some actions also are not performed for the same amount of time during play as they are when animals are not playing. Our definition centers on what animals *do* when they play, or the structure of play.

Play activity that crosses species borders: One miniature horse playing chase with several dogs, all of whom live at a rescue center in Texas. The horse is a certified therapy animal who visits hospitals.
Courtesy Animal Image.

What this all means is that when animals play they use actions that are used in such activities as predation (hunting), reproduction (mating), and aggression. Full-blown threatening and submitting occur only rarely, if ever, during play. Behavior patterns that are used in antipredatory behavior also are observed in play. This is especially so in prey animals such as ungulates (deer, elk, moose, gazelles) who run about in unpredictable zig-zag patterns during play. These actions may be changed in their form and intensity and combined in a wide variety of unpredictable sequences. For example, in polecats, coyotes, and American black bears, biting in play fighting is inhibited when compared to biting in real fighting. Clawing in bears is also inhibited and less intense. Play in bears also is nonvocal, and biting and clawing are directed to more parts of the body of another individual during play than during aggression.

Play sequences may also be more variable and less predictable because individuals are mixing actions from a number of different contexts. Because there are more actions for individuals to choose from it is not surprising that sequences are more variable. "More variable" simply means that it is more difficult to predict which actions will follow one another during play than, for example, during predation or aggression. During play in dogs, coyotes, or wolves one might see the following sequence: biting, chasing, wrestling, body slamming, wrestling, mouthing, chasing, lunging, biting, and wrestling, whereas during aggression it would be more likely to see threatening, chasing, lunging, attacking, biting, wrestling, and then one individual submitting to the other.

One reason why studying play has been so difficult is because play is a hodge-podge of many different activities, and it takes a lot of time to learn about the details of this fascinating behavior. In my own studies, I can spend hours conducting frame-by-frame analyses of 5 minutes of play caught on video.

Dog Minds and Dog Play
Alexandra Horowitz

A Great Dane, at its shoulders the height of a small horse, spots his target across the lawn: a six-pound Chihuahua almost hidden in the high grasses. With one languorous leap, his ears perked, the Dane arrives in front of the trembling Chihuahua. He lowers his head and bows to the little dog, raising his rear end up in the air, and wagging his tail. Instead of fleeing, the Chihuahua mirrors this pose in return, and she leaps onto the head of the Dane, embracing his nose with her tiny paws. They begin to play.

My video camera catches it all. With their play bows, these improbable dog playmates joined my other research subjects as data in my search to understand the mind of the domestic dog (*Canis familiaris*). How does this social interaction succeed? How does the little dog know that the big dog wants to play? What—if anything—does the Great Dane, or the Chihuahua, need to understand about the perspective of each other, for play to erupt and succeed?

While much science takes place in controlled environments, ethologists believe that careful observation of the natural behavior of animals, including domestic animals, can also reveal intriguing, sound data about the cognition of animals. The natural environment of a domesticated animal like a dog is among dogs and people; as a result, I pursued research into what they understand about each other, not by bringing a dog into the laboratory, but by looking at pairs of dogs interacting in dog parks near San Diego, California.

Any interaction a dog has with other dogs—straining on leashes toward each other on the sidewalk; while running in fields; competing with others for food— could potentially provide information about what each dog knows about other dogs. I examined how dogs negotiate during rough-and-tumble play, which involves rambunctious wrestling, biting, leaping, and chasing. This kind of play is a fertile source because it requires coordination: Each dog needs to tailor his play to match the other, or he risks losing a playmate or even being attacked.

Such coordination might involve the ability to take the perspective of the other animal: What in humans we call having a "theory of mind." Theory of mind is the explanation of the behavior of others by reference to their mental states— particularly belief, desire, and knowledge. Do the Great Dane and the Chihuahua understand something about each other's minds?

My research indicated that yes, they do. How to determine if an animal has a theory of mind has been much debated among psychologists. In my research I identified behaviors that are tailored to the attitude or perspective of others— and, thus, might establish if an animal has a theory of other animals' minds. In particular, I focused on communications—the play signals—and the opportunity to use or manipulate attention. For instance, did a dog desiring play ensure the other dog was watching before play-bowing, or did he start playing outright?

After hundreds of hours of observation, 39 play bouts were translated into a sequential listing of each dog's posture and actions, through slow motion videotape playback, often frame (1/30th of a second)-by-frame. An astonishing wealth of brief behaviors and glances is visible in slow playback, but invisible in "real time." What I found is that these dogs used attention skillfully in communicating. The dogs

(continued)

Dog Minds and Dog Play (continued)

signaled requests for play almost exclusively to present, attentive audiences, and when attention was away, they moved to get attention, persistently and creatively. They used attention-getters which were tailored to their audiences: more forceful (a nip) if looking away or playing with others, more subtle (running into the dog's field of vision) if less distracted. These behaviors indicate that the dogs understood that in order to communicate successfully, one must take the attention—the mind—of the recipient into account: a kind of "rudimentary" theory of mind.

Careful observation of animals' natural behavior is an exciting and realistic method for understanding their cognition. You may have an animal in your own household whose behavior you can watch over time: What you will find will be unexpected—if you look closely.

See also Cognition—*Fairness in Monkeys*
Cognition—*Theory of Mind*
Cooperation
Emotions—*Emotions and Cognition*
Empathy
Play—*Social Play Behavior and Social Morality*

You can see some of the features of play in my field notes from research on play in domestic dogs:

> Jethro bounds towards Zeke, stops immediately in front of him, crouches on his forelimbs, wags his tail, barks, and immediately lunges at him, bites his scruff and shakes his head rapidly from side to side, works his way around to his backside and mounts him, jumps off, does a rapid bow, lunges at his side and slams him with his hips, leaps up and bites his neck, and runs away. Zeke takes off in wild pursuit of Jethro and leaps on his back and bites his muzzle and then his scruff, and shakes his head rapidly from side to side. Suki bounds in and chases Jethro and Zeke and they all wrestle with one another. They part for a few minutes, sniffing here and there and resting. Then, Jethro walks slowly over to Zeke, extends his paw toward Zeke's head, and nips at his ears. Zeke gets up and jumps on Jethro's back, bites him, and grasps him around his waist. They then fall to the ground and mouth wrestle. Then they chase one another and roll over and play. Suki decides to jump in and the three of them frolic until they're exhausted.

It is important to note that on no occasion did play between Jethro and Zeke escalate into aggression. Of course, dogs aren't the only animals who love to play and repeatedly seek out opportunities to play. Young cats, chimpanzees, foxes, bears, and rats love to play to exhaustion. When a potential playmate doesn't respond to a play invitation these "wired kids" often turn to another individual or to their own tails.

Animals who play do not spend a lot of time engaged in this activity. The amount of time and energy that young mammals devote to various types of play is usually less than 10% of their total time and energy budgets. In most species play occurs mainly during infant and juvenile life. Adults do engage in social play but less so than the young of their species. Primatologists Marc Hauser and Richard Wrangham have observed wild adult male chimpanzees engage in vigorous play. I have seen adult coyotes chase one another and play to exhaustion, but less so than indefatigable youngsters.

Play and Emotions

When animals play they are having fun. They experience joy and happiness. The rhythm, dance, and spirit of animals at play is incredibly contagious; just seeing animals playing can stimulate play in others. Animals who are sick or stressed usually do not play. Thus, play may a useful measure of an individual's well-being.

One way to get animals (including humans) to do something is to make it enjoyable. Studies of the chemistry of play support the claim that play is fun. Neurochemicals in the brain such as dopamine (and perhaps serotonin and norepinephrine) are important in the regulation of play. Rats show an increase in dopamine activity when anticipating the opportunity to play, and they enjoy being playfully tickled.

Charles Darwin, whose theory of natural selection revolutionized the study of biology, had much to say about a lot of different topics in zoology and botany, and he also reflected on many different aspects of animal behavior. He was an unabashed advocate for the idea that animals have rich and deep emotional lives and perhaps even a sense of social morality. Darwin also had no doubts that animals loved to play. In his book *The Descent of Man and Selection in Relation to Sex* he wrote that: "Happiness is never better exhibited than by young animals, such as puppies, kittens, lambs, etc., when playing together, like our own children." In the same book Darwin also wrote: "It is a significant fact, that the more the habits of any particular animal are studied by a naturalist, the more he attributes to reason, and the less to unlearnt instincts."

Darwin also demonstrated that there is evolutionary continuity among different species, and that anatomical, behavioral, cognitive, emotional, and moral variations among different species are differences in *degree* rather than differences in *kind*. Darwin's well-accepted view argues that there shades of gray and not absolute black–white differences among different animals and between nonhumans and humans. Thus, there isn't a void in cognitive abilities or emotional lives among different animals but rather a continuum. If something is so for humans there's no reason to assume that it isn't so for other animals. Human brains do not look much like dog or rat brains, but there's no doubt that all of these creatures have organs that sit in their craniums and we call these organs brains. The same can be said for their digestive systems and how their brains and stomachs work. Using Darwin's arguments, there is little reason to believe that the neurochemical basis of joy in dogs would differ substantially from that of rats, cats, chimpanzees, or humans.

How Animals Tell One Another, "I Want to Play with You!"

To learn about the dynamics of play it is essential to pay attention to subtle details that are otherwise lost in superficial analyses. Dogs and other animals keep track of what is happening when they play, so we need to do this also. My studies of play are based on careful observation and analyses of videotape. I watch tapes of play one frame at a time to see what the animals are doing and how they exchange information about their intentions and desires to play. This is tedious work, and some of my students who were excited about studying dog play had second thoughts after watching the same video frames over and over again. But when they then were able to go out and watch dogs play and understand what was happening, they came to appreciate that while studying play can be hard work it's well worth the effort.

By studying play we can learn about what may be going on in an individual's minds, what she is thinking about, what she wants, and what she is likely to do during a social encounter. During play, behavior patterns that are observed in mating may be intermixed in unpredictable and varying sequences with actions that are used during fighting, looking for prey, and avoiding being eaten. Because of the intermingling of actions from different contexts, it is important to ask such questions as, "How do animals know that they are playing?," How do they communicate their desires or intentions to play or to continue to play?," and "How is the play mood maintained?"

Because there is a chance that various behavior patterns that are performed during ongoing social play can be misinterpreted, individuals need to tell others, "I want to play with you," "this is still play no matter what I am going to do to you," or "this is still play regardless of what I just did to you." An agreement to play, rather than to fight, mate, or engage in predatory activities, can be negotiated in various ways. Individuals may use various behavior patterns called *play signals* or *play markers* to initiate play or to maintain and prevent termination of a play mood by "punctuating" play sequences with these actions when it is likely that a particular behavior may have been, or will be, misinterpreted. There also are auditory (play sounds such as play panting), olfactory (play odors), and tactile (touch) play signals.

Pumpernickel in a classic play bow stance.
Courtesy of Alexandra Horowitz.

To illustrate how animals communicate their intentions and desires to play, let's consider domesticated dogs and their wild relatives. One action that is very common in play among canids is the *bow*. A dog asks another to play by bowing—crouching on her forelimbs, raising her hind end in the air, and often barking and wagging her tail as she bows. Bows occur almost exclusively in the context of social play. The bow is a highly ritualized and stereotyped movement. Bows are a Modal Action Pattern (MAP). Bows stimulate recipients to engage or to continue to engage in social play. Bows occur throughout play sequences, but most commonly are performed at the beginning or toward the middle of playful encounters.

Bows help to establish a "play mood." I discovered that play bows are always less variable (more stereotyped or "fixed") when performed at the beginning, rather than in the middle of, ongoing play sequences. The more variable form of the bow after play has begun might occur because there is less of a need to communicate that "this is still play" while animals are playing than there is when trying to initiate a new play interaction.

To solve the problems that might be caused by confusing play with mating or fighting, many species have evolved signals such as the bow that function to establish and maintain a play mood. In a long-term and continuing study of social play, I also found that play signals in infant canids (domestic dogs, wolves, and coyotes) were used *nonrandomly*, especially when biting accompanied by rapid side-to-side shaking of the head was performed. These youngsters were intentionally using bows with a purpose in mind. I learned not only that bows are used right at the beginning of play to tell another dogs, "I want to play with you," but they're also used right before biting accompanied by rapid side-to-side head shaking as if to say, "I'm going to bite you hard, but it's still in play," and right after vigorous biting as if to say, "I'm sorry I just bit you so hard, but it was play." Biting accompanied by rapid side-to-side shaking of the head is performed during serious aggressive and predatory encounters and can easily be misinterpreted if its meaning is not modified by a play signal. Bows serve as punctuation, an exclamation point, to call attention to what the dog wants to tell his friend. Bows also reduce the likelihood of aggression in African wild dogs. Dogs and their wild relatives rapidly learn how to play fairly, and their response to play bows seems to be innate.

Play signals are an example of what ethologists call "honest signals." There is little evidence that social play is a manipulative activity. Play signals are rarely used to deceive others in canids or other species. My own long-term studies indicate that deceptive signaling is so rare that I cannot recall more than a few occurrences in thousands of play sequences. Cheaters are unlikely to be chosen as play partners because others can simply refuse to play with them and can choose others. Infant coyotes who mislead others into playing so that they dominate them have difficulty getting other young coyotes to play with them. The message is clear: Don't bow if you don't want to play.

In domestic dogs there is little tolerance for noncooperative cheaters. Cheaters may be avoided or chased from play groups. There seems to be a sense of what is right, wrong, and fair. While studying dog play on a beach in San Diego, California, Alexandra Horowitz observed a dog she called Up-ears enter into a play group and interrupt the play of two other dogs, Blackie and Roxy. Up-ears was chased out of the group, and when she returned, Blackie and Roxy stopped playing and looked off toward a distant sound. Roxy began moving in the direction of the sound, and Up-ears ran off following their line of sight. Roxy and Blackie immediately began playing once again. Even in rats, fairness and trust are important in the dynamics of playful interactions. Sergio Pellis, a psychologist at the University of Lethbridge in Canada, discovered that sequences of rat play consist of individuals assessing and monitoring one another and then fine-tuning and changing their own behavior to maintain the play mood. When the rules of play are violated, when fairness breaks down, so does play.

Role-Reversing and Self-Handicapping

Individuals also engage in role-reversing and self-handicapping to maintain the play mood. Each can serve to uphold the agreement to play that is needed for play to occur. *Self-handicapping* happens when an individual performs behavior patterns that might compromise herself. For example, a coyote might not bite her play partner as hard as she can, or she might not play as vigorously as she can. Inhibiting the intensity of a bite during play helps to maintain the play mood. I once picked up a 22-day-old coyote only to have him bite through my thumb with his needle-sharp teeth. The fur of young coyotes is very thin and an intense bite results in high-pitched squeals and much pain to the recipient. An intense bite is a play stopper. In adult wolves, a bite can generate as much as 1,500 pounds of pressure per square inch, so there is a good reason to inhibit its force. Years ago I foolishly tried to show a captive adult male wolf, Lupey, where his bone was by pointing toward it, and he immediately showed me that he knew where it was by clasping his mouth over my extended forearm and squeezing ever so gently. I wore Lupey's teeth marks for 2 weeks, but he didn't even break the skin. Lucky me.

Red-neck wallabies also engage in self-handicapping by adjusting their play to the age of their play partner. When a partner is younger, the older animal adopts a defensive, flat-footed posture, and pawing rather than sparring occurs. In addition, the older players are more tolerant of its partner's tactics and take the initiative in prolonging play interactions.

Role-reversing occurs when a dominant animal performs an action during play that would not normally occur during real aggression. Basically, there is a breakdown in dominance relationships during play, and dominant individuals do not try to intimidate or to immobilize subordinate animals. For example, a dominant wolf might not voluntarily roll over on his back during fighting, but would do so while playing. In some instances, role-reversing and self-handicapping might occur together. Thus, a dominant wolf might roll over while playing with a subordinate animal and inhibit the intensity of a bite.

Self-handicapping and role-reversing, similar to using specific play invitation signals and gestures, or altering behavioral sequences, might serve to signal an individual's intention to continue to play. In this way there can be mutual benefits to each individual player because of their agreeing to play and not fight or mate.

Why Play? What Are the Functions of Play?

A big question and one that is very difficult to answer is, "Why has play evolved or why is play adaptive?" Despite risking injury and using energy that might be needed for growth or to get food associated with play (these are called *costs of play*), many individuals, especially youngsters, persistently seek it out and play to exhaustion. What benefits might outweigh the costs of play?

Patagonian Gray Fox (Dusicyon griseus) *pups playing at their den, at about 10 weeks of age. Young foxes often play until exhaustion.*
© Joe McDonald / Visuals Unlimited.

Evolutionary (or functional) explanations are often tied to analyses of what individuals do when they play. Although many researchers agree that play is important in the development of social skills, locomotor skills, or in the development of cognitive skills that support motor performance, solid evidence is scant, and opinions are divided. One important theme from recent comparative research is that there does not appear to be only one function of play across the diverse species in which individuals play. For example, play-fighting was once considered important in learning fighting skills that would be used in adulthood or for physical training. Whereas in some species play may be important for the development of certain skills and physical (aerobic and anaerobic) training, in others this might not be the case. Thus, play-fighting does not appear to be important in the development of motor training for fighting skills in laboratory rats, but research has shown that play may be important in the development of motor training, cognitive/motor training, or in the development of other social skills in other mammals, including humans. Play also may serve a number of functions simultaneously, for example, socialization, exercise, practice, or cognitive development. Play can fulfill different functions in individual species and animals who differ in age and sex.

No matter what the functions of play may be, it is thought by many researchers that play is "brain food" because it provides important nourishment for brain growth and helps to rewire the brain, increasing the connections between neurons in the cerebral cortex. Play is not an idle waste of time.

A number of researchers have suggested that play hones cognitive skills including logical reasoning and behavioral flexibility—the ability to make appropriate choices in changing and unpredictable environments. Marek Spinka, Ruth Newberry, and I suggested that play allows animals to develop flexible responses to unexpected events. Play increases the versatility of movement used to recover from sudden shocks, such as a loss of balance and falling over, and to enhance the ability of animals to cope emotionally with unexpected stressful situations. The jury is still out about why most animals play. Spinka, Newberry, and I presented 24 predictions about why animals play. Most await detailed study.

Fine-Tuning Play on the Run: Why Cooperate and Play Fairly?

Play is a voluntary activity. Individuals who don't want to play, don't play.

Individuals of different species fine-tune on-going play sequences to maintain a play mood and to prevent play from escalating into real aggression. Detailed analyses of film show that in canids there are subtle and fleeting movements and rapid exchanges of eye contact that suggest that players are exchanging information on the run, from moment to moment, to make certain everything is all right, that this is still play. Dogs and their wild relatives communicate with one another (and with us) using their face, eyes, ears, tail, body, and various gaits and vocalizations. They combine facial expressions and tail positions with different types of barks to produce a large number of detailed messages about what they want or how they feel. Dogs and wolves have more than a dozen facial expressions, a wide variety of tail positions, and numerous vocalizations, postures, and gaits, so you can see that the number of possible combinations of all of these modes of communication are staggering. They are able to send detailed messages about what they want and what they intend to do.

Playtime generally is safe time. Transgressions and mistakes are forgiven, and apologies are accepted by others, especially when one player is a youngster who is not yet a competitor for social status, food, or mates. Individuals must cooperate with one another when they play, they must negotiate agreements to play. The highly cooperative nature of play has evolved in many species.

Why might animals continually keep track of what they and others are doing and modify and fine-tune play on the run, while they are playing? Why might they try hard to share one anothers' intentions? Why do animals carefully use play signals to tell others that they really want to play and not try to dominate them? Why do they engage in self-handicapping and role-reversing? Why do animals behave fairly? By "behave fairly" I mean that animals often have social expectations when they engage in various sorts of social encounters, the violation of which constitutes being treated unfairly because of a lapse in social etiquette.

During early development there is a small window of time when young mammals can play without being responsible for their own well-being. Parents and other adults feed them and protect them from intruders. This time period is generally referred to as the *socialization period* for this is when species-typical social skills are learned most rapidly. It is important for individuals to engage in some play, and there is a premium for playing fairly if one is to be able to play at all. If individuals do not play fairly, they may not be able to find willing play partners. In coyotes, for example, youngsters are hesitant to play with an individual who does not play fairly or with an individual whom they fear. In many species, individuals also show play partner preferences, and it is possible that these preferences are based on the trust that specific individuals place in one another. Because social play cannot occur in the absence of cooperation or fairness, it might be a "foundation of fairness."

During social play, while individuals are having fun in a relatively safe environment, they learn social ground rules that are acceptable to others—how hard they can bite, how roughly they can interact—and how to resolve conflicts. There is a premium on playing fairly and trusting others to do so as well. There are codes of social conduct that regulate actions that are and are not permissible, and the existence of these codes is related to the evolution of social morality. What could be a better atmosphere in which to learn social skills than during social play, where there are few penalties for transgressions? Individuals might also generalize codes of conduct learned in playing with specific individuals to other group members and to other situations such as sharing food sharing, defending resources, grooming, and giving care.

Does It Feel Good to Be Fair?

It is important to consider the possibility that it feels good to be fair to others, to cooperate with them and treat them fairly, and to forgive them for their mistakes and shortcomings. We know that animals and humans share many of the same emotions and same chemicals that play a role in the experience and expressions of emotions such as joy and pleasure. If being nice feels good, then that is a good reason for being nice. It's also a good way for a pattern of behavior to evolve and to remain in an animal's repertoire.

Recent research on humans, using functional magnetic resonance imaging (fMRI) by Princeton University's James Rilling and his colleagues, has shown that the brain's pleasure centers are strongly activated when people cooperate with one another, and that we might be wired to be nice to one another. Perhaps this is also so for animals. Rilling's research is very important because it shows that there is a strong neural basis for human cooperation, that it feels good to cooperate, and that being nice is rewarding in social interactions and might be a stimulus for fostering cooperation and fairness. This sort of noninvasive research is precisely what is needed on other animals. We really do not know much about the neural bases of being nice and fair in any animals, even in our primate relatives. Science can help us along, and it will not be long before we can use fMRI to study fairness. fMRI has already been used by Rilling and others to study mother–infant interactions and other types of social behavior in primates.

Wild Justice: Playing Fairly, Being Nice, and Minding Manners

Studies of play are thus related to the evolution of fairness and morality. Explanations of social play rely on such notions as trusting, niceness, fairness, forgiving, apologizing, and perhaps justice—behavioral attributes that underlie human social morality and moral agency. Dogs and other animals can be just—democratic—and honest. Recent research on nonhuman primates has shown that punishment and apology play important roles in maintaining cooperation. Given what we know about play, I do not find this at all surprising. During play animals need to cooperate with one another and play fairly. Each individual has to trust that others will "play by the rules" and not cheat or try to beat them up, after "telling" them that they want to play.

I believe that social morality, in this case behaving fairly, being nice, and minding manners, is an adaptation that is shared by many mammals. Behaving fairly evolved because it helped young animals acquire social and other skills needed as they mature into adults. Without social play, individuals and social groups would lose out. Morality evolved because it's adaptive and because it helps animals, including humans, survive and thrive in particular environments. This idea might sound radical or outlandish, but there is no reason to assume that social morality is unique to humans. "Uncooperative play" is in fact impossible, an oxymoron, and so it is likely that natural selection weeds out cheaters, those who do not play by the accepted and negotiated rules.

Future Research: The Importance of Social Carnivores

While there have been many studies of cooperative behavior in animals, none has considered the details of social play, the requirement for cooperation and reciprocity, and its possible role in the evolution of social morality. Group-living animals, in which there are a variety of complex social interactions among individuals and in which individuals assess social relationships, may provide many insights into animal morality. In many social groups, individuals establish social hierarchies and develop and maintain tight social bonds that help to regulate social behavior. Individuals coordinate their behavior—some mate, some hunt, some defend resources, some accept subordinate status—to achieve common goals and to maintain social stability.

In social groups, the solidarity and cohesiveness depends upon individuals agreeing that certain rules regulate their behavior.
Courtesy of Corbis.

Let's consider pack-living wolves, exemplars of highly developed cooperative and coordinated behavior. For a long time researchers thought pack size was regulated by available food resources. Wolves typically feed on such prey as elk and moose, each of which is larger than an individual wolf. Hunting such large ungulates successfully takes more than one wolf, so it made sense to postulate that wolf packs evolved because of the size of wolves' prey. Defending food might also be associated with pack-living. However, long-term research by the field biologist David Mech showed that pack size in wolves was regulated by *social* and not food-related factors. Mech discovered that the number of wolves who could live together in a coordinated pack was governed by the number of wolves with whom individuals could closely bond (*social attraction factor*), balanced against the number of individuals from whom an individual could tolerate competition (*social competition factor*). Codes of conduct and packs broke down when there were too many wolves. Whether or not the breakup of packs was due to individuals behaving unfairly is unknown, but this would be a valuable topic for future research in wolves and other social animals.

In social groups, individuals often learn what they can and cannot do, and the group's solidarity and cohesiveness depends upon individuals agreeing that certain rules regulate their behavior. At any given moment individuals know their place or role and that of other group members. As a result of lessons in social cognition and empathy that are offered in social play, individuals learn what is "right" or "wrong"—what is acceptable to others—the result of which is the development and maintenance of a social group that operates efficiently.

Animal Morality and Human Morality

Mammalian social play is a good choice for a behavior to study in order to learn more about the evolution of fairness and social morality even in humans. This is not to say that animal morality is the same as human morality. If one is a "good" Darwinian, it is premature to claim that *only* humans can be empathic and moral beings.

So, where are we? It all boils down to the idea that social play may be a unique category of behavior because inequalities in play are tolerated more than in other social situations. Play can't occur if individuals choose not to engage in the activity, and the equality (or symmetry) needed for play to continue makes it different from other forms of seemingly cooperative behavior such as hunting and caregiving. This sort of symmetry, or egalitarianism, is thought to be a precondition for the evolution of social morality in humans. But we really do not know much about where human morality came from.

It is clear that morality and virtue did not suddenly appear in the evolutionary epic beginning with humans. Although fair play in animals may be a rudimentary form of social morality, it still could be a forerunner of more complex and more sophisticated human moral systems. It is self-serving anthropocentrism to claim that we're the *only* moral beings in the animal kingdom. The origins of virtue, egalitarianism, and morality are more ancient than our own species. But, we will never learn about animal morality if we close the door on the possibility that it exists.

Animals can have fun, be nice to one another, be fair, and honor the trust that each has for the other. In order to play, they have to do just this because most, if not all, individuals in a group will benefit from adopting this behavioral strategy, and group stability may also be fostered. Numerous mechanisms (play invitation signals, variations in the sequencing of actions performed during play when compared to other contexts, self-handicapping, role-reversing) have evolved to facilitate the initiation and maintenance of social play in numerous mammals—to keep others engaged—so that agreeing to play fairly and the resulting benefits of doing so can be readily achieved.

You can easily observe much about play behavior, especially in dogs. Visit your local dog park and watch play groups of dogs and how they go about having fun with their friends, how they dance in play. You can enjoy their frolicking and at the same time learn a lot about the social dynamics of social play. And carefully join in, but only if you know the dogs well. Get down on all fours and bow and lunge; there is no need for words, but maybe a bark will help. Dogs know what play signals given by humans mean. Nicola Rooney and her colleagues at the University of Southampton in the United Kingdom discovered that bowing and lunging by humans increased play in dogs, but that lunging was more effective. You should not try these positions with dogs who are unfamiliar to you, because some dogs are not all that tolerant of unfamiliar humans who want to play with them and who get too close without the proper introduction or greeting.

Studying play is exciting, challenging, and fun. In the last few years, the scientific study of play behavior has undergone many significant changes. New information is forcing people to give up old ideas and set ways of thinking about play. Because social play is a widespread phenomenon, especially among mammals, it offers a wonderful opportunity for comparative research on intentionality, communication, information sharing, fairness, and morality.

See also Cognition—*Fairness in Monkeys*
Cognition—*Social Cognition in Primates
  and Other Animals*
Cognition—*Theory of Mind*
Communication—*Modal Action Patterns*
Communication—Vocal—*Communication
  in Wolves and Dogs*
Cooperation
Emotions—*Emotions and Affective Experiences*

Emotions—*Emotions and Cognition*
Empathy
Play—*Dog Minds and Dog Play*

Further Resources

Bekoff, M. 2002. *Minding Animals: Awareness, Emotions, and Heart*. New York: Oxford University Press.
Bekoff, M. 2002. *Virtuous nature*. New Scientist, 13 July, 34–37.
Bekoff, M. & Byers, J. A. (Eds.) 1998. *Animal play: Evolutionary, Comparative, and Ecological perspectives*. New York: Cambridge University Press.
Burghardt, G. M. 2004. *The Genesis of Play*. Cambridge, MA: MIT Press.
Fagen, R. M. 1981. *Animal Play Behavior*. New York: Oxford University Press.
Power, T. G. 2000. *Play and Exploration in Children and Animals*. Hillsdale, NJ: Lawrence Erlbaum Associates.
Spinka, M., Newberry, R. C., and Bekoff, M. 2001. *Mammalian play: Training for the unexpected*. Quarterly Review of Biology, 76, 141–168.

Marc Bekoff

■ Predatory Behavior
Ghost Predators and Their Prey

Predators, such as tigers, wolves and lions, all have the capacity to shape ecosystems because species that live there will become prey if they do not avoid these meat eaters effectively. The ways in which prey species, such as deer and elk, avoid predators have striking effects far beyond their own immediate realm because their local ecology is shaped by ways in which prey navigate the landscape. For instance, the formation of large groups by prey enhances detection of predators, but it can also result in trampled vegetation, clustering in sensitive habitats that harbor insects or birds, or perhaps even increased competition and aggression among group members for access to limited food. But, how do prey species respond when the predators are gone? Does their behavior and ecology change with the loss of predators?

Two general frameworks have been developed to examine how prey respond when predators have been lost, each forming extreme points along a continuum. At one end is the situation where historic predators, such as saber-toothed cats, cave lions, and short-faced bears, have been extinct for a long time, and their prey now exist in the absence of effective predation. The extinct carnivores are referred to as *ghost predators* because, despite their absence, they are expected to exert effects on prey species that are still living. The most striking example of this are pronghorn, an endemic North American ungulate. In the Pleistocene epoch, the American cheetah was likely to be the major predator of this fleet animal, who can run at almost 100 kph (62 mph). However, the American cheetah has been extinct for more than 10,000 years, yet pronghorn still maintain the skeletal and muscular development now that they did when cheetahs roamed the central prairies. Pronghorn also form large groups to enhance their ability to detect predators, even where predators are absent and when fawns do not exist. John Byers has argued that pronghorn are overbuilt and still exhibit behavior that derive from ghost predators.

At the opposite end of the continuum are much more recent losses of carnivores. For example, during last 125 years in the vast areas between the Canadian border and central Mexico, grizzly bears and wolves have been extirpated from more than 95% of the areas where they once existed. As a consequence, some of the prey species such as elk, bison, and moose have now been released from a predation pressure that had been in place for thousands of years. Whether these predators should be considered ghost predators is not as clear. But what is clear is that these losses are part of a global biodiversity crisis that spans the planet.

Some interesting present-day conservation conundrums are raised when previously extinct or even new predators either re-enter or colonize an area. When humans first crossed the Bering Land Bridge and entered North America from Asia more than 12,000 years ago, waves of extinction followed with some 70% of the fauna larger than 45 kg (99 lb) becoming extinct. Among these were mammoths, camels, and horses. Some have argued that these massive extinctions were due to a lack of behavioral adaptations of prey to these new human predators, although others suggest climate change was responsible.

In a modern conservation context, the loss of recent predators and efforts to restore them raise questions of concern to society. Will prey species remember the previously vanquished predators? Does effective antipredator behavior remain within the behavioral repertoire of prey? And, if prey do not remember the predators or they cannot readjust rapidly, then are they themselves likely doomed for local extinction? These are the types of questions that conservation biologists, wildlife managers, big game hunters, ranchers, the American public, and the American government had to confront in deciding what the responses of wolf restoration might be like in Yellowstone National Park and its surrounding lands that comprise the Greater Yellowstone Ecosystem.

Studies of elk and moose in the Yellowstone region have now shown that prey lose their fear of predators in just a few generations after the predators have been extirpated. But, once populations of predators such as wolves or grizzly bear have been restored, prey do respond effectively. Moose mothers, for instance, after having lost just a single calf to wolves, respond with heightened vigilance to the howls of wolves, and they tend to leave the immediate area. In other words, at least some species of prey have the capacity, through experience, to develop very rapid responses to thwart predators effectively. Whether such flexibility exists in the behavioral repertoires of prey to respond to predators that have been extinct for thousands of years is a topic that invites additional study.

See also Predatory Behavior—*Predator–Prey Communication*

Further Resources

Berger, J., J. E. Sweonson, & I. Lill-Persson. 2001. *Recolonizing carnivores and naïve prey; conservation lessons from lessons from Pleistocene extinctions*. Science, 291, 1036–1039.
Byers, J. A. 1997. *American Pronghorn: Social Adaptations and Ghosts of Predators Past*. Chicago: University of Chicago Press.

Joel Berger

■ Predatory Behavior
Orb-Web Spiders

The spiders have always caught the attention of many people because of their beautiful web designs. These architectural buildings assume multiple shapes, and their location is diverse. Many scientists start working with spiders just because they believe that these animals can be manipulated easily, but quickly find out the fascinating creatures they really are.

Spider webs are the result of a preprogrammed behavior: The same species of spider builds the same type of web. The orb webs are geometrical and can be considered an engineering work of balance. The orb web-building spiders belong to one of the following families: Araneidae, Tetragnathiidae and Uloboridae. Uloboridae spiders have a special silk, called cribellate-silk. These spiders have two structures for coiling the silk: a plate in the abdomen or opisthosoma (calamistrum) and special hairs in the last pair of legs or legs IV (cribellum).

Orb Weavers

Araneidae and Tetragnathiidae build orb webs with a sticky or viscose silk, produced by aggregates and flagelliforme or coronate silk glands. The orb web has the structure you can see in the figure. The building behavior is stereotypical and takes the following steps:

1. The spider makes the threads to build the frame of the web;
2. It builds the radii in an alternative order;
3. It builds the auxiliary spiral and later the definitive threads of the sticky or catching spiral.

The typical orb web shows variations according to the location within the habitat and can present different inclinations, from vertical to horizontal. Some species can build different type of webs throughout their development, as in the case of the *retreat*, a structure made at the end of the orb-webbing behavior. As an example, *Metepeira gressa* builds an orb web with a retreat and a barrier web. The retreat is useful for the spider to hide and protect itself from the rain, sun, or predators, and it will also contain the egg sacs (usually two or three). Webs can exhibit decorations, as in the spider *Argiope* which presents a web with a silken band leading in a zigzag manner, with a possible function of camouflaging the spider against sun radiations and predators.

Only the juveniles and the females are capable of building orb webs. When reaching adulthood, males lose the ability to build orb webs. Most adult males change their habits after their last molt, leaving their webs and becoming wanderers looking for females; as adults, they rarely catch prey again. Many males die soon after copulating with as many females as they can. The legendary cannibalism by females is, in fact, not very frequent in nature, and the few reports coincide with laboratory cases. In the field we observed males occupying abandoned juvenile or female webs. These observations made us ask the question of why the males do this. In experiments carried out in 1988

Web of Metepeira gressa.
Courtesy of Carmen Viera.

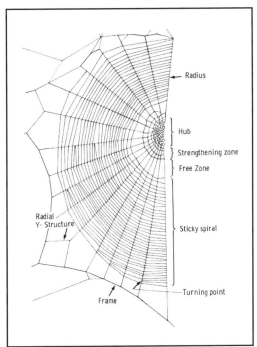

Structure web of Araneus diadematus.
Courtesy of Carmen Viera.

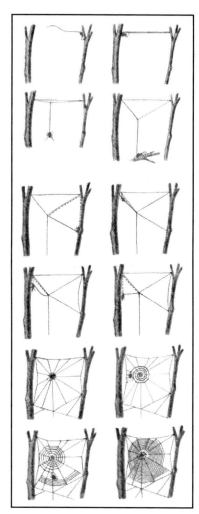

Steps of web building.
Courtesy of Carmen Viera.

with *M. gressa* under laboratory conditions, Fernando Costa and I observed that adult males are incapable of building orb webs for capturing prey, but they can capture prey in juvenile or female conspecific webs. In these experiments, we compared the predatory behavior of males in both female-abandoned and juvenile-abandoned webs. The results showed that males are able to capture prey in both types of webs, but are less successful in female webs. Perhaps the persistence of sexual pheromone in female webs motivate males to continue to search for females, rather than to capture.

The Web as a Trap

The spiders are obligatorily carnivorous and need to capture living prey to survive. Most of their prey are insects, but other spiders and little vertebrates can be captured depending on the type and size of the web. Sticky silk works for catching some prey, but the web is not all that is required to be successful; the spider must also actively attack. *M. gressa*, as well as other araneids, capture prey with the "sit and wait" strategy, hiding in the retreat which is connected to the hub of the orb web by signal threads, much like a fisherman patiently awaiting the tightening of his line.

Hungry Spiders

A predator is limited by the lack of prey. According to David Wise (1993), within a season, the spatial foraging pattern of spiders appears to have been molded by prey shortages. Then, the sit-and-wait strategy of many spiders is an adaptation to a shortage of prey, and when the spiders change foraging sites they often locate in habitats of high prey availability.

Experiments under laboratory conditions allow us to control certain parameters like temperature and humidity. But, just as important, we can control food provision, producing starved and/or well-fed spiders. This method is useful for observing capturing behavior. In experiments in the field, it is rather difficult to know the history of each spider, its physiological conditions, number of molts or reproduction stage. These factors can modify the behavioral responses of the spider. However, studies in the field allow us to take a look into the life of these mysterious creatures, observing different behaviors without any restriction or manipulation.

Taking into account these facts, I made both types of experiments—at the laboratory and in the field—to find out the predatory behavior pattern in *M. gressa*. I found out that these spiders are always hungry and eat all the prey that are offered, but the time of approaching each prey can vary widely with the condition of the spider. The spider responds quickly if it has been starved and slowly if it has been well fed. In fact, the time it takes to capture is an important parameter for evaluating capture success, allowing the comparison of the responses on different type of prey. Susan Riechert in 1982 proved experimentally that spiders capture more prey than what they need for consumption. Such behavior could be useful to control populations of insects considered to be pests.

Prey Discrimination

The same species do not always use the same strategy for capturing different kinds of prey. I found that *M. gressa* uses three types of strategies elicited by three different types of prey. The capturing behavior consists of a sequence of units which can be grouped into three phases or stages: detection, immobilization and carry–ingestion. The *detection phase* includes locating the prey, turning toward the prey, and grasping it with the front legs. The *immobilization phase* consists of pulling the prey with the chelicerae and biting it (venom injection) and/or wrapping the prey with silk. The last phase (carry–ingestion) is the transportation of the controlled prey, depending on the prey size, using either the jaws or hanging it from the spinnerets with the help of the leg IV, and finally feeding. Other behavioral categories used are: self-grooming, short and long bites, pause, touch prey, prey retrieve and displacements. The units of behavior must be selected carefully and each one must be descriptive, but must not include functional explanations. The number or frequency of units used by the spider and the time spent in capturing can indicate the difficulties found by the individual in completing this behavior. Also, the type of immobilization used indicates that the spider can determine the potential danger of the prey. Michael Robinson and José Olazarri were the first researchers to establish quantitative capture behavior in 1969, and their description of behavioral units is still used today by specialists. I observed spider predatory behavior elicited by ants (*Acromyrmex*), domestic flies (*Musca domestica*) and mealworm larvae (*Tenebrio*) reared at

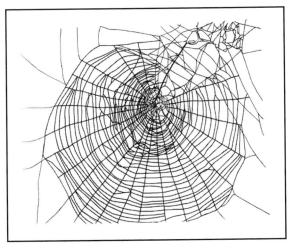

Web of M. gressa *with the retreat. Drawing was traced from a photograph.*
Courtesy of Carmen Viera.

the laboratory. Both ants and flies are well-known spider prey, which have two absolutely different defensive tactics, but the third type of prey was unknown to the spider because it is highly improbable for a mealworm to meet a spider in natural conditions. In this way I could test the potential of the spider to capture a "new type of prey" without any experience. The inexperienced spiders were obtained from breeding spiders under laboratory conditions and, since their emergence, were fed with pieces of *Drosophila*. Experience can improve capture success, so for this reason I designed the predatory experiments to offer prey to the groups of spiders in an alternate sequence in order to standardize the experiments.

I observed three types of capture strategies that indicate the complex variation of these captures. The easiest

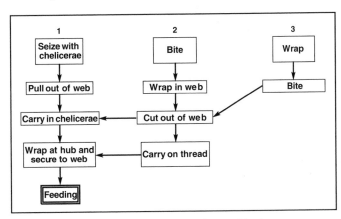

Predatory stategies in orb weavers (Araneidae).
Courtesy of Carmen Viera.

Metepeira gressa *wrapping ant during the capture.*
Courtesy of Carmen Viera.

capture was the fly and the greatest difficulty was shown when capturing the ant, as was expected. The ant is a dangerous prey with strong jaws and formic acid secretion, so the spider must capture and defend itself at the same time, using its wrapping unit for immobilizing the prey. This unit has advantages for the spider: It is less dangerous to immobilize because the spider will then not be in close contact with the chemical and mechanical prey defenses of the ant, and it allows the spider to capture prey that get caught in the web just for a brief time before escaping. The responses found with the mealworm were unexpected, because the mealworm has no defensive mechanism and presents no risk for the spider. With this last prey, the capture was unusually long and implied several different units being used at the same time, because the larvae broke the web, and the spider needed to repair it and capture at the same time. Losing the web is a very serious problem for the spider because without the web it is unable to capture and even to stay alive. The successful capture of the mealworm shows us the potential for spiders to adapt to new situations, and proves the plasticity of the capturing behavior. Hilton Japyassú & Carmen Viera (2002) have confirmed in the field the previous results on spider prey discrimination, but added another variable, the prey size. In order to evaluate the influence of prey size on the behavioral sequence, we offered three sizes of each type of prey. Each spider captured only one prey item and was observed only once. We found that the relation of prey size to spider size is very important when choosing the capture strategy.

Recapitulating

Spiders represent a group of carnivores that are easy to study because of their abundance and the simplicity of their requirements for surviving under laboratory conditions. Experiments made in the field with web spiders present some problems with the identification of individuals, and some difficulties in determining individual life history. However, in several sites the spiders live naturally adapted to human environment and this helps us in the understanding of their mysterious ways. Stereotypical behaviors can be useful in the work of identification and in finding the relationship between the studied species and other taxa. The plasticity found in some of the studied spider behaviors can be considered important in terms of adaptation and relevant in expanding the knowledge of the evolution trends of capturing behavior.

Metepeira gressa *capturing Tenebrio larvae.*
Courtesy of Carmen Viera.

Further Resources

Foelix, R.F. 1982. *Biology of Spiders.* Cambridge, MA: Harvard University Press.

Japyassú, H. F. & Viera, C. 2002. *Predatory plasticity* in Nephilengys cruentate (Araneae, Tetragnathidae): *Relevance for phylogeny reconstruction.* Behaviour, 139, 529–544.

Wise, D. H. 1993. *Spider in Ecological Webs*. Cambridge, MA: Cambridge University Press.

Witt, P. N. & Rovner, J. S. 1982. *Spider Communication. Mechanisms and Ecological Significance*. Princeton, NJ: Princeton University Press.

Carmen Viera

■ Predatory Behavior
Praying Mantids

The praying mantids are a group of fascinating, predatory insects that make their living as generally sedentary, ambush predators. Their striking morphology, unusually mobile head, large compound eyes, piercing gaze, and the remarkably fast strike of their spined, raptorial forelegs have captured peoples' imaginations for thousands of years. In spite of that, however, until recently, little was known about this animal's behavior, and most of what was believed was based on misconceptions, myths, and legends. However, recent work, such as that reviewed in the book, *The Praying Mantids*, has revealed that this animal holds unusual promise as a model system to study a variety of complex behaviors across a number of levels of analyses—from functional neurophysiology to ecology.

As unique as mantids are in appearance and behavior, it is surprising that they are members of the order Dictyoptera which includes, besides Mantodea (mantids), the Blattodea (cockroaches) and the Isoptera (termites). Mantids and cockroaches apparently diverged from a primitive cockroach-like Paleozoic ancestor (543 to 248 million years ago). Fossil mantids date from at least as early as the Upper Cretaceous period (about 70 million years ago), so mantids were probably well established during the Tertiary period (around 60 million years ago).

The approximately 2,000 species of mantid are divided into eight families. The Metallyticidae are rare blue or green metallic colored insects restricted to the Malayan region. Chaeteessidae exist now only in South America. They do not have the strong hook-shaped spine on the end of their foreleg as do other mantids, and their other foreleg spines are thin and delicate. The neotropical Mantoida are the smallest mantids, similar in appearance to Chaetessa. Amorphoscelidae are also small mantids that range from the Mediterranean and Asia to Africa and Australia. They are characterized by tiny, cone-shaped extensions behind their eyes, and a short pronotum (the upper body plate of the first section of the thorax, which is frequently elongated in mantids). Eremiaphilidae are stout, ground-dwelling mantids that live in desert and semidesert regions of northern Africa and Asia. They have a globe-shaped head, slightly protruding eyes, small antennae, a short trapezoidal pronotum with a bumpy surface, strong predatory forelegs armed with short spines, and mesothoracic (middle) and metathoracic (hind) legs that are long and thin which make them fast runners. The Empusidae are slender, unusual-looking mantids that are found in southern Europe, Africa, Madagascar, and western Asia up to China. They are distinguished by a number of characteristics including comb-like projections on both sides of the male's antennae, an elongated point on top of their head, and, frequently, lobes on their meso- and metathoracic legs. The Hymenopodidae are found in all tropical countries except Australia. This group includes the spectacular flower mimics such as the orchid mantid.

The remaining roughly 360 genera, including 80% of the species, are grouped in the family Mantidae. This is a very diverse group, including both the more common looking mantids and most of the grass and bark mimics. Virtually all research on mantids has been

This is an adult female Sphodromantis lineola *eating an anole lizard. Mantids will capture and eat any prey that meets a certain set of perceptual criteria. These include small items (e.g., flies) and larger animals such as this lizard, or other mantids.* Reproduced with the kind permission of the photographer, Mark Gonka.

done on members of this group, and there is still only a small amount of comparative behavioral or physiological data across species or families.

What Mantids Eat

Mantids are opportunistic predators that prey primarily, but not exclusively, on arthropods and their larvae. As generalized predators they may capture and eat beetles, bees, wasps, crickets, grasshoppers, caterpillars, or butterflies among other insects. In addition to these relatively smaller creatures, however, mantids have been known to capture and eat much larger prey such as other, same-sized mantids, and newts, lizards, frogs, small birds, small turtles or mice. We also know that under conditions in which prey or water is scarce, they will eat fruit and, as young nymphs, pollen. These alternate food sources can tide them over until water or prey is available. This fact also demonstrates that, contrary to popular belief, mantids *will* eat something other than live prey.

How Mantids Identify Prey

Being a generalized predator has both advantages and disadvantages. The former includes the fact that the predator is not limited to one specific type of prey. If one type is in short supply, another can be eaten. The disadvantage is that the predator has to be able to identify rapidly and reliably any moving object that may appear in its visual field as a potential meal, even if it has never seen the object before. This is a challenge for an animal with a brain that can fit comfortably on the head of pin. Surprisingly, however, mantids accomplish this task as do much larger predators such as toads, frogs, and salamanders. In fact, they employ a cognitive strategy that is commonly used by humans.

When you decide if a particular meal is fit to eat, you do not have in mind a picture of one invariant acceptable meal (i.e., a *template*) to which you try to match the meal in front of you. Rather, you weigh a variety of stimuli (odor, color, taste, texture) associated with the meal and if, *all together*, the stimuli are acceptable, you eat. The entire set of stimuli associated with a given meal defines one *exemplar* (or example) of the category *acceptable meal*. The category can be understood as a theoretical, perceptual envelope that includes all of the various acceptable combinations and permutations of the original key stimulus parameters (odor, color, taste, texture), and it may include particular stimulus combinations associated with meals as diverse as sushi, pizza, borscht, and goulash.

A mantid (via the functional neuroanatomy of its visual system) does the same thing when it decides what an acceptable meal is: It identifies an object as an *exemplar* of the category *acceptable meal* by assessing each of an identified set of specific stimulus parameters. If a moving object creates an image on the mantid's retinae (i.e., a retinal image) that meets a sufficient number of these parameters, the mantid will identify the moving object as prey and attempt to capture it.

"Sexual Cannibalism": Is It Really "Sexual" or Is It Just Predatory Behavior?

Frederick R. Prete

One seemingly grisly behavior in which mantids engage is so-called "sexual cannibalism" by females. This involves an adult female mantid eating a male who is captured during or just after copulation. Although this event occurs very rarely, and then only in certain species and when other food is scarce, the idea that it is a real phenomenon and a regular part of mantid mating behavior stubbornly persists.

There are a number of psychological, social, scientific and political reasons that have fueled the belief that female-on-male cannibalism is a part of mantid mating behavior (see especially the 1992 article by Prete & Wolfe). Of these, the following three have been particularly influential: First, the male mantis can, indeed, mate even after a devastating injury, such as having its head devoured by the hungry female with which he is mating; second, people have mistakenly thought that fly-sized prey is the normal fare for mantids and that capturing large prey, in this case a female capturing a nearly same-sized male, is an "abnormal" behavior and in need of a special explanation (such as "sexual cannibalism"). Third, it is often argued that the male's brain serves as an "inhibitory" influence on the mating reflex and decapitation "releases" mating behavior from the brain's inhibition.

Regarding the first reason, although it is true that a male with much of the front part of his body gone can still mate, this has nothing to do with the organization of mantids per se. As in many other insects, the mantid's central nervous system is distributed along the length of its body in bundles of cells (ganglia) that are connected by a pair of ventral nerve cords. Hence, the sensory neurons, motor neurons, and many of the integrating interneurons necessary to control any given body segment are contained in the ganglion located in that segment. Hence, all of the neural machinery necessary to run the legs and genitalia of many insects will work fairly well even if the more anterior portion of the insect is gone. For instance, you can experimentally train a single cockroach leg and its associated ganglion to avoid a noxious stimulus even if the rest of the roach has been disposed of. Or, the adult female abdomen of the giant cecropia silk moth, without its head or thorax, can still emit pheromones, attract males, and lay eggs. Even a headless female mantis can lay a perfectly formed ootheca (egg case). The fact that a male mantid can mate without a head is no demonstration that he has evolved the ability to both mate and serve as a nutritious meal for its mate, as is often suggested. Further, when a female captures and eats a male it is usually long before he has had a chance to mate with her.

Regarding the second reason, we now know that mantids regularly capture very large prey using the prey–recognition algorithm explained in the text. Catching large prey, such as a male mantis, a newt, or a mouse, needs no other explanation than the female was hungry. It has nothing to do with mating behavior, per se. Further, recent historical analyses of the literature on mantid behavior have rediscovered that in certain species, mantids of both sexes and of all ages—sexually mature or immature—may eat other mantids. Cannibalism is not limited to the mating event in any way.

(continued)

"Sexual Cannibalism": Is It Really "Sexual" or Is It Just Predatory Behavior? (continued)

Finally, regarding the idea that the mantid's brain (or, more correctly, the ganglia in its head) inhibit the male's mating reflex has held sway primarily because it appeared first in an article (in 1935) and then in a very influential book (in 1963) written by Ken Roeder, a pioneer in the study of the neural control of animal behavior. However, although most credit Roeder with the idea, he actually got it from Etienne Rabaud who published it in a French journal in 1916. (Roeder put Rabaud's article in his bibliography, but did not credit him with the hypothesis.) In 1916, of course, people had a much more simplistic idea about how behavior is controlled than we do now. The idea that the mantid's brain inhibits male mating, and once removed allows it to occur, simply no longer fits with our understanding of the complex sensory integration and nuanced control that is a part of all animals' behaviors. After all, if a farmer decapitates a chicken, it runs around for a few minutes before it dies. No one, however, would argue that the chicken's brain simply "inhibits" running behavior and decapitation frees the chicken to perform its final behavioral display.

The idea of a retinal image is best understood by thinking of it as a "shadow" cast by an object onto the mantid's large, compound eyes. Such images are measured in terms of visual angle rather than absolute object size. Visual angle is that angle made by two theoretical lines emanating from the same point on the animal's eye to the edges of the object being viewed.

Most of the research in this area has been done using adult, female mantis *Sphodromantis lineola*, so this will serve as the example here. When an object enters *S. lineola's* visual field, the mantid's visual system simultaneously assesses the following parameters. The first is the overall size of the retinal image—is it a compact image (i.e., roughly round or oval, like that made by a lady beetle or bumblebee)? Ideally, the image would subtend between five and twelve degrees of visual angle. However, if the image is elongated (i.e., like that made by a long beetle, another mantid, or a lizard), the image must be moving parallel, but not perpendicular, to its long axis and, again, its leading edge should be between five and twelve degrees. Next, the image must be darker than the background against which it is moving, and it must be moving at an acceptable speed (approximately 35–85 deg/second for female *S. lineola*). The image's speed allows the mantid to gauge how far away the object is. Next, the image must move a sufficient distance across the mantid's retinae and should pass through the center/lower center of its visual field. These parameters help the mantis to assess the location of the object in three-dimensional space. Finally, image direction is assessed. Images that move from dorsal to ventral (i.e., down) across the retinae are more likely to be categorized as prey because, from the mantid's perspective, a downward moving retinal image represents an object that is moving toward it.

What the mantid is doing when it assesses these stimulus parameters is employing a computational algorithm or "rule of thumb" which allows it to easily determine if a retinal image represents something that can reasonably be classified as prey. This approach is used by humans and computers every day to simplify complex decision making in analogous situations. The mantid uses the prey-identification algorithm to solve what would otherwise be the impossible task of having memorized each and every conceivable prey item that it might encounter (from flies to newts) and then trying to figure out if a given retinal image matches any of the dozens of image templates stored in its memory. The algorithm approach works reliably; the latter would be impossible for the mantid.

How Mantids Capture Their Prey

When potential prey moves into the mantid's visual field, the mantid may orient toward it or visually track it. These behaviors serve to move the object's retinal image onto the center/lower center of the compound eyes where it can be viewed by an area of the retinae called the *acute zone* analogous to the vertebrate's fovea. The visual details of the object can then be assessed. Although generally an ambush predator, if the object satisfies the stimulus parameters that define prey, the mantid may slowly stalk or chase it. The decision to do so is based primarily on the movement characteristics of the object's retinal image. An otherwise prey-like object whose retinal image moves very slowly will be perceived as being too far away to catch and the mantid may attempt to get closer before striking.

Once an object is identified as prey, and it is within catching distance, the mantid will attempt a capture. The prey capture sequence consists of two distinct and separate behavioral components. The first is the strike, a rapid extension and grasping movement made by the raptorial forelegs. Depending on the perceived distance to the prey, the strike may be accompanied by a lunge, a displacement of the mantid's body effected by the meso- and metathoracic legs.

Mantid prey identification is based on the assessment of certain key stimulus parameters of an object's retinal image. For adult female Sphodromantis lineola, *an object is most likely classified as prey if its image or the leading edge of its image subtends 5–12 deg of visual angle (A), if the image moves parallel to its long axis (B), if the image is darker than the background (C), if the image moves quickly, moves a sufficient distance across the retinae, and passes through the retinae's acute zones (which view the center/lower center of the visual field) (D), and if the image moves down across the retinae (E).*
Courtesy of Frederick Prete.

Just before the strike occurs, the mantid may make some deliberate, final postural adjustments to maximize the probability that the capture will be successful. For instance, if the prey item is high overhead relative to the mantid, it will pitch its body upward and backward, and extend its middle legs to close the gap slightly between itself and the prey. If the prey is low, for instance, on the stem on which the mantid is perched, depending on the distance to the prey, the mantid may lean forward by extending its rear legs slightly.

The strike, itself, is very fast. For adult female *S. lineola*, the elapsed time from the beginning of foreleg movement to the time that the forelegs initially contact the prey is only 40–220 milliseconds; The elapsed time from the beginning of the strike until the forelegs close around the prey is a brief 80–260 milliseconds.

As noted, depending upon relative prey distance, the strike may be accompanied by a lunge. Lunge distance and direction is variable depending on the prey's relative position, and it allows the mantid to capture prey within a large three dimensional area in front,

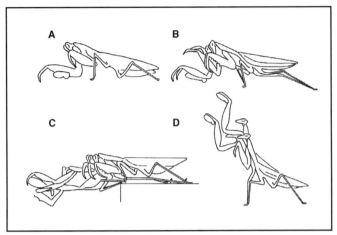

These are composite tracings made from high speed (200 frames/sec) video recordings of adult female Spodromantis lineola *catching an adult cricket. These demonstrate the variability of the prey-capture sequence. The first two panels depict low strikes during which there was no measurable lunge (A) and only a slight downward and forward lunge (B). Panel C depicts a low strike accompanied by a dramatic forward directed lunge. Note that the mantid's rear tarsi (feet) lost their grip and slid forward during the lunge. Panel D depicts two high strikes by the same mantid, each of which began from exactly the same initial position. In the higher of the two strikes, the mantid's body displacement was primarily upward. In the lower of the two, the body displacement was primarily forward.*
Reprinted from Prete & Cleal 1996, with the kind permission of S. Karger AG, Basel.

above, and to either side of where it is standing. During "low strikes" (i.e., strikes at prey near the level of the mantid's tarsi, or "feet"), lunge velocity is correlated with prey distance: When prey is farther away, mantids lunge faster (as fast as 295 mm/sec). So, the elapsed time from the beginning of the strike/lunge sequence to the time that the prey is initially contacted by the forelegs is relatively constant across prey distances. However, when lunging for prey that is high in the air ("high" strikes) the lunge is considerably slower and not correlated with prey distance. This difference is due to the differences between the postures assumed just before the strike/lunge sequence begins. During low strikes, the mantid's body displacement is primarily forward, and the lunge is powered by extensions of the meso- and metathoracic legs which, together, can generate considerable force. However, high strikes begin with the mesothoracic legs more extended than in low strikes, so they cannot generate the same forces or body speed during the lunge.

When strikes are directed toward prey that is to the right or left of the mantid's midline, it will turn toward the prey as it lunges by extending the legs on the side opposite the prey and flexing those on the same side. In these cases, the hind legs generate the power that moves the mantid forward during the lunge and the middle legs act to steer the lunge to one side.

Once the prey is captured, the mantis flexes its meso- and metathoracic legs and returns to its initial standing position. It then begins eating whatever it has caught—be it fly, grasshopper, lizard, or another mantid—by biting off pieces of the live prey with its powerful jaws.

The speed, variability, and accuracy of the overall prey–capture sequence demonstrates that the underlying central nervous system motor programs that direct it must take into account and integrate a variety of ongoing sensory inputs, including visual information about prey location and speed, and proprioceptive information indicating the relative positions of the mantid's body parts. In fact, the information processing that underlies a successful prey–capture sequence by a mantid is fundamentally as complicated as that orchestrating a similar movement sequence by a much larger animal, for instance, a soccer player driving to catch a ball.

See also Reproductive Behavior—*Sexual Cannibalism*

Further Resources

Beckman, N. & Hurd, L. E. 2003. *Pollen feeding and fitness in praying mantids: The vegetarian side of a tritrophic predator.* Environmental Entomology, 32, 881–885.

Prete, F. R., & Cleal, K. 1996. *The predatory strike of free ranging praying mantises,* Sphodromantis lineola (*Burr.*). *I: Strikes in the midsagittal plane.* Brain, Behavior and Evolution, 48, 173–190.

Prete, F. R. & Kral, K. 2004. *In the mind of a hunter: The visual world of the praying mantis.* In: *Complex Worlds from Simpler Nervous Systems* (Ed. by F. R. Prete), pp. 75–115. Cambridge, MA: MIT Press.

Prete, F. R., Lum, H. & Grossman, S. P. 1992. *Non-predatory ingestive behaviors of the praying mantids* Tenodera aridifolia sinensis (*Sauss.*) *and* Sphodromantis lineola (*Burm.*). Brain, Behavior and Evolution, 39, 124–132.

Prete, F. R., & Mahaffey, R. J. 1993. *Appetitive responses to computer generated visual stimuli by the praying mantis,* Sphodromantis lineola (*Burm.*).Visual Neurosciences, 10, 669–679.

Prete, F. R., Wells, H., Wells, P. & Hurd, L. E. (Eds.) 1999. *The Praying Mantids.* Baltimore, MA: Johns Hopkins University Press.

Prete, F. R. & Wolfe, M. M. 1992. *Religious supplicant, seductive cannibal, or reflex machine? In search of the praying mantis.* Journal of History of Biology, 25, 91–136.

Frederick R. Prete